Studies in Neuroscience, Psychology and Behavioral Economics

Series editors

Martin Reuter, Bonn, Germany
Christian Montag, Ulm, Germany

More information about this series at http://www.springer.com/series/11218

Nikolai Axmacher · Björn Rasch
Editors

Cognitive Neuroscience of Memory Consolidation

Springer

Editors
Nikolai Axmacher
Ruhr-University Bochum
Bochum
Germany

Björn Rasch
University of Fribourg
Fribourg
Switzerland

ISSN 2196-6605 ISSN 2196-6613 (electronic)
Studies in Neuroscience, Psychology and Behavioral Economics
ISBN 978-3-319-45064-3 ISBN 978-3-319-45066-7 (eBook)
DOI 10.1007/978-3-319-45066-7

Library of Congress Control Number: 2016961668

Printed on acid-free paper

This Springer imprint is published by Springer Nature
The registered company is Springer International Publishing AG
The registered company address is: Gewerbestrasse 11, 6330 Cham, Switzerland

Contents

Part I Conceptual Questions of Memory Consolidation

Cellular and Systems Consolidation of Declarative Memory 3
Lisa Genzel and John T. Wixted

Mechanisms of Memory Consolidation and Transformation 17
Melanie J. Sekeres, Morris Moscovitch and Gordon Winocur

The Medial Prefrontal Cortex is a Critical Hub in the Declarative Memory System 45
Guillén Fernández

Consolidation of Episodic Memory: An Epiphenomenon of Semantic Learning 57
Sen Cheng

Part II Memory Consolidation During Off-Line Periods and the Role of Sleep

The Impact of Sleep Deprivation on Molecular Mechanisms of Memory Consolidation in Rodents 75
Judith C. Kreutzmann, Jennifer C. Tudor, Christopher C. Angelakos and Ted Abel

Sleep and Odor Memory Consolidation in Non-human Animal Models 87
Donald A. Wilson, Kacper Kondrakiewicz and Dylan C. Barnes

The Effect of Sleep on Multiple Memory Systems 105
Monika Schönauer and Steffen Gais

A Role of Sleep in Forming Predictive Codes 117
Karsten Rauss and Jan Born

Emotional Memory Consolidation During Sleep 133
Tony J. Cunningham and Jessica D. Payne

Daily Life Experiences in Dreams and Sleep-Dependent Memory Consolidation 161
Jean-Baptiste Eichenlaub, Sydney S. Cash and Mark Blagrove

Is Dreaming Related to Sleep-Dependent Memory Consolidation? 173
Michael Schredl

Part III Mechanisms of Memory Consolidation on a Systems Physiology Level

Neuronal Oscillations and Reactivation Subserving Memory Consolidation 185
Til Ole Bergmann and Bernhard P. Staresina

The Role of Sleep Spindles in Sleep-Dependent Memory Consolidation 209
Elizabeth A. McDevitt, Giri P. Krishnan, Maxim Bazhenov and Sara C. Mednick

Hippocampal Sharp Wave/Ripple Complexes—Physiology and Mechanisms 227
Nikolaus Maier and Richard Kempter

Replay in Humans—First Evidence and Open Questions 251
Hui Zhang, Lorena Deuker and Nikolai Axmacher

Cortico-Hippocampal Circuits for Memory Consolidation: The Role of the Prefrontal Cortex 265
Lisa Genzel and Francesco P. Battaglia

Part IV Modulation of Memory Consolidation

Stress and Memory Consolidation 285
Shira Meir Drexler and Oliver T. Wolf

Electric Stimulation to Improve Memory Consolidation During Sleep 301
Diana Campos-Beltrán and Lisa Marshall

Memory Manipulation During Sleep: Fundamental Advances and Possibilities for Application 313
Lucia M. Talamini

Scents and Reminiscence: Olfactory Influences on Memory Consolidation in the Sleeping Human Brain 335
Laura K. Shanahan and Jay A. Gottfried

Reinforcing Language Learning During Sleep 347
Thomas Schreiner, Mick Lehmann and Björn Rasch

Part V Clinical Translation

Reconsolidation and Trauma Memory 369
Henrik Kessler, Simon E. Blackwell and Aram Kehyayan

Sleep-Related Interventions to Improve Psychotherapy 381
Christoph Nissen, Marion Kuhn, Elisabeth Hertenstein
and Nina Landmann

Accelerated Long-Term Forgetting in Epilepsy—and Beyond 401
John Baker and Adam Zeman

Introduction

Memory is a fundamental trait underlying adaptive behavior. Only if previous experiences can be stored and used to influence and alter future behavior, adaptation with respect to the environment and environmental changes is feasible. Thus, the capacity to store memories exists in the vast majority of living organisms, starting from simple cell forms to highly complex organisms including animals and humans. And as the ability to store memories is also the prerequisite to develop consciousness, self-awareness, and personality, memory and its underlying mechanisms have fascinated researchers ever since.

The formation of memories is typically separated into different phases. The initial phase always requires the encoding of information by the organism. And as we still do not have any other means to identify the successful formation of a memory, some types of retrieval processes are also necessarily involved. In this book, we specifically focus on the time period between encoding and retrieval: What mechanisms are involved *after* the successful encoding of information?

One major advancement in understanding the mechanisms underlying long-term formation of memories was the proposition of a consolidation process. In 1900, Müller and Pilzecker proposed that after successful encoding, a physiological process "perseverates" and stabilizes encoded memory representations (Lechner et al. 1999; Müller 1900), gradually rendering them less susceptible to future interferences caused, e.g., by new learning. Since then, tons of evidence in favor of the consolidation of memories has been provided by very different research areas, including the fact that long-term memory retention initially depends on the synthesis of new proteins, whereas it does not any more after a certain period of consolidation (Dudai 2004; Kandel 2001). In addition, temporally graded amnesia was observed in patients with hippocampal damage (Squire and Wixted 2011). Furthermore, pharmacological studies in animals show that certain manipulations are only effective shortly after encoding, but not after a period of consolidation (McGaugh and Izquierdo 2000). Based on these studies, McGaugh (2000) proposed in his seminal review paper "a century of consolidation," that there are probably different waves of memory consolidation after their encoding, initially including cellular stabilization processes but involving also large-scale changes in the organization of the memory trace (i.e., systems consolidation, see Dudai 2004). And even after the successful consolidation of a memory, retrieving a memory can under certain circumstances render these memories again instable and susceptible to interfering influences, requiring a period of reconsolidation in order to persist (Nader and Hardt 2009).

In spite of the overwhelming experimental support for consolidation processes from neuroscientists, cognitive psychologists have been quite skeptical whether such a process indeed exists. On the one side, the skepticism was due to the failure to find consistent evidence that memories become more resistant to interference by new learning with time (see Wixted 2004, for an overview). Thus, forgetting by either interference, decay, or a combination of both were (and still are) the prominent concepts in cognitive psychology instead of memory consolidation. On the other side, cognitive memory researchers did not have the appropriate methodological approaches to study this process, as the cognitive processes underlying memory consolidation cannot be directly observed. The bridging of this gap started with the emergence of studies in the cognitive neurosciences that combine cognitive testing with neuroimaging recording and stimulation technique during off-line periods such as sleep. In these studies, several important concepts, including, for example, the functional role of brain oscillations and spontaneous memory reactivation during off-line periods, have been put forward to characterize the physiological processes of memory consolidation and to link them to putative behavioral measures of consolidation. The goal of this book was to give an overview of the state of the art of this endeavor, based on the contributions from leading cognitive neuroscientists who summarize recent advances in the exciting research field of the cognitive neuroscience of memory consolidation.

The book is organized into five parts: In the first part, conceptual questions of memory consolidation and its effect on the reorganization of memory systems are presented. Chapters in the second part describe more closely the processes of memory consolidation in animals and humans, emphasizing the role of sleep. Contributions in the third part describe the mechanisms of memory consolidation on the level of systems physiology or neural networks, focusing on oscillations and replay. In the fourth part, several factors are introduced which modulate memory consolidation during waking and sleep. Finally, the last part translates concepts of memory consolidation to clinical populations.

Part I: Conceptual Questions of Memory Consolidation

More specifically, in the first part, Genzel and Wixted provide an overview of the concepts and empirical foundations of cellular and systems consolidation. With respect to cellular consolidation, they introduce synaptic long-term potentiation (LTP) as a putative neural mechanism as well as the concepts of synaptic and behavioral "tagging." On the system level, they discuss a specific role of sleep and sleep-related oscillations, replay and scaling as well as prior knowledge on system consolidation processes. Adopting an alternative perspective on memory consolidation (called the "Trace Transformation Theory"), Sekeres, Moscovitch, and Winocur argue that the hippocampus is uniquely suited to promote pattern separation and is required to support a very specific, context-dependent form of memory. By contrast, neocortical areas—in particular in medial prefrontal cortex and anterior cingulate cortex—may only support memory for a context-independent, generalized representation of an event, its "gist." In this framework, they also describe how memory traces may be reorganized following their reactivation and how this could be related to memory reconsolidation. In the next chapter, Fernandez capitalizes on the important role of the medial prefrontal cortex in memory consolidation: Lesion experiments in animals, neuropsychological studies in patients, and neuroimaging research in healthy participants provide converging evidence that while newly acquired memory traces are dependent on the hippocampus, consolidation transfers this role to the medial prefrontal cortex. This is further explained by the relevance of the medial prefrontal cortex for the control of schema-congruent knowledge that is extracted from individual episodes during memory consolidation. Finally, Cheng introduces a third alternative to the standard consolidation theory and the multiple trace theory: He assumes that episodic memory traces remain always hippocampus-dependent. However, he further argues that semantic learning exploits hippocampal traces, and that autobiographical memory may be based to different degrees on hippocampal episodic memory or neocortical semantic information. As a result, consolidated information may appear hippocampus-independent when it is largely based on such semantic information.

Part II: Memory Consolidation During Off-Line Periods and the Role of Sleep

In the second part on the role of sleep in memory consolidation, Kreutzmann, Tudor, Angelakos, and Abel review which signaling pathways related to the consolidation of hippocampus-dependent memories are impaired by sleep deprivation after learning in rodents. They discuss that the timing of sleep deprivation is crucial and highlight its effects on molecular mediators of synaptic plasticity as well as on the expression and translation of genes that are relevant for learning and memory. Remaining in non-human animal models, Wilson, Kondrakiewicz, and Barnes specifically discuss the consolidation of odor memories and the influence of sleep on this process. They first review the anatomy and physiology of the olfactory system in both vertebrates and invertebrates and describe different types of odor memory. Afterward, they discuss how processing in the piriform cortex changes during different sleep stages and how their prominent features affect odor memory consolidation, considering various vertebrate and invertebrate species. Turning to humans, Schönauer and Gais describe evidence that sleep benefits the consolidation of different memory systems, including declarative, procedural, and emotional memory. After summarizing studies showing the effects of sleep on each of these systems separately and the presumed mechanism underlying this effect—reactivation—they discuss how memory systems interact and how consolidation may affect this interaction. In the next chapter, Rauss and Born emphasize that not all memories are equally consolidated during sleep, but rather those which are relevant for future behavior. They propose that memory consolidation during sleep mainly serves to optimize predictive coding of upcoming events, i.e., generating expectations about regularities in the world. They demonstrate how previous studies (e.g., on gist abstraction and language learning during sleep) can be interpreted within this novel framework and discuss open questions for future research. As emotional events are also highly relevant for future behavior, Cunningham and Payne first describe that the amygdala plays a critical role for facilitating consolidation of emotional memories over long time periods. Next, they review the role of sleep (particularly REM sleep) for the consolidation of emotional memories and discuss the physiological processes supporting this relationship, such as high levels of amygdala activity and acetylcholine. If memory consolidation is facilitated by sleep, is it related to dreaming? In their chapter, Eichenlaub, Cash, and Blagrove discuss which events during daily live are incorporated into dream contents. They first review studies indicating that in particular very recent events from the 1–2 days before a night of sleep and those around 5–7 days earlier are incorporated, and then describe that emotionally laden events often become part of dreams. Finally, they critically discuss arguments for and against the view that dreaming is related to memory consolidation. Along similar lines, Schredl reviews studies suggesting that cueing during REM sleep may increase the likelihood that dream contents are related to the events associated with the cues before sleep. Furthermore, he describes evidence that waking contents are incorporated into dreams, and then reports results that dreaming may improve subsequent performance, suggestive of a functional role of dreams. Finally, he reviews studies about possible positive effects of lucid dreaming and suggests future research topics in this area.

Part III: Mechanisms of Memory Consolidation on a Systems Physiology Level

The third part specifically focusses on neural mechanisms of memory consolidation during sleep and wakefulness on a systems physiology level. Bergmann and Staresina capitalize on the role of neural oscillations—in particular, slow oscillations, sleep spindles, and ripples—for memory consolidation. Summarizing work in both animals and humans, they describe both the physiological mechanisms and the putative functional role of these oscillations for hippocampal–neocortical interactions. They then discuss if and how these oscillations support the reactivation of previously acquired memory traces during sleep. Focusing on sleep spindles,

McDevitt, Krishnan, Bazhenov, and Mednick review their putative physiological basis and discuss how sleep spindles are related to the consolidation of different memory systems. They first describe correlational studies and then pharmacological manipulations showing that sleep spindles are functionally related to consolidation processes during sleep. In their chapter on hippocampal ripples and sharp waves, Maier and Kempter review their putative functional role for memory consolidation and discuss different models on their physiological generation. Their conclusion is paradigmatic for the active yet still emerging field of neural oscillation research: "Despite the huge amount of available data on SWRs and their likely physiological relevance, the basic mechanisms underlying this phenomenon remain largely enigmatic, both *in vitro* and *in vivo*." Hippocampal ripples are typically associated with memory reactivations or "replay." In their chapter, Zhang, Deuker, and Axmacher review the current evidence for replay of stimulus-specific memory traces, or "engram patterns," in humans. They describe evidence both for the existence and for the behavioral relevance of replay in fMRI experiments and then again emphasize the large remaining gaps in our understanding of replay: What is the role of sleep for replay, how is it related to hippocampal ripples, and is it actually causally relevant? Turning back to the system level, the chapter by Genzel and Battaglia capitalizes on the role of the prefrontal cortex in sleep-related memory consolidation. They describe the relationship between sharp wave-ripple complexes and replay, both in the hippocampus and in the neocortex, and discuss their relationship to synaptic plasticity. In particular, they provide evidence that replay in prefrontal cortex is related to hippocampal replay and associated with specific processes at the behavioral, neural oscillation, and neuromodulation level.

Part IV: Modulation of Memory Consolidation

The fourth part deals with important modulators of memory consolidation during wakefulness and sleep. In the first chapter of this part, Meir-Drexler and Wolf emphasize the role of stress for memory, which strongly depends on the memory phase: While stress impairs memory retrieval, it typically improves memory consolidation. The authors discuss the inverse effects of different stress levels and durations and review how extreme stress—trauma—may induce symptoms of posttraumatic stress disorder such as intrusions and flashbacks. In the next chapter, Campos-Beltrán and Marshall focus on the application of weak electric fields for modulating memory consolidation. They first describe how finite element modeling can be used to estimate the effects of artificially applied electric fields on cognitive and physiological processing. They then summarize results from meta-analyses on the impact of these electric fields on declarative and procedural memory consolidation in humans and animals. The contribution of specific pathways in memory consolidation and/or brain rhythms is presented based on studies using optogenetic stimulation. Talamini then describes how the causal relevance of consolidation-related neural processes can be tested. She first summarizes previous studies which either modulated sleep-related oscillations via electric stimulation or used olfactory or auditory cues to enhance reactivation of specific memory traces. She then describes very recent approaches that target reactivating cues to specific phases of slow oscillations during sleep. These approaches enhance the effects of stimulation on memory consolidation and may be relevant for various applications during health and disease. More specifically, Shanahan and Gottfried discuss previous studies using olfactory cueing to probe memory consolidation. After reviewing studies that investigated the role of cueing on consolidation of declarative memories, they describe the effect of cueing on emotional memory consolidation and other consolidation-related processes such as creative problem solving. Finally, they review studies that used other cueing modalities (such as auditory cueing) and describe burning questions for future research. Following up on auditory cueing, Schreiner, Lehmann, and Rasch describe the potential of sleep and targeted memory reactivation during sleep for language learning. They first summarize studies providing evidence for a benefit of sleep for language learning, and then describe how auditory cues like words presented during sleep can improve central aspects of language learning. They finish with a discussion of the potential oscillatory mechanisms of successful memory reactivation during sleep.

Part V: Clinical Translation

The last part focusses on the clinical aspects of memory consolidation. In particular, reconsolidation might open a window of opportunity to destabilize already consolidated, but non-adaptive memories. Interfering with reconsolidation is clinically beneficial when memories are pathological, as in the case of posttraumatic stress disorder (PTSD). This novel translational approach is described by Kessler, Blackwell, and Kehyayan. They first review research on reconsolidation and then summarize evidence that PTSD can be conceptualized as a disorder of memory. Then, they describe novel studies showing that interference with reconsolidation may indeed reduce the core symptoms of PTSD, intrusions, and flashbacks. While memory traces of traumatic events should be weakened, it would be helpful to augment the influence of experiences made during psychotherapeutic treatment. Novel research has shown that sleep after therapy plays an important role here. These studies and their underlying rationale are discussed by Nissen, Kuhn, Hertenstein, and Landmann. The authors suggest that psychotherapy can be conceptualized as a learning process which either strengthens or reorganizes existing memory traces, and describe putative neural correlates of this learning process. They distinguish between the effects of different sleep stages on either the strengthening or the reorganization of memories. Finally, they discuss whether manipulations of sleep-related consolidation processes after psychotherapy may further augment the therapeutic impact. The final chapter deals with the phenomenon of "accelerated long-term forgetting," which may occur in temporal lobe epilepsy patients. Baker and Zeman were among the first to describe accelerated long-term forgetting. In their chapter, they review the development of this concept, suggest methodological guidelines to measure accelerated long-term forgetting in a clinical setting, and discuss whether early or later consolidation stages are affected. Finally, they describe the dependence of accelerated long-term forgetting on sleep and its occurrence in different clinical conditions, and consider various underlying mechanisms.

Together, these chapters provide compelling evidence that the field of memory consolidation is actively flourishing, and we thank all authors for their insightful contributions. The concept of memory consolidation bridges a wide range of research areas from cellular and systems physiology via cognitive neuroscience and psychology to clinical applications. With further conceptual and methodological developments, we anticipate more exciting findings to emerge in the near future, which will eventually lead to a deeper understanding of this fundamental yet complex phenomenon.

References

Dudai Y (2004) The neurobiology of consolidations, or, how stable is the engram? Ann Rev Psychol 55:51–86

Kandel ER (2001) The molecular biology of memory storage: a dialogue between genes and synapses. Science (New York, NY) 294(5544):1030–1038

Lechner HA, Squire LR, Byrne JH (1999) 100 years of consolidation–remembering Müller and Pilzecker. Learn Mem (Cold Spring Harbor, NY) 6(2):77–87

McGaugh, JL (2000) Memory—a century of consolidation. Science (New York, NY) 287(5451): 248–251

McGaugh JL, Izquierdo I (2000) The contribution of pharmacology to research on the mechanisms of memory formation. Trends pharmacol sci 21(6):208–210

Müller G, Pilzecker A (1900) Experimentelle Beiträge zur Lehre vom Gedächtnis. Z. Psychol. Suppl. 1:1–300

Nader K, Hardt O (2009) A single standard for memory: the case for reconsolidation. Nature reviews. Neurosci. 10(3):224–234

Squire LR, Wixted JT (2011) The cognitive neuroscience of human memory since h.m. Ann rev of neurosci. 34:259–288

Wixted JT (2004) The psychology and neuroscience of forgetting. Ann Rev Psychol 55(1):235–269

Part I
Conceptual Questions of Memory Consolidation

Cellular and Systems Consolidation of Declarative Memory

Lisa Genzel and John T. Wixted

Abstract For memories to last consolidation has to occur, with this chapter referring to both cellular consolidation and systems consolidation. Cellular consolidation takes place in the hours after learning, stabilizing the memory trace—a process that likely involves structural changes in hippocampal neurons. Systems consolidation refers to a more protracted process by which memories eventually become independent of the hippocampus as they are established in cortical neurons. Both forms of consolidation may serve to render memories less vulnerable to forgetting. Although generally treated separately, these two forms of consolidation are presumably closely related. In this chapter, we will provide an overview of both cellular and systems consolidation and how they interact. Further, we will discuss effects of novelty, sleep and previous knowledge on consolidation.

Keywords Memory · Consolidation · Synaptic · System · Hippocampus

Introduction

The modern idea that memories require time to consolidate has a long history. In 1900, the German experimental psychologists Georg Müller and Alfons Pilzecker published a monograph in which they proposed a new theory of memory and forgetting, one that included—for the first time—a role for consolidation. According to Müller and Pilzecker's (1900) view, consolidation consists of a physiological process

L. Genzel (✉)
CCNS, University of Edinburgh, Edinburgh, UK
e-mail: lgenzel@ed.ac.uk

L. Genzel
Donders Institute, Nijmegen, The Netherlands

J.T. Wixted
Department of Psychology, University of California, San Diego, La Jolla, CA 92093, USA

N. Axmacher and B. Rasch (eds.), *Cognitive Neuroscience of Memory Consolidation*, Studies in Neuroscience, Psychology and Behavioral Economics, DOI 10.1007/978-3-319-45066-7_1

that perseverates and eventually renders the memory trace less vulnerable to interference caused by new learning (Wixted and Cai 2013).

Although Müller and Pilzecker (1900) are credited with conceiving of the concept, the main impetus for the study of consolidation can be traced to Patient HM. Following bilateral medial temporal lobe resection to control his epileptic seizures, HM was unexpectedly left with a profound case of anterograde amnesia (i.e., the inability to form new memories from that point on) despite retaining normal perceptual and intellectual functioning, including normal working memory capacity (Scoville and Milner 1957). Critically, HM also exhibited temporally graded retrograde amnesia (Squire 2009). That is, memories that were formed prior to surgery were also impaired, and the degree of impairment was greater for memories that had been formed just prior to surgery than for memories that had been formed well before. Although memories of up to 3 years prior to his surgery appeared to be somewhat impaired, HM's older memories were apparently intact (Scoville and Milner 1957). This result suggested that medial temporal lobe structures are involved in the maintenance of memories for a limited period of time after the memory is formed. In other words, memories consolidate in that sense as well.

Memory consolidation is now a multifaceted concept. At a minimum, it refers to both cellular consolidation and systems consolidation. Cellular consolidation takes place in the hours after learning, stabilizing the memory trace—a process that likely involves structural changes in hippocampal neurons. Systems consolidation refers to a more protracted process by which memories eventually become independent of the hippocampus as they are established in cortical neurons. Both forms of consolidation may serve to render memories less vulnerable to forgetting. Although generally treated separately, these two forms of consolidation are presumably closely related and are perhaps best conceptualized as different stages of the consolidation process that Müller and Pilzecker (1900) conceived of more than a century ago.

Cellular Consolidation

Hebb postulated that when two neurons repeatedly fire together, they become more likely to fire together again in the future. The mechanism underlying this durable change in the coordinated firing propensities of two neurons is termed cellular consolidation. Investigations into the mechanisms of cellular consolidation have used a wide array of model systems, ranging from Aplysia to the mammalian hippocampus. Since the early 1970s, these investigations have led to a series of insights, beginning with the seminal discovery of long-term-potentiation (LTP, Bliss and Lomo 1973) and continuing with our still growing understanding of the role of CREB and plasticity-related immediate early gene expression (Bailey et al. 2015).

LTP refers to a long-lasting increase in synaptic strength following high-frequency stimulation of the pre-synaptic neuron. While there are many types of LTP (for review see Bailey et al. 2015), the classic form is NMDA receptor dependent LTP: in response to high-frequency stimulation, the excitatory neurotransmitter glutamate is released from the presynaptic neuron and binds to post-synaptic AMPA receptors, depolarizing the post-synaptic neuron (causing it to fire) and opening a channel in the post-synaptic NMDA receptors. The open NMDA channel results in an influx of calcium ions into the post-synaptic neuron, which, in turn, induces a molecular cascade of phosphorylation (Fig. 1). Autonomously phosphorylated (and thus active) CaMKII and PKC phosphorylate existing AMPA receptors, increasing the conductance of the receptors already in the synapse, and triggering the insertion of additional AMPA receptors into synapse.

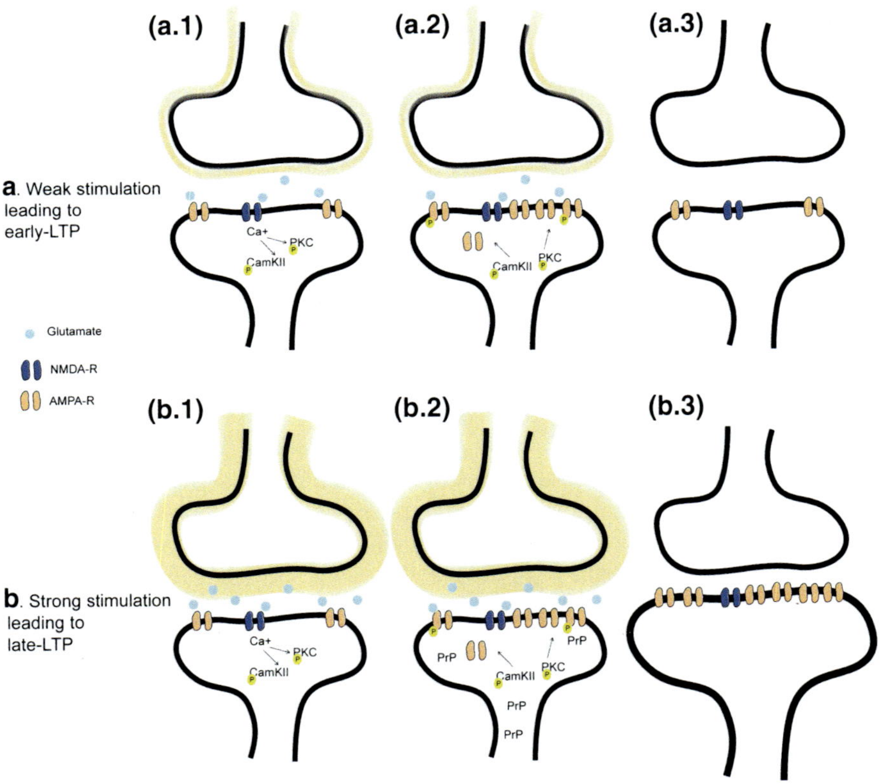

Fig. 1 Cellular consolidation. *a.1* Weak stimulation of a pre-synaptic neuron leads to the release of Glutamate and activation of AMPA and NMDA receptors in the post-synaptic neuron. *a.2* The resulting Ca+ influx activates CamKII and PKC, which in turn phosphorylate existing AMPA receptors and integrate new ones as well. *a.3* If not stabilized by protein production, this potentiation—early LTP—lasts only a few hours after which the synapse returns to its original state. *b.1* After strong stimulation, causing more Glutamate release, the same processes occur initially. *b.2* Additionally, plasticity-related proteins (*PRP*) are produced. *b.3* This leads to a long lasting, structural (e.g., larger) and functional change of the synapse. If a strong stimulation (as in **b**) occurs on one synapse but a weak stimulation (as in **a**) occurs on a nearby synapse of the same neuron, the weakly simulated synapse (**a**) can hijack some of the PRPs of the other synapse (**b**). This leads to the long lasting strengthening of the weakly stimulated synapse **a**

These events are not dependent on protein synthesis and result in early-LTP, which refers to an increase in synaptic strength that will degrade in minutes (or a few hours at most) if not stabilized to late-LTP. Unlike early-LTP, which is independent of protein synthesis, late-LTP requires gene transcription and protein synthesis in the postsynaptic cell. The processes that lead to late-LTP will only occur if the tetanic stimulation of the pre-synaptic neuron is sufficiently strong. Many cytoplasmic and nuclear molecules are needed to ultimately result in the protein synthesis and morphological changes observed in late-LTP. Most significant among these is the transcription factor CREB, discovered by Eric Kandel. Together, these cascades lead to the stabilization of the new AMPA receptors in the cell membrane so that long-lasting synaptic potentiation is achieved. The importance of NMDA-receptor dependent LTP for memory was shown in a seminal study by Richard Morris, in which the inhibition of NMDA-receptors in the hippocampus prevented the induction of LTP and led to a memory deficit in the spatial watermaze task (Morris et al. 1982).

Typically, the processes associated with early-LTP and late-LTP are synapse-specific in that they unfold in the stimulated spine—but not nearby spines—of the postsynaptic neuron. However, Frey and Morris (1998) proposed the "synaptic tagging and capture hypothesis" to model how early-LTP that is destined to degrade at one synapse can be transformed to late-LTP by the strong tetanisation of a different, nearby synapse on the same neuron. Weak tetanisation (too weak to induce late-LTP but strong enough to induce early LTP) at synapse A theoretically leads to the setting of a "tag", which may, for example, consist of the introduction of new AMPA receptors in the synapse. On its own, this "tag" will not become stabilized because the biochemical cascade leading to protein synthesis at that synapse will not occur. However, strong tetanisation shortly before or after the weak event at a separate synapse of the same cell will lead to the setting of its own "tag" and also to the synthesis of plasticity-related-proteins (PRP). These PRPs now not only stabilize the "tag" of the strongly tetanised synapse but can also be "hijacked" by the tag of the weakly tetanised synapse, which is thus stabilized as well (Redondo and Morris 2011). In a series of experiments, they and others went on to show that dopamine plays an important role for the persistence of memory by inducing PRPs (Redondo and Morris 2011; Rossato et al. 2009; Wang et al. 2010).

Synaptic tagging and capture was subsequently translated to *behavioural* tagging. Behavioural tagging is said to occur when a weak and otherwise transient memory is transformed into a more durable memory when it occurs close in time to other behaviourally relevant experiences that provide PRPs (Moncada et al. 2015). For example, Wang et al. (2010) showed that exposure to a novel event shortly before or after a hippocampal-dependent, weak, spatial encoding experience, allowed the originally weak memory to last much longer. In parallel electrophysiological and behavioural experiments, they showed that this memory enhancement was dependent on dopamine and protein synthesis. For a tagging and capture effect to be seen, both events or tasks have to rely on the same brain area, contain overlapping neuronal populations and the events have to occur in close temporal proximity, usually within a 1–2 h window surrounding the weak event (Moncada et al. 2015; Wang et al. 2010).

In conceptually related behavioural tagging work, Dunsmoor et al. (2015) found that weak memories of incidentally presented items from a semantic category (e.g., animals or tools) could be retroactively and selectively strengthened if other items from that same category were made emotionally salient shortly thereafter by pairing them with shock (an amygdala-dependent Pavlovian fear-conditioning task). The retroactive enhancement of the weakly encoded items was not observed on an immediate recognition test but only emerged following a period of consolidation. This finding is consistent with prior work showing that the amygdala can influence the post-training consolidation processes in the hippocampus (McGaugh 2004) (see also chapters by Cunningham and Payne and by Meir Drexler and Wolf) but is the first to show that this effect can be selective (occurring only for previously encoded items that are related to the emotionally conditioned items).

On the surface, these findings are surprising because, usually, two tasks or experiences that occur in close proximity will show detrimental retroactive interference effects on memory for the first event instead of a strengthening of memory. In other words, on the surface, synaptic tagging almost seems to deny retroactive interference, but it doesn't really. There are times when subsequent memories interfere and times when they enhance, but the details matter. It seems to be the case that enhancement occurs when one memory is weak and the other is strong. In that case, the weak memory is enhanced even if the strong event occurs shortly after the weak event. Under other conditions, such as when two strong memory events occur in succession, interference might occur. Studies indicate that competition for protein resources between different learning tags is one of the main factors that give rise to memory interference (Moncada et al. 2015).

Systems Consolidation

Memories initially dependent on the hippocampus are thought to become less dependent on the hippocampus over time and to instead rely more on cortical representations. This process of memory reorganization is termed systems consolidation. Initial evidence for systems consolidation came from patients with hippocampal lesions (Scoville and Milner 1957), and this evidence was later confirmed by experimental studies using artificially induced hippocampal lesions in animals (for review see Squire et al. 2015; Zola-Morgan et al. 1994); in addition to an impairment in the encoding of new memories, these subjects displayed a retrograde memory deficit with a very characteristic temporal gradient: more recent memories were lost while older memories remained intact. Over the past decades, a multitude of studies have investigated this phenomenon in an effort to uncover its underlying mechanisms. The weight of evidence suggests that the hippocampus initially binds the details of our daily experiences that are initially recorded by independent neocortical regions. However, it only serves a temporary role (Morris 2006). The hippocampus is thought to establish connections between these neocortical regions, allowing the newly learned information to be assimilated into existing neocortical networks without causing interference and at the same time extracting the salient information and compressing the memory when necessary (Battaglia et al. 2012; Frankland and Bontempi 2005). These memories are not "transferred" from the hippocampus to the cortex; instead the memory engrams in the cortex are already established during the encoding experience and only need to be linked together to enable retrieval without hippocampal assistance.

Lesburguères et al. (2011) presented initial evidence for an AMPA- and NMDA receptor dependent "tagging process" in the cortex during encoding, which was crucial for the progressive hippocampal-driven rewiring of cortical networks supporting remote memory storage. In a seminal study Cowansage et al. (2014) then went on to show that when such neural ensembles in the retrosplenial cortex are activated by optogenetic techniques, memory retrieval can occur even with hippocampal inactivation at a time point when sensory cues are not sufficient for memory retrieval without hippocampal involvement (i.e., when the memory is still hippocampal dependent). Both of these studies used tasks that were novel for the animal. With previous knowledge of the task, cortical representations during encoding become even more important. For example, after rats have learned a map of flavour-location associations over a period of weeks to months, a new paired-associate can be learned and integrated into the known map in one single trial and induce plasticity-related gene expression in the prefrontal cortex (Tse et al. 2011). Further, this previous knowledge, most likely represented in an extended cortical network, now allows for systems consolidation to occur at a much more rapid pace. Classically, weeks to months are needed for a memory to become independent of the hippocampus, but when previous knowledge in form of a schema is present, this process can be completed in 24–48 h (Tse et al. 2007). These studies, together with human experiments (van Buuren et al. 2014; van Kesteren et al. 2010a; b; 2012; Wagner et al. 2015), suggest that the prefrontal cortex has a special importance when new information is learned in the context of previous knowledge, perhaps by binding the information distributed across other brain areas (see also chapters by Fernandez and by Genzel and Battaglia). While in this case schemas allowed for rapid consolidation, initially, during learning, the hippocampus was still needed (Bethus et al. 2010; Tse et al. 2007).

Conceivably, when cortical schemas are extensive enough and are harnessed during encoding, the hippocampus may not even be needed during initial learning. For example, during rapid word learning, words are *fast mapped* onto new concepts, an important learning mechanism during vocabulary building in childhood. During "normal", explicit word learning, lists are presented to be "learned" by the subjects. In contrast, during *fast mapping*, subjects are not explicitly asked to learn a new word; instead, the word is introduced in context with a known item and its meaning is apparent through inference (Coutanche and Thompson-Schill 2015). For example, instead of presenting a picture of a new animal with the name of the animal, the subject would be shown the new animal together with a known animal and asked "Is the tail of X pointing down?" Studies have shown that this type of learning leads to rapid integration into cortical networks (Coutanche and Thompson-Schill 2014) and may not need the hippocampus even during the encoding experience (Sharon et al. 2011); however the latter finding is still controversial because efforts to replicate it have not been successful (Greve et al. 2014).

We are usually awake when we learn something new, but systems consolidation may take place mainly during sleep (see chapters by Schönauer and Gais, by Rauss and Born and by Kreutzmann and colleagues), most likely because, during sleep, no new experiences can interfere with the process. Two mechanisms have been proposed to act during sleep: memory replay and synaptic scaling. These mechanisms are thought to act together to enable the extraction of salient features and integration into cortical networks (Genzel et al. 2014). "Replay," is the reactivation of patterns of network activity that had occurred during previous experience and is thought to lead to potentiation of relevant synaptic connections in the cortex (see also chapter by Zhang, Deuker and Axmacher). "Scaling" refers to "…sleep homeostatically but nonspecifically regulating synaptic weights to improve the signal-to-noise ratio of memory traces" (Tononi and Cirelli 2006, 2014). The combined "push–pull" action of replay on the one hand ("push" equals potentiating "important" traces) and scaling on the other ("pull" equals weakening irrelevant traces) may together aid the construction and updating of memory networks in the cortex (Diekelmann and Born 2010; Genzel et al. 2014; Lewis and Durrant 2011).

Non-REM sleep (NREM) is especially important for systems consolidation of memories, with replay occurring throughout NREM and scaling becoming more dominant during deeper NREM also known as slow wave sleep (for the role of REM sleep in memory consolidation see Genzel et al. 2015c). Different oscillations have been shown to play specific roles in these processes (see also chapters by Bergmann and Staresina and by Maier and Kempter). Replay is initiated by a slow oscillation (0.5–1 Hz, seen as K-complex in the surface EEG) in the prefrontal cortex, which travels to the medial temporal lobe, where it is followed by a sharp-wave-ripple (100–200 Hz) in the hippocampus (Fig. 2). During the sharp-wave-ripple replay can be measured in the hippocampus and prefrontal cortex; this replay is then followed by a sleep spindle (13–16 Hz) (see also chapter by McDevitt and colleagues) deafferenting the prefrontal cortex from the hippocampus perhaps to enable integration into pre-existing networks (Genzel et al. 2014; Peyrache et al. 2009, 2011; Sullivan et al. 2014). Interestingly, motor cortex replay after motor sequence learning is seen later on in this oscillatory sequence, with replay occurring during, not before, the spindle (Ramanathan et al. 2015), even though this type of learning has been shown to involve the hippocampus (Genzel et al. 2015a; Schendan et al. 2003). Perhaps these two types of replay represent different mechanisms, or perhaps the delay is caused by the time that is needed for the information to travel across the cortex (Buzsaki 2015; Genzel and Robertson 2015). Replay can be measured in many brain areas (hippocampus (Wilson and McNaughton 1994), striatum (Pennartz et al. 2004), VTA (Gomperts et al. 2015), olfactory/prefrontal/visual/motor cortex (Barnes and Wilson 2014; Ji and Wilson 2007; Peyrache et al. 2009; Ramanathan et al. 2015; Yang et al. 2014) and disrupting sharp-wave-ripples and thus replay during sleep leads to a deficit in hippocampal led consolidation (Ego-Stengel and Wilson 2010; Girardeau et al. 2009), providing further evidence for the importance of sleep in systems consolidation.

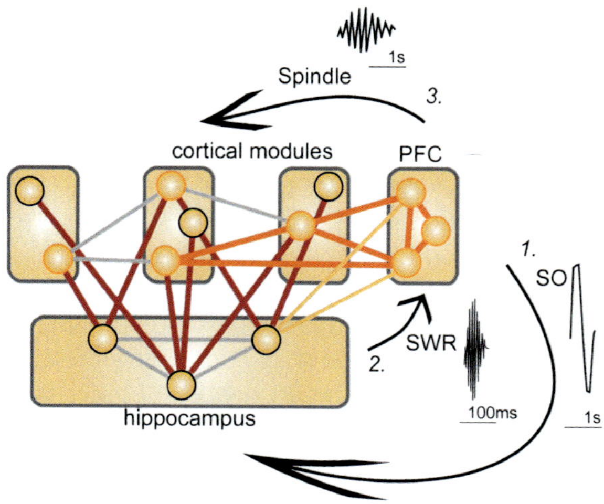

Fig. 2 Systems consolidation during sleep. Memory replay during NREM sleep is initialized by a slow oscillation (*SO*) traveling from the prefrontal cortex (*PFC*) to the medial temporal lobe and the hippocampus (*1*). There it is followed by a sharp-wave-ripple (*SWR*) and the reactivation of hippocampal and PFC neural ensembles of new memories (*2*). Subsequently, the sleep spindle can be seen in the cortex (*3*), deafferenting the PFC from the hippocampus, and accompanied by memory replay events farther along the cortex. Adapted from Genzel and Robertson (2015)

After replay strengthens important memory traces during early stages of NREM, scaling in the cortex is thought to occur during delta-waves (1–4 Hz) during deeper NREM.

Neural replay and related consolidation processes may preferentially unfold during periods in which no new information is being actively encoded (Mednick et al. 2011). NREM sleep is obviously one such period, but neural replay has also been found to occur during periods of quiet wake (Karlsson and Frank 2009). Indeed, in various kinds of learning tasks, post-learning wakeful resting has yielded effects on the consolidation of memory that are similar to the effects associated with NREM sleep (Dewar et al. 2012; Tambini et al. 2010). There seem to be three types of sharp-wave-ripple related memory replay, during the task, quiet rest and sleep; each contributing to memory but perhaps in a slightly different way (Dupret et al. 2010). Replay during task execution has been related to working memory (Jadhav et al. 2012) and replay during quiet rest seems to stabilise the hippocampal memory trace (Dupret et al. 2010). Some evidence suggests that only during sleep does replay occur in the cortex as well as the hippocampus (Peyrache et al. 2009; Ramanathan et al. 2015). This systems-wide replay during sleep may be due to increased cross-brain connectivity seen during light NREM in comparison to wake (Spoormaker et al. 2011). Then again, other evidence suggests that coordinated reactivation between the hippocampus and cortex may also occur during the awake state. For example, using fMRI, Tambini et al. (2010) found that following an associative learning task, enhanced hippocampal-cortical interactions occurred during subsequent rest, and the magnitude of resting correlations across subjects predicted individual differences in later associative memory for the previously learned items. Thus, exactly how consolidation processes differ between the sleep and awake states remains an open question. Nevertheless, at a minimum, it seems reasonable to suppose that most systems consolidation occurs during sleep, if no other reason than much of the awake state involves the active encoding of new information (Buzsaki 1989).

Interaction of Cellular and Systems Consolidation

Most discussions of cellular and systems consolidation treat them separately, as if they are independent from each other. Further, in mammals, cellular consolidation is also often used as a synonym for consolidation occurring in the hippocampus, since it is the classic brain area used for investigation of this process. Of course, the picture is more complex (Mednick et al. 2011). Only those memories initially stabilized in the hippocampus via cellular consolidation will survive long enough for systems consolidation to occur in the following sleep periods (Dupret et al. 2010). Furthermore, cellular consolidation is needed in the cortex directly after encoding (Lesburgueres et al. 2011; Tse et al. 2011) as well as later on for systems consolidation to be effective. Instead of viewing cellular and systems consolidation as separate entities, we need to focus more on their interactive dynamics. Interestingly, while a minimum amount of cellular consolidation in the hippocampus is needed for later systems consolidation, too much of the former can actually inhibit the latter. Very novel events lead to very large dopamine release in the hippocampus via pathways from the VTA and LC, enabling very strong cellular consolidation. This seems to "tag" memories to remain hippocampal with its more detailed memory representation and inhibits systems consolidation to occur in later sleep phases (Genzel et al. 2015b). In humans this form of memory is known as flashbulb memory.

Although the standard model holds that all declarative memories eventually become independent of the hippocampus (Fig. 3a), some memories, even though they may undergo systems consolidation, never become fully hippocampal independent. For example, while spatial memory learned in the watermaze shows signatures of cortical consolidation (Genzel et al. 2015b), most findings indicate that the hippocampus is always needed for retrieval (for review see Squire et al. 2015). This may be due to navigational issues during swimming, since similar dry land tasks usually become independent of the hippocampus and if a schema is present even at a rapid time-scale (Tse et al. 2007) (Fig. 3b). In other well-controlled animal studies, the temporal gradient of retrograde amnesia is also not always observed even on tasks that do not have an obvious spatial navigation aspect (e.g. context fear conditioning, Broadbent and Clark 2013), however the reasons for the empirical variability are not well understood. The transformation theory (former Multiple Trace Theory) argues that detail-rich, episodic memories

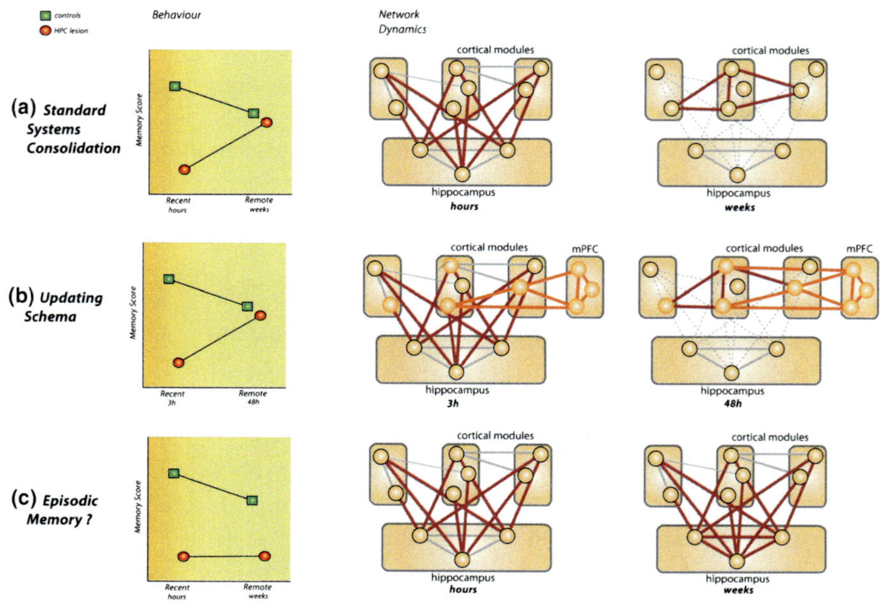

Fig. 3 Systems consolidation. **a** According to the standard model, the classic retrograde consolidation gradient in hippocampal lesioned animals shows that while memories are initially hippocampal dependent over time (weeks to months) the cortical network is eventually strengthened to be sufficient for memory recall. **b** When a schema, in form of a cortical network of relevant information, is present, this process occurs at a rapid rate (within 48 h). **c** Some memories never seem to become hippocampus independent. The transformation theory proposes that these represent episodic memories that always rely on the detailed representation in the hippocampus. Adapted from Squire et al. (2015)

remain dependent on the hippocampus (Nadel and Moscovitch 1997) (Fig. 3c) (see also chapters by Sekeres, Moscovitch and Winocur and by Cheng). However, this view remains controversial because patients with bilateral damage limited to the hippocampus generally do not exhibit the profound and selective loss of episodic memories that has been strikingly apparent in patients with more extensive cortical damage, such as Patient K.C. (Squire et al. 2015). Further, just because a memory can be retrieved when the hippocampus is lesioned, does not mean that the hippocampus is not involved when intact (Axmacher et al. 2009). Technical and methodological issues may also play a role. Studies have shown that using inactivation methods with different time scales (optogenetics, pharmacology, permanent lesions) leads to different results (Goshen et al. 2011; Otchy et al. 2015), which may be due to a time-lag in compensatory mechanisms (Goshen et al. 2011) or negative effects of transient manipulations on downstream circuits (Otchy et al. 2015).

Conclusion

The idea that memories consolidate began as a simple concept: the physiological processes associated with encoding perseverate for a limited period of time, thereby rendering the memory trace more resistant to retroactive interference than it otherwise would be (Müller and Pilzecker1900). After more than a century of research, one thing has become abundantly clear: consolidation is not a simple process. Our understanding of how consolidation works—and our awareness of how much we still do not know about it—have both increased enormously. Many of the intricate details of the cellular consolidation process have now been deciphered, but critical details are still largely unknown. In particular, how those cellular processes trigger the later processes that theoretically underlie systems consolidation—namely, neural replay and the associated exchange of information between the hippocampus and neocortex—remain mysterious. Fortunately, the tools needed to advance our understanding of what remains to be discovered are becoming more powerful than one might have hoped (or even imagined) only a few years ago. It therefore seems safe to assume that the remaining secrets of the memory consolidation process will be exposed sooner rather than later.

References

Axmacher N, Draguhn A, Elger C, Fell J (2009) Memory processes during sleep: beyond the standard consolidation theory. Cell Mol Life Sci 66(14):2285–2297

Bailey CH, Kandel ER, Harris KM (2015) Structural components of synaptic plasticity and memory consolidation. Cold Spring Harb Perspect Biol 7(7):a021758

Barnes DC, Wilson DA (2014) Slow-wave sleep-imposed replay modulates both strength and precision of memory. J Neurosci 34(15):5134–5142

Battaglia FP, Borensztajn G, Bod R (2012) Structured cognition and neural systems: from rats to language. Neurosci Biobehav Rev 36(7):1626–1639

Bethus I, Tse D, Morris RG (2010) Dopamine and memory: modulation of the persistence of memory for novel hippocampal NMDA receptor-dependent paired associates. J Neurosci 30 (5):1610–1618

Bliss TV, Lomo T (1973) Long-lasting potentiation of synaptic transmission in the dentate area of the anaesthetized rabbit following stimulation of the perforant path. J Physiol 232(2):331–356

Broadbent NJ, Clark RE (2013) Remote context fear conditioning remains hippocampus-dependent irrespective of training protocol, training-surgery interval, lesion size, and lesion method. Neurobiol Learn Mem 106:300–308

Buzsaki G (1989) Two-stage model of memory trace formation: a role for "noisy" brain states. Neuroscience 31(3):551–570

Buzsaki G (2015) Hippocampal sharp wave-ripple: a cognitive biomarker for episodic memory and planning. Hippocampus 25(10):1073–1188

Coutanche MN, Thompson-Schill SL (2014) Fast mapping rapidly integrates information into existing memory networks. J Exp Psychol Gen 143(6):2296–2303

Coutanche MN, Thompson-Schill SL (2015) Rapid consolidation of new knowledge in adulthood via fast mapping. Trends Cogn Sci 19(9):486–488

Cowansage KK, Shuman T, Dillingham BC, Chang A, Golshani P, Mayford M (2014) Direct reactivation of a coherent neocortical memory of context. Neuron 84(2):432–441

Dewar M, Alber J, Butler C, Cowan N, Della Sala S (2012) Brief wakeful resting boosts new memories over the long term. Psychol Sci 23(9):955–960

Diekelmann S, Born J (2010) The memory function of sleep. Nat Rev Neurosci 11(2):114–126

Dunsmoor JE, Murty VP, Davachi L, Phelps EA (2015) Emotional learning selectively and retroactively strengthens memories for related events. Nature 520(7547):345–348

Dupret D, O'Neill J, Pleydell-Bouverie B, Csicsvari J (2010) The reorganization and reactivation of hippocampal maps predict spatial memory performance. Nat Neurosci 13(8):995–1002

Ego-Stengel V, Wilson MA (2010) Disruption of ripple-associated hippocampal activity during rest impairs spatial learning in the rat. Hippocampus 20(1):1–10

Frankland PW, Bontempi B (2005) The organization of recent and remote memories. Nat Rev Neurosci 6(2):119–130

Frey U, Morris RGM (1998) Synaptic tagging: implications for late maintenance of hippocampal long-term potentiation. Trends Neurosci 21(5):181–188

Genzel L, Robertson EM (2015) To replay, perchance to consolidate. PLoS Biol 13(10):e1002285

Genzel L, Kroes MCW, Dresler M, Battaglia FP (2014) Light sleep vs. slow wave sleep in memory consolidation: a question of global vs. local processes? Trends Neurosci 37(1):10–19

Genzel L, Dresler M, Cornu M, Jager E, Konrad B, Adamczyk M, Friess E, Steiger A, Czisch M, Goya-Maldonado R (2015a) Medial prefrontal-hippocampal connectivity and motor memory consolidation in depression and schizophrenia. Biol Psychiatry 77(2):177–186

Genzel L, Rossato JI, Jacobse J, Morris RG (2015b) Differential consolidation induced by novelty and sleep associated with contrasting behavioural expression of hippocampal and cortical memory traces. SfN Poster 535:10

Genzel L, Spoormaker VI, Konrad BN, Dresler M (2015c) The role of rapid eye movement sleep for amygdala-related memory processing. Neurobiol Learn Mem 122:110–121

Girardeau G, Benchenane K, Wiener SI, Buzs ki G, Zugaro MB (2009) Selective suppression of hippocampal ripples impairs spatial memory. Nat Neurosci 12(10):1222–1223

Gomperts SN, Kloosterman F, Wilson MA (2015) VTA neurons coordinate with the hippocampal reactivation of spatial experience. Elife 14(4):05360

Goshen I, Brodsky M, Prakash R, Wallace J, Gradinaru V, Ramakrishnan C, Deisseroth K (2011) Dynamics of retrieval strategies for remote memories. Cell 147(3):678–689

Greve A, Cooper E, Henson RN (2014) No evidence that 'fast-mapping' benefits novel learning in healthy older adults. Neuropsychologia 60:52–59

Jadhav S, Kemere PC, German PW, Frank LM (2012) Awake hippocampal sharp-wave ripples support spatial memory. Science 336:1454–1458

Ji D, Wilson MA (2007) Coordinated memory replay in the visual cortex and hippocampus during sleep. Nat Neurosci 10(1):100–107

Karlsson MP, Frank LM (2009) Awake replay of remote experiences in the hippocampus. Nat Neurosci 12(7):913–918

Lesburgueres E, Gobbo OL, Alaux-Cantin S, Hambucken A, Trifilieff P, Bontempi B (2011) Early tagging of cortical networks is required for the formation of enduring associative memory. Science 331(6019):924–928

Lewis PA, Durrant SJ (2011) Overlapping memory replay during sleep builds cognitive schemata. Trends Cogn Sci 15(8):343–351

McGaugh JL (2004) The amygdala modulates the consolidation of memories of emotionally arousing experiences. Annu Rev Neurosci 27(1):1–28

Mednick SC, Cai DJ, Shuman T, Anagnostaras S, Wixted JT (2011) An opportunistic theory of cellular and systems consolidation. Trends Neurosci 34(10):504–514

Moncada D, Ballarini F, Viola H (2015) Behavioral tagging: a translation of the synaptic tagging and capture hypothesis. Neural Plast 2015:650780

Morris RG (2006) Elements of a neurobiological theory of hippocampal function: the role of synaptic plasticity, synaptic tagging and schemas. Eur J Neurosci 23(11):2829–2846

Morris RG, Garrud P, Rawlins JN, O'Keefe J (1982) Place navigation impaired in rats with hippocampal lesions. Nature 297(5868):681–683

Müller GE, Pilzecker A (1900) Experimentelle Beiträge zur Lehre vom Gedächtnis. Psychol Ergänzungsband (Experimental contributions to the science of memory) 1:1–300

Nadel L, Moscovitch M (1997) Memory consolidation, retrograde amnesia and the hippocampal complex. Curr Opin Neurobiol 7(2):217–227

Otchy TM, Wolff SB, Rhee JY, Pehlevan C, Kawai R, Kempf A, Gobes SM, Olveczky BP (2015) Acute off-target effects of neural circuit manipulations. Nature 528(7582):358–363

Pennartz CM, Lee E, Verheul J, Lipa P, Barnes CA, McNaughton BL (2004) The ventral striatum in off-line processing: ensemble reactivation during sleep and modulation by hippocampal ripples. J Neurosci 24(29):6446–6456

Peyrache A, Khamassi M, Benchenane K, Wiener SI, Battaglia FP (2009) Replay of rule-learning related neural patterns in the prefrontal cortex during sleep. Nat Neurosci 12(7):919–926

Peyrache A, Battaglia FP, Destexhe A (2011) Inhibition recruitment in prefrontal cortex during sleep spindles and gating of hippocampal inputs. Proc Natl Acad Sci 108(41):17207–17212

Ramanathan DS, Gulati T, Ganguly K (2015) Sleep-dependent reactivation of ensembles in motor cortex promotes skill consolidation. PLoS Biol 13(9):e1002263

Redondo RL, Morris RGM (2011) Making memories last: the synaptic tagging and capture hypothesis. Nat Rev Neurosci 12(1):17–30

Rossato JI, Bevilaqua LR, Izquierdo I, Medina JH, Cammarota M (2009) Dopamine controls persistence of long-term memory storage. Science 325(5943):1017–1020

Schendan HE, Searl MM, Melrose RJ, Stern CE (2003) An fMRI study of the role of the medial temporal lobe in implicit and explicit sequence learning. Neuron 37(6):1013–1025

Scoville W, Milner B (1957) Loss of recent memory after bilateral hippocampal lesions. J Neurol Neurosurg Psychiat 20(11):11–23

Sharon T, Moscovitch M, Gilboa A (2011) Rapid neocortical acquisition of long-term arbitrary associations independent of the hippocampus. Proc Natl Acad Sci 108(3):1146–1151

Spoormaker VI, Czisch M, Maquet P, Jancke L (2011) Large-scale functional brain networks in human non-rapid eye movement sleep: insights from combined electroencephalographic/functional magnetic resonance imaging studies. Philos Trans A Math Phys Eng Sci 369(1952):3708–3729

Squire LR (2009) The legacy of patient H.M. for neuroscience. Neuron 61(1):6–9

Squire LR, Genzel L, Wixted JT, Morris RG (2015) Memory consolidation. Cold Spring Harb Perspect Biol 7(8):a021766

Sullivan D, Mizuseki K, Sorgi A, Buzsaki G (2014) Comparison of sleep spindles and theta oscillations in the hippocampus. J Neurosci 34(2):662–674

Tambini A, Ketz N, Davachi L (2010) Enhanced brain correlations during rest are related to memory for recent experiences. Neuron 65(2):280–290

Tononi G, Cirelli C (2006) Sleep function and synaptic homeostasis. Sleep Med Rev 10(1):49–62

Tononi G, Cirelli C (2014) Sleep and the price of plasticity: from synaptic and cellular homeostasis to memory consolidation and integration. Neuron 81(1):12–34

Tse D, Langston RF, Kakeyama M, Bethus I, Spooner PA, Wood ER, Witter MP, Morris RG (2007) Schemas and memory consolidation. Science 316(5821):76–82

Tse D, Takeuchi T, Kakeyama M, Kajii Y, Okuno H, Tohyama C, Bito H, Morris RG (2011) Schema-dependent gene activation and memory encoding in neocortex. Science 333 (6044):891–895

van Buuren M, Kroes MC, Wagner IC, Genzel L, Morris RG, Fernandez G (2014) Initial investigation of the effects of an experimentally learned schema on spatial associative memory in humans. J Neurosci 34(50):16662–16670

van Kesteren MT, Fernandez G, Norris DG, Hermans EJ (2010a) Persistent schema-dependent hippocampal-neocortical connectivity during memory encoding and postencoding rest in humans. Proc Natl Acad Sci U S A 107(16):7550–7555

van Kesteren MT, Rijpkema M, Ruiter DJ, Fernandez G (2010b) Retrieval of associative information congruent with prior knowledge is related to increased medial prefrontal activity and connectivity. J Neurosci 30(47):15888–15894 van Kesteren MT, Ruiter DJ, Fernandez G, Henson RN (2012) How schema and novelty augment memory formation. Trends Neurosci 35(4):211–219

Wagner IC, van Buuren M, Kroes MC, Gutteling TP, van der Linden M, Morris RG, Fernandez G (2015) Schematic memory components converge within angular gyrus during retrieval. Elife 4

Wang SH, Redondo RL, Morris RG (2010) Relevance of synaptic tagging and capture to the persistence of long-term potentiation and everyday spatial memory. Proc Natl Acad Sci U S A 107(45):19537–19542

Wilson MA, McNaughton BL (1994) Reactivation of hippocampal ensemble memories during sleep. Science 265:676–679

Wixted JT, Cai DJ (2013) Memory consolidation. Oxford University Press, New York

Yang G, Wan Lai CS, Cicgen J, Ma L, Li W, Gan WB (2014) Sleep promotes branch-specific formation of dendritic spines after learning. Science 344(6188):1174–1178

Zola-Morgan S, Squire LR, Ramus SJ (1994) Severity of memory impairment in monkeys as a function of locus and extent of damage within the medial temporal lobe memory system. Hippocampus 4(4):483–495

Mechanisms of Memory Consolidation and Transformation

Melanie J. Sekeres, Morris Moscovitch and Gordon Winocur

Abstract Memory consolidation is a dynamic process occurring over the lifetime of a memory, yet the underlying mechanisms are not well understood. The hippocampus is considered to be a critical structure for the acquisition, initial storage, and retrieval of a memory, but there is considerable debate over the continuing role of the hippocampus in representing a memory as it ages. Studies in rodents and humans both point towards a reorganization of hippocampus-dependent memory traces in the cortex over time, but when and how long it takes these large-scale network changes to occur is uncertain. In this chapter, we address how a memory that is initially dependent on the hippocampus becomes represented in the cortex, independently of the hippocampus. We also discuss how the quality of the memory changes (transforms) as the trace reorganizes over time, with a focus on hippocampal-cortical interactions as described by Trace Transformation Theory (TTT), and consider the degree to which evidence related to the mechanistic basis of memory consolidation in rodents applies to complex human memory. We conclude that theories like TTT

M.J. Sekeres (✉) · M. Moscovitch (✉) · G. Winocur (✉)
Rotman Research Institute, Baycrest, Toronto, ON M6A 2E1, Canada
e-mail: melanie_sekeres@baylor.edu

M. Moscovitch
e-mail: momos@psych.utoronto.ca

G. Winocur
e-mail: gwinocur@research.baycrest.org

M. Moscovitch
Department of Psychology, Baycrest, Toronto, ON M6A 2E1, Canada

M.J. Sekeres · M. Moscovitch · G. Winocur
Department of Psychology, University of Toronto, Toronto, ON M5S 3G3, Canada

G. Winocur
Department of Psychiatry, University of Toronto, Toronto, ON M5S 3G3, Canada

M.J. Sekeres
Department of Psychology and Neuroscience, Baylor University, Waco, TX 76798, USA

G. Winocur
Department of Psychology, Trent University, Peterborough, ON K9J 7B8, Canada

N. Axmacher and B. Rasch (eds.), *Cognitive Neuroscience of Memory Consolidation*, Studies in Neuroscience, Psychology and Behavioral Economics, DOI 10.1007/978-3-319-45066-7_2

provide a new approach to thinking about consolidation as an ongoing and interactive process involving the hippocampus, mPFC, and other brain regions.

Keywords Memory · Consolidation · Transformation · Reconsolidation · Hippocampus · Medial prefrontal cortex · Episodic · Semantic · Context fear

Abbreviations

aCC	Anterior cingulate cortex
CREB	Cyclic-AMP response element binding protein
IEG	Immediate early gene
L-LTP	Late phase long-term potentiation
mPFC	Medial prefrontal cortex
MTL	Medial temporal lobe
MTT	Multiple Trace Theory
PFC	Prefrontal cortex
PRP	Plasticity-related protein
SCT	Standard Consolidation Theory
TGRA	Temporally-graded retrograde amnesia
TTT	Trace Transformation Theory
fMRI	Functional magnetic resonance imaging

Introduction

The foundation for modern ideas of memory consolidation goes back over a century (Ribot 1882). Müller and Pilzecker (1900) provided the first experimental evidence that memories require a certain amount of time to become permanent, but other early researchers suggested that a memory trace does not simply become imprinted and fixed in its original form. Rather, it was proposed that the process is an evolving one and that with each subsequent recall or reactivation, the 'consolidated' memory is updated, eventually resulting in a modified version that is quite different from the original (Bartlett 1932). In 1903, Burnham had already summarized both the tension, and complementarity, between the two views. "*The fixing of an impression depends on a physiological process. It takes time for an impression to become so fixed that it can be reproduced after a long interval; for it to become part of the permanent store of memory considerable time may be necessary. This we may suppose is not merely a process of making a permanent impression upon the nerve cells, but also a process of association, of organization of the new impressions with the old ones*" (Burnham 1903, p. 128). This debate has dominated memory research for the past several decades, as researchers attempt to understand (1) how a memory trace forms, (2) how a memory trace changes over time, and (3) the physiological mechanisms underlying the change, or transformation, of a memory over time.

The process of consolidation is generally discussed in binary terms of (1) *cellular (or synaptic) consolidation*: rapid post-translational changes in local synapse efficiency and structural modifications of local neuronal networks in the hippocampus[1] following memory acquisition; and (2) *systems consolidation*: a prolonged process following acquisition during which the memory trace reorganizes and distributes in cortical regions beyond the hippocampus (see also Chapters by Nissen and colleagues). Cellular consolidation and systems consolidation tend to be studied separately but it is important to keep in mind that they are part of one continuous and dynamic process (see also Chapter by Genzel and Wixted).

Within the field of systems level memory consolidation, it is often said that memories 'transfer' out of the hippocampus into the cortex, or 'form representations' in certain brain regions, but researchers are often vague when defining these terms. It is unlikely that the physical substrate of a memory is transported from one brain region to another the way a piece of mail is delivered from one address to the next. So how does a memory trace that was initially dependent on the hippocampus become 'represented' in the cortex, independently of the hippocampus? How does the quality of the memory change (transform) as the trace reorganizes over time? Do the same molecular cascades that result in physical changes in the local neuronal network underlying cellular consolidation extend to systems consolidation in humans and other species? In this chapter, we address these issues, with a focus on hippocampal-cortical interactions as described by Trace Transformation Theory (TTT), and consider the degree to which evidence from the mechanistic basis of memory consolidation in rodents translates to complex human memory. Consideration of these issues will help identify questions that are still outstanding in the field of memory consolidation and transformation.

How Does a Memory Trace Form? Mechanisms of Cellular Consolidation

For an experienced event to become stored as a long-term memory, it must form a physical 'memory trace' in the brain, in the form of synaptic connections within a set of neurons (Cajal 1894; Lorenté de No 1934; Hebb 1949, but see Gallistel and Matzel 2013). For this to occur, an experienced event must induce neuronal depolarization and an influx of intracellular Ca^{2+}, which initiates a downstream molecular cascade that results in transcription and translation of plasticity-related proteins (PRPs). These PRPs induce structural and functional changes in local neuronal networks, resulting in new, remodeled, or strengthened synaptic connections. Activity within these synaptic contacts leads to the development of a cell

[1]For this review, we will focus primarily on the type of memory that depends on the hippocampus for initial acquisition, including episodic memory in humans and episodic-like contextual and spatial memory in rodents.

assembly (Hebb 1949) of interconnected neurons known as a 'memory engram' (Semon 1923; Lashley 1950; Schacter et al. 1978; see Dudai 2012; Tonegawa et al. 2015; Josselyn et al. 2015 for review). These changes, which occur within minutes to hours of an experience, require a period of quiescence in order to stabilize. Interruption of this process via protein-synthesis inhibition, or interference from new learning events, can disrupt the stabilization process, leading to incomplete consolidation.

One specific transcription factor, cyclic-AMP response element binding protein (CREB), has been characterized as an essential 'memory gene' (Bourtchuladze et al. 1994) regulating the expression of many PRPs implicated in cellular consolidation, including growth factors, structural proteins, signal transduction proteins, and other transcription factors (see Alberini 2009; Sekeres et al. 2012b; Kandel et al. 2014 for review). The threshold for late-phase long-term potentiation (L-LTP), the electrophysiological correlate of long-term memory (Bliss and Collingridge 1993), is lowered in the presence of elevated CREB activity (Lonze and Ginty 2002). The enhanced neuronal excitability and plasticity induced by CREB-mediated transcription produces structural changes in neuronal morphology of activated cells, including growth of dendritic spines (the site of excitatory synaptic neurotransmission) (Barco and Marie 2011), and the formation of new synaptic connections (Martin and Kandel 1996). Prevention of these processes in the hippocampus during the post-acquisition phase disrupts subsequent consolidation at a systems level (Restivo et al. 2009b; Vetere et al. 2011; Cole et al. 2012). Conversely, enhancing the intrinsic excitability and availability of CREB in the CA1 or the dentate gyrus of the hippocampus allows for memory consolidation, even under weak learning conditions that do not otherwise support memory formation (Restivo et al. 2009a; Sekeres et al. 2010, 2012a).

For a weakly encoded memory to undergo cellular consolidation, it must be supported by de novo protein synthesis (Davis and Squire 1984). Due to the high intrinsic excitability and plasticity induced in neurons with elevated CREB function, enhancing intracellular CREB levels may similarly promote consolidation of a weakly encoded memory by artificially increasing the availability of PRPs during the post-encoding consolidation window (Barco et al. 2002). The synaptic tagging and capture hypothesis (Frey and Morris 1997, 1998) makes predictions that are consistent with these observations. It proposes that a weakly potentiated synapse can capture available PRPs produced by subsequent strong inputs to a nearby synapse. These newly available proteins provide the necessary plasticity factors required for synaptic remodeling and cellular consolidation of memories in the hippocampus (Wang et al. 2010; for review see Wang and Morris 2010). Similar phenomena, presumably dependent on the same underlying processes, have been observed in humans. Dunsmoor et al. (2015) showed that memory for neutral objects was enhanced if other objects from the same category were later paired with shock. This retroactive enhancement was observed only for weakly encoded items, and then only after a period of consolidation.

Together, these findings highlight the critical role of CREB in hippocampus-dependent memory consolidation. Similar enhancements have been demonstrated in multiple brain regions in rodents, including the amygdala (Josselyn et al. 2001), insular cortex (Sano et al. 2014), and the retrosplenial cortex (Czajkowski et al. 2014). The high intrinsic excitability and plasticity induced in neurons with enhanced CREB function are thought to bias the selection of these neurons for memory encoding (Han et al. 2007). According to the memory allocation paradigm proposed by Silva et al. (2009), items presented in close succession may be encoded by overlapping populations of neurons. Due to the high excitability induced by the initial consolidation phase (Zhou et al., 2009), a subsequently presented item or context is likely to re-activate a large subset of this same population of excitable neurons, and thus, be linked during retrieval (Silva et al. 2009; Rogerson et al. 2014; Cai et al. 2016).

The neuronal allocation phenomenon may partially account for the temporal contiguity effect observed in humans, in which items experienced in close temporal proximity to each other during memory acquisition tend to be bound together at retrieval (Landauer 1975; Moscovitch 1992; Howard and Kahana 1999; Sadeh et al. 2015; see Davachi and DuBrow 2015 for review). Intracranial recordings from within the medial-temporal lobe revealed neuronal coding of temporal links between events and their reinstatement at retrieval (Gelbard-Sagiv et al. 2008; Paz et al. 2010; Manning et al. 2011). Together, this evidence supports the idea that an overlapping population of neurons may be recruited during encoding of contiguous items.

The protein synthesis-dependent processes observed in rodents may underlie human memory consolidation but, due to limitations in studying this phenomenon in vivo, cellular consolidation is not well characterized in humans. One functional magnetic resonance imaging (fMRI) study of episodic memory (Ben-Yakov and Dudai 2011) suggests that the immediate post-encoding period shows a spike in hippocampal activity following the offset of a short movie clip. These investigators proposed that the increase in hippocampal activity reflects the initiation of synaptic consolidation for the event, similar to that seen in hippocampal replay in rodents following spatial learning (Foster and Wilson 2006). Presenting stimuli during this phase interferes with the purported consolidation process and leads to poorer subsequent memory. This interpretation, however, is difficult to assess due the poor spatial resolution of fMRI, and the sluggish hemodynamic response underlying neural activity.

Another limitation in extending the cellular consolidation literature to human memory processing is that investigations of cellular consolidation in rodents typically involve single trial learning, such as tone or context fear conditioning, or conditioned taste aversion. One exception is spatial learning which occurs over multiple spaced trials. These types of learning are unlike those used in most human memory experiments, which typically involve many different trials in rapid succession, leaving no time for cellular consolidation processes to occur for individual trials. If a similar consolidation process occurs in humans, it is unlikely that the temporal dynamics would be so rapid that each item or trial within a multi-trial acquisition session undergoes its own consolidation process. From this perspective, the subsequently presented information could actually be seen as interfering stimuli that render it difficult for any information to become consolidated at all. Similarly, in everyday life, humans rarely have an opportunity to rest quietly in the absence of any other stimuli while a newly acquired memory undergoes cellular consolidation and stabilizes. Despite this reality, somehow we manage to form strong and long-lasting memories. Precisely how this is accomplished in humans remains unclear.

One hypothesis relates to the architecture of the hippocampus. The hippocampus is thought to encode items in sparse, orthogonal synaptic connections which allow for unique representations (pattern separation) of distinct memories (Leutgeb et al. 2007; Bakker et al. 2008). Sadeh et al. (2013, 2016; see also Hardt et al. 2013) recently proposed that episodic memories that maintain a ‘recollective’ quality, are represented in such a way in the hippocampus, and as a result, are relatively resistant to interference from subsequently presented items. Information that is not represented in this manner, such as familiarity-based memories, will be more susceptible to interference. Familiarity-based memories, which are thought to be represented in the perirhinal cortex, lack the pattern separation capabilities of those controlled by the hippocampus. As a result, a subsequently presented similar item may activate the same synaptic contacts, resulting in disruption of the previously acquired item and its recall. Consistent with this view, memory loss in patients (Winocur and Weiskrantz 1976) and in animals (Winocur and Mills 1970; Winocur et al. 2012) with hippocampal dysfunction is exacerbated under conditions of high interference.

How Does a Memory Trace Change Over Time? Reorganization and Distribution of the Memory Network

Systems consolidation is well studied in rodents and humans, but our understanding of the physiological mechanisms underlying this gradual process is limited. As noted above, the dynamic cellular consolidation process occurs within a narrow window following memory acquisition. Once this window closes, the memory trace stabilizes and is less susceptible to protein synthesis inhibition, pharmacological disruption, or interference from new learning (Nader and Hardt 2009). The memory trace then transitions into a much more prolonged process of systems consolidation, in which it begins to form new synaptic connections within neuronal networks throughout the brain (distributed memory traces) (Dudai 2012).

Leading modern theories of memory consolidation agree that the hippocampus and cortex both form traces upon memory acquisition (McClelland et al. 1995; Nadel and Moscovitch 1997; Squire 2004; Winocur and Moscovitch 2011). Where they differ is in whether the hippocampus has a continuing role in memory storage and retrieval as the cortical traces are consolidated over time. They also disagree on the nature of information represented within the cortical traces. For example, declarative memories, that can be consciously and explicitly recalled, are viewed differently in discussions of memory consolidation. Declarative memory is traditionally subdivided into episodic (memory for a unique event occurring within a precise spatio-temporal context) and semantic (fact-based information related to an event, but lacking precise contextual details) components (Tulving 1972).

The Standard Consolidation Theory (SCT) treats both episodic and semantic memories equivalently in this regard. According to SCT, the hippocampus plays a time-limited role in the temporary storage of declarative memory (Squire 2004). Initially, memory is primarily supported by the hippocampus, but over time, the memory reorganizes and forms new traces in the cortex (see Chapter by Genzel and Wixted). As cortical representations strengthen, hippocampal involvement weakens. Eventually, the memory disengages from the hippocampus, and is represented, in its original form, in cortex (Alvarez and Squire 1994; McClelland et al. 1995; Frankland and Bontempi 2005). Early support for this position came from lesion studies of human medial temporal lobe (MTL) patients (Scoville and Milner 1957), monkeys (Zola-Morgan and Squire 1990), and rodents (Winocur 1990; Kim and Fanselow 1992; Squire 1992). Across species, a pattern of temporally-graded retrograde amnesia (TGRA) was observed in which memories acquired long before hippocampal damage were preserved, while memories acquired just prior to hippocampal damage were lost.

Upon further inspection, investigators noted that the preserved remote memories following MTL damage were qualitatively different from similarly aged memories in individuals with intact hippocampi (Nadel and Moscovitch 1997; Fujii et al. 2000; Corkin 2002; Rosenbaum et al. 2000, 2005). Surviving memories were more semantic in nature, and lacked perceptual, temporal, and contextual details—qualities that allow an individual to mentally re-experience a memory. There is a large body of neuropsychological evidence for TGRA for semanticized or schematic memory in MTL amnesic patients. Older semantic memories tend to be retained following MTL damage, but detailed autobiographical episodic memories tend to be lost regardless of how long ago they were formed. This suggests that the hippocampus is not required for the storage and retrieval of semantic memories (Steinvorth et al. 2005; St-Laurent et al. 2009).

The dissociation between preserved remote semantic memory and temporally extensive amnesia for episodic memory cannot be explained by SCT. In an attempt to reconcile shortcomings of SCT, Nadel and Moscovitch (1997) proposed the Multiple Trace Theory (MTT). This position holds that cortical memory traces are extractions of common elements, from repeated activations over time, which become integrated into existing schematic knowledge networks. Memory reactivations also allow for the formation of multiple, distributed traces of the precise contextual and perceptual details of the memory in the hippocampus. These details continue to depend on the hippocampus for their storage and retrieval.

Expanding upon MTT, Moscovitch (2007), Winocur and Moscovitch (2011) developed the Trace Transformation Theory (TTT), in which they theorized that the cortical version of the memory that develops over time is an extraction of the schematic, or gist-like features of the memory (see also Chapter by Cheng). As schematic memories lack many of the unique contextual and perceptual details of the original experience, they may be recovered without the hippocampus. To retrieve episodically detailed information, however, the hippocampus is thought to be required, regardless of the age of the memory. Critically, TTT holds that, in the intact brain, the detailed, hippocampus-dependent version of the memory, and the generalized or semanticized cortical version of the memory, can co-exist. The original memory and the transformed memory representations are in dynamic flux or interaction, and the conditions at retrieval influence which version of the memory is expressed. If the schematic version is sufficient to support retrieval in a given situation, the cortical version will be engaged. If retrieval of more intricate contextual details is required, the hippocampus is recruited, but the cortical version may still be engaged. Thus, the *nature* of the targeted memory mediates the reliance on the hippocampus over time. Data from our group support this idea in both healthy humans and rodents (Sekeres et al. 2015, 2016b) in which recently acquired, perceptually detailed episodic memories (or context-specific memories in rodents) highly engage the hippocampus. As the memory ages and loses specificity and detail, hippocampal activity declines. Under these conditions, areas in the medial prefrontal cortex (mPFC) become increasingly active, suggesting a shift towards cortical activity as the memory generalizes (see also Chapters by Fernandez and by Genzel and Battaglia). In line with the predictions of TTT, as long as the retrieved memory retains perceptual detail (or context-precision in rodents), the hippocampus continues to be similarly active at both recent and remote time points.

Parallels to this pattern are found in the animal literature. As in MTL patients, rodents with hippocampal lesions show TGRA for context memory (episodic-like memory). Lesions performed soon after acquisition (i.e. 1 day) abolish the recently acquired memory, whereas hippocampal lesions performed long after acquisition (i.e. 1 month) spare the remote memory (Kim and Fanselow 1992). Similar to humans, when the retrieved memory was probed, it became clear that the hippocampus-independent version of the memory was qualitatively different from one that involves the hippocampus. When tested at short delays, control rats expressed fear memory in the original context, but not in a novel context (Winocur et al. 2007), whereas lesioned rats did not express the fear memory in either context; at long delays, however, lesioned rats exhibited fear in both contexts, suggesting that the memories at each delay differed fundamentally from each other. This observation (see also Wiltgen and Silva 2007; Goshen et al. 2011; Einarsson et al. 2014), led to the proposal that the hippocampus-independent version of the memory is a generalized memory which lacks context-specificity. Interestingly, Winocur and colleagues observed that control animals also exhibited this generalized memory when tested in the novel context after a long delay.

There are indications in the literature that the context-specific memory may co-exist with the transformed generalized memory. For example, Winocur et al. (2009) used a reactivation paradigm (i.e. briefly replacing the rat back in the conditioning chamber) to show that a stable consolidated memory may be returned to a labile state during which it is susceptible to disruption (see also Nader et al. 2000; Sara 2000; Debiec et al. 2002; see also Chapter by Kessler, Blackwell and Kehyayan). Winocur et al. (2009) found that the brief reminder experience reinstated (1) context-specificity and (2) hippocampal-dependency to the fear memory. Following reactivation, lesions to the hippocampus abolished memory for the fear response in both contexts, but did not impair retrieval of a non-reactivated remote memory, suggesting that the cortical representation of the memory continues to support retrieval of the remote memory. Together, these observations were taken as strong support for the notion within TTT that the context-general, hippocampus-independent memory dominates at a remote time point, but the context-specific memory continues to remain represented in the hippocampus. The latter can be re-engaged following reactivation, suggesting the two representations can co-exist. These findings, it should be noted, are not compatible with SCT, which argues that once represented in the cortex, the memory can no longer be returned to a hippocampus-dependent state.

Studies of rodents point to the anterior cingulate cortex (aCC) of the mPFC as a region that is involved in remote memory (Bontempi et al. 1999; Restivo et al. 2009b; Einarsson and Nader 2012). Increased expression of the immediate early gene (IEG) c-Fos, a commonly used marker of neuronal activity, (Greenberg and Ziff 1983) is observed in the aCC following retrieval of remote context memory. Conversely, inactivation of the aCC at a remote time point results in decreased freezing during context memory testing (Frankland et al. 2004), and decreased IEG expression in the aCC (Goshen et al. 2011). Interestingly, Einarsson et al. (2014) found that pharmacological inactivation of both the aCC and the hippocampus disrupted retrieval of the context memory, whereas inactivation of either aCC or hippocampus alone did not impair retrieval, suggesting that, at the remote time point, either structure can support the memory representation. Together these findings suggest that contextually-detailed memories continue to rely on the hippocampus, whereas over time a generalized memory trace develops in the aCC which can also support memory retrieval under certain conditions.

In humans, as in rodents, damage to the mPFC impairs episodic memory. There is no evidence, however, to suggest that it affects remote memories more than recent ones, though no-one has investigated this systematically. Deficits also extend to semantic memory, as might be expected if the mPFC is implicated in representing gist and schemas. Here, too, there is some dispute as to whether episodic memory is affected more than semantic memory (Gilboa et al. 2002; Dalla-Barba and La Corte 2015). Damage to mPFC in humans, however, does not lead merely to a loss of memory as one might infer on the basis of animal studies, but also to confabulation, a severe form of memory distortion in which the individual reports patently false memories, without any intention to deceive (Moscovitch 1989, 1995a, b; Gilboa et al. 2002; Nieuwenhuis and Takashima 2011; Hebscher et al. 2015). There is debate in the literature as to the nature of the deficits underlying this disorder, but a storage failure is not among the leading candidates (Ghosh et al. 2014). Instead, the deficits seem to result from corrupted or over-inclusive schemas which impair memory encoding and search, likely combined with poor monitoring of retrieved memories, resulting in failure to satisfy the criteria or goals of the memory task (Moscovitch and Winocur 2002; Gilboa et al. 2006; Ciaramelli and Ghetti 2007;

Moscovitch et al. 2016). These deficits may also impair the individual's ability to inhibit competing memories that are not relevant for the task at hand (Schnider 2008).

Reconciling the human studies with the rodent data, the most parsimonious explanation is that mPFC is implicated in processing or representing schemas, which guide perception, memory encoding and retrieval, and also provide a template against which retrieved memories can be compared to ensure that only plausible responses are emitted. Although in humans, damage to such a mechanism can manifest itself as confabulation, in non-humans, lacking verbal report, the deficit will just be one of impaired memory. In both cases, the deficit will appear when memory search and retrieval are primarily schema-dependent and strategic. When cues are sufficiently strong to specify the target memory, no deficits will be evident. This explanation may account for the relatively good performance on recognition or cued recall tasks under some conditions in humans, and on recently learned tasks in rodents, where the cues or contexts are at their most potent and specific, as compared to when the memory is more distant (Moscovitch et al. 2016).

A Case for the Transformation of Spatial Memory

Since O'Keefe and Nadel (1978) published their seminal book "The Hippocampus as a Cognitive Map", one type of spatial memory was postulated to depend on the hippocampus's ability to form and maintain a cognitive map, namely an allocentric (viewpoint-independent) representation that captured the configuration of the environment (O'Keefe and Nadel 1978; Morris et al. 1982; Morris 1984). Spatial memories that are non-allocentric, but dependent on egocentric coordinates, routes or specific landmarks, could be formed and retained without the hippocampus.

Given the wealth of evidence supporting cognitive map theory, it was surprising to discover that, with sufficient time and practice since acquisition, humans with hippocampal damage exhibit accurate allocentric spatial memories of familiar neighborhoods. This observation was first noted in MTL lesion patients with extensive episodic memory impairment (Milner et al. 1968; Zola-Morgan et al. 1986; Beatty et al. 1987; Teng and Squire 1999; Rosenbaum et al. 2000; Corkin 2002). Patients could navigate normally in their neighborhoods and pass tests of mental navigation that were diagnostic of cognitive maps, such as finding the next shortest route to a goal when the shortest one was blocked, or the shortest route "as the crow flies" between two locations (vector mapping). Navigation along major routes was normal, though it was impaired along smaller, side streets (Maguire et al. 2006). They could even draw accurate maps of their neighborhoods and floor maps of their homes (Beatty et al. 1987), though they were not as detailed as those of controls (Rosenbaum et al. 2000, 2004). Later, functional neuroimaging studies of healthy young adults corroborated the conclusion that performance on these tests of mental navigation did not activate the hippocampus if the memory was acquired over a year ago and had become a familiar environment, but did activate the hippocampus if the memory was more recent (Rosenbaum et al. 2005; Hirshhorn et al. 2012). Further investigation suggested that this preserved spatial ability relied on a schematized topographical representation, not unlike a skeletal cognitive map, that could be supported by extra-hippocampal representations. Detailed, perceptually-rich internal representation of the environment, even if acquired long ago, continued to require the hippocampus (Rosenbaum et al. 2000, 2004; Hirshhorn et al. 2012).

This way of thinking about remote spatial memory in humans was consistent with novel findings of spared remote spatial memory in rodents with hippocampal lesions. Winocur and colleagues reared rats in a 'village' environment for several months prior to lesioning the rats' hippocampus. Such prolonged pre-morbid experience allowed for the development of a map of the environment from multiple perspectives, similar to the way individuals learn the layout of their home and neighborhood. Hippocampal lesions performed after the development of this map did not impair the rat's ability to navigate along highly familiar routes in the village (Winocur et al. 2005b). This paralleled the finding in MTL lesion patients. When major spatial cues within the environment were re-configured, or when previous routes were blocked, however, lesioned rats exhibited significant spatial memory impairments. This suggests that rodents were relying on a schematic representation of the environment that differed from that of controls; they lacked a detailed, cohesive allocentric representation which would allow them to adapt and re-map to accommodate changes in the environment (Winocur et al. 2010). Together, these findings support the idea that both contextually-detailed and schematic components of spatial memory may develop for well-learned spatial environments, and support the position that spatial memory undergoes a similar transformation as other episodic memories during systems consolidation. In line with TTT, non-transformed spatial memories always depend on the hippocampus, and a result, remain vulnerable to disruption following hippocampal damage (Winocur et al. 2005a, 2013). Despite the similarities with respect to detail, there is a discrepancy between the human and animal findings in that humans with hippocampal lesions, unlike rats, seem able to adjust better to spatial changes, such as blocking routes. Future research will determine whether the nature of the underlying extra-hippocampal representation differs between rats and humans, or whether this difference in performance arises because humans' greater intelligence enables them to compensate better than rats in operating on an impoverished, and fundamentally different, representation than the one mediated by the hippocampus.

IEG expression studies in rodents showed that the aCC emerges as a key extra-hippocampal region supporting remote spatial memory (Bontempi et al. 1999; Frankland et al. 2004; Maviel et al. 2004; Teixeira et al. 2006). Reports of reduced hippocampal activity in rodents during remote spatial memory retrieval was initially taken as support for SCT (Frankland and Bontempi 2005). Recent evidence, however, is consistent with the TTT as indicated above. The evidence suggests that the remote spatial memory that is represented is different from the initial memory represented in the hippocampus; by comparison to the detailed memory in the hippocampus, the memory that develops in the mPFC is more schematic (Tse et al.

2007, 2011; Richards et al. 2014). As with context memory, it is not a 'transfer' of the memory trace that takes place, but rather, the development of a distributed network in the cortex consisting of multiple traces of the schematic elements of the spatial memory which, in the absence of the hippocampus, can support retrieval of the schematic spatial memory.

Studies of the effects of mPFC lesions on spatial memory in humans are rare. In the one reported study, participants were asked to travel from one familiar location to another, but erred in doing so, and sometimes got lost. The deficit seems to have arisen from an inability to keep distracting or intruding spatial information at bay, such as following inappropriate routes triggered by cues in the environment while navigating to a destined location, rather than from a loss of remotely learned spatial information (Ciaramelli 2008). As result, the patient is sidetracked from the intended path.

Evidence for Systems Consolidation and Memory Transformation in the Healthy Rodent Brain

The fact that context-specificity of a memory and hippocampal dependence can be restored by reminders speaks to the dynamic interaction between the hippocampus and extra-hippocampal structures in memory retention and retrieval. It also suggests that some vestige of the original specific memory is likely retained by the hippocampus, and contributes to remote memory performance. Evidence in support of this interpretation comes from investigations using fast temporal and precise spatial resolution to identify engram cells thought to be the physical storage site of a specific memory (Tayler et al. 2013; Denny et al. 2014; Josselyn et al. 2015; Tonegawa et al. 2015). Previous studies with rapid optogenetic inactivation of tagged engram cells suggest that the hippocampus may be the default retrieval structure, but when it is unavailable, the cortical version of the memory is expressed (Goshen et al. 2011). These studies provide strong support for the idea that, in the healthy brain, a specific context memory may continue to be supported by specific cell assemblies. Although, over time, the memory network may reorganize and distribute in the cortex, activation of the original cell assembly can result in expression of the context memory (Goshen et al. 2011; Liu et al. 2012; Ramirez et al. 2013). While these cells continue to play a critical role in the storage and retrieval of the memory, stimulation of these cell assemblies alone likely does not activate the entire memory trace. Rather, stimulation may induce activity in other parts of the network, which together support retrieval of the memory.

Recent evidence in mice supports this position. Using an inducible transgenic mouse model, Denny et al. (2014) were able to tag hippocampal neurons active during acquisition of a context fear memory. Selectively silencing, via optogenetic inhibition, this neuronal network at the time of remote memory retrieval abolished the freezing response, indicating impaired memory for the conditioning context (Denny et al. 2014). In light of the previous discussion regarding distributed memory representations, it is puzzling that the extra-hippocampal neuronal assembly would not be sufficient to support retrieval of the context memory in the absence of the hippocampal engram cells. These results suggest that in the healthy brain, rapid inactivation of a normally functional part of the memory network disrupts the coordination of the entire network. In the case of hippocampal lesions, or slow-acting pharmacological inhibitors, the long temporal lag between inactivation and retrieval may allow the network sufficient time to adapt and to compensate for the hippocampal disruption (Goshen et al. 2011). From this perspective, so long as it exists, the original hippocampal neuronal assembly formed during encoding may be the default region which coordinates the rest of the memory network, supporting retrieval of the original, detailed version of the memory (Lee et al. 2016). When this original neuronal assembly is no longer accessible, if the brain has sufficient time to compensate, other components of the memory network can come online to support retrieval. The retrieved version, however, may be a more generalized representation.

Limitations to Rodent Models of Memory Consolidation

To date, few studies have attempted to characterize the brain-wide remote memory network in rodents (Bontempi et al. 1999; Wheeler et al. 2013). Studying the mechanistic basis of consolidation in animals using IEG expression as a marker of neuronal activity is a time-consuming and labor intensive process, and therefore it is more practical to study changes at the neuronal level within a limited number of regions (but see Vousden et al. 2015 and Ye et al. 2016 for novel brain-wide imaging approaches). Importantly, in the analyses reported in these papers, the hippocampus emerges as a crucial hub linking several regions even when the memory is remote. While the mPFC is a major hub in the remote memory network, in rodents, it is only one node of a larger retrieval network involving regions, including the hippocampus and posterior cingulate cortex, that have been identified in the human recollection network (Rugg and Vilberg 2013; Wheeler et al. 2013). Further research is needed to understand how damage to key nodes of this network changes activity and functional connectivity during recent and remote memory retrieval (see Vetere et al. 2015 for preliminary investigation silencing key nodes of the rodent functional connectome). Although novel gene expression techniques, in vivo Ca^{2+} imaging, and high resolution fluorescence microscopy in rodent models provide valuable insight into neural dynamics of consolidation, they are limited in their ability to visualize brain-wide changes in activity over *multiple retrieval events within the same animal*. fMRI approaches in humans allow us to overcome this drawback, and emphasize the importance of considering changes in the overall retrieval network over time. This approach also allows for a more nuanced investigation of changes to the quality of memory as remote memory networks reorganize.

Evidence for Systems Consolidation and Memory Transformation in the Healthy Human Brain

Prior to the development of intact brain neuroimaging, our knowledge of the role of the hippocampus and the mPFC in human memory was based largely on loss-of-function lesion studies in patients. As noted earlier, beginning with H.M., studies of the effects of MTL lesions suggested that remote memories were spared, but memories acquired just before the lesion and subsequent to it were impaired (Scoville and Milner 1957; Penfield and Milner 1958). More careful observation indicated that only semantic memories and gist-like memories of autobiographical events followed that pattern, whereas richly-detailed, episodic memories were impaired across the lifetime (Sanders and Warrington 1971; Nadel and Moscovitch 1997; Moscovitch et al. 2005, 2006, 2016). These findings in patients, however, are complicated by the fact that damage often extended to adjacent medial temporal regions (Squire and Bayley 2007; Squire and Wixted 2011), contributing to the debate over the continuing role of different MTL regions in remote memory.

Modern functional neuroimaging techniques allow the use of multivariate analyses to identify changes in brain-wide patterns of activity underlying memory encoding and retrieval over time. In the healthy individual, hippocampal activity declines after 1 week as episodic memory recollection fades (Viskontas et al. 2009), but remote autobiographical episodic memories which retain their vividness and perceptual detail continue to be associated with high hippocampal activity (Addis et al. 2004; Gilboa et al. 2004; Sheldon and Levine 2013). Vividly retrieved remote autobiographical memories are also correlated with increased activation of pre-frontal cortical regions, particularly the ventromedial prefrontal cortex, supporting it as a candidate region for remote episodic memory processing in humans (Bonnici et al. 2012). Testing autobiographical memory is a valuable way of assessing the nuanced quality of human memory, but verifying the accuracy of memory details remains a problem. To address this limitation, researchers have used functional neuroimaging during encoding and retrieval of film clips of everyday events (Ben-Yakov and Dudai 2011; St-Laurent et al. 2014, 2016) as a means of inferring how brain networks change as the quality of naturalistic episodic memory changes over time (Furman et al. 2012; Sekeres et al. 2015, 2016b). This approach retains control over the conditions during encoding, as well as the ability to assess the accuracy of memory content.

Behavioral studies have confirmed that perceptual, contextual, and central schematic elements that together make up human episodic memories decline at different rates over time, with perceptual and contextual details declining more rapidly (Thorndyke 1977; Bahrick 1984). As the hippocampus is critical for the representation of those perceptual and contextual details, our group used memory for film clips to determine how the network of brain activity changes as different elements of episodic memory are lost over time (Sekeres et al. 2016a). A fMRI study of healthy young adults revealed how the retrieval network reorganizes as the quality and content of memory for events in the film clips changes (transforms) over the course of one week. In line with the predictions of TTT, we found that (1) immediately after encoding, retrieval of perceptually-detailed memory for events in the film clips highly engaged the hippocampus; (2) memory for perceptual details declines over time, whereas memory for the central story elements is retained. This is accompanied by a reduction of hippocampal activity and an increase in mPFC activity during retrieval of the 7 day old memory; (3) vivid and perceptually detailed retrieval of the film clips highly engages the hippocampus at both immediate and 7d delayed time points. Vivid retrieval of the 7 day old memory, however, was also supported by strong activity in the mPFC, consistent with the idea that the memory becomes distributed and supported by a cortical network over time, but also continues to depend on the hippocampus for retrieval. It remains unclear if the mPFC is playing a supportive or redundant role during retrieval of the vivid memory. These results provide further support for the idea that both the hippocampally-dependent detailed version of a memory, and the cortically-dependent schematic version can co-exist (Sekeres et al. 2015, 2016b). In the latter regard, the evidence is consistent with that obtained in rodents. It may be that these parallel representations interact with each other when intact, but also build in compensatory representations that can be accessed by different cues and can operate under different task demands.

Dudai and colleagues used a short documentary to test time-dependent changes in the quality of naturalistic episodic memory. In line with our findings, they found that the hippocampus continued to be recruited during accurate recall of memory details in the weeks and months following encoding; as recognition accuracy for events in the film decreased over time, hippocampal activity also decreased, though accuracy was still correlated with the extent of hippocampal activation (Mendelsohn et al. 2010; Furman et al. 2012). Together, these findings provide converging evidence for the time-dependent reorganization of the memory network which shifts towards frontal activity as the memory ages and loses precision. The quality, rather than the age, of the retrieved memory appears to mediate hippocampal engagement.

What Are the Mechanisms of Systems Consolidation?

Investigations over the past several decades have significantly advanced our understanding of the molecular mechanisms supporting cellular consolidation (Kandel et al. 2014). Comparatively little work of this nature has focused on mechanisms underlying systems consolidation, and remodeling of the distributed memory network.

One mechanism possibly implicated in remodeling is reconsolidation, a process, as described above, in which memory reactivation makes the memory trace temporarily labile, and vulnerable to disruption or alteration, and in need of further consolidation (re-consolidation) if it is to be retained (Misanin et al. 1968, Nader et al. 2000; Sara 2000). The few available studies in rodents suggest that similar molecular mechanisms are involved in this process as in initial consolidation. Debiec et al. (2002) were the first to demonstrate that remote memory undergoes a similar protein-synthesis dependent systems reconsolidation process in the hippocampus. Here, an infusion of the protein-synthesis inhibitor anisomycin into the hippocampus following reactivation of a remote context memory disrupted the restabilization of the memory, and abolished freezing when re-tested in the context. In the absence of reactivation, however, anisomycin had no effect on the subsequent retrieval of the context memory.

Several recent studies involving rodents have investigated the possibility of *enhancing* systems consolidation and using the same PRPs that boost cellular consolidation. Shema et al. (2011) found that virally increasing PKMζ, a protein implicated in the maintenance of LTP (Ling et al. 2002), in the insular cortex 6 days after conditioned taste aversion training (at a time when the window for protein-synthesis dependent cellular consolidation has closed) enhanced subsequent memory even though the memory had not undergone reactivation. It is surprising that, in the absence of direct reactivation, increasing PKMζ would enhance the retrieval of a consolidated memory; it is possible, however, that enhancing cortical plasticity at this time facilitated systems consolidation, which is presumed to be an ongoing process that, over time, continues to form multiple memory traces in the cortex.

To directly test how enhancing PRPs during initial consolidation may have enduring effects on systems consolidation, Sekeres et al. (2012a) virally over-expressed the transcription factor CREB in the dentate gyrus of the mouse hippocampus prior to context fear conditioning. This served to potentiate plasticity during the initial cellular consolidation phase. One month later, the memory was tested in the original context, and in a novel context. Importantly, using this transient viral expression technique, CREB levels were elevated during memory acquisition (within the window for cellular consolidation), but had returned to basal endogenous levels several days later. When tested one month later, mice conditioned with high hippocampal CREB continued to show robust, context-specific memory, suggesting that increasing plasticity during acquisition facilitated the cellular consolidation of the context memory. It is likely that these neurons were re-engaged during remote memory testing, leading to the retrieval of persistent, context-specific remote memory.

This finding does not argue against the development of schematic, or context-general memory traces in the cortex, but suggests that the hippocampus-dependent version of the memory dominates at retrieval. Similarly, in line with Debiec et al.'s report (2002) of protein synthesis-dependent systems reconsolidation, over-expression of CREB in the hippocampus just prior to reactivation of a remote context memory enhanced reconsolidation of context-specific fear memory. This suggests that not only can a reminder bring a generalized remote memory back to a hippocampus-dependent, context-specific state, but the same molecular mechanisms underlying initial cellular consolidation also support systems consolidation and reconsolidation. Future studies are needed to determine the downstream effects (i.e. synaptic remodeling) that support this facilitation of systems consolidation.

In the absence of explicit retrieval or reactivation, which may re-initiate synaptic consolidation or reconsolidation processes throughout the brain, how does a physical trace of a memory form in the cortex? One proposed mechanism is the 'active consolidation in sleep' hypothesis. This position proposes that offline hippocampal reactivation or replay during slow wave sleep or rest (Diekelmann et al. 2011; see Atherton et al. 2015 for review) results in multiple traces distributed throughout the hippocampus and cortex. This replay or reactivation initiates new waves of synaptic consolidation that gradually support physical changes in the morphological neuronal network structure (including dendritic growth and the formation or remodeling of dendritic spines). It also promotes changes in synaptic strength and efficiency in neurotransmission within the cortical neuronal networks. Consistent with TTT, studies in humans suggest that these sleep-dependent changes can be accompanied by a transformation of the initial memory trace from one that is context-specific to one that is more schematic, retaining the gist but not the context (Cairney et al. 2011; Lewis and Durrant 2011, see Chapters by Schönauer and Gais as well as Rauss and Born).

These ongoing or recurrent waves of synaptic consolidation can be considered 'subroutines of systems consolidation' (Dudai 2012; Dudai et al. 2015), and require that the memory be reactivated in order to undergo systems consolidation across a distributed network. Accordingly, not all memories will undergo systems consolidation or transformation. A memory that is never, or rarely, reactivated will not have undergone sufficient waves of synaptic consolidation to allow the formation of multiple distributed traces in the cortex. This is not to say that a memory undergoes the synaptic consolidation process every single time it is reactivated or replayed, as there are boundaries limiting the conditions that will initiate the reconsolidation process, such as novelty, and the strength and the age of the memory (Dudai 2012; Finnie and Nader 2012), as well as the context in which the memory is reactivated (Hupbach et al. 2008).

This explanation does provide an attractive mechanism for MTT and TTT, and is also in line with the growing evidence that memories, especially contextually-bound memories, continue to be represented in their original neural ensembles. This does not argue against the parallel development of other traces in mPFC (see also Chapters by Fernandez as well as Genzel and Battaglia). These cortical traces are likely not randomly formed, but rather, integrate into existing schematic memory networks represented in the mPFC, updating the general knowledge network. Accordingly, newly acquired memories that are consistent with existing schemas may be rapidly consolidated in the mPFC, where they may be supported independently of the hippocampus (Tse et al. 2007; van Kesteren et al. 2012, 2013; Richards et al. 2014; Ghosh and Gilboa 2014).

Studies on reconsolidation in humans are few, and predominantly behavioral, though the results have neurobiological implications and are consistent with the above hypotheses (see Chapter by Kessler, Blackwell and Kehyayan). Schiller and her colleagues (2010) have shown that reactivating a fear-conditioned response can promote extinction of the fear response if the extinction procedure is administered during a temporal window of about 6 h after reactivation, when the reactivated memory is labile.

Other studies, however, have shown that memory reactivation and subsequent reconsolidation also can lead to alteration of the existing memory trace, rather than its elimination. Hupbach, Nadel and their colleagues showed that reactivating memory for a set of objects alters that memory if another set of objects is presented shortly after reactivation. The memory for the initial items is not reduced, but the new items are incorporated into it, leading participants to (mistakenly) recall the new items along with the old. In this way, reconsolidation can serve as a mechanism for memory updating at retrieval. Importantly, Hupbach et al. (2008) noted that such updating only occurs if memories are reactivated in the initial spatial context.

Chan et al. (2009) showed that reactivating a memory of an event, makes individuals more prone to the misinformation effect, namely the incorporation of false information, delivered after the event, into an eyewitness account. Such alterations of an existing memory can occur even if the memory had been acquired days earlier (Chan and LaPaglia 2013). These observation are reinforced by studies of reconsolidation, updating, and false memory conducted by St. Jacques and Schacter (2013) using a novel paradigm in which participants toured a museum while wearing a camera that recorded the event. At test, they were presented with photos taken during their tour, along with photos taken from a different tour of the same museum. Subsequent recognition memory was better for those photos that matched highly reactivated memories, but importantly, reactivation also increased false memories of the novel photos. A follow-up fMRI study (St. Jacques et al. 2013), showed that highly reactivated true memories were associated with increased activity in the retrosplenial, parahippocampal and inferior temporal cortices. These areas are associated with contextual reinstatement, particularly involving spatial context and scene construction, and are consistent with Hupbach et al. (2008) findings. Importantly, false recognition of the novel photos was associated with activation of the anterior hippocampus and vmPFC, regions identified by TTT as implicated in memory transformation (Winocur et al. 2010; Winocur and Moscovitch 2011; Kroes and Fernández 2012).

Timeline of Systems Consolidation

Studies in rodents and humans both point towards a reorganization and distribution of the memory network over time, but *when* and *how long* it takes these large-scale network changes to occur is uncertain. One major limitation to this field of study is the broad temporal range over which systems consolidation can occur (Varela et al. 2016). Using fMRI in humans, we see evidence that large-sale reorganization of declarative memory networks can be detected only 24-hrs following associative memory acquisition (Takashima et al. 2009; Ritchey et al. 2015), and one week following encoding of complex episodic memory for film clips (Sekeres et al. 2015, 2016b). Other naturalistic studies find that reorganization of the autobiographical memory network follows a similar pattern extending over a much longer period of months and years (Nieuwenhuis and Takashima 2011; Maguire 2014). Moreover, if we take the time-dependent effects of damage to the hippocampus as a marker of systems consolidation, it seems that the process could last years. It is likely that the reorganization of memory networks begins early, and continues over the lifetime of the memory. This process may be dependent on memory reactivation, but there is evidence that similar network changes may be detected for very old memories that had not been recently reactivated (Bonnici et al. 2012).

In rodents, hippocampal neurons activated during context memory acquisition are reactivated when the context memory is retrieved 2 days later; however, fewer hippocampal cells show this overlap when the memory is retrieved 2 weeks later, supporting the idea of a pruning of activated hippocampal engram cells and a reorganization and broadening of the memory network over time (Tayler et al. 2013). It is likely that this two-week old memory has already begun the transformation process, and has formed traces in the mPFC. As a result, fewer of the original hippocampal neurons will be engaged as the extra-hippocampal regions mediating the reorganized memory trace become increasingly active during remote memory retrieval.

Several studies have investigated how preventing morphological modifications to memory networks at various stages during systems consolidation affects remote memory in rodents. Formation of a remote memory is associated with dendritic spine growth in the aCC, which is prevented if the hippocampus is lesioned one day after memory encoding (Restivo et al. 2009b). This suggests that the cortical plasticity supporting remote memory formation may be driven by interactions with the hippocampus occurring early in the consolidation process (Vetere et al. 2011). If spine growth is suppressed in the hippocampus, recent context memory consolidation is similarly impaired (Cole et al. 2012). There is evidence for the development and clustering of dendritic spines on pyramidal motor neurons following motor task learning (Xu et al. 2009; Fu et al. 2012), and Tonegawa and colleagues have proposed that a similar spine clustering may support the development of the engram following episodic memory acquisition (Govindarajan et al. 2006). To date, very little is known about the dynamics of plasticity mechanisms in humans, and while this spine growth and remodeling has not been studied in vivo, it suggests one potential neural mechanism supporting changes in large-scale memory networks in the human brain.

Conclusion: Transforming the Concept of Consolidation

The term ‘consolidation’ is commonly used to describe the process by which memories become represented in the brain and available for retrieval, but it is clear from the dictionary definition of the term consolidation, ‘to make firm’, that it is a misnomer. Cellular consolidation, systems consolidation, and reconsolidation are all part of a dynamic, non-linear process. A recent review by Dudai and colleagues proposed that consolidation is an outdated term usefully employed to describe a range of memory-related processes, but should possibly be reconsidered. "*The term consolidation… is well-rooted in the memory literature and therefore deserves not to be reconsolidated even in systems level discussion, but research… indicates that in the memory dictionary, its translation is ongoing transformation, not fixation*" (Dudai et al. 2015, pg. 28). Research in the field of memory consolidation and transformation has scratched the surface in determining how memory 'traces' are formed, transformed and 'represented' throughout the brain. The terminology and ideas behind these concepts will likely become more refined as the development of genetic technology, *in vivo* functional neuroimaging, and computational modeling advance our understanding of the mechanisms of systems level memory consolidation. Whatever these new developments reveal, it is clear that there is no returning to traditional notions that memory consolidation marks the end of hippocampal processing after memories become represented permanently in extra-hippocampal structures. Rather, theories like TTT provide a new direction for thinking about consolidation as an ongoing and interactive process in the hippocampus and mPFC, and also across broader memory networks in the brain.

Acknowledgements The preparation of this chapter was supported by a grant to MM and GW from the Canadian Institutes for Health Research (CIHR—grant # MGP 6694). Research by the authors reported in this paper was supported by CIHR grant MGP 6694 and by grants from the Natural Sciences and Engineering Research Council to GW (# A8181) and MM (# A8347). MJS was supported by a CIHR post-doctoral fellowship.

References

Addis DR, Moscovitch M, Crawley AP, McAndrews MP (2004) Recollective qualities modulate hippocampal activation during autobiographical memory retrieval. Hippocampus 14:752–762

Alberini CM (2009) Transcription factors in long-term memory and synaptic plasticity. Physiol Rev 89:121–145

Alvarez P, Squire LR (1994) Memory consolidation and the medial temporal lobe: a simple network model. Proc Natl Acad Sci 91:7041–7045

Atherton LA, Dupret D, Mellor JR (2015) Memory trace replay: the shaping of memory consolidation by neuromodulation. Trends Neurosci 38:560–570

Bahrick HP (1984) Semantic memory content in permastore: fifty years of memory for Spanish learned in school. J Exp Psychol Gen 113:1–29

Bakker A, Kirwan CB, Miller M, Stark CE (2008) Pattern separation in the human hippocampal CA3 and dentate gyrus. Science 319(5870):1640–1642

Barco A, Marie H (2011) Genetic approaches to investigate the role of CREB in neuronal plasticity and memory. Mol Neurobiol 44:330–349

Barco A, Alarcon JM, Kandel ER (2002) Expression of constitutively active CREB protein facilitates the late phase of long-term potentiation by enhancing synaptic capture. Cell 108:689–703

Bartlett FC (1932) Remembering: an experiment in experimental and social psychology. Cambridge, University Press, London

Beatty WW, Salmon DP, Bernstein N, Butters N (1987) Remote memory in a patient with amnesia due to hypoxia. Psychol Med 17:657–665

Ben-Yakov A, Dudai Y (2011) Constructing realistic engrams: poststimulus activity of hippocampus and dorsal striatum predicts subsequent episodic memory. J Neurosci 31:9032–9042

Bliss TV, Collingridge GL (1993) A synaptic model of memory: long-term potentiation in the hippocampus. Nature 361:31–39

Bonnici HM, Chadwick MJ, Lutti A, Hassabis D, Weiskopf N, Maguire EA (2012) Detecting representations of recent and remote autobiographical memories in vmPFC and hippocampus. J Neurosci 32:16982–16991

Bontempi B, Laurent-Demir C, Destrade C, Jaffard R (1999) Time-dependent reorganization of brain circuitry underlying long-term memory storage. Nature 400:671–675

Bourtchuladze R, Frenguelli B, Blendy J, Cioffi D, Schutz G, Silva AJ (1994) Deficient long-term memory in mice with a targeted mutation of the cAMP-responsive element-binding protein. Cell 79:59–68

Burnham WH (1903) Retroactive amnesia: illustrative cases and a tentative explanation. Am J Psychol 118–132

Cai DJ, Aharoni D, Shuman T, Shobe J, Biane J, Song W, Wei B, Veshkini M, La-Vu M, Lou J et al (2016) A shared neural ensemble links distinct contextual memories encoded close in time. Nature Adv Online Publ. doi:10.1038/nature17955

Cairney SA, Durrant SJ, Musgrove H, Lewis PA (2011) Sleep and environmental context: interactive effects for memory. Exp Brain Res 214(1):83–92

Cajal SRY (1894) The croonian lecture: La fine structure des centres nerveux. Proc R Soc Lond 88:444–468

Chan JC, LaPaglia JA (2013) Impairing existing declarative memory in humans by disrupting reconsolidation. Proc Natl Acad Sci 110:9309–9313

Chan JC, Thomas AK, Bulevich JB (2009) Recalling a witnessed event increases eyewitness suggestibility the reversed testing effect. Psychol Sci 20:66–73

Ciaramelli E (2008) The role of ventromedial prefrontal cortex in navigation: a case of impaired way finding and rehabilitation. Neuropsychologia 46(7):2099–2105

Ciaramelli E, Ghetti S (2007) What are confabulators' memories made of? A study of subjective and objective measures of recollection in confabulation. Neuropsychologia 45:1489–1500

Cole CJ, Mercaldo V, Restivo L, Yiu AP, Sekeres MJ, Han J-H, Vetere G, Pekar T, Ross PJ, Neve RL (2012) MEF2 negatively regulates learning-induced structural plasticity and memory formation. Nat Neurosci 15:1255–1264

Corkin S (2002) What's new with the amnesic patient H.M.? Nat Rev Neurosci 3:153–160

Czajkowski R, Jayaprakash B, Wiltgen B, Rogerson T, Guzman-Karlsson MC, Barth AL, Trachtenberg JT, Silva AJ (2014) Encoding and storage of spatial information in the retrosplenial cortex. Proc Natl Acad Sci 111:8661–8666

Dalla Barba G, La Corte V (2015) A neurophenomenological model for the role of the hippocampus in temporal consciousness. Evidence from confabulation. Frontiers Behav Neurosci 9

Davachi L, DuBrow S (2015) How the hippocampus preserves order: the role of prediction and context. Trends Cogn Sci 19(2):92–99

Davis HP, Squire LR (1984) Protein synthesis and memory: a review. Psychol Bull 96:518–559

Debiec J, LeDoux JE, Nader K (2002) Cellular and systems reconsolidation in the hippocampus. Neuron 36:527–538

Denny CA, Kheirbek MA, Alba EL, Tanaka KF, Brachman RA, Laughman KB, Tomm NK, Turi GF, Losonczy A, Hen R (2014) Hippocampal memory traces are differentially modulated by experience, time, and adult neurogenesis. Neuron 83:189–201

Diekelmann S, Büchel C, Born J, Rasch B (2011) Labile or stable: opposing consequences for memory when reactivated during waking and sleep. Nat Neurosci 14:381–386

Dudai Y (2012) The restless engram: consolidations never end. Annu Rev Neurosci 35:227–247

Dudai Y, Karni A, Born J (2015) The consolidation and transformation of memory. Neuron 88:20–32

Dunsmoor JE, Murty VP, Davachi L, Phelps EA (2015) Emotional learning selectively and retroactively strengthens memories for related events. Nature 520:345–348

Einarsson EO, Nader K (2012) Involvement of the anterior cingulate cortex in formation, consolidation, and reconsolidation of recent and remote contextual fear memory. Learn Mem 19:449–452

Einarsson EO, Pors J, Nader K (2014) Systems reconsolidation reveals a selective role for the anterior cingulate cortex in generalized contextual fear memory expression. Neuropsychopharmacology 40:480–487

Finnie PS, Nader K (2012) The role of metaplasticity mechanisms in regulating memory destabilization and reconsolidation. Neurosci Biobehav Rev 36:1667–1707

Foster DJ, Wilson MA (2006) Reverse replay of Behavioral sequences in hippocampal place cells during the awake state. Nature 440:680–683

Frankland PW, Bontempi B (2005) The organization of recent and remote memories. Nat Rev Neurosci 6:119–130

Frankland PW, Bontempi B, Talton LE, Kaczmarek L, Silva AJ (2004) The involvement of the anterior cingulate cortex in remote contextual fear memory. Science 304:881–883

Frey U, Morris RG (1997) Synaptic tagging and long-term potentiation. Nature 385:533–536

Frey U, Morris R (1998) Weak before strong: dissociating synaptic tagging and plasticity-factor accounts of late-LTP. Neuropharmacology 37:545–552

Fu M, Yu X, Lu J, Zuo Y (2012) Repetitive motor learning induces coordinated formation of clustered dendritic spines in vivo. Nature 483:92–95

Fujii T, Moscovitch M, Nadel L (2000) Memory consolidation, retrograde amnesia, and the temporal lobe. Handbook of neuropsychology, 2nd ed., vol 2. Memory and its disorders. Elsevier Science Publishers, pp 223–250

Furman O, Mendelsohn A, Dudai Y (2012) The episodic engram transformed: Time reduces retrieval-related brain activity but correlates it with memory accuracy. Learn Mem 19:575–587

Gallistel CR, Matzel LD (2013) The neuroscience of learning: beyond the Hebbian synapse. Annu Rev Psychol 64:169–200

Gelbard-Sagiv H, Mukamel R, Harel M, Malach R, Fried I (2008) Internally generated reactivation of single neurons in human hippocampus during free recall. Science 322(5898):96–101

Ghosh VE, Gilboa A (2014) What is a memory schema? A historical perspective on current neuroscience literature. Neuropsychologia 53:104–114

Ghosh VE, Moscovitch M, Colella BM, Gilboa A (2014) Schema representation in patients with ventromedial PFC lesions. J Neurosci 34:12057–12070

Gilboa A, Moscovitch M, Baddeley A, Kopelman M, Wilson B (2002) The cognitive neuroscience of confabulation: a review and a model. Handb Mem Disord 2:315–342

Gilboa A, Winocur G, Grady CL, Hevenor SJ, Moscovitch M (2004) Remembering our past: functional neuroanatomy of recollection of recent and very remote personal events. Cereb Cortex 14:1214–1225

Gilboa A, Alain C, Stuss DT, Melo B, Miller S, Moscovitch M (2006) Mechanisms of spontaneous confabulations: a strategic retrieval account. Brain 129:1399–1414

Goshen I, Brodsky M, Prakash R, Wallace J, Gradinaru V, Ramakrishnan C, Deisseroth K (2011) Dynamics of retrieval strategies for remote memories. Cell 147:678–689

Govindarajan A, Kelleher RJ, Tonegawa S (2006) A clustered plasticity model of long-term memory engrams. Nat Rev Neurosci 7:575–583

Greenberg ME, Ziff EB (1983) Stimulation of 3T3 cells induces transcription of the c-fos proto-oncogene. Nature 311(5985):433–438

Han JH, Kushner SA, Yiu AP, Cole CJ, Matynia A, Brown RA, Neve RL, Guzowski JF, Silva AJ, Josselyn SA (2007) Neuronal competition and selection during memory formation. Science 316:457–460

Hardt O, Nader K, Nadel L (2013) Decay happens: the role of active forgetting in memory. Trends Cogn Sci 17(3):111–120

Hebb DO (1949) The organization of behavior: a neuropsychological approach. Wiley

Hebscher M, Barkan-Abramski M, Goldsmith M, Aharon-Peretz J, Gilboa A (2015) Memory, decision-making, and the Ventromedial Prefrontal Cortex (vmPFC): the roles of subcallosal and posterior orbitofrontal cortices in monitoring and control processes. Cerebral Cortex, bhv220

Hirshhorn M, Grady C, Rosenbaum RS, Winocur G, Moscovitch M (2012) Brain regions involved in the retrieval of spatial and episodic details associated with a familiar environment: an fMRI study. Neuropsychologia 50:3094–3106

Howard MW, Kahana MJ (1999) Contextual variability and serial position effects in free recall. J Exp Psychol Learn Mem Cogn 25(4):923

Hupbach A, Hardt O, Gomez R, Nadel L (2008) The dynamics of memory: context-dependent updating. Learn Mem 15:574–579

St. Jacques PL, Schacter DL (2013) Modifying memory selectively enhancing and updating personal memories for a museum tour by reactivating them. Psychol sci 24:537–543

St. Jacques PL, Olm C, Schacter DL (2013) Neural mechanisms of reactivation-induced updating that enhance and distort memory. Proc Nat Acad Sci 110:19671–19678

Josselyn SA, Shi C, Carlezon WA Jr, Neve RL, Nestler EJ, Davis M (2001) Long-term memory is facilitated by cAMP response element-binding protein overexpression in the amygdala. J Neurosci 21:2404–2412

Josselyn SA, Köhler S, Frankland PW (2015) Finding the engram. Nat Rev Neurosci 16:521–534

Kandel ER, Dudai Y, Mayford MR (2014) The molecular and systems biology of memory. Cell 157:163–186

Kim JJ, Fanselow MS (1992) Modality-specific retrograde amnesia of fear. Science 256:675–677

Kroes MC, Fernández G (2012) Dynamic neural systems enable adaptive, flexible memories. Neurosci Biobehav Rev 36:1646–1666

Landauer TK (1975) Memory without organization: properties of a model with random storage and undirected retrieval. Cogn Psychol 7(4):495–531

Lashley KS (1950) In search of the engram. In: Society for experimental biology, symposium IV: physiological mechanisms in animal behavior, pp 454–482 (Academic Press)

Lee JQ, Zelinski EL, McDonald RJ, Sutherland RJ (2016) Heterarchic reinstatement of long-term memory: a concept on hippocampal amnesia in rodent memory research. Neurosci Biobehav Rev 71:154–166

Leutgeb JK, Leutgeb S, Moser M-B, Moser EI (2007) Pattern separation in the dentate gyrus and CA3 of the hippocampus. Science 315(5814):961–966

Lewis PA, Durrant SJ (2011) Overlapping memory replay during sleep builds cognitive schemata. Trends Cogn Sci 15(8):343–351

Ling DS, Benardo LS, Serrano PA, Blace N, Kelly MT, Crary JF, Sacktor TC (2002) Protein kinase Mζ is necessary and sufficient for LTP maintenance. Nat Neurosci 5:295–296

Liu X, Ramirez S, Pang PT, Puryear CB, Govindarajan A, Deisseroth K, Tonegawa S (2012) Optogenetic stimulation of a hippocampal engram activates fear memory recall. Nature 484:381–385

Lonze BE, Ginty DD (2002) Function and regulation of CREB family transcription factors in the nervous system. Neuron 35:605–623

Lorenté de No R (1934) Studies on the structure of the cerebral cortex II: continuation of the study if the ammonic system. J Psychol Neurol 46:113–177

Maguire EA (2014) Memory consolidation in humans: new evidence and opportunities. Exp Physiol 99:471–486

Maguire EA, Nannery R, Spiers HJ (2006) Navigation around London by a taxi driver with bilateral hippocampal lesions. Brain 129:2894–2907

Manning JR, Polyn SM, Baltuch GH, Litt B, Kahana MJ (2011) Oscillatory patterns in temporal lobe reveal context reinstatement during memory search. Proc Natl Acad Sci 108(31):12893–12897

Martin KC, Kandel ER (1996) Cell adhesion molecules, CREB, and the formation of new synaptic connections. Neuron 17:567–570

Maviel T, Durkin TP, Menzaghi F, Bontempi B (2004) Sites of neocortical reorganization critical for remote spatial memory. Science 305:96–99

McClelland JL, McNaughton BL, O'Reilly RC (1995) Why there are complementary learning systems in the hippocampus and neocortex: insights from the successes and failures of connectionist models of learning and memory. Psychol Rev 102:419

Mendelsohn A, Furman O, Dudai Y (2010) Signatures of memory: brain coactivations during retrieval distinguish correct from incorrect recollection. Front Behav Neurosci 4:18

Milner B, Corkin S, Teuber H-L (1968) Further analysis of the hippocampal amnesic syndrome: 14-year follow-up study of HM. Neuropsychologia 6:215–234

Misanin JR, Miller RR, Lewis DJ (1968) Retrograde amnesia produced by electroconvulsive shock after reactivation of a consolidated memory trace. Science 160:554–555

Morris R (1984) Developments of a water-maze procedure for studying spatial learning in the rat. J Neurosci Methods 11:47–60

Morris R, Garrud P, Rawlins J, O'Keefe J (1982) Place navigation impaired in rats with hippocampal lesions. Nature 297:681–683

Moscovitch M (1989) Confabulation and the frontal systems: strategic versus associative retrieval in neuropsychological theories of memory. In: Roediger HL, Craik FIM (eds) Varieties of memory and consciousness: essays in honour of Endel Tulving. Lawrence Erlbaum Associates, Hillsdale, pp 133–160

Moscovitch M (1992) A neuropsychological model of memory and consciousness. Neuropsychol Mem 2:5–22

Moscovitch M (1995a) Confabulation. In: Schacter DL (Ed) Memory distortions: how minds, brains, and societies reconstruct the past. Harvard University Press, Cambridge, MA, US, pp 226–251

Moscovitch M (1995b) Recovered consciousness: a hypothesis concerning modularity and episodic memory. J Clin Exp Neuropsychol 17:276–290

Moscovitch M, Winocur G (2002) The frontal cortex and working with memory. Principles of frontal lobe function, pp 188–209

Moscovitch M, Rosenbaum RS, Gilboa A, Addis DR, Westmacott R, Grady C, McAndrews MP, Levine B, Black S, Winocur G, Nadel L (2005) Functional neuroanatomy of remote episodic, semantic and spatial memory: a unified account based on multiple trace theory. J Anat 207:35–66

Moscovitch M, Nadel L, Winocur G, Gilboa A, Rosenbaum RS (2006) The cognitive neuroscience of remote episodic, semantic and spatial memory. Curr Opin Neurobiol 16:179–190

Moscovitch M (2007) Why the engram is elusive: memory does not exist until it is recovered. In: Tulving E, Roediger HL III, Dudai Y, Fitzpatrick SM (eds) Science of memory: concepts. Oxfod, UK: Oxford University Press, pp 17–22

Moscovitch M, Cabeza R, Winocur G, Nadel L (2016) Episodic memory and beyond: the hippocampus and neocortex in transformation. Annu Rev Psychol 67:105–134

Müller GE, Pilzecker A (1900) Experimentelle beiträge zur lehre vom gedächtniss (Vol 1). JA Barth

Nadel L, Moscovitch M (1997) Memory consolidation, retrograde amnesia and the hippocampal complex. Curr Opin Neurobiol 7:217–227

Nader K, Hardt O (2009) A single standard for memory: the case for reconsolidation. Nat Rev Neurosci 10:224–234

Nader K, Schafe GE, Le Doux JE (2000) Fear memories require protein synthesis in the amygdala for reconsolidation after retrieval. Nature 406:722–726

Nieuwenhuis IL, Takashima A (2011) The role of the ventromedial prefrontal cortex in memory consolidation. Behav Brain Res 218:325–334

O'Keefe J, Nadel L (1978) The hippocampus as a cognitive map, vol 3. Clarendon Press, Oxford

Paz R, Gelbard-Sagiv H, Mukamel R, Harel M, Malach R, Fried I (2010) A neural substrate in the human hippocampus for linking successive events. Proc Natl Acad Sci 107(13):6046–6051

Penfield W, Milner B (1958) Memory deficit produced by bilateral lesions in the hippocampal zone. AMA Arch Neurol Psychiatry 79:475–497

Ramirez S, Liu X, Lin P-A, Suh J, Pignatelli M, Redondo RL, Ryan TJ, Tonegawa S (2013) Creating a false memory in the hippocampus. Science 341:387–391

Restivo L, Tafi E, Ammassari-Teule M, Marie H (2009a) Viral-mediated expression of a constitutively active form of CREB in hippocampal neurons increases memory. Hippocampus 19:228–234

Restivo L, Vetere G, Bontempi B, Ammassari-Teule M (2009b) The formation of recent and remote memory is associated with time-dependent formation of dendritic spines in the hippocampus and anterior cingulate cortex. J Neurosci 29:8206–8214

Ribot TA (1882) Diseases of memory, an essay in the positive psychology, vol 43

Richards BA, Xia F, Santoro A, Husse J, Woodin MA, Josselyn SA, Frankland PW (2014) Patterns across multiple memories are identified over time. Nat Neurosci 17:981–986

Ritchey M, Montchal ME, Yonelinas AP, Ranganath C (2015) Delay-dependent contributions of medial temporal lobe regions to episodic memory retrieval. eLife 4, e05025

Rogerson T, Cai DJ, Frank A, Sano Y, Shobe J, Lopez-Aranda MF, Silva AJ (2014) Synaptic tagging during memory allocation. Nat Rev Neurosci 15:157–169

Rosenbaum RS, Priselac S, Kohler S, Black SE, Gao F, Nadel L, Moscovitch M (2000) Remote spatial memory in an amnesic person with extensive bilateral hippocampal lesions. Nat Neurosci 3:1044–1048

Rosenbaum RS, McKinnon MC, Levine B, Moscovitch M (2004) Visual imagery deficits, impaired strategic retrieval, or memory loss: disentangling the nature of an amnesic person's autobiographical memory deficit. Neuropsychologia 42:1619–1635

Rosenbaum RS, Kohler S, Schacter DL, Moscovitch M, Westmacott R, Black SE, Gao F, Tulving E (2005) The case of K.C.: contributions of a memory-impaired person to memory theory. Neuropsychologia 43:989–1021

Rugg MD, Vilberg KL (2013) Brain networks underlying episodic memory retrieval. Curr Opin Neurobiol 23:255–260

Sadeh T, Ozubko JD, Winocur G, Moscovitch M (2013) How we forget may depend on how we remember. Trends Cogn Sci 18(1):26–36

Sadeh T, Moran R, Goshen-Gottstein Y (2015) When items 'pop into mind': Variability in temporal-context reinstatement in free-recall. Psychon Bull Rev 22(3):779–790

Sadeh T, Ozubko JD, Winocur G, Moscovitch M (2016) Forgetting patterns differentiate between two forms of memory representation. Psychol Sci 0956797616638307

Sanders HI, Warrington EK (1971) Memory for remote events in amnesic patients. Brain 94:661–668

Sano Y, Shobe JL, Zhou M, Huang S, Shuman T, Cai DJ, Golshani P, Kamata M, Silva AJ (2014) CREB regulates memory allocation in the insular cortex. Curr Biol 24:2833–2837

Sara SJ (2000) Retrieval and reconsolidation: toward a neurobiology of remembering. Learn Mem 7:73–84

Schacter DL, Eich JE, Tulving E (1978) Richard Semon's theory of memory. J Verbal Learn Verbal Behav 17:721–743

Schiller D, Monfils M-H, Raio CM, Johnson DC, LeDoux JE, Phelps EA (2010) Preventing the return of fear in humans using reconsolidation update mechanisms. Nature 463:49–53

Schnider A (2008) The confabulating mind: how the brain creates reality. Oxford University Press, New York

Scoville WB, Milner B (1957) Loss of recent memory after bilateral hippocampal lesions. J Neurol Neurosurg Psychiatry 20:11–21

Sekeres MJ, Neve RL, Frankland PW, Josselyn SA (2010) Dorsal hippocampal CREB is both necessary and sufficient for spatial memory. Learn Mem 17:280–283

Sekeres MJ, Mercaldo V, Richards B, Sargin D, Mahadevan V, Woodin MA, Frankland PW, Josselyn SA (2012a) Increasing CRTC1 function in the dentate gyrus during memory formation or reactivation increases memory strength without compromising memory quality. J Neurosci 32:17857–17868

Sekeres MJ, Sargin D, Ross PJ, Josselyn SA (2012b) The role of CREB and CREB Co-activators in memory formation. Memory Mechanisms in Health and Disease: Mechanistic Basis of Memory pp 171–194 (World Scientific)

Sekeres MJ, Anderson JAE, Winocur G, Moscovitch M, Grady C (2015) The neural correlates of episodic memory transformation in humans. Soc Neurosci 2016, Abstract 18.03

Sekeres MJ, Bonasia K, St-Laurent M, Pishdadian S, Winocur G, Grady C, Moscovitch M (2016a) Recovering and preventing loss of detailed memory: differential rates of forgetting for detail types in episodic memory. Learn Memory 23:72–82

Sekeres MJ, Winocur G, Moscovitch M, Anderson JAE, Pishdadian S, Wojtowicz JM, St-Laurent M, McAndrews MP, Grady C (2016b) Changes in patterns of neural activity underlie a time-dependent transformation of memory across species (submitted)

Semon R (1923) Mnemic psychology (1909). George Allen and Unwin, Ltd, London

Sheldon S, Levine B (2013) Same as it ever was: vividness modulates the similarities and differences between the neural networks that support retrieving remote and recent autobiographical memories. Neuroimage 83:880–891

Shema R, Haramati S, Ron S, Hazvi S, Chen A, Sacktor TC, Dudai Y (2011) Enhancement of consolidated long-term memory by overexpression of protein kinase Mzeta in the neocortex. Science 331:1207–1210

Silva AJ, Zhou Y, Rogerson T, Shobe J, Balaji J (2009) Molecular and cellular approaches to memory allocation in neural circuits. Science 326:391–395

Squire LR (1992) Memory and the hippocampus: a synthesis from findings with rats, monkeys, and humans. Psychol Rev 99:195–231

Squire LR (2004) Memory systems of the brain: a brief history and current perspective. Neurobiol Learn Mem 82:171–177

Squire LR, Bayley PJ (2007) The neuroscience of remote memory. Curr Opin Neurobiol 17:185–196

Squire LR, Wixted JT (2011) The cognitive neuroscience of human memory since HM. Annu Rev Neurosci 34:259

Steinvorth S, Levine B, Corkin S (2005) Medial temporal lobe structures are needed to re-experience remote autobiographical memories: evidence from H.M. and W.R. Neuropsychologia 43:479–496

St-Laurent M, Moscovitch M, Levine B, McAndrews MP (2009) Determinants of autobiographical memory in patients with unilateral temporal lobe epilepsy or excisions. Neuropsychologia 47:2211–2221

St-Laurent M, Moscovitch M, Jadd R, McAndrews MP (2014) The perceptual richness of complex memory episodes is compromised by medial temporal lobe damage. Hippocampus 24:560–576

St-Laurent M, Moscovitch M, McAndrews MP (2016) The retrieval of perceptual memory details depends on right hippocampal integrity and activation. Cortex (under review)

Takashima A, Nieuwenhuis IL, Jensen O, Talamini LM, Rijpkema M, Fernández G (2009) Shift from hippocampal to neocortical centered retrieval network with consolidation. J Neurosci 29:10087–10093

Tayler KK, Tanaka KZ, Reijmers LG, Wiltgen BJ (2013) Reactivation of neural ensembles during the retrieval of recent and remote memory. Curr Biol 23:99–106

Teixeira CM, Pomedli SR, Maei HR, Kee N, Frankland PW (2006) Involvement of the anterior cingulate cortex in the expression of remote spatial memory. J Neurosci 26:7555–7564

Teng E, Squire LR (1999) Memory for places learned long ago is intact after hippocampal damage. Nature 400:675–677

Thorndyke PW (1977) Cognitive structures in comprehension and memory of narrative discourse. Cogn Psychol 9:77–110

Tonegawa S, Pignatelli M, Roy DS, Ryan TJ (2015) Memory engram storage and retrieval. Curr Opin Neurobiol 35:101–109

Tse D, Langston RF, Kakeyama M, Bethus I, Spooner PA, Wood ER, Witter MP, Morris RG (2007) Schemas and memory consolidation. Science 316:76–82

Tse D, Takeuchi T, Kakeyama M, Kajii Y, Okuno H, Tohyama C, Bito H, Morris RG (2011) Schema-dependent gene activation and memory encoding in neocortex. Science 333:891–895

Tulving E (1972) Episodic and semantic memory. Organization of memory. Academic, London 381(4)

van Kesteren MT, Ruiter DJ, Fernández G, Henson RN (2012) How schema and novelty augment memory formation. Trends Neurosci 35(4):211–219

van Kesteren MT, Beul SF, Takashima A, Henson RN, Ruiter DJ, Fernández G (2013) Differential roles for medial prefrontal and medial temporal cortices in schema-dependent encoding: from congruent to incongruent. Neuropsychologia 51(12):2352–2359

Varela C, Weiss S, Meyer R, Halassa M, Biedenkapp J, Wilson MA, Goosens KA, Bendor D (2016) Tracking the time-dependent role of the hippocampus in memory recall using DREADDs. PLoS ONE 11:e0154374

Vetere G, Restivo L, Cole CJ, Ross PJ, Ammassari-Teule M, Josselyn SA, Frankland PW (2011) Spine growth in the anterior cingulate cortex is necessary for the consolidation of contextual fear memory. Proc Natl Acad Sci 108:8456–8460

Vetere G, Kenney JW, Tran L, Wheeler A, Josselyn S, Frankland PW (2015) In vivo and in silico interrogation of a fear memory network. Soc Neurosci 2016, Abstract 534.17

Viskontas IV, Carr VA, Engel SA, Knowlton BJ (2009) The neural correlates of recollection: hippocampal activation declines as episodic memory fades. Hippocampus 19:265–272

Vousden DA, Epp J, Okuno H, Nieman BJ, van Eede M, Dazai J, Ragan T, Bito H, Frankland PW, Lerch JP (2015) Whole-brain mapping of Behaviorally induced neural activation in mice. Brain Struct Funct 220:2043–2057

Wang S-H, Morris RG (2010) Hippocampal-neocortical interactions in memory formation, consolidation, and reconsolidation. Annu Rev Psychol 61:49–79

Wang S-H, Redondo RL, Morris RG (2010) Relevance of synaptic tagging and capture to the persistence of long-term potentiation and everyday spatial memory. Proc Natl Acad Sci 107:19537–19542

Wheeler AL, Teixeira CM, Wang AH, Xiong X, Kovacevic N, Lerch JP, McIntosh AR, Parkinson J, Frankland PW (2013) Identification of a functional connectome for long-term fear memory in mice. PLoS Comput Biol 9:e1002853

Wiltgen BJ, Silva AJ (2007) Memory for context becomes less specific with time. Learn Mem 14:313–317

Winocur G (1990) Anterograde and retrograde amnesia in rats with dorsal hippocampal or dorsomedial thalamic lesions. Behav Brain Res 38:145–154

Winocur G, Mills JA (1970) Transfer between related and unrelated problems following hippocampal lesions in rats. J Comp Physiol Psychol 73(1):162

Winocur G, Moscovitch M (2011) Memory transformation and systems consolidation. J Int Neuropsychol Soc 17:766–780

Winocur G, Weiskrantz L (1976) An investigation of paired-associate learning in amnesic patients. Neuropsychologia 14(1):97–110

Winocur G, Moscovitch M, Caruana DA, Binns MA (2005a) Retrograde amnesia in rats with lesions to the hippocampus on a test of spatial memory. Neuropsychologia, 43(11):1580–1590

Winocur G, Moscovitch M, Fogel S, Rosenbaum RS, Sekeres M (2005b) Preserved spatial memory after hippocampal lesions: effects of extensive experience in a complex environment. Nat Neurosci 8:273–275

Winocur G, Moscovitch M, Sekeres M (2007) Memory consolidation or transformation: context manipulation and hippocampal representations of memory. Nat Neurosci 10:555–557

Winocur G, Frankland PW, Sekeres M, Fogel S, Moscovitch M (2009) Changes in context-specificity during memory reconsolidation: selective effects of hippocampal lesions. Learn Mem 16:722–729

Winocur G, Moscovitch M, Rosenbaum RS, Sekeres M (2010) An investigation of the effects of hippocampal lesions in rats on pre- and postoperatively acquired spatial memory in a complex environment. Hippocampus 20:1350–1365

Winocur G, Becker S, Luu P, Rosenzweig S, Wojtowicz JM (2012) Adult hippocampal neurogenesis and memory interference. Behav Brain Res 227:464–469

Winocur G, Sekeres MJ, Binns MA, Moscovitch M (2013) Hippocampal lesions produce both nongraded and temporally graded retrograde amnesia in the same rat. Hippocampus, 23(5): 330–341

Xu T, Yu X, Perlik AJ, Tobin WF, Zweig JA, Tennant K, Jones T, Zuo Y (2009) Rapid formation and selective stabilization of synapses for enduring motor memories. Nature 462:915–919

Ye L, Allen WE, Thompson KR, Tian Q, Hsueh B, Ramakrishnan C et al.(2016). Wiring and molecular features of frefrontal ensembles representing distinct experiences. Cell (in press)

Zhou Y, Won J, Karlsson MG, Zhou M, Rogerson T, Balaji J, Neve R, Poirazi P, Silva AJ (2009) CREB regulates excitability and the allocation of memory to subsets of neurons in the amygdala. Nat Neurosci 12:1438–1443

Zola-Morgan SM, Squire LR (1990) The primate hippocampal formation: evidence for a time-limited role in memory storage. Science 250:288–290

Zola-Morgan S, Squire LR, Amaral D (1986) Human amnesia and the medial temporal region: enduring memory impairment following a bilateral lesion limited to field CA1 of the hippocampus. J Neurosci 6:2950–2967

The Medial Prefrontal Cortex is a Critical Hub in the Declarative Memory System

Guillén Fernández

Abstract What enables us to acquire and use our knowledge? The classical declarative memory system with the hippocampus at its core appears not sufficient to explain knowledge acquisition and retrieval satisfactorily. Recent evidence suggests an extension of this classical model by assigning the medial prefrontal cortex a particular, yet not fully defined role in long-term memory. This chapter will integrate data derived from experiments with rodents and humans providing the basis for an extended declarative memory system that includes the medial prefrontal cortex. Here, I discuss how the medial prefrontal cortex, interacting with the medial temporal lobe, posterior representational areas and specific subcortical structures, may help obtain, integrate and apply our knowledge for long-term usage.

Keyword Declarative memory · Knowledge · Hippocampus · Medial prefrontal cortex

The last sixty years have brought us important insight into the brain circuits that underlie our ability to remember our past. The seminal work of Scoville and Milner (1957) has revealed the critical and specific role, the medial temporal lobe, with the hippocampus at its core, plays in episodic memory. Following bilateral hippocampal resection Patient H.M. was able to remember remote experiences normally, but not events experienced shortly before or any time after his surgery. This pattern of results showed that episodic memories are initially hippocampally dependent, but with time they can be retrieved independently of the hippocampus, a transition termed systems-level consolidation (Squire and Alvarez 1995). This process assigns a linking role to the hippocampus that is necessary to bind distributed neocortical representations together into one coherent episode during initial encoding and during reinstatement when remembering recent events (e.g., Tanaka et al. 2014). However, with time the network of neocortical representations

G. Fernández (✉)
Donders Institute for Brain, Cognition and Behavior,
Radboud University Medical Center, Nijmegen, The Netherlands
e-mail: g.fernandez@donders.ru.nl

N. Axmacher and B. Rasch (eds.), *Cognitive Neuroscience of Memory Consolidation*, Studies in Neuroscience, Psychology and Behavioral Economics, DOI 10.1007/978-3-319-45066-7_3

becomes stabilized so that remote episodes can be remembered without the binding function of the hippocampus (e.g., Takashima et al. 2009). This stabilization appears to be the consequence of off-line processes based on replaying neural activation patterns that occurred during the initial encoding causing Hebbian plasticity (Hebb 1949) in neocortical networks (Schwindel and McNaughton 2011).

In this classical model of systems-level consolidation, the intricate interaction between the medial temporal lobe and representational brain areas in posterior neocortex form the basic infrastructure of episodic memory. However, recent evidence suggests an extension of this classical model by assigning the medial prefrontal cortex a particular, yet not fully defined role in long-term memory. This chapter will integrate data derived from experiments with rodents, human lesion studies and functional neuroimaging studies providing the basis for an extended declarative memory system that includes the medial prefrontal cortex.

The first empirical evidence for a role of the medial prefrontal cortex in long-term memory retrieval was revealed by Bontempi and his colleagues in 1999. They assessed regional neural activity in mice during memory retrieval and contrasted different study-test delays, a recent (5 days) and a remote (25 days) condition. Single-shot functional brain imaging based on (14C) 2-deoxyglucose uptake was used to evaluate how regional brain activity underlying memory retrieval is changing with time. Their results showed that hippocampal activity during memory retrieval decreased from recent to remote test, while medial prefrontal and posterior brain activity increased. This pattern of results is in line with the classical model of systems-level consolidation in which the hippocampus has a time-limited role in memory retrieval and that neocortical circuits are sufficient for remote memory retrieval. However, the medial prefrontal cortex is not known as a representational area as they are occurring in posterior brain regions and thus, this report appeared to extend the classical model by adding first evidence for a higher-order prefrontal area specifically involved in remote memory retrieval that may have a binding function integrating representations in posterior brain regions.

Frankland et al. (2004) extended these findings by providing first causative evidence for the critical role of the medial prefrontal cortex in remote memory retrieval that went beyond correlative evidence. Calmodulin-dependent protein kinase II (CaMKII) is essential for long-term memory formation in neocortex (Frankland et al. 2001) and they showed that a loss-of-function mutation of the CaMKII gene abolished medial prefrontal activation during a remote memory test and impaired remote retrieval. They found a similar pattern of results when reversibly inactivating the medial prefrontal cortex in normal mice. Most critically, these manipulations did not disrupt recent memory retrieval revealing the specific and essential role the medial prefrontal cortex plays in remote memory retrieval. More recently, an integrated set of experiments that combined RNA sequencing, microscopy and electrophysiology revealed in the medial prefrontal cortex learning-related increases in transcription, expansion of the synaptic active zone, enhanced dendritic spine plasticity, and increased synaptic efficacy (Bero et al. 2014). Moreover, this study provided also a mechanistic account for this medial prefrontal remodeling, by optogenetic inhibition of excitatory neurons during learning, revealing that this inhibition impairs associative memory formation.

Well-controlled lesion data in experimental animals confirmed a double dissociation between medial temporal and medial prefrontal contributions to recent and remote memory retrieval. Takehara et al. (2003) ablated selectively the medial prefrontal cortex or hippocampus in rats at various time points following learning. When hippocampal lesions were made one day after learning, rats were clearly impaired whereas rats whose medial prefrontal cortex was lesioned at the same time point performed almost normally. With an increasing delay between learning and ablation, hippocampal lesions showed progressively less effect on memory performance so that after a four week delay no impairment could be detected anymore. In contrast, medial prefrontal lesions led to progressively more severe impairments so that animals could not retain the conditioned response if the lesion was made four weeks after learning. This pattern of results was replicated by Quinn et al. (2008), who showed that neurotoxic lesions of the medial prefrontal cortex or the hippocampus made one or 200 days following memory formation led to distinct impairments at test. Hippocampal lesions led to impaired performance for recently formed memories while medial prefrontal lesions disrupted mostly memories encoded 200 days earlier. Also lidocaine inactivation one day or one month after learning led to a clear double dissociation with medial prefrontal inactivation leading to a specific remote memory deficit that could not simply be explained by memory strength or content (Ding et al. 2008). Thus, the circuitry necessary for memory retrieval is reorganized in a way that the role of the hippocampus is time limited, as predicted by the classical systems-level theory, and this role appears to be taken over by the medial prefrontal cortex when retrieving consolidated long-term memories.

What is happening in the medial prefrontal cortex during this delay? Successful formation of medial prefrontal memory traces requires NMDA receptor-dependent processes during the first week after learning (Takehara-Nishiuchi et al. 2006). Such NMDA receptor-dependent processes are essential for Hebbian plasticity during memory consolidation (Shimizu et al. 2000) and thus may mediate the consolidation of medial prefrontal networks underlying long-term memory representations. Prerequisite for Hebbian plasticity is (repeated) neural activity that represents the information to be encoded into a durable memory trace. Neural replay during post-encoding offline states like rest or sleep has been shown to be related to successful memory stabilization (Schwindel and McNaughton 2011; see also chapter by Zhang, Deuker and Axmacher), allowing optimal information integration and storage by strengthening neural connections within representations. This process appears to be orchestrated by the hippocampus through reactivating and reinstating pattern of neural activity as phase sequences, which are compressed in time in all representational modules, enabling the establishment of stabilized connections (Euston et al. 2007). In addition to representational modules in posterior brain areas, hippocampally orchestrated replay has been also observed in prefrontal, in particular medial prefrontal areas during rest and sleep (Peyrache et al. 2009). During post-encoding rest, neural firing in the medial prefrontal cortex becomes selective for the acquired memories over a period of several weeks after learning without continued learning (Takehara-Nishiuchi and McNaughton 2008). Thus, hippocampally induced, post-encoding replay may stabilize specific memory traces in the medial prefrontal cortex via NMDA receptor-dependent Hebbian plasticity allowing long-term storage of memories.

This classical model with two memory systems with distinct time constants, a fast hippocampally centered and a slow neocortically centered memory system (McClelland et al. 1995), has been extended based on studies manipulating the relationship between new memories and already preexisting knowledge (McClelland 2013; see also chapter by Cheng). Rats that have acquired an associative knowledge structure, also called 'schema' about a complex environment learned over several iterations, can acquire new associative memories within this environment based on single-trial learning and these congruent memories become already hippocampally independent within 24 h (Tse et al. 2007). A mechanistic follow-up study (Tse et al. 2011) showed that learning of new associative memories that fit preexisting knowledge leads to increased immediate early gene synthesis in the medial prefrontal cortex. Moreover, pharmacological blockade of that area prevented new learning as well as recall of both remotely and recently consolidated memories. These schema studies have prompted a revision of the classical consolidation theory by incorporating rapid consolidation of putative linking nodes in medial prefrontal areas if new associative information can be readily assimilated in already existing knowledges structures. An alternative, perhaps also complementary interpretation assigns a computation to the medial prefrontal cortex that resolve conflicts between newly acquired information and already existing knowledge (Richards et al. 2014).

The pattern of results reviewed so far suggests an intricate interaction between the hippocampus and the medial prefrontal cortex in memory, especially during memory consolidation and remote memory retrieval, even if it is concerned with spatial memories (Vieira and Korzus 2015). However, rodent hippocampal–mPFC connections are unidirectional and not evenly distributed over the long axis of the hippocampus (Vertes et al. 2007; Hoover and Vertes 2012) and hence, an indirect communication appears necessary to enable a bidirectional information exchange between the medial prefrontal cortex and the hippocampus. Thalamic midline structures like the nucleus reuniens are reciprocally connected to both the medial prefrontal cortex and the hippocampus as well as to posterior representational areas (Aggleton and Brown 1999; van der Werf et al. 2002; Vertes et al. 2006, 2007; Hoover and Vertes 2012; Cassel et al. 2013; Wheeler et al. 2013). Indeed, the nucleus reuniens acts as a relay station connecting the medial prefrontal cortex and the hippocampus during memory encoding essential for the generalization of memory attributes (Xu and Südhof 2013). This mid thalamic relay appears also to be critical for offline consolidation that occurs after encoding. For instance, temporally specific and transient inactivation of the nucleus reunions in rats after training impaired subsequent memory retrieval (Davoodi et al. 2011). Thus, one can readily formulate a heuristic model in which the midline thalamus serves as a relay between the medial prefrontal cortex and medial temporal lobe enabling consolidation of memories rather generalized over several specific episodes, and initial imaging data in humans appears to translate this findings to humans (Thielen et al. 2015).

Human imaging studies appear to align well to studies with rodents. Takashima et al. (2006) performed a prospective memory consolidation study. On the first day, volunteers studied a large set of real-world photographs of landscapes. After a short nap they learned a second, smaller set of photographs before going into an MRI scanner where they performed an old-new recognition memory test in which they were instructed to distinguish old items (from the first and second set) and new items. This sequence of pre-scan study phase and scanned recognition memory test was repeated on the next day and on days 30 and 90. Thus, the authors were able to compare retrieval-related activity that became progressively more remote with retrieval-related brain activity of photographs studied just prior to scanning. In this way they revealed that hippocampal activity associated with confident, correct recognition decreased over the course of 90 days while medial prefrontal activity increased; a pattern of results in line with Bontempi et al. (1999). Also remote memories encoded in real-life and tracked by a wearable camera showed after a five-month delay and compared to a 36-hour delay this decreased hippocampal and increased medial prefrontal activity during successful recognition memory (Milton et al. 2011). Using functional MRI and multivoxel pattern analysis to detect representations of specific memories Bonnici et al. (2012) compared activity pattern while volunteers remembered specific events that were experienced recently (two weeks old) or remotely (10 years old). They revealed that the medial prefrontal cortex contained information about both recent and remote memories, but representations of remote memories were stronger, showing that a medial prefrontal consolidation has occurred for these memories.

Sleep appears to be critical for this transition from hippocampal to medial prefrontal retrieval processes. Gais et al. (2007) asked volunteers to memorize a list of word pairs in the evening before they were allowed to sleep normally or were sleep deprived (see also chapter by Schönauer and Gais). Memory was tested two days or six months after this initial study phase. They showed that post-learning sleep led to increased retrieval-related activity in the hippocampus and connectivity between the hippocampus and medial prefrontal cortex when tested two days after study. In line with the proposed transition, six months after study, the medial prefrontal cortex was more strongly activated when retrieving word-pairs that were studied before sleep as opposed to word-pairs that were studied prior to sleep deprivation. Thus, the initial post-learning sleep has long-lasting consequences for memory retrieval by augmenting medial prefrontal processes at retrieval. Very similar results regarding the medial prefrontal cortex were found when using emotional items. Again, initial post-learning sleep had a long-term effect at a six-month test on emotional memory retrieval with increased medial prefrontal activity and connectivity (Sterpenich et al. 2009). These results show that sleep during the first night after an experience appears to be critical for medial prefrontal involvement in long-term memory retrieval.

Brain processes during sleep that are responsible for this transition are difficult to disclose. Probing brain activity using functional MRI during sleep is experimentally challenging and requires unconventional methods in data analysis, because volunteers are not performing a well-controlled cognitive task that allows conventional model-based data analysis. Regardless, van Dongen et al. (2011) acquired functional MRI data while volunteers took a nap reaching shallow sleep (stage I & II) following a study session in which they learned a series of face-location associations. Functional connectivity between the face-representing area in the fusiform gyrus and the medial prefrontal cortex during sleep stages I and II predicted retention of associations as assessed in a post-nap cued-recall memory test. This increased connectivity may represent spontaneous replay supporting systems-level consolidation including the above described transition to medial prefrontal circuits. A second study (van Dongen et al. 2012) used auditory stimuli to induce replay during deep sleep in a temporarily and stimulus specific way as done in a behavioral study before (Rudoy et al. 2009). Volunteers studied a series of sound-object-location associations before "going to bed" in the MRI scanner. When reaching deep sleep stages characterized by slow-wave EEG, sounds used during study, and control sounds, were played to the subjects via headphones without awakening them. Replaying study sounds compared to control sounds induced specific changes in regional brain activity and connectivity. In addition, during a post sleep test connectivity between an object representational area in the posterior parahippocampal cortex and a medial prefrontal region was correlated with the positive effect of replaying on retention of associations Fig. 1.

To my knowledge, no clear cut double dissociation for remote and recent memory retrieval performance between patients with medial prefrontal and medial temporal lesions has been published so far. Several studies investigated memory retrieval after medial prefrontal lesions as they can typically occur after ruptures of aneurysms of the anterior communicans artery. In contrast to controlled lesions in experimental animals, these lesions are usually limited to very ventral aspects,

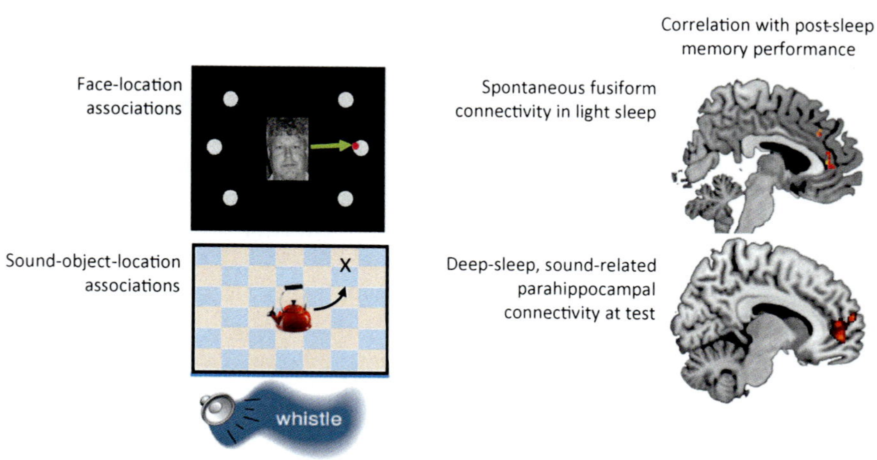

Fig. 1 Probing memory consolidation during sleep. The *left hand panel* depicts examples of stimulus material used by van Dongen et al. (2011, 2012) during a pre-sleep study task. The maps on the *right* highlight brain areas that have increased functional connectivity with particular representational areas that positively correlate with post-sleep memory performance

keeping large parts of the medial prefrontal cortex intact. Regardless, these patients show substantial retrograde amnesia (O'Connor and Lafleche 2004) and deficits in access to knowledge or monitoring of knowledge retrieval in two domains semantic and social information (Kan et al. 2010; Stolk et al. 2015; Hebscher et al. 2016). Moreover, medial prefrontal lesions alter an early event-related potential component during remote memory retrieval thought to be related to an automatic (pre-conscious) process (Gilboa et al. 2009).

Functional imaging studies comparing retrieval related activity associated with remotely and recently learned information as described above are commonly confounded by differences in task difficulty or memory performance. Thus, differences in medial prefrontal activity might be related to higher task difficulty instead of remoteness. However, rapid consolidation based on congruency with prior knowledge (schema effect) goes along with better task performance and thus, parallel schema and remoteness effects are not readily explained by task difficulty or a performance account. And indeed retrieval of schema congruent, as opposed to schema incongruent information leads to stronger medial prefrontal activity and connectivity with posterior representational areas. To show this, van Kesteren et al. (2010a) asked volunteers to memorize triplets of visual motives, written words describing an object made of fabric and a tactile experience of an actual fabric before going into the scanner. Half of the triplets were congruent with prior, real-world knowledge in terms of word-fabric associations, the other half incongruent, while visual motives were randomly assigned to the two conditions. Subsequent memory for the visual motives and their associations was better for motives in congruent than for incongruent triplets and associated with pronounced medial prefrontal involvement. Also retrieval of object-location associations that define a mental schema based on a large visuo-spatial layout, conceptually similar to the event arena used for rodents, is associated with medial prefrontal activity. Moreover, Brod et al. (2015) corroborated these findings, by using a set-up that induces an artificial, hierarchic knowledge structure. They showed that enhanced medial prefrontal activity was associated with successful retrieval of schema-congruent compared to schema-incongruent information.

While schema-congruent retrieval appears associated with medial prefrontal computations, formation of such memory appears also processed preferably there. In two studies using very distinct associative information as study material that is either congruent or incongruent with general or education-related knowledge van Kesteren and co-workers found two very similar interactions between the factors 'schema congruency' and 'subsequent memory' (2013, 2014) (Fig. 2). In both studies, the subsequent memory effect for schema-congruent information was larger in the medial prefrontal cortex, while the subsequent memory effect for schema-incongruent information was larger in the medial temporal lobe. Thus, processes underlying memory formation appear balancing between two routes into long-term memory. A medial prefrontal route predominates when to be learned information fits well with preexisting knowledge and a medial temporal route predominates when information is characterized by novelty relative to already

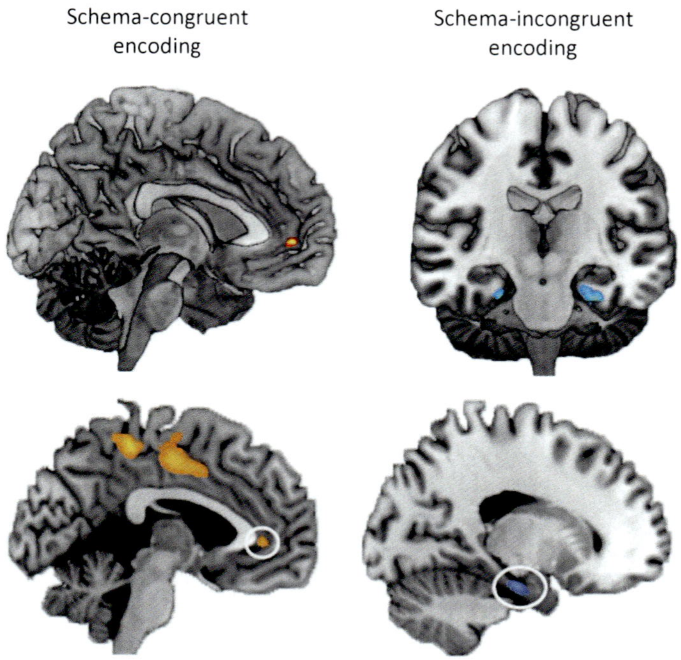

Fig. 2 Probing memory formation. Maps highlight brain areas where the factors of 'subsequent memory' and 'schema congruency' interact. *Warm colors* depict larger subsequent memory effects for schema-congruent information and *cold colors* for schema-incongruent information from two different tasks (adapted from van Kesteren et al. 2013, 2014)

existing knowledge. This conclusion is confirmed by a massive prospective study tracking the formation of schemas over the course of 300 days showing that learning new, related information that fit well into the already acquired knowledge was linked to a medial prefrontal effect (Sommer, in press). Thus, during memory formation the medial prefrontal cortex and the medial temporal lobe appear to interact, enabling the formation of memories by integrating overlapping information between events or between preexisting knowledge and a related experience providing the bases of conceptual knowledge or understanding (Kumaran et al. 2009; van Kesteren et al. 2010b; Zeithamova et al. 2012; Bein et al. 2014, Liu et al., in press). However, the representation of underlying knowledge might not be stored in the medial prefrontal cortex, but in connected posterior brain areas like the angular gyrus that combine underlying perceptual features with a schema that structures those features into one memory of conceptual representation (Wagner et al. 2015).

Recent neuropsychological evidence in patients with medial prefrontal lesions confirms that this brain region carries an essential role in schema memory. It has been repeatedly reported that medial prefrontal lesions can lead to confabulations, the erroneous production of objectively false memories (e.g., Turner et al. 2008). Damaged medial prefrontal cortex may cause irregular schema activation and thereby causing confabulations (Ghosh et al. 2014). Moreover, schematic memory can also underlie memory failures when causing false memories, by activating an entire conceptual network that contains semantically related information. When applying the false memory paradigm of Roediger and McDermott (1995), patients with medial prefrontal lesions had less false memories, showing that the medial prefrontal cortex is not necessarily the location where schemas are represented, but it appears to control the impact of conceptual networks or schemas when exposed to related information (Warren et al. 2014). This controlling function has also been confirmed by another study that probed schema-based operations that do not require memory formation but rather schema usage (Ghosh et al. 2014). In that study, patients and controls had to decide whether words were related to a semantic schema previously described. Shortly thereafter, they repeated the task for another schema, but with a word list that included some lures, words congruent to the first schema. Medial prefrontal patients with confabulations showed a deficit in rejecting lures and in identifying words that were actually related to the schema as being related. Thus, neuropsychological data in brain damaged patients show that the medial prefrontal cortex is not the site where schemas are represented, but where they are processed enabling integration or assimilation of new information and schema activation or reinstatement—results nicely in line with neuroimaging findings.

Summing up, I propose that the evidence linking the medial prefrontal cortex to memory, as reviewed here, warrants an extension of the declarative memory system with an additional hub. One can conclude that there is a medial prefrontal—hippocampal interaction, presumably mediated by midline structures of the thalamus that are, under certain circumstances, critical for memory formation, consolidation and retrieval. Medial prefrontal involvement appears enabled by or most needed for when there is overlap between new information to be encoded or consolidated with already existing knowledge, enabling integration in existing mental schemas and generalization across specific episodes. The memory dependent on the medial prefrontal cortex appears associative in nature, but rather less episodic or vivid. Thus, it does not fall easily on one side of the semantic-episodic divide.

Certainly, there are as many accounts for medial prefrontal processes as disciplines in cognitive neuroscience and thus, we do not have full understanding yet of what kind of computation(s) is or are actually performed. Also, we do not have insight into how different medial prefrontal regions including medial aspects of the orbital frontal cortex and the anterior cingulate cortex contribute uniformly or differently to memory and other cognitive operations. Different connectivity patterns might help delineating functional modules (Piray et al. 2016). Thus, the extension of the declarative memory system, as proposed here, can be regarded as a heuristic model paving the way for further studies clarifying the intricate interaction between the medial temporal lobe, the medial prefrontal cortex, posterior representational areas as well as specific subcortical structures that jointly enable us to acquire and apply a vast amount of knowledge.

References

Aggleton JP, Brown MW (1999) Episodic memory, amnesia, and the hippocampal-anterior thalamic axis. Behav Brain Sci 22:425–444

Bein O, Reggev N, Maril A (2014) Prior knowledge influences on hippocampus and medial prefrontal cortex interactions in subsequent memory. Neuropsychologia 64:320–330

Bero AW, Meng J, Cho S, Shen AH, Canter RG, Ericsson M, Tsai LH (2014) Early remodeling of the neocortex upon episodic memory encoding. Proc Nat Acad Sci U S A 111:11852–11857

Bonnici HM, Chadwick MJ, Lutti A, Hassabis D, Weiskopf N, Maguire EA (2012) Detecting representations of recent and remote autobiographical memories in vmPFC and hippocampus. J Neurosci 32:16982–16991

Bontempi B, Laurent-Demir C, Destrade C, Jaffard R (1999) Time-dependent reorganization of brain circuitry underlying long-term memory storage. Nature 400:671–675

Brod G, Lindenberger U, Werkle-Bergner M, Shing YL (2015) Differences in the neural signature of remembering schema-congruent and schema-incongruent events. Neuroimage 117:358–366

Cassel JC, Pereira de Vasconcelos A, Loureiro M, Cholvin T, Dalrymple-Alford JC, Vertes RP (2013) The reuniens and rhomboid nuclei: neuroanatomy, electrophysiological characteristics and behavioral implications. Prog Neurobiol 111:34–52

Davoodi FG, Motamedi F, Akbari E, Ghanbarian E, Jila B (2011) Effect of reversible inactivation of reuniens nucleus on memory processing in passive avoidance task. Behav Brain Res 221:1–6

Ding HK, Teixeira CM, Frankland PW (2008) Inactivation of the anterior cingulate cortex blocks expression of remote, but not recent, conditioned taste aversion memory. Learn Mem 15: 290–293

Euston DR, Tatsuno M, McNaughton BL (2007) Fast-forward playback of recent memory sequences in prefrontal cortex during sleep. Science 318:1147–1150

Frankland PW, O'Brien C, Ohno M, Kirkwood A, Silva AJ (2001) Alpha-CaMKII-dependent plasticity in the cortex is required for permanent memory. Nature 411:309–313

Frankland PW, Bontempi B, Talton LE, Kaczmarek L, Silva AJ (2004) The involvement of the anterior cingulate cortex in remote contextual fear memory. Science 304:881–883

Gais S, Albouy G, Boly M, Dang-Vu TT, Darsaud A, Desseilles M, Rauchs G, Schabus M, Sterpenich V, Vandewalle G, Maquet P, Peigneux P (2007) Sleep transforms the cerebral trace of declarative memories. Proc Natl Acad Sci U S A 104:18778–18783

Gilboa A, Alain C, He Y, Stuss DT, Moscovitch M (2009) Ventromedial prefrontal cortex lesions produce early functional alterations during remote memory retrieval. J Neurosci 29:4871–4881

Ghosh VE, Moscovitch M, Melo Colella B, Gilboa A (2014) Schema representation in patients with ventromedial PFC lesions. J Neurosci 34:12057–12070

Hebb D (1949) The organization of behavior. Wiley, New York

Hebscher M, Barkan-Abramski M, Goldsmith M, Aharon-Peretz J, Gilboa A (2016) Memory, decision-making, and the ventromedial prefrontal cortex (vmPFC): the roles of subcallosal and posterior orbitofrontal cortices in monitoring and control processes. Cereb Cortex 26 (12):4590–4601

Hoover WB, Vertes RP (2012) Collateral projections from nucleus reuniens of thalamus to hippocampus and medial prefrontal cortex in the rat: a single and double retrograde fluorescent labeling study. Brain Struct Funct 217:191–209

Kan IP, Larocque KF, Lafleche G, Coslett HB, Verfaellie M (2010) Memory monitoring failure in confabulation: evidence from the semantic illusion paradigm. J Int Neuropsychol Soc 16:1006–1017

Kumaran D, Summerfield JJ, Hassabis D, Maguire EA (2009) Tracking the emergence of conceptual knowledge during human decision making. Neuron 63:889–901

Liu ZX, Grady C, Moscovitch M (in press) Effects of prior-knowledge on brain activation and connectivity during associative memory encoding. Cerebral Cortex

McClelland JL (2013) Incorporating rapid neocortical learning of new schema-consistent information into complementary learning systems theory. J Exp Psychol Gen 142:1190–11210

McClelland JL, McNaughton BL, O'Reilly RC (1995) Why there are complementary learning systems in the hippocampus and neocortex: insights from the successes and failures of connectionist models of learning and memory. Psychol Rev 102:419–457

Milton F, Muhlert N, Butler CR, Smith A, Benattayallah A, Zeman AZ (2011) An fMRI study of long-term everyday memory using SenseCam. Memory 19:733–744

O'Connor MG, Lafleche GM (2004) Retrograde amnesia in patients with rupture and surgical repair of anterior communicating artery aneurysms. J Int Neuropsychol Soc 10:221–229

Peyrache A, Khamassi M, Benchenane K, Wiener SI, Battaglia FP (2009) Replay of rule-learning related neural patterns in the prefrontal cortex during sleep. Nat Neurosci 12:919–926

Piray P, Toni I, Cools R (2016) Human choice strategy varies with anatomical projections from ventromedial prefrontal cortex to medial striatum. J Neurosci 36:2857–2867

Quinn JJ, Ma QD, Tinsley MR, Koch C, Fanselow MS (2008) Inverse temporal contributions of the dorsal hippocampus and medial prefrontal cortex to the expression of long-term fear memories. Learn Mem 15:368–372

Richards BA, Xia F, Santoro A, Husse J, Woodin MA, Josselyn SA, Frankland PW (2014) Patterns across multiple memories are identified over time. Nat Neurosci 17:981–986

Roediger HL, McDermott KB (1995) Creating false memories remembering words not presented in lists. J Exp Psychol Learn Mem Cogn 4:803–814

Rudoy JD, Voss JL, Westerberg CE, Paller KA (2009) Strengthening individual memories by reactivating them during sleep. Science 326:1079

Schwindel CD, McNaughton BL (2011) Hippocampal-cortical interactions and the dynamics of memory trace reactivation. Prog Brain Res 193:163–177

Scoville WB, Milner B (1957) Loss of recent memory after bilateral hippocampal lesions. J Neurol Neurosurg Psychiatry 20:11–21

Shimizu E, Tang YP, Rampon C, Tsien JZ (2000) NMDA receptor-dependent synaptic reinforcement as a crucial process for memory consolidation. Science 290:1170–1174

Sommer T (in press) The emergence of knowledge and how it supports the memory for novel related information. Cerebal Cortex

Squire LR, Alvarez P (1995) Retrograde amnesia and memory consolidation: a neurobiological perspective. Curr Opin Neurobiol 5:169–177

Sterpenich V, Albouy G, Darsaud A, Schmidt C, Vandewalle G, Dang Vu TT, Desseilles M, Phillips C, Degueldre C, Balteau E, Collette F, Luxen A, Maquet P (2009) Sleep promotes the neural reorganization of remote emotional memory. J Neurosci 29:5143–5152

Stolk A, D'Imperio D, di Pellegrino G, Toni I (2015) Altered communicative decisions following ventromedial prefrontal lesions. Curr Biol 25:1469–1474

Takehara K, Kawahara S, Kirino Y (2003) Time-dependent reorganization of the brain components underlying memory retention in trace eyeblink conditioning. J Neurosci 23: 9897–9905

Takashima A, Petersson KM, Rutters F, Tendolkar I, Jensen O, Zwarts MJ, McNaughton BL, Fernández G (2006) Declarative memory consolidation in humans: a prospective functional magnetic resonance imaging study. Proc Nat Acad Sci U S A 103:756–761

Takehara-Nishiuchi K, Nakao K, Kawahara S, Matsuki N, Kirino Y (2006) Systems consolidation requires postlearning activation of NMDA receptors in the medial prefrontal cortex in trace eyeblink conditioning. J Neurosci 26:5049–5058

Takehara-Nishiuchi K, McNaughton BL (2008) Spontaneous changes of neocortical code for associative memory during consolidation. Science 322:960–963

Takashima A, Nieuwenhuis IL, Jensen O, Talamini LM, Rijpkema M, Fernández G (2009) Shift from hippocampal to neocortical centered retrieval network with consolidation. J Neurosci 29:10087–10093

Tanaka KZ, Pevzner A, Hamidi AB, Nakazawa Y, Graham J, Wiltgen BJ (2014) Cortical representations are reinstated by the hippocampus during memory retrieval. Neuron 84: 347–354

Thielen JW, Takashima A, Rutters F, Tendolkar I, Fernández G (2015) Transient relay function of midline thalamic nuclei during long-term memory consolidation in humans. Learn Mem 22:527–531

Tse D, Langston RF, Kakeyama M, Bethus I, Spooner PA, Wood ER, Witter MP, Morris RG (2007) Schemas and memory consolidation. Science 316:76–82

Tse D, Takeuchi T, Kakeyama M, Kajii Y, Okuno H, Tohyama C, Bito H, Morris RG (2011) Schema-dependent gene activation and memory encoding in neocortex. Science 333:891–895

Turner MS, Cipolotti L, Yousry TA, Shallice T (2008) Confabulation: damage to a specific inferior medial prefrontal system. Cortex 44:637–648

van der Werf YD, Witter MP, Groenewegen HJ (2002) The intralaminar and midline nuclei of the thalamus. Anatomical and functional evidence for participation in processes of arousal and awareness. Brain Res Rev 39:107–140

van Dongen EV, Takashima A, Barth M, Fernández G (2011) Functional connectivity during light sleep is correlated with memory performance for face-location associations. Neuroimage 57:262–270

van Dongen EV, Takashima A, Barth M, Zapp J, Schad LR, Paller KA, Fernández G (2012) Memory stabilization with targeted reactivation during human slow-wave sleep. Proc Natl Acad Sci U S A 109:10575–10580

van Kesteren MT, Rijpkema M, Ruiter DJ, Fernández G (2010a) Retrieval of associative information congruent with prior knowledge is related to increased medial prefrontal activity and connectivity. J Neurosci 30:15888–15894

van Kesteren MT, Fernández G, Norris DG, Hermans EJ (2010b) Persistent schema-dependent hippocampal-neocortical connectivity during memory encoding and postencoding rest in humans. Proc Natl Acad Sci U S A 107:7550–7555

van Kesteren MT, Beul SF, Takashima A, Henson RN, Ruiter DJ, Fernández G (2013) Differential roles for medial prefrontal and medial temporal cortices in schema-dependent encoding: from congruent to incongruent. Neuropsychologia 51:2352–2359

van Kesteren MT, Rijpkema M, Ruiter DJ, Morris RG, Fernández G (2014) Building on prior knowledge: schema-dependent encoding processes relate to academic performance. J Cogn Neurosci 26:2250–2261

Vertes RP, Hoover WB, Do Valle AC, Sherman A, Rodriguez JJ (2006) Efferent projections of reuniens and rhomboid nuclei of the thalamus in the rat. J Comp Neuro 499:768–796

Vertes RP, Hoover WB, Szigeti-Buck K, Leranth C (2007) Nucleus reuniens of the midline thalamus: link between the medial prefrontal cortex and the hippocampus. Brain Res Bull 71:601–609

Vieira PA, Korzus E (2015) CBP-Dependent memory consolidation in the prefrontal cortex supports object-location learning. Hippocampus 25:1532–1540

Wagner IC, van Buuren M, Kroes MC, Gutteling TP, van der Linden M, Morris RG, Fernández G (2015) Schematic memory components converge within angular gyrus during retrieval. eLIFE 4:e09668

Warren DE, Jones SH, Duff MC, Tranel D (2014) False recall is reduced by damage to the ventromedial prefrontal cortex: implications for understanding the neural correlates of schematic memory. J Neurosci 34:7677–7682

Wheeler AL, Teixeira CM, Wang AH, Xiong X, Kovacevic N, Lerch JP, McIntosh AR, Parkinson J (2013) Frankland PW identification of a functional connectome for long-term fear memory in mice. PLoS Comput Biol 9:e1002853

Xu W, Südhof TC (2013) A neural circuit for memory specificity and generalization. Science 339:1290–1295

Zeithamova D, Dominick AL, Preston AR (2012) Hippocampal and ventral medial prefrontal activation during retrieval-mediated learning supports novel inference. Neuron 75:168–179

Consolidation of Episodic Memory: An Epiphenomenon of Semantic Learning

Sen Cheng

Abstract Two hypotheses dominate the literature on systems consolidation of episodic memory: the transfer hypothesis and the transformation hypothesis. The former postulates a transfer of episodic memory from a fast-learning hippocampus to a slow-learning neocortex. The latter postulates that only the hippocampus genuinely stores episodic memories, and that systems consolidation arises due to multiple memory traces in the hippocampus and a transformation of episodic into semantic memories. While both hypotheses are supported by some evidence, they are contradicted by other, and hence remain controversial. Here, I suggest a new account of systems consolidation. It is based on the transformation hypothesis and introduces two modifications. First, episodic and semantic memory differ in their representational format, which is optimized for different purposes: Rigid sequences for episodic memories and flexible representations for semantic memory. Second, multiple memory traces in the hippocampus are not required to account for the temporal gradient of retrograde amnesia, if there is any. To this end, slow semantic learning from episodic memories suffices. The main hypothesis that I propose in this chapter is that systems consolidation is an epiphenomenon of semantic learning from episodic memory.

Keywords Episodic memory · Semantic memory · Hippocampus · CRISP · Sequence analysis

Introduction

Müller and Pilzecker (1900) coined the term "consolidation" more than 100 years ago, but we still do not fully understand why consolidation occurs and what happens to memories during this process. In the current literature, consolidation might

S. Cheng (✉)
Institute for Neural Computation and Mercator Research Group "Structure of Memory", Ruhr Universität Bochum, 44801 Bochum, Germany
e-mail: sen.cheng@ruhr-uni-bochum.de

N. Axmacher and B. Rasch (eds.), *Cognitive Neuroscience of Memory Consolidation*, Studies in Neuroscience, Psychology and Behavioral Economics, DOI 10.1007/978-3-319-45066-7_4

refer to several different phenomena (see also chapter by Genzel and Wixted). Synaptic consolidation refers to protein-synthesis-dependent late-phase long-term potentiation (Frey and Morris 1997; Bramham and Messaoudi 2005; Dudai 2012) and consolidation in procedural learning refers to a stabilization of motor memory (Brashers-Krug et al. 1996) or an improvement in performance during a rest period post-training (Walker et al. 2002). The contemporary term "systems consolidation" most closely matches the phenomenon Müller and Pilzecker originally studied: The process by which an initially labile memory is made stable such that it becomes robust to interference or brain injury (Rosenbaum et al. 2008; Dudai 2012). Here, I focus on the systems consolidation of episodic memory, the memory of personally experienced events. About 60 years ago, it was discovered that episodic memories cannot be stored without a functioning hippocampus (anterograde amnesia) (Scoville and Milner 1957). For memories that were stored before the hippocampal insult, the impact is less severe on remote memories than on more recent memories (graded retrograde amnesia) (Zola-Morgan and Squire 1990; Kim and Fanselow 1992; Rosenbaum et al. 2008).

Theoretical Accounts of Consolidation: Transfer Versus Transformation

Due to their influence on the field two hypotheses stand out from among the many hypotheses about systems consolidation. According to the *transfer hypothesis*, the memory contents are transferred from the hippocampal memory system to a neocortical memory system during systems consolidation (Buzsaki 1989; Alvarez and Squire 1994; McClelland et al. 1995). Systems consolidation is necessary because a neural network cannot store memories both rapidly and stably over time because of catastrophic interference (McClelland et al. 1995). With two complementary learning systems (CLS), episodic memories can be stored quickly in the hippocampus, where plasticity is rapid, and then gradually transferred to the neocortex, where memories are encoded more slowly, but also more stably. The memory contents are hypothesized to remain similar, if not identical, during and after consolidation. The transfer hypothesis is supported, for instance, by findings in imaging studies that hippocampal activation during retrieval is lower for more remote than for recent memories (Bontempi et al. 1999; Takashima et al. 2009; see chapters by Genzel and Wixted, by Fernandez and by Genzel and Battaglia).

However, there are challenges for the transfer hypothesis (Nadel and Moscovitch 1997; Winocur and Moscovitch 2011; see chapter by Sekeres, Moscovitch and Winocur). For instance, there is currently little experimental support for a fundamental difference in the time scales of plasticity in the hippocampus and neocortex (Cheng 2013). On the contrary, some studies find rapid neocortical learning on the timescale of single trials (Sharon et al. 2011). The most serious challenge, however, is the long temporal gradient of retrograde amnesia. In amnesics, whose brain damage was limited to the hippocampus, the gradient extends 15 years or more in some cases (Squire et al. 1989) and retrograde amnesia appears to be not graded at all in other cases (Steinvorth et al. 2005). It is difficult to conceive why the transfer process would last more than a decade or the entire lifetime of an individual. Moreover, such prolonged storage in the hippocampal memory system contradicts the core assumption of the transfer hypothesis that memory storage in the hippocampus is not stable (Buzsaki 1989).

An alternative view of systems consolidation is the *transformation hypothesis* (Winocur and Moscovitch 2011). It holds that genuine episodic memories are stored only in the hippocampus and that their recall requires the hippocampus regardless of the age of the memory (Nadel and Moscovitch 1997). This prediction is supported by fMRI studies (Nadel et al. 2000; Ryan et al. 2001; Harand et al. 2012). Multiple Trace Theory (MTT) suggests that each retrieval gives rise to a new copy of the memory trace. So older episodic memories, which have been retrieved more frequently, will be represented by more memory traces than recent memories. Therefore, partial hippocampal lesions are less likely to affect remote than recent memories because remote memories have more widely distributed representations in the network. In short, MTT can account for the long (Squire et al. 1989) or flat gradient (Steinvorth et al. 2005) gradient in retrograde amnesia, which are difficult to reconcile with the transfer hypothesis.

The transformation hypothesis additionally holds that episodic memories are transformed into semantic memories (Cermak 1984; Nadel and Moscovitch 1998; Winocur et al. 2010; Winocur and Moscovitch 2011). When episodic memories are reactivated, there are some deviations in details. Those central features that overlap in different memory traces, the gist, can be identified by neocortex and stored as semantic memory.[1] In other words, during consolidation there is a "loss of detailed, contextual features." (Winocur and Moscovitch 2011, p. 767). Such a process has been observed in animal experiments (Wiltgen and Silva 2007). Once a semantic memory has been established, the same information is located in two different sites: once in the hippocampus and once in the neocortex. Hence, remote auto-biographical information could be recalled from semantic memory in the neocortex, even if the hippocampus were completely lesioned. This would make it appear as if some episodic memory were stored outside the hippocampus.

A central element of the transformation hypothesis is the distinction between episodic memory and semantic memory, which the hypothesis relies on to account for the different functions of hippocampus and neocortex and to account for systems consolidation. We therefore review the evidence for a dissociation between episodic and semantic memory in the following.

[1]Interestingly, in the most widely cited article that champions the transfer hypothesis, McClelland et al. (1995) propose that the hippocampus stores single experiences whereas the function of neocortex is "discovering structure". These ideas are reminiscent of the transformation hypothesis, but the computational models in McClelland et al. assume that the contents of both episodic and semantic memory are very similar, if not identical.

The Distinction Between Episodic and Semantic Memory

Semantic memory is thought to be the memory of general facts (Quillian 1966; Collins and Quillian 1969; Tulving 1972), which differs from the memory of particular episodes in episodic memory. When Tulving (1972) first conceptualized episodic memory as memory of personally-experienced events, he hypothesized that episodic memories could be distinguished on the basis of the information that they contain. He suggested that "[e]pisodic memory receives and stores information about temporally dated episodes or events, and temporal-spatial relations among these events." (Tulving 1972, p. 385) This criterion later came to be known as the what-where-when (Griffiths et al. 1999) or WWW criterion (Suddendorf and Busby 2003). Tulving (1972) thought that, on this basis, episodic memories could be distinguished from other memory systems such as semantic or procedural memory. However, Tulving (1985) later realized that the WWW criterion was neither necessary nor sufficient to define episodic memory. For instance, we frequently know what happened where and when (e.g., such as details about Abraham Lincoln's assassination) without having personally experienced the event. Tulving therefore developed a new definition that focused on the subjective experience during the retrieval of episodic memories. He proposed that anoetic, noetic, and autonoetic awareness characterize the recall of procedural, semantic and episodic memory, respectively. Roughly speaking, anoetic consciousness lacks awareness of the content of the memory, noetic consciousness includes awareness of the content, and auto-noetic consciousness includes the awareness that one self has been part of the recalled event. The experience during retrieval of episodic memories was likened to mental time travel into the past (Tulving 1985; Suddendorf and Corballis 1997), a virtual reliving of the past.

Notwithstanding the lack of a clear definition between the two types of memory, numerous findings suggest that two separate memory systems exist. Patients with lesions of the medial temporal lobe (MTL) have a severe impairment of episodic memory (Bayley et al. 2008; Rosenbaum et al. 2008), but their semantic memory appears to be largely spared (Hirano and Noguchi 1998; Schmolck et al. 2002) and they can learn new semantic information (Vargha-Khadem et al. 1997; Bayley et al. 2008; Sharon et al. 2011). By contrast, patients with semantic dementia suffer from a severe loss of semantic memory, while their episodic memory is relatively spared (Snowden et al. 1994; Hodges and Patterson 2007). Further dissociations have been observed in neuroimaging studies. Semantic memory activates the left inferior prefrontal cortex and left posterior temporal areas (Wiggs et al. 1998), whereas autobiographical and episodic memory retrieval involves increased activity in the medial frontal cortex, middle temporal, temporopolar and right prefrontal areas (Maguire 2001; Gilboa 2004). Taken together, these findings suggest to some authors that episodic memory is a separate memory system from semantic memory.

However the above conclusion is not uncontroversial. Some authors have argued that the conclusion is based on studies that have focused only on activation difference where the experimental paradigms and statistical methods were designed to dissociate the neural correlates of the two types of memories (Rajah and McIntosh 2005; Burianova and Grady 2007). Other studies, found that amnesics have semantic retrieval deficits in addition to profound impairments in episodic memory (Squire and Zola 1998; Kopelman and Kapur 2001; Manns et al. 2003), suggesting that the MTL is important for both semantic and episodic retrieval. Furthermore, the left prefrontal cortex, which is thought to be part of the semantic memory system, is also involved during episodic memory retrieval (Martin and Chao 2001; Wagner et al. 2001). Damage to left prefrontal cortex has been found to impair free recall and context memory task performance (Stuss et al. 1996; Mangels 1997).

As a result, alternative views have emerged. The unitary system view suggests that a single declarative memory system subserves both episodic and semantic memory (Baddeley 1984; Rajah and McIntosh 2005; Burianova and Grady 2007). Specifically, the encoding of to-be-remembered material is always contextual (Baddeley 1984), but at retrieval, memories may, or may not, become decontextualized depending on the task demands (Westmacott and Moscovitch 2003; Rajah and McIntosh 2005; Yassa and Reagh 2013). What is called episodic memory, they posit, is simply a memory that is more contextual than one that is called semantic. Indeed, semantic memory is rarely entirely context-free, but rather may involve some contextual and episodic components (Westmacott and Moscovitch 2003; Gilboa 2004), and episodic retrieval is not free of general world knowledge (Levine et al. 2002; Gilboa 2004). Similarly, Mayes and Roberts (2001) suggest that episodic memory is the result of a synergy between perceptual and semantic aspects of objects and spatio-temporal contextual features. In a similar vein, Klein (2013) proposed that one and the same memory trace could be retrieved as either episodic or semantic memory, depending on whether the retrieval was associated with auto-noetic or noetic consciousness, respectively.

Greenberg and Verfaellie (2010) suggested that memory exists on a continuum, and that episodic and semantic memory merely represent two extremes of the same type of memory. As a consequence, intermediate manifestations of that memory also exist and episodic memory cannot clearly be separated from semantic memory. A number of neuroimaging and neuropsychological studies reported commonalities in neural activations across tasks involving semantic and episodic memories (Rajah and McIntosh 2005; Burianova and Grady 2007). For example, in an event-related functional magnetic resonance imaging (fMRI) study, Burianova et al. (2010) showed subjects pictures to cue three types of memory retrieval: autobiographical, episodic, and semantic. In every retrieval condition, they found that three regions (left hippocampus, left lingual gyrus, and right caudate nucleus) showed a common pattern of connectivity. Along similar lines, the semantic system identified by Binder et al. (2009) in a meta analysis of 120 functional neuroimaging studies overlaps to a large extent with the core network of the episodic memory system (Maguire 2001; Svoboda et al. 2006).

The Component-Structure of Episodic Memory

In the following, I suggest a novel account of systems consolidation, which largely adopts the transformation hypothesis, but deviates from it in two significant points. First, in contrast to previous approaches that were based on memory content or subjective experience, I suggest that episodic and semantic memory differ in their representational formats. Each representational format is optimized for a different purpose. The sequential character of one-time experiences involving arbitrary objects and modalities requires that episodic memories are represented by neuronal sequences, whereas capturing propositions about particular objects and the relationships between them requires flexible representational structures in semantic memory. Second, multiple memory traces in hippocampus are not required to account for systems consolidation. Slow semantic learning from episodic memories alone can account for a temporal gradient of retrograde amnesia, if there is any. However, slow learning is not due to slow plasticity in neocortex as the transfer hypothesis would have it, but rather due to the nature of semantic information.

Perhaps episodic memory and its consolidation are so difficult to understand because episodic memory in humans is a complex compound phenomenon. If so, then the first step to understanding its nature and neuronal basis is to dissociate episodic memory into its constitutive components. Adopting this view, I suggest here that the following components can be distinguished in episodic memory: the subjective experience, scenario generation, episodic memory traces,[2] semantic representations, and semantic information. Except for the distinction between semantic representation and semantic information, these components were introduced previously (Cheng et al. 2016). For the sake of brevity, the subjective experience in episodic memory is not discussed here, since in my view it does not add to our understanding of systems consolidation. Nevertheless, the nature of the subjective experience during retrieval can be incorporated into the framework for episodic memory proposed here (Cheng et al. 2016).

Semantic Representations and Information

Semantic memory, as used in the field, is probably an umbrella term that subsumes several concepts that are quite different from each other (Renoult et al. 2012). For this reason, I use the more specific terms "semantic representation" and "semantic information" in the following.

Semantic representations are needed in an episodic memory to account for the finding that episodic memory is not detailed, but contains only gist information

[2] I use the term "episodic memory trace" to refer to the quintessential information that is stored in the brain about a particular episode and "episodic memory" to refer to the compound phenomenon that involves the episodic memory trace, but also other components and processes.

(Roediger and McDermott 1995; Koutstaal and Schacter 1997; Oliva 2005). I regard the gist as the information content from the most abstract level that is relevant in that one episode. The relevant level is established dynamically for each episode depending on the context and attention (Graham et al. 2000). In some cases, the most abstract level could in fact be quite detailed, a level that in most other cases would not enter into the episodic memory traces. For instance, after going for a leisurely walk around a lake, we might only remember a path, trees and water, but not the types of trees, the shape of their leaves, or state of bloom. However, we are more likely to remember these details, if we had been given the task to study the trees around the lake. I propose that episodic memory includes only gist information, because it stores a semantic representation of the input, not the rich sensory input itself. The semantic representation is more abstract and combines various objects to compound entities. For instance, in most instances we represent a car as a single entity rather than a collection of tires, windows, doors and other parts. Note that each of the parts is in itself also a semantic representation. Some might object that possessing a semantic representation does not qualify as having a semantic memory and they might be right, but that is a different debate. One more important feature of semantic representations is that they are not only categorical, e.g. a woman, but can also represent particular objects or persons, e.g., my mother. My suggestion that semantic representations are stored in episodic memory is similar Teyler and DiScenna's (1986) suggestion that episodic memory stores indices of the items rather than the items themselves. Semantic representations are probably localized in sensory areas of the neocortex (Huth et al. 2012).

We cannot only represent objects and people, but also associate properties to them and store the relationships between them. This information I call semantic information. It can involve symbolic rules involving categories, e.g. "police cars are blue" or "policemen carry guns", or idiosyncratic properties of and relationships between particular objects, as long as they are consistent across different episodes, such as, for example, the name of a particular person (Quian Quiroga et al. 2005) or information of the kind "The Eiffel Tower is in Paris" or "my car is red". As these examples indicate, semantic information is about items of which we have developed semantic representations. Semantic information are general facts that are usually extracted from multiple experiences and/or apply to more than one instance (Maguire and Frith 2004). This slow learning accords well with findings in computational studies that, during training, information has to be presented multiple times to a neural network interleaved with previously learned information (e.g., McClelland et al. 1995). The structure of semantic information can be quite different, e.g. hierarchical, sequential, or map-like. So its representational format must support these different structures. Since semantic information is also discovered slowly, the discovered structure might change over time. Therefore the representation must be flexible and support changes in the organization of the information. The association areas of the neocortex are likely specialized for these purposes (Svoboda et al. 2006).

Sequential Episodic Memory Traces

Episodic memories are stored incidentally without a special trigger (Tulving 1972). In the vast majority of cases, the potential importance of a particular episodic memory is not discernible at the time of storage and the usefulness of the stored information might only become apparent in the future. Since episodes unfold in time, it has been suggested that episodes are sequences of events and that they are represented by sequences of neuronal patterns (Cheng 2013; Werning and Cheng 2014; Cheng and Werning 2016). Several prior studies have used sequences to model episodic memory (Jensen and Lisman 1996; Levy 1996), but this idea has not been adopted widely in the field. I suggest that the episodic memory trace forms the backbone of episodic memory, providing the sequential ordering of events to which the items involved in the events can be associated to. Since episodic memory traces are linked to specific, experienced episodes, the traces have to be stored rapidly after a single experience (Ennaceur and Delacour 1988). The episodic memory trace is the quintessential component of episodic memory. While other components involved in episodic memory serve important functions in other cognitive processes, episodic memory traces exclusively contribute to episodic memory. The reverse is also true. Removing the episodic memory trace from the system, by removing the hippocampus, abolishes episodic memory.

The hippocampal circuitry appears to be specialized for rapidly storing sequences of neuronal patterns (Cheng 2013). The core of this specialization might be the recurrent network in CA3 that is able to generate intrinsic sequences (Azizi et al. 2013), and the feedforward projections probably serve to store the associations between simultaneous inputs from the entorhinal cortex (Neher et al. 2015). Evidence from a number of species suggest that the hippocampus is involved in storing sequence memory, for instance in humans (Curran 1997; Schendan et al. 2003), monkey (Kimble and Pribram 1963), and rats (Agster et al. 2002; Fortin et al. 2002).

Intriguingly, hippocampal neurons generate different types of sequential activity intrinsically (Buhry et al. 2011). In the offline state, i.e., when the animal sits quietly or is asleep, neurons fire spikes in temporal sequences of lengths between 50 ms and 400 ms (Lee and Wilson 2002). These offline neuronal sequences correlate with the order, in which neurons were active at an earlier time in the online state (Nádasdy et al. 1999; Mehta et al. 2002). It was therefore suggested that sequential neural activity in the offline state is a replay of sequential activity during prior experience. Replay has been studied mostly in rodent hippocampus, but has also been found during free recall of movie sequences in humans (Gelbard-Sagiv et al. 2008).

Scenario Generation

The retrieval of episodic memories is not a passive process, in which information is simply found in its original form, but an active process, in which information about a past episode is (re-)generated (Dudai and Carruthers 2005; Michaelian 2011). We also call this mnemonic simulation (Cheng and Werning 2016) or scenario generation (Cheng et al. 2016). During retrieval, a scenario is generated that more or less accurately represents a past episode. A scenario is based on the sequences and associations stored in the episodic memory traces. The gist information in the semantic representations can be expanded upon with details stored in the semantic system. Furthermore, semantic information might be used to fill in gaps or resolve inconsistencies in episodic memory, which might arise through inattention, encoding, storage, or retrieval errors (Cheng et al. 2016). These processes occur automatically, even if subjects strive to accurately retrieve a past experience, and might account for the unreliable nature of human episodic memory (Loftus et al. 1978; Zaragoza and Lane 1994; Loftus and Pickrell 1995; Roediger and McDermott 1995; Schacter 2002). Scenario generation is different from the concept of "scene construction", which Hassabis and Maguire (2007) introduced. The latter refers to scenes that are static spatial arrangements while scenarios extend in space as well as time. Another difference is that scenes are isolated entities, whereas scenarios can be embedded to allow constructions of larger narratives and reflection (Suddendorf 2013).

Evidence for the Component Nature of Episodic Memory

Some evidence for the proposed component nature of episodic memory comes from developmental studies (Suddendorf and Redshaw 2013). Even though adult-like episodic memory develops only by the age of about four in children (Nelson 1993; Hayne and Imuta 2011; Scarf et al. 2013), children as young as 2½ years old appear to store and retrieve episodic memory traces (Bauer et al. 1998; Wenner and Bauer 1999). These latter studies found that children were able to re-enact a complex sequence of actions after observing it only once. Importantly, the sequence was arbitrary, and therefore the children could not use semantic information about the natural sequential order of events to reproduce it. If, however, there are so-called enabling relations between the elements of the sequence, even younger toddlers (in these studies, 16 months olds) could re-enact the sequence. One interpretation of these results is that children are first able to reproduce a sequence based on semantic information, then develop the ability to store and retrieve arbitrary sequences as episodic memory traces, and later develop the capacity to construct nested scenarios and to distill more complex semantic information. The final stages of the development of episodic memory continue well after age 4, when increases in working memory capacity, among others factors, lead to the construction of more complex episodic memories and larger narratives (Suddendorf and Redshaw 2013).

Novel Account of Systems Consolidation

Putting it all together, I propose the following account of systems consolidation. Episodic memory traces are the core component of episodic memory. They are represented as temporal sequences of neural activity that are stored only in the hippocampus. The hippocampal circuitry is optimized for rapidly storing associations between arbitrary items in temporal sequences. Structure hidden in specific episodic experiences can be extracted and stored as semantic information in the neocortex, which is optimized for discovering and representing structured information. This is consistent with observations that episodic memory greatly facilitates semantic learning (Gabrieli et al. 1988; Kitchener et al. 1998; Bayley et al. 2008). Since semantic information can be extracted from episodic memory, it might be possible in some instances to retrieve the same information from both systems. This would be more likely for more remote memories, since there was more time for semantic learning. So it would then appear as if remote autobiographical information becomes independent of the hippocampus, because it can be retrieved as semantic information from neocortex (Cermak and O'Connor 1983; Cermak 1984; Kirwan et al. 2008). In this sense, systems consolidation is an epiphenomenon of slow semantic learning that is driven by episodic memory.

The passage of time is not the sole or decisive factor for the extraction of semantic information. Whether and when semantic information is extracted from episodic memory likely depends on a range of factors, including the number of retrievals (Cermak and O'Connor 1983; Nadel and Moscovitch 1998), the salience and importance of the information (Westmacott and Moscovitch 2003; Westmacott et al. 2004) and the difficulty of extracting the information. This diversity of factors might account for the diversity in experimental findings. While some studies suggest that memories of remote episodes never become independent of the hippocampus (Steinvorth et al. 2005), others find that they do (Kirwan et al. 2008).

Slow learning of semantic information does not imply that neocortical plasticity is in itself slow, only that discovering and/or storing structured information takes time. Nor do I wish to imply that semantic information can only be learned from episodic memory. Plasticity in neocortex alone is sufficient to support semantic learning (Vargha-Khadem et al. 1997; Sharon et al. 2011). Nevertheless, episodic memory might aid semantic learning for several reasons. First, stored experiences can be reactivated for interleaved learning (McClelland et al. 1995; Kali et al. 2004). Second, episodic memory traces can connect temporally discontiguous information (Staresina and Davachi 2009; Pyka and Cheng 2014). Third, the temporal structure of inputs can be exploited to extract semantic information (Wiskott and Sejnowski 2002; Franzius et al. 2011). Since the first and third cases have been modeled and discussed more extensively in the literature, I focus on the second function below.

Learning Associations Across Time Gaps

We recently studied a model for learning an association between two temporally discontiguous pieces of information (Pyka and Cheng 2014). Associations bridging a time gap occur, for instance, in trace conditioning, which has been found to depend on the hippocampus in human (McGlinchey-Berroth et al. 1997), rabbit (Solomon et al. 1986), rat (McEchron et al. 1998; Weiss et al. 1999), and mouse (Tseng et al. 2004). In addition, functional magnetic resonance imaging revealed increased hippocampal activation during trace as compared to delay conditioning (Cheng et al. 2008). In our model, we adopted a two layer neural network as a model of the cortico-hippocampal system. The key property of the model was that axonal conduction delays were heterogeneous, in particular, the delays between cortex and hippocampus were longer than delays within neocortex or within the hippocampus. If neocortex was isolated, it could not store associations between two patterns that were separated by more than 60 ms, because of the narrow time window imposed by spike-timing dependent plasticity. With the addition of the hippocampus, the combined network could store associations across a temporal separation of up to 140 ms. Intriguingly, after repeated retrievals of the association in the neocortico-hippocampal loop, neocortex alone was sufficient to retrieve the associations, i.e., the association was consolidated.

While semantic learning in general is likely to require much more sophisticated algorithms, this simple model offers at least two important insights. First, consolidation in the neocortico-hippocampal loop changes the information stored. One might think that the association was transferred from hippocampus to the neocortex, but actually it is rather an extraction of information, since the information about the temporal relationship, which might be an important aspect of episodic memory, is lost after consolidation. Second, systems consolidation in this model occurs as a byproduct of repeated retrieval in the absence of any dedicated consolidation mechanism.

Conclusion

I have proposed a new account of systems consolidation based on a recently proposed theory of the component-structure of episodic memory, in which episodic memory traces and semantic information are combined to generate scenarios during episodic memory retrieval. The key idea in this new account is that episodic memory traces are represented as somewhat rigid temporal sequences that can be flexibly associated with arbitrary items, while semantic information consists of flexible relationships between somewhat rigid items. These suggestions offer an explanation for the special role of the hippocampus in systems consolidation and are well supported by experimental evidence. To put the main points of this chapter succinctly: episodic memory traces are stored, episodic memory is generated, semantic information is learned, and systems consolidation is an epiphenomenon of semantic learning from episodic memory.

Acknowledgements I thank Jing Fang and Richard Görler for helpful comments on the manuscript. This work was supported by the German Federal Ministry of Education and Research (BMBF), grant 01GQ1506, and by the Stiftung Mercator.

References

Agster KL, Fortin NJ, Eichenbaum H (2002) The hippocampus and disambiguation of overlapping sequences. J Neurosci 22:5760–5768

Alvarez P, Squire L (1994) Memory consolidation and the medial temporal lobe: a simple network model. Proc Natl Acad Sci U S A 91:7041–7045

Azizi AH, Wiskott L, Cheng S (2013) A computational model for preplay in the hippocampus. Front Comput Neurosci 7:161

Baddeley AD (1984) Neuropsychological evidence and the semantic/episodic distinction. Behav Brain Sci 7:238

Bauer PJ, Hertsgaard La, Dropik P, Daly BP (1998) When even arbitrary order becomes important: developments in reliable temporal sequencing of arbitrarily ordered events. Memory 6: 165–198

Bayley PJ, O'Reilly RC, Curran T, Squire LR (2008) New semantic learning in patients with large medial temporal lobe lesions. Hippocampus 18:575–583

Binder JR, Desai RH, Graves WW, Conant LL (2009) Where is the semantic system? A critical review and meta-analysis of 120 functional neuroimaging studies. Cereb Cortex 19:2767–2796

Bontempi B, Laurent-Demir C, Destrade C, Jaffard R (1999) Time-dependent reorganization of brain circuitry underlying long-term memory storage. Nature 400:671–675

Bramham CR, Messaoudi E (2005) BDNF function in adult synaptic plasticity: the synaptic consolidation hypothesis. Prog Neurobiol 76:99–125

Brashers-Krug T, Shadmehr R, Bizzi E (1996) Consolidation in human motor memory. Nature 382:252–255

Buhry L, Azizi AH, Cheng S (2011) Reactivation, replay, and preplay: how it might all fit together. Neural Plast 2011:203462

Burianova H, Grady CL (2007) Common and unique neural activations in autobiographical, episodic, and semantic retrieval. J Cogn Neurosci 19:1520–1534

Burianova H, McIntosh AR, Grady CL (2010) A common functional brain network for autobiographical, episodic, and semantic memory retrieval. Neuroimage 49:865–874

Buzsaki G (1989) Two-stage model of memory trace formation: a role for "noisy" brain states. Neuroscience 31:551–570

Cermak LS (1984) The episodic-semantic distinction in amnesia. In: Squire LR, Butters N (eds) Neuropsychology of memory. Guilford Press, New York, pp 55–62

Cermak LS, O'Connor M (1983) The anterograde and retrograde retrieval ability of a patient with amnesia due to encephalitis. Neuropsychologia 21:213–234

Cheng S (2013) The CRISP theory of hippocampal function in episodic memory. Front Neural Circuits 7:88

Cheng S, Werning M (2016) What is Episodic Memory if it is a Natural Kind? Synthese 193:1345–1385

Cheng DT, Disterhoft JF, Power JM, Ellis DA, Desmond JE (2008) Neural substrates underlying human delay and trace eyeblink conditioning. Proc Natl Acad Sci U S A 105:8108–8113

Cheng S, Werning M, Suddendorf T (2016) Dissociating memory traces and scenario construction in mental time travel. Neurosci Biobehav Rev 60:82–89

Collins AM, Quillian MR (1969) Retrieval time from semantic memory. J Verbal Learn Verbal Behav 8:240–247

Curran T (1997) Higher-order associative learning in amnesia: evidence from the serial reaction time task. J Cogn Neurosci 9:522–533

Dudai Y (2012) The restless engram: consolidations never end. Annu Rev Neurosci 35:227–247

Dudai Y, Carruthers M (2005) The Janus face of Mnemosyne. Nature 434:567

Ennaceur A, Delacour J (1988) A new one-trial test for neurobiological studies of memory in rats. 1: Behavioral data. Behav Brain Res 31:47–59

Fortin NJ, Agster KL, Eichenbaum HB (2002) Critical role of the hippocampus in memory for sequences of events. Nat Neurosci 5:458–462

Franzius M, Wilbert N, Wiskott L (2011) Invariant object recognition and pose estimation with slow feature analysis. Neural Comput 23:2289–2323

Frey U, Morris RGM (1997) Synaptic tagging and long-term potentiation. Nature 385:533–536

Gabrieli JD, Cohen NJ, Corkin S (1988) The impaired learning of semantic knowledge following bilateral medial temporal-lobe resection. Brain Cogn 7:157–177

Gelbard-Sagiv H, Mukamel R, Harel M, Malach R, Fried I (2008) Internally generated reactivation of single neurons in human hippocampus during free recall. Science 322:96–101

Gilboa A (2004) Autobiographical and episodic memory—one and the same? Evidence from prefrontal activation in neuroimaging studies. Neuropsychologia 42:1336–1349

Graham KS, Simons JS, Pratt KH, Patterson K, Hodges JR (2000) Insights from semantic dementia on the relationship between episodic and semantic memory. Neuropsychologia 38:313–324

Greenberg DL, Verfaellie M (2010) Interdependence of episodic and semantic memory: evidence from neuropsychology. J Int Neuropsychol Soc 16:748–753

Griffiths D, Dickinson A, Clayton N (1999) Episodic memory: What can animals remember about their past? Trends Cogn Sci 3:74–80

Harand C, Bertran F, La Joie R, Landeau B, Mézenge F, Desgranges B, Peigneux P, Eustache F, Rauchs G (2012) The hippocampus remains activated over the long term for the retrieval of truly episodic memories. PLoS ONE 7:e43495

Hassabis D, Maguire EA (2007) Deconstructing episodic memory with construction. Trends Cogn Sci 11:299–306

Hayne H, Imuta K (2011) Episodic memory in 3- and 4-year-old children. Dev Psychobiol 53:317–322

Hirano M, Noguchi K (1998) Dissociation between specific personal episodes and other aspects of remote memory in a patient with hippocampal amnesia. Percept Mot Skills 87:99–107

Hodges JR, Patterson K (2007) Semantic dementia: a unique clinicopathological syndrome. Lancet Neurol 6:1004–1014

Huth AG, Nishimoto S, Vu AT, Gallant JL (2012) A continuous semantic space describes the representation of thousands of object and action categories across the human brain. Neuron 76:1210–1224

Jensen O, Lisman JE (1996) Hippocampal CA3 region predicts memory sequences: accounting for the phase precession of place cells. Learn Mem 3:279–287

Kali S, Dayan P, Káli S, Dayan P (2004) Off-line replay maintains declarative memories in a model of hippocampal-neocortical interactions. Nat Neurosci 7:286–294

Kim J, Fanselow M (1992) Modality-specific retrograde amnesia of fear. Science 256:675–677

Kimble DP, Pribram KH (1963) Hippocampectomy and behavior sequences. Science 139:824–825

Kirwan CB, Bayley PJ, Galvan VV, Squire LR (2008) Detailed recollection of remote autobiographical memory after damage to the medial temporal lobe. Proc Natl Acad Sci 105:2676–2680

Kitchener EG, Hodges JR, McCarthy R (1998) Acquisition of post-morbid vocabulary and semantic facts in the absence of episodic memory. Brain 121:1313–1327

Klein SB (2013) Making the case that episodic recollection is attributable to operations occurring at retrieval rather than to content stored in a dedicated subsystem of long-term memory. Front Behav Neurosci 7:3

Kopelman MD, Kapur N (2001) The loss of episodic memories in retrograde amnesia: single-case and group studies. Philos Trans R Soc Lond B Biol Sci 356:1409–1421

Koutstaal W, Schacter DL (1997) Gist-based false recognition of pictures in older and younger adults. J Mem Lang 37:555–583

Lee AK, Wilson MA (2002) Memory of sequential experience in the hippocampus during slow wave sleep. Neuron 36:1183–1194

Levine B, Svoboda E, Hay JF, Winocur G, Moscovitch M (2002) Aging and autobiographical memory: dissociating episodic from semantic retrieval. Psychol Aging 17:677–689

Levy WBW (1996) A sequence predicting CA3 is a flexible associator that learns and uses context to solve hippocampal-like tasks. Hippocampus 6:579–590

Loftus EF, Pickrell JE (1995) The formation of false memories. Psychiatr Ann 25:720–725

Loftus EF, Miller DG, Burns HJ (1978) Semantic integration of verbal information into a visual memory. J Exp Psychol Hum Learn Mem 4:19–31

Maguire EA (2001) Neuroimaging studies of autobiographical event memory. Philos Trans R Soc Lond B Biol Sci 356:1441–1451

Maguire EA, Frith CD (2004) The brain network associated with acquiring semantic knowledge. Neuroimage 22:171–178

Mangels JA (1997) Strategic processing and memory for temporal order in patients with frontal lobe lesions. Neuropsychology 11:207–221

Manns J, Hopkins R, Squire L (2003) Semantic memory and the human hippocampus. Neuron 38:127–133

Martin A, Chao LL (2001) Semantic memory and the brain: structure and processes. Curr Opin Neurobiol 11:194–201

Mayes AR, Roberts N (2001) Theories of episodic memory. Philos Trans R Soc Lond B Biol Sci 356:1395–1408

McClelland JJL, McNaughton BLB, O'Reilly RRC (1995) Why there are complementary learning systems in the hippocampus and neocortex: insights from the successes and failures of connectionist models of learning and memory. Psychol Rev 102:419–457

McEchron MD, Bouwmeester H, Tseng W, Weiss C, Disterhoft JF (1998) Hippocampectomy disrupts auditory trace fear conditioning and contextual fear conditioning in the rat. Hippocampus 8:638–646

McGlinchey-Berroth R, Carrillo MC, Gabrieli JDE, Brawn CM, Disterhoft JF (1997) Impaired trace eyeblink conditioning in bilateral, medial-temporal lobe amnesia. Behav Neurosci 111:873–882

Mehta MR, Lee AK, Wilson MA (2002) Role of experience and oscillations in transforming a rate code into a temporal code. Nature 417:741–746

Michaelian K (2011) Generative memory. Philos Psychol 24:323–342

Müller GE, Pilzecker A (1900) Experimentelle Beiträge zur Lehre vom Gedächtnis. Zeitschrift für Psychol Ergänzungsband 1:1–300

Nádasdy Z, Hirase H, Czurkó A, Csicsvari J, Buzsáki G (1999) Replay and time compression of recurring spike sequences in the hippocampus. J Neurosci 19:9497–9507

Nadel L, Moscovitch M (1997) Memory consolidation, retrograde amnesia and the hippocampal complex. Curr Opin Neurobiol 7:217–227

Nadel L, Moscovitch M (1998) Hippocampal contributions to cortical plasticity. Neuropharmacology 37:431–439

Nadel L, Samsonovich A, Ryan L, Moscovitch M (2000) Multiple trace theory of human memory: computational, neuroimaging, and neuropsychological results. Hippocampus 10:352–368

Neher T, Cheng S, Wiskott L (2015) Memory storage fidelity in the hippocampal circuit: the role of subregions and input statistics. PLoS Comput Biol 11:e1004250

Nelson K (1993) The psychological and social origins of autobiographical memory. Psychol Sci 4:7–14

Oliva A (2005) Gist of the scene. In: Itti L, Rees G, Tsotsos JK (eds) Neurobiology of attention. Elsevier Academic Press, London, pp 251–256

Pyka M, Cheng S (2014) Pattern association and consolidation emerges from connectivity properties between cortex and hippocampus. PLoS ONE 9:e85016

Quian Quiroga R, Reddy L, Kreiman G, Koch C, Fried I (2005) Invariant visual representation by single neurons in the human brain. Nature 435:1102–1107

Quillian MR (1966) Semantic memory. Bedford, MA

Rajah MN, McIntosh AR (2005) Overlap in the functional neural systems involved in semantic and episodic memory retrieval. J Cogn Neurosci 17:470–482

Renoult L, Davidson PSRR, Palombo DJ, Moscovitch M, Levine B (2012) Personal semantics: at the crossroads of semantic and episodic memory. Trends Cogn Sci 16:550–558

Roediger HL, McDermott KB (1995) Creating false memories: remembering words not presented in lists. J Exp Psychol Learn Mem Cogn 21:803–814

Rosenbaum RS, Moscovitch M, Foster JK, Schnyer DM, Gao F, Kovacevic N, Verfaellie M, Black SE, Levine B (2008) Patterns of autobiographical memory loss in medial-temporal lobe amnesic patients. J Cogn Neurosci 20:1490–1506

Ryan L, Nadel L, Keil K, Putnam K, Schnyer D, Trouard T, Moscovitch M (2001) Hippocampal complex and retrieval of recent and very remote autobiographical memories: evidence from functional magnetic resonance imaging in neurologically intact people. Hippocampus 11: 707–714

Scarf D, Gross J, Colombo M, Hayne H (2013) To have and to hold: episodic memory in 3- and 4-year-old children. Dev Psychobiol 55:125–132

Schacter DL (2002) The seven sins of memory: how the mind forgets and remembers. Houghton Mifflin, Boston

Schendan HE, Searl MM, Melrose RJ, Stern CE (2003) An FMRI study of the role of the medial temporal lobe in implicit and explicit sequence learning. Neuron 37:1013–1025

Schmolck H, Kensinger Ea, Corkin S, Squire LR (2002) Semantic knowledge in patient H.M. and other patients with bilateral medial and lateral temporal lobe lesions. Hippocampus 12:520–533

Scoville WB, Milner B (1957) Loss of recent memory after bilateral hippocampal lesions. J Neurol Neurosurg Psychiatry 20:11–21

Sharon T, Moscovitch M, Gilboa A (2011) Rapid neocortical acquisition of long-term arbitrary associations independent of the hippocampus. Proc Natl Acad Sci U S A 108:1146–1151

Snowden J, Griffiths H, Neary D (1994) Semantic dementia: autobiographical contribution to preservation of meaning. Cogn Neuropsychol 11:265–288

Solomon PR, Vander Schaaf ER, Thompson RF, Weisz DJ (1986) Hippocampus and trace conditioning of the rabbit's classically conditioned nictitating membrane response. Behav Neurosci 100:729–744

Squire LR, Zola SM (1998) Episodic memory, semantic memory, and amnesia. Hippocampus 8:205–211

Squire LR, Haist F, Shimamura AP (1989) The neurology of memory: quantitative assessment of retrograde amnesia in two groups of amnesic patients. J Neurosci 9:828–839

Staresina BP, Davachi L (2009) Mind the gap: binding experiences across space and time in the human hippocampus. Neuron 63:267–276

Steinvorth S, Levine B, Corkin S (2005) Medial temporal lobe structures are needed to re-experience remote autobiographical memories: evidence from H.M. and W.R. Neuropsychologia 43:479–496

Stuss DT, Craik FIM, Sayer L, Franchi D, Alexander MP (1996) Comparison of older people and patients with frontal lesions: evidence from world list learning. Psychol Aging 11:387–395

Suddendorf T (2013) The gap: the science of what separates us from other animals. Basic Books, New York

Suddendorf T, Busby J (2003) Mental time travel in animals? Trends Cogn Sci 7:391–396

Suddendorf T, Corballis MC (1997) Mental time travel and the evolution of the human mind. Genet Soc Gen Psychol Monogr 123:133–167

Suddendorf T, Redshaw J (2013) The development of mental scenario building and episodic foresight. Ann N Y Acad Sci 1296:135–153

Svoboda E, McKinnon MC, Levine B (2006) The functional neuroanatomy of autobiographical memory: a meta-analysis. Neuropsychologia 44:2189–2208

Takashima A, Nieuwenhuis ILC, Jensen O, Talamini LM, Rijpkema M, Fernandez G (2009) Shift from hippocampal to neocortical centered retrieval network with consolidation. J Neurosci 29:10087–10093

Teyler TJ, DiScenna P (1986) The hippocampal memory indexing theory. Behav Neurosci 100:147–154

Tseng W, Guan R, Disterhoft JF, Weiss C (2004) Trace eyeblink conditioning is hippocampally dependent in mice. Hippocampus 14:58–65

Tulving E (1972) Episodic and semantic memory. In: Tulving E, Donaldson W (eds) Organization of memory. Academic Press Inc., New York, pp 381–402

Tulving E (1985) Memory and consciousness. Can J Psychol 26:1–26

Vargha-Khadem F, Gadian D, Watkins K, Connelly A, Van Paesschen W, Mishkin M (1997) Differential effects of early hippocampal pathology on episodic and semantic memory. Science 277:376–380

Wagner AD, Paré-Blagoev EJ, Clark J, Poldrack RA (2001) Recovering meaning. Neuron 31: 329–338

Walker MP, Brakefield T, Morgan A, Hobson AA, Stickgold R (2002) Practice with sleep makes perfect: sleep-dependent motor skill learning. Neuron 35:205–211

Weiss C, Bouwmeester H, Power JM, Disterhoft JF (1999) Hippocampal lesions prevent trace eyeblink conditioning in the freely moving rat. Behav Brain Res 99:123–132

Wenner JA, Bauer PJ (1999) Bringing order to the arbitrary: one- to two-year-olds' recall of event sequences. Infant Behav Dev 22:585–590

Werning M, Cheng S (2014) Is episodic memory a natural kind?—A defense of the sequence analysis. In: Bello P, Guarini M, McShane M, Scassellati B (eds) Proceedings of the 36th annual conference of the cognitive science society. Cognitive Science Society, Austin, TX, pp 964–969

Westmacott R, Moscovitch M (2003) The contribution of autobiographical significance to semantic memory. Mem Cognit 31:761–774

Westmacott R, Black SE, Freedman M, Moscovitch M (2004) The contribution of autobiographical significance to semantic memory: evidence from Alzheimer's disease, semantic dementia, and amnesia. Neuropsychologia 42:25–48

Wiggs CL, Weisberg J, Martin A (1998) Neural correlates of semantic and episodic memory retrieval. Neuropsychologia 37:103–118

Wiltgen BJ, Silva AJ (2007) Memory for context becomes less specific with time. Learn Mem 14:313–317

Winocur G, Moscovitch M (2011) Memory transformation and systems consolidation. J Int Neuropsychol Soc 17:766–780

Winocur G, Moscovitch M, Bontempi B (2010) Memory formation and long-term retention in humans and animals: convergence towards a transformation account of hippocampal-neocortical interactions. Neuropsychologia 48:2339–2356

Wiskott L, Sejnowski TJ (2002) Slow feature analysis: unsupervised learning of invariances. Neural Comput 14:715–770

Yassa Ma, Reagh ZM (2013) Competitive trace theory: a role for the hippocampus in contextual interference during retrieval. Front Behav Neurosci 7:1–13

Zaragoza MS, Lane SM (1994) Source misattributions and the suggestibility of eyewitness memory. J Exp Psychol Learn Mem Cogn 20:934–945

Zola-Morgan S, Squire L (1990) The primate hippocampal formation: evidence for a time-limited role in memory storage. Science 250:288–290

Part II
Memory Consolidation During Off-Line Periods and the Role of Sleep

The Impact of Sleep Deprivation on Molecular Mechanisms of Memory Consolidation in Rodents

Judith C. Kreutzmann, Jennifer C. Tudor, Christopher C. Angelakos and Ted Abel

Abstract Loss of sleep leads to deleterious effects on cognitive function, with memory being significantly impaired when rodents are sleep deprived during the first 5–6 h following learning. Recent findings suggest that lack of sleep during this specific time period after learning alters molecular signaling pathways critical for the formation of memories. Several molecular mechanisms underlying impaired cognition caused by sleep deprivation have been described. In this review, we will discuss the proposed molecular mechanisms underlying memory consolidation affected by sleep deprivation, which includes timing of sleep deprivation and the molecular signaling pathways involved.

Keywords Sleep loss · Memory · Cognition · Neuronal plasticity · Hippocampus · Mouse · cAMP · PKA · Transcription · Translation · Protein synthesis · Gene expression

Introduction

The common perception that individuals can function properly on little sleep without any consequences has been rebutted by a variety of studies in both humans and animals, thereby demonstrating that proper mental function requires sufficient sleep (Banks and Dinges 2007). The deleterious effects of sleep loss on cognitive function have extensively been studied, and include memory impairments

J.C. Kreutzmann · J.C. Tudor · C.C. Angelakos · T. Abel (✉)
Department of Biology, University of Pennsylvania, 10-133 Smilow Center for Translational Research, 3400 Civic Center Boulevard, Building 421, Philadelphia, PA 19104, USA
e-mail: abele@sas.upenn.edu

J.C. Kreutzmann
Groningen Institute for Evolutionary Life Sciences, University of Groningen, Groningen, The Netherlands

N. Axmacher and B. Rasch (eds.), *Cognitive Neuroscience of Memory Consolidation*, Studies in Neuroscience, Psychology and Behavioral Economics, DOI 10.1007/978-3-319-45066-7_5

(for review see Walker and Stickgold 2004; Walker 2008; Prince and Abel 2013; Kreutzmann et al. 2015) (see also Chapter by Schönauer and Gais).

A brain region crucial for the formation of spatial, contextual, and declarative memories that is sensitive to sleep loss is the hippocampus. Several studies demonstrate that hippocampus-dependent memory is impaired when animals are sleep deprived during the immediate 5–6 h following learning (e.g. Graves et al. 2003; Palchykova et al. 2006). These data suggest that the timing of when the sleep deprivation (SD) occurs in relation to learning is a critical component of memory impairment induced by SD. Because insufficient sleep is a common problem in our society, it is crucial to elucidate and understand the molecular and cellular mechanisms of impaired memory caused by SD in order to identify novel therapeutic approaches that can counteract the adverse consequences.

In this review, we will outline several molecular signaling mechanisms underlying hippocampus-dependent memory consolidation, and discuss how molecular targets of sleep loss are affected when SD occurs. Finally, we will briefly summarize the findings, and discuss how the identification of molecular pathways could potentially provide new drug targets to improve upon the deleterious effects of SD on memory consolidation.

Molecular Mechanisms of Memory Consolidation

Memory formation consists of three main stages: acquisition, consolidation, and retrieval. Memory begins with *acquisition*, also known as encoding, during which an animal learns to make an association between a context and experience. Then during *consolidation*, memory is strengthened from its labile state to a more fixed one so that it is available for future recall. Finally, *retrieval* is the stage during which an animal recalls information and adjusts its behavior accordingly (Nadel et al. 2012). There is strong evidence demonstrating that SD interferes with critical signaling cascades during memory consolidation leading to memory impairments (e.g., Smith and Rose 1996, 1997; Graves et al. 2003; Hairston et al. 2005; Palchykova et al. 2006; Yang et al. 2008; Li et al. 2009; Havekes et al. 2014; Prince et al. 2014). These will be discussed in the following sections.

cAMP-PKA Signaling Cascade as Key Modulator During Memory Consolidation

Research has shown that several molecular signaling pathways are critical for the consolidation of hippocampus-dependent memories, such as the Cyclic adenosine monophosphate (cAMP)—Protein Kinase A (PKA) signaling cascade. Learning generates a transient increase of calcium (Ca^{2+}), thereby elevating activity levels of

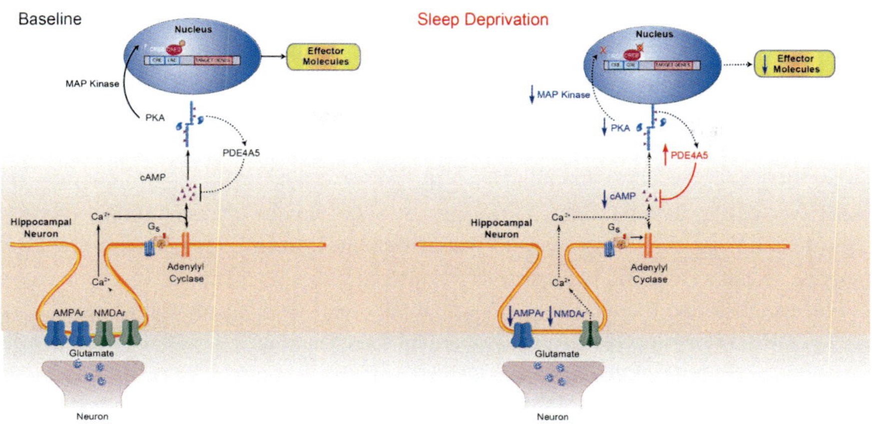

Fig. 1 Schematic overview of hippocampal signaling pathways whose modulation by sleep deprivation (SD) may contribute to the effects of SD on memory formation. **a** Signaling pathways under baseline conditions. **b** SD has been reported to reduce glutamatergic signaling which in turn attenuates cAMP signaling and CREB-mediated gene transcription. All of these molecular events are shown in a single connected pathway in order to demonstrate how the effects of SD could potentially interact to impact learning and memory. *Dashed black lines* and *black arrows pointing down* indicate attenuation of the signaling pathway. *Red lines* and *upward pointing arrows* indicate an increase of the signaling pathway (Figure modified after Havekes et al. 2012)

Ca^{2+}-dependent adenylyl cyclase in pyramidal neurons (Xia and Storm 2012). This enzyme enhances the production of the second messenger cAMP, which in turn activates downstream targets important for memory consolidation, such as PKA, hyperpolarization-activated cyclic nucleotide-gated channels and exchange proteins activated by cAMP, as well as multiple kinases, including calmodulin-dependent protein kinase (CAMKII), extracellular signal-regulated kinase (ERK1/2) and mitogen activated protein kinase (MAPK) (Ahmed and Frey 2005; Prince and Abel 2013). Activation of these downstream targets facilitates the phosphorylation of transcription factors including cAMP response element binding protein (CREB), which up-regulates gene expression for proteins critically involved in memory consolidation (see Fig. 1) (Xia and Storm 2012; Prince and Abel 2013; Kreutzmann et al. 2015). Inhibiting these critical signaling elements in turn, for example via SD, has been suggested to impair memory consolidation.

Critical Time Windows During Memory Consolidation

Memory consolidation following training requires the activation of molecular signaling pathways at precise time points (Abel and Lattal 2001). Previous time course studies have shown that inhibiting signaling elements within a certain time frame post-learning can impair the consolidation of memories (Hernandez and Abel 2011). There are three peaks in cAMP magnitude at 0.5, 3, and 6 h following training in the inhibitory avoidance learning task (Bernabeu et al. 1997). Manipulation of cellular signaling pathways at these specific time "windows" have been shown to impair memory consolidation. Protein synthesis inhibition or PKA inhibition either immediately or at 4 h after contextual fear conditioning disrupts memory consolidation. This disruption was not seen by manipulating other time points after training (Bourtchouladze et al. 1998). These studies demonstrate that cAMP signaling within hours of training is required for memory consolidation. Sleep loss during these critical time windows may in turn interfere with signaling patterns and lead to deficits in memory consolidation.

To further define the time window in which sleep loss affects fear conditioning, Graves and colleagues investigated whether five hours of SD either immediately following training (0–5 h), or with a delay (5–10 h), leads to memory impairments in a single-trial Pavlovian conditioning task. The results showed that 5 h of immediate SD post-training selectively impaired hippocampus-dependent contextual fear memory, while SD 5–10 h following training had no impact on memory consolidation (Graves et al. 2003). Elaborating on this, Prince and co-workers were able to narrow down the sensitive time window for the effects of SD on memory consolidation (Prince et al. 2014). The authors found that 3 h of SD immediately after training did not affect the object location memory (0–3 h), however memory was impaired when SD began one hour post-training (1–4 h) (Prince et al. 2014). Altogether, these studies suggest that molecular signaling pathways must occur within a critical window of memory consolidation after information acquisition for the information to be available for retrieval.

Sleep Deprivation Impairs Molecular Targets Important for Memory Consolidation

Sleep Deprivation Impairs CAMP-PKA Signaling and Influences Downstream Targets

Given that cAMP and PKA cascade signaling are key elements underlying memory consolidation, it is possible that SD attenuates the signaling of these molecular targets, thereby disrupting associated cognitive processes. Several studies looked at the effect of disruptive sleep on the signaling pathways underlying memory consolidation (Graves et al. 2001; Hellman et al. 2010; Luo et al. 2013). SD reduced cAMP signaling, which was mediated by elevated Phosphodiesterase 4 (PDE4) activity, an enzyme responsible for degradation of hippocampal cAMP. Here, protein expression of the PDE4 isoform PDE4A5 was particularly increased following SD. Inhibiting PDE4 signaling during the SD-period with the PDE4 inhibitor Rolipram in turn rescued cAMP signaling, as well as SD-induced hippocampus-dependent memory consolidation (Vecsey et al. 2009). In concert with this finding, a transient increase in cAMP using a virally-delivered pharmacogenetic approach was sufficient to prevent memory deficits associated with sleep loss (Havekes et al. 2014). The authors showed that transiently increasing cAMP levels during the course of SD rescued hippocampus-dependent spatial memory deficits, thereby implicating the fundamental role of hippocampal cAMP in cognitive processes (Havekes et al. 2014).

Disruption of the cAMP-PKA signaling cascade in the hippocampus can have detrimental effects on downstream targets of this pathway, such as altered phosphorylation of serine 845 in the Glutamate A1 receptor subunit induced by PKA activity (Vyazovskiy et al. 2008; Ravassard et al. 2009; Hagewoud et al. 2010b). Attenuated phosphorylation of this site alters glutamate receptor subunit expression at the cell membrane, which, to some extent, influences neuronal excitability (Kreutzmann et al. 2015). The most prominent glutamate receptor family, which has been described to be involved in learning and memory, is the ionotrophic receptor family, which contains the N-Methyl-D-aspartate (NMDA) receptors and the α-Amino-3-Hydroxy-5-Methyl-4-Isoxazolepropionic acid (AMPA) receptors (Watkins and Jane 2006). By mediating Ca^{2+} influx, glutamate receptors play an essential role at all three stages of memory and can be regarded as vital regulators of synaptic plasticity processes (Hernandez and Abel 2011). A number of studies directly addressed the effect of SD on these receptors, and found that, for instance, SD for five days reduced intracellular Ca^{2+} influx in the hippocampus, which, in turn, altered NMDA receptor subunit composition and surface expression (Chang et al. 2012). SD has also been shown to directly modify NMDA receptor surface expression. Sleep-deprived rats express higher levels of cytoplasmic NMDA receptor NR1 and NR2A subunits compared to their controls (Chen et al. 2006; Ravassard et al. 2009; Chang et al. 2012).

These experiments provide evidence that the cAMP-PKA pathway and its downstream targets are key elements for the observed functional deficits in behavior and plasticity following the disruption of sleep (see Fig. 1). Inaccurate cascade signaling may alter coinciding gene expression and protein synthesis, which influences the consolidation of memory. These alterations will be discussed in the next section.

Sleep Deprivation Impairs Gene Expression

In order to illuminate molecular targets of SD, various research groups have assessed gene expression in response to SD using microarray or RNA Sequencing, which allow the examination of thousands of transcripts at once. Moreover, these gene-expression studies allow the identification of gene classes whose expressions are altered following SD (Cirelli 2002; Mackiewicz et al. 2007, 2009; Wang et al. 2010).

Gene-expression studies investigating changes in response to SD have observed contradictory findings between brain regions, as well as within the hippocampus. Using genome-wide microarray, Vecsey and colleagues found that five hours of SD caused up- or down-regulation in the expression of 533 genes, including many which have previously been implicated to be affected by SD, such as the immediate early gene activity-regulated cytoskeleton-associated protein (*Arc*) Arc/Arg3.1, *c-Fos*, Heterogeneous nuclear ribonucleoprotein D-like (*Hnrpdl*) or RNA binding motif protein 3 (*Rbm3*) (Vecsey et al. 2012). The study did not confirm previously reported changes in *Homer-1a* or *Zif268/Egr1/NGFI-A* as markers of SD (Conti et al. 2007; Maret et al. 2007; Wang et al. 2010). These inconsistencies may be attributed to method or length of SD. Nevertheless, bioinformatic analysis from this study revealed that up-regulated clusters of genes following acute SD include genes involved in ATP binding, GTP signaling, negative regulation of kinase activity, positive regulation of transcription, nucleosome/chromatin assembly, and unfolded protein response. Conversely, genes involved in translation, RNA-binding/processing, ubiquitination and negative regulation of transcription were down-regulated in the hippocampus after 5 h of SD. Two and a half hours of recovery sleep reversed the extensive changes in hippocampal gene expression initiated by SD (Vecsey et al. 2012).

Gene expression after longer periods of SD has also been examined in the hippocampus. For instance, Taishi and colleagues sleep deprived rats for eight hours and observed increases in Arc, with Brain-Derived Neurotrophic Factor (BDNF) remaining unchanged (Taishi et al. 2001). In another study, 24 h of SD led to selective changes in several transcripts encoding proteins involved in synaptic plasticity in the rat hippocampus (Conti et al. 2007). However, Guzmán-Marín and colleagues found that 48 h of SD caused a reduction in BDNF, CREB, CaMK II, and Synapsin I gene expression (Guzmán-Marín et al. 2006). These contradictory findings may be attributed to differences in the duration and method of SD. However, it cannot be discounted that SD affects multiple genes and their expression dynamics. The interpretation of these results is further complicated by the recent discovery that SD impairs translation.

Sleep Deprivation Impairs Translation

The process of translation can be divided in three main phases: initiation, elongation and termination. Initiation is the rate-limiting step to translation (Kapp and Lorsch 2004; Gal-Ben-Ari et al. 2012), during which the target mRNA is recruited to the ribosome. The translation factor, eukaryotic initiation factor 4E (eIF4E), is required for translation of most mRNAs as it binds to the 5' terminal cap of mRNA. eIF4E forms a complex with the scaffolding protein eukaryotic initiation factor 4G (eIF4G), as well as other factors. This interaction between eIF4E and eIF4G is controlled and modulated by the mammalian target of rapamycin (mTOR) signaling pathway (Proud 2007). mTOR is a gene that is crucially involved in the regulation of protein synthesis during memory consolidation (Hoeffer and Klann 2010). While increased levels of this translational regulator seem to improve memory (Dash et al. 2006), SD has been shown to decrease total and phosphorylated mTOR protein expression (Vecsey et al. 2012). This suggests that SD reduces translation required for memory consolidation. By using a non-radioactive in vivo protein translation assay, Tudor and colleagues were able to show that five hours of SD impaired protein synthesis in the hippocampus. This was due to attenuated levels of mTORC1-mediated phosphorylation of eIF4E-binding protein 2, as well as reduced eIF4E and eIF4G interaction in the mouse hippocampus. When hippocampal protein synthesis was restored to non-sleep-deprived levels using viral approaches, hippocampus-dependent memory deficits associated with sleep loss were prevented (Tudor et al. 2016). The study demonstrates that mTORC1-eIF4E-binding protein 2-regulated translation is a key mediator in SD-induced memory impairments.

Translation elongation is the process by which the ribosome moves along the mRNA and forms the new polypeptide chain. Eukaryotic elongation factor 2 (eEF2) catalyzes the ribosomal translocation reaction of peptidyl-tRNAs from the A-site to the P-site of the ribosome, resulting in the movement of ribosomes along the mRNA. eEF2 is target of the eEF2 kinase, which is regulated by Ca^{2+}/Calmodulin, mTORC1 and ERK signaling. Phosphorylation of eEF2 by this kinase converts eEF2 to an inactive form that does not bind to the ribosome and extenuates its function, leading to a temporary inhibition of overall translation (Ryazanov and Spirin 1990; Proud 2007). A recent study by Grønli and colleagues showed that eight hours of SD decreased phosphorylation of eIF4E in the dentate gyrus, but not in the CA regions of the hippocampus. eEF2 phosphorylation, conversely, was elevated in both hippocampal regions, as well as the prefrontal cortex (Grønli et al. 2012). Although the exact function of phosphorylation is unclear, eIF4E phosphorylation has often been associated with enhanced translation rates (Gingras et al. 1999; Scheper et al. 2002). Phosphorylation of eEF2, in turn, seems to inhibit its binding to the ribosome, thereby slowing down elongation of the polypeptide chain and reducing the rate of general protein synthesis (Gal-Ben-Ari et al. 2012; Grønli et al. 2012). These data support the finding that SD-induced hippocampus-dependent memory impairments are due to altered protein synthesis initiation, and ultimately support the hypothesis that one function of sleep is to facilitate protein synthesis necessary for memory consolidation.

Conclusion

Sleep loss leads to profound changes in the molecular and cellular mechanisms underlying memory consolidation, including cAMP-PKA signaling, gene expression, receptor localization, and translation. These wide-ranging molecular and cellular changes suggest that therapeutic intervention could be difficult. Nevertheless, several studies described in this review have shown that interventions aimed at a single specific process are sufficient to prevent the memory deficits caused by SD.

One question that is yet to be resolved is how cAMP-PKA signaling and protein synthesis signaling are interrelated during SD. Both of these pathways have been shown to be sufficient to cause memory deficits with SD. Although it is conceivable that altered cAMP and mTOR signaling occur in parallel and independent of one another during SD, these two pathways have points of intersection. Understanding of the temporal dynamics of these signaling pathways during SD and how these pathways intersect could be important for novel therapeutic treatment for SD-induced cognitive impairments.

It is also unknown whether the myriad of processes affected by SD are hippocampus-specific. The hippocampus has been described to store memories for a short time before transferring them to the cerebral cortex for long-term storage (McClelland et al. 1995; Squire et al. 2004) (see Chapter by Genzel and Wixted). Therefore, future research should aim at investigating how the molecular mechanisms described here apply to neural circuitry.

Insufficient sleep is an epidemic in our modern society, which can lead to physical and mental health deterioration. Continued research for a better mechanistic understanding of how SD affects memory, and memory consolidation specifically, is essential to the development of future drug treatments and public health.

Acknowledgements Research in the Abel lab is supported by the National Institute of Health (P50 AG 017628, RO1 MH 099544 and R21 MH 102679) and the Defense Advanced Research Projects Agency (W911NF1010093).

References

Abel T, Lattal KM (2001) Molecular mechanisms of memory acquisition, consolidation and retrieval. Curr Opin Neurobiol 11:180–187

Ahmed T, Frey JU (2005) Plasticity-specific phosphorylation of CaMKII, MAP-kinases and CREB during late-LTP in rat hippocampal slices in vitro. Neuropharmacology 49:477–492

Banks S, Dinges DF (2007) Behavioral and physiological consequences of sleep restriction. J Clin Sleep Med 3:519–528

Bernabeu R, Bevilaqua L, Ardenghi P, Bromberg E, Schmitz P, Bianchin M, Izquierdo I, Medina JH (1997) Involvement of hippocampal cAMP/cAMP-dependent protein kinase signaling pathways in a late memory consolidation phase of aversively motivated learning in rats. Proc Natl Acad Sci USA 94:7041–7046

Bourtchouladze R, Abel T, Berman N, Gordon R, Lapidus K, Kandel ER (1998) Different training procedures for contextual memory in mice can recruit either one or two critical periods for memory consolidation that require protein synthesis and PKA. Learn Mem 5:365–374

Chang HM, Liao WC, Sheu JN, Chang CC, Lan CT, Mai FD (2012) Sleep deprivation impairs Ca^{2+} expression in the hippocampus: Ionic imaging analysis for cognitive deficiency with TOF-SIMS. Microsc Microanal 18:425–435

Chen C, Hardy M, Zhang J, LaHoste GJ, Bazan NG (2006) Altered NMDA receptor trafficking contributes to sleep deprivation-induced hippocampal synaptic and cognitive impairments. Biochem Biophys Res Commun 340:435–440

Cirelli C (2002) How sleep deprivation affects gene expression in the brain: a review of recent findings. J Appl Physiol 92:394–400

Conti B, Maier R, Barr AM, Morale MC, Lu X, Sanna PP, Bilbe G, Hoyer D, Bartfai T (2007) Region-specific transcriptional changes following the three antidepressant treatments electro convulsive therapy, sleep deprivation and fluoxetine. Mol Psychiatry 12:167–189

Dash PK, Orsi SA, Moore AN (2006) Spatial memory formation and memory-enhancing effect of glucose involves activation of the tuberous sclerosis complex-Mammalian target of rapamycin pathway. J Neurosci 26:8048–8056

Gal-Ben-Ari S, Kenney JW, Ounalla-Saad H, Taha E, David O, Levitan D, Gildish I, Panja D, Pai B, Wibrand K, Simpson TI, Proud CG, Bramham CR, Armstrong JD, Rosenblum K (2012) Consolidation and translation regulation. Learn Mem 19:410–422

Graves L, Pack A, Abel T (2001) Sleep and memory: a molecular perspective. Trends Neurosci 24:237–243

Graves LA, Heller EA, Pack AI, Abel T (2003) Sleep deprivation selectively impairs memory consolidation for contextual fear conditioning. Learn Mem 10:168–176

Gingras AC, Raught B, Sonenberg N (1999) eIF4 initiation factors: effectors of mRNA recruitment to ribosomes and regulators of translation. Annu Rev Biochem 68:913–963

Grønli J, Dagestad G, Milde AM, Murison R, Bramham CR (2012) Post-transcriptional effects and interactions between chronic mild stress and acute sleep deprivation: regulation of translation factor and cytoplasmic polyadenylation element-binding protein phosphorylation. Behav Brain Res 235:251–262

Guzmán-Marín R, Ying Z, Suntsova N, Methippara M, Bashir T, Szymusiak R, Gomez-Pinilla F, McGinty D (2006) Suppression of hippocampal plasticity-related gene expression by sleep deprivation in rats. J Physiol 575:807–819

Hagewoud R, Havekes R, Novati A, Keijser JN, Van der Zee EA, Meerlo P (2010b) Sleep deprivation impairs spatial working memory and reduces hippocampal AMPA receptor phosphorylation. J Sleep Res 19:280–288

Hairston IS, Little MT, Scanlon MD, Barakat MT, Palmer TD, Sapolsky RM, Heller HC (2005) Sleep restriction suppresses neurogenesis induced by hippocampus-dependent learning. J Neurophysiol 94:4224–4233

Havekes R, Vecsey CG, Abel T (2012) The impact of sleep deprivation on neuronal and glial signaling pathways important for memory and synaptic plasticity. Cell Signal 24:1251–1260

Havekes R, Bruinenberg VM, Choi JHK, Baumann A, Meerlo P, Abel T (2014) Transiently increasing cAMP levels selectively in hippocampal excitatory neurons during sleep deprivation prevents memory deficits caused by sleep loss. J Neurosci 34:15715–15721

Hellman K, Hernandez P, Park A, Abel T (2010) Genetic evidence for a role for protein kinase A in the maintenance of sleep and thalamocortical oscillations. Sleep 33:19–28

Hernandez PJ, Abel T (2011) A molecular basis for interactions between sleep and memory. Sleep Med Clin 6:71–84

Hoeffer CA, Klann E (2010) mTOR signaling: at the crossroads of plasticity, memory and disease. Trends Neurosci 33:67–75

Kapp LD, Lorsch JR (2004) The molecular mechanics of eukaryotic translation. Annu Rev Biochem 73:657–704

Kreutzmann JC, Havekes R, Abel T, Meerlo P (2015) Sleep deprivation and hippocampal vulnerability: changes in neuronal plasticity, neurogenesis and cognitive function. Neuroscience 309:173–190

Li S, Tian Y, Ding Y, Jin X, Yan C, Shen X (2009) The effects of rapid eye movement sleep deprivation and recovery on spatial reference memory of young rats. Learn Behav 37:246–253

Luo J, Phan TX, Yang Y, Garelick MG, Storm DR (2013) Increases in cAMP, MAPK activity, and CREB phosphorylation during REM sleep: implications for REM sleep and memory consolidation. J Neurosci 33:6460–6468

Mackiewicz M, Shockley KR, Romer MA, Galante RJ, Zimmerman JE, Naidoo N, Baldwin DA, Jensen ST, Churchill GA, Pack AI (2007) Macromolecule biosynthesis: a key function of sleep. Physiol Genomics 31:441–457

Mackiewicz M, Zimmerman JE, Shockley KR, Ga Churchill, Pack AI (2009) What are microarrays teaching us about sleep? Trends Mol Med 15:79–87

Maret S1, Dorsaz S, Gurcel L, Pradervand S, Petit B, Pfister C, Hagenbuchle O, O'Hara BF, Franken P, Tafti M (2007) Homer1a is a core brain molecular correlate of sleep loss. Proc Natl Acad Sci USA 104: 20090–20095

McClelland JL, McNaughton BL, O'Reilly RC (1995) Why there are complementary learning systems in the hippocampus and neocortex: insights from the successes and failures of connectionist models of learning and memory. Psychol Rev 102:419–457

Nadel L, Hupbach A, Gomez R, Newman-Smith K (2012) Memory formation, consolidation and transformation. Neurosci Biobehav Rev 36:1640–1645

Palchykova S, Winsky-Sommerer R, Meerlo P, Dürr R, Tobler I (2006) Sleep deprivation impairs object recognition in mice. Neurobiol Learn Mem 85:263–271

Prince TM, Abel T (2013) The impact of sleep loss on hippocampal function. Learn Mem 20:558–569

Prince TM, Wimmer M, Choi J, Havekes R, Aton S, Abel T (2014) Sleep deprivation during a specific 3-hour time window post-training impairs hippocampal synaptic plasticity and memory. Neurobiol Learn Mem 109:122–130

Proud CG (2007) Signalling to translation: how signal transduction pathways control the protein synthetic machinery. Biochem J 403:217–234

Ravassard P, Pachoud B, Comte JC, Mejia-Perez C, Scoté-Blachon C, Gay N, Claustrat B, Touret M, Luppi PH, Salin PA (2009) Paradoxical (REM) sleep deprivation causes a large and rapidly reversible decrease in long-term potentiation, synaptic transmission, glutamate receptor protein levels, and ERK/MAPK activation in the dorsal hippocampus. Sleep 32:227–240

Ryazanov AG, Spirin AS (1990) Phosphorylation of elongation factor 2: a key mechanism regulating gene expression in vertebrates. New Biologist 2:843–850

Scheper GC, van Kollenburg B, Hu J, Luo Y, Goss DJ, Proud CG (2002) Phosphorylation of eukaryotic initiation factor 4E markedly reduces its affinity for capped mRNA. J Biol Chem 277:3303–3309

Smith C, Rose GM (1996) Evidence for a paradoxical sleep window for place learning in the morris water maze. Physiol Behav 59:93–97

Smith C, Rose GM (1997) Post-training paradoxical sleep in rats is increased after spatial learning in the Morris water maze. Behav Neurosci 111:1197–1204

Squire LR, Stark CE, Clark RE (2004) The medial temporal lobe. Annu Rev Neurosci 27:279–306

Taishi P, Sanchez C, Wang Y, Fang J, Harding JW, Krueger JM (2001) Conditions that affect sleep alter the expression of molecules associated with synaptic plasticity. Am J Physiol Regul Integr Comp Physiol 281:R839–845

Tudor JC, Davis EJ, Peixoto L, Wimmer ME, van Tilborg E, Park AJ, Poplawski SG, Chung CW, Havekes R, Huang J, Gatti E, Pierre P, Abel T (2016) Sleep deprivation impairs memory by attenuating mTORC1-dependent protein synthesis. Sci Signal 9: ra41

Vecsey CG, Baillie GS, Jaganath D, Havekes R, Daniels A, Wimmer M, Huang T, Brown KM, Li XY, Descalzi G, Kim SS, Chen T, Shang YZ, Zhuo M, Houslay MD, Abel T (2009) Sleep deprivation impairs cAMP signaling in the hippocampus. Nature 461:1122–1125

Vecsey CG, Peixoto L, Choi JH, Wimmer M, Jaganath D, Hernandez PJ, Blackwell J, Meda K, Park AJ, Hannenhalli S, Abel T (2012) Genomic analysis of sleep deprivation reveals translational regulation in the hippocampus. Physiol Genomics 44:981–991

Vyazovskiy VV, Cirelli C, Pfister-Genskow M, Faraguna U, Tononi G (2008) Molecular and electrophysiological evidence for net synaptic potentiation in wake and depression in sleep. Nat Neurosci 11:200–208

Walker MP, Stickgold R (2004) Sleep-dependent learning and memory consolidation. Neuron 44:121–133

Walker MP (2008) Sleep-dependent memory processing. Harvard Rev Psychiatry 16:287–298

Wang H, Liu Y, Briesemann M, Yan J (2010) Computational analysis of gene regulation in animal sleep deprivation. Physiol Genomics 42:427–436

Watkins JC, Jane DE (2006) The glutamate story. Br J Pharmacol 147:S100–S108

Xia Z, Storm DR (2012) Role of signal transduction crosstalk between adenylyl cyclase and MAP kinase in hippocampus-dependent memory. Learn Mem 19:369–374

Yang RH, Hu SJ, Wang Y, Zhang WB, Luo WJ, Chen JY (2008) Paradoxical sleep deprivation impairs spatial learning and affects membrane excitability and mitochondrial protein in the hippocampus. Brain Res 1230:224–232

Sleep and Odor Memory Consolidation in Non-human Animal Models

Donald A. Wilson, Kacper Kondrakiewicz and Dylan C. Barnes

Abstract Odor perception is a memory based phenomenon involving the synthesis of odorant molecular components into a configural odor object percept, e.g., coffee. In addition to the synthesis of molecular features, odors are commonly associated with visual objects, hedonic valence and context. The piriform cortex serves as a major locus for the memory of these associations. Recent work from several labs has demonstrated that the unique physiology of the piriform cortex during sleep plays a critical role in the strength and acuity of odor memory consolidation. The anatomically simple olfactory pathway from nose to higher-order cortex makes olfaction an ideal system for studying the role of sleep in memory consolidation. Sleep also plays a role in odor memory in invertebrates, suggesting sleep-dependent odor memory consolidation is a wide-spread phenomenon. This chapter reviews recent findings in sleep-dependent odor memory consolidation in animal models.

Keywords Piriform cortex · Memory consolidation · Odor memory · Olfactory system · Slow wave sleep

D.A. Wilson (✉)
Department of Child and Adolescent Psychiatry,
New York University School of Medicine, New York, NY, USA
e-mail: DWilson@NKI.RFMH.ORG

D.A. Wilson
Emotional Brain Institute, Nathan Kline Institute
for Psychiatric Research, Orangeburg, NY, USA

K. Kondrakiewicz
Department of Neurophysiology, Nencki Institute
of Experimental Biology PAS, Warsaw, Poland

D.C. Barnes
Department of Biology, University of Oklahoma, Norman, OK, USA

N. Axmacher and B. Rasch (eds.), *Cognitive Neuroscience of Memory Consolidation*, Studies in Neuroscience, Psychology and Behavioral Economics, DOI 10.1007/978-3-319-45066-7_6

Introduction

Sleep is a behavioral and physiological state expressed in almost all animals—vertebrates and invertebrates—examined to date (Cirelli 2009; Vorster and Born 2015). Sleep plays a critical role in a variety of brain functions including neural plasticity (Guzman-Marin et al. 2006; Aton et al. 2009; Walker 2009), waste clearance (Xie et al. 2013), and homeostatic control (Tononi and Cirelli 2006). As one part of these functions, sleep serves as an important physiological window for memory consolidation. Through sleep-dependent changes in neuromodulatory tone, circuits underlying memory storage change responsiveness to ongoing sensory inputs, modify temporal patterns of neural activity, and alter functional connectivity within and across brain regions. This combination of changes allows for replay, strengthening, and transfer of memories in a relatively interference free context. Through these sleep-dependent processes, memory consolidation can occur, allowing relatively accurate, long-term storage of recent experiences.

Evidence for sleep-dependent memory consolidation exists for a variety of neural systems including neocortex, hippocampus and several subcortical areas (Wilson and McNaughton 1994; Ji and Wilson 2007; Lansink et al. 2008; Peyrache et al. 2009; Popa et al. 2010; Valdes et al. 2015). Both rapid eye movement (REM) sleep and non-REM (or slow-wave) sleep have been linked to memory consolidation. In general, sleep-dependent memory consolidation is believed to involve spontaneous reactivation of circuits that were activated during initial encoding of an experience; essentially a form of memory replay. This replay allows repeated activation of synapses within neural ensembles representing the to-be-stored information, effectively strengthening them in a Hebbian, long-term potentiation-dependent manner (Walker and Stickgold 2006; Cirelli 2013; Frank and Cantera 2014). This replay also allows weakening of synapses not critical for the stored information, allowing them to reset and become available for strengthening during later experiences (Tononi and Cirelli 2006). In addition, due to the enhancement of long-range functional connectivity during slow-wave sleep (Volgushev et al. 2006; Popa et al. 2010), coherent activity across distant brain regions allows sleep-dependent transfer of information from one region to another, which may further strengthen consolidation. Finally, during slow-wave sleep (SWS) most circuits are isolated from outside sensory inputs due to state-dependent sensory gating, which primarily occurs in the thalamus (McCormick and Bal 1994). It is hypothesized that this sensory gating protects memory ensembles from outside interference during replay, enabling precise and accurate memory storage.

Direct evidence in support of a critical role for sleep in memory consolidation comes from a wide range of work. Training followed by sleep results in better memory retrieval than training not followed by sleep (e.g., Binder et al. 2012; Wooden et al. 2014). Activity in neural ensembles during encoding of recently experienced events is recapitulated during post-training sleep (replay) (Wilson and McNaughton 1994; Ji and Wilson 2007; Peyrache et al. 2009; Deuker et al. 2013; Barnes and Wilson 2014b). Slow-wave activity and other sleep features within brain regions that were active during training are enhanced during post-training sleep (Huber et al. 2004; Hanlon et al. 2009) (see also chapter by Zhang, Deuker and Axmacher), and artificially enhancing the magnitude of slow-waves during post-training sleep, for example with trans-cranial stimulation, enhances memory (Marshall et al. 2004; Antonenko et al. 2013). Finally, although currently correlational, a variety of disorders associated with memory dysfunction (see also chapter by Campos Beltran and Marshall), such as dementia and depression, have sleep disruption as a major disease component (Bellivier et al. 2015; Miller 2015; Murphy and Peterson 2015).

In this review, we focus on the role of sleep in odor memory consolidation (see also chapter by Nissen and colleagues). The olfactory system has a variety of characteristics that make it potentially very different from the thalamocortical systems most commonly studied. In addition, it is a relatively simple system, with the primary sensory cortex just two synapses from the outside world. This relative simplicity and the resulting ability to precisely control patterns of learned input, may make it an ideal system for the study of sleep-dependent memory consolidation. This chapter will focus on non-human animal models, while another chapter in this volume will focus on human models (see chapter by Shanahan and Gottfried).

Olfactory System Anatomy and Physiology

The olfactory system is an evolutionarily ancient and well conserved system for sensing and perceiving volatile chemicals. In both vertebrates and invertebrates, sensory transduction is mediated by a large family of olfactory receptors (~400 in humans, ~1000 in mice, ~40 in fruitflies, ~1200 in pigs) expressed in olfactory sensory neurons. Olfactory sensory neurons project directly into the forebrain where they synapse within the olfactory bulb (antennal lobe in invertebrates). Sensory neurons expressing the same receptor converge onto structures called glomeruli where they synapse with second order neurons. Glomeruli receive input from sensory neurons all expressing the same receptor. Thus, activation of a population of olfactory sensory neurons results in a spatiotemporal pattern of activity within the olfactory bulb/antennal lobe. Given that different odorant molecules activate different combinations of receptors, each odorant activates a specific pattern of olfactory bulb activity. The olfactory system does not have a receptor for coffee or lavender, which are actually composed of complex mixtures of molecules. Rather, olfactory receptors appear to bind submolecular features, and it is the job of higher order olfactory regions, such as the primary olfactory cortex (piriform cortex) to merge these features into perceptual objects (Gottfried 2010; Wilson and Sullivan 2011). This synthesis of odorant features into objects is a memory-dependent pattern recognition process (Haberly 2001; Wilson and Sullivan 2011). At the same time, the piriform cortex is processing patterned input from the olfactory bulb, it also receives inputs from other brain regions (Fig. 1), conveying information associated with that stimulus (e.g., hedonics, context, reward, etc.;

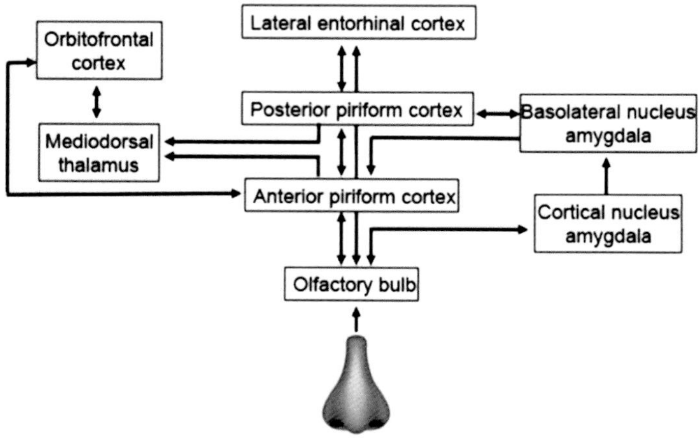

Fig. 1 Schematic of major (though not all) early components of the mammalian olfactory pathway. Note that the primary olfactory cortex (piriform cortex) can be anatomically and functionally divided into anterior and posterior sections. Also note that many, but not all connections are reciprocal, with very strong cortical feedback to the olfactory bulb

Cleland and Linster 2003). Piriform cortex, therefore serves as an associative cortex, linking odors to their non-olfactory contexts (Haberly 2001). In many invertebrates, the mushroom bodies, which receive direct input from the antennal lobes serve a similar function (Guven-Ozkan and Davis 2014).

As in other sensory systems, in addition to feed-forward flow from receptors to cortex, there are also extensive feedback olfactory pathways. The olfactory bulb receives feedback from all cortical areas to which it projects, including the anterior olfactory nucleus, piriform cortex and lateral entorhinal cortex, among others. Interestingly, the majority of this feedback to the olfactory bulb targets inhibitory interneurons in the glomerular layer and granule cell layers (Luskin and Price 1983). Most of these neurons, such as the olfactory bulb granule cells, undergo lifelong neurogenesis (Lepousez et al. 2013). Survival of adult-born olfactory bulb granule cells is experience-dependent and contributes to odor memory (Gheusi et al. 2000; Moreno et al. 2009). As we will discuss below, sleep plays an important role in adult-born granule cell fate.

Odor Memory

As mentioned above memory and neural plasticity are involved not only in learning that a specific odor signals a specific outcome (associative memory), but also in odor perception itself (perceptual memory) (Barkai and Wilson 2014). Distributed ensembles of piriform cortical neurons encode spatiotemporal patterns of olfactory bulb input (common co-occurrences of olfactory receptor activation) and store templates of those patterns by Hebbian adjustments of synapses connecting the neurons within those ensembles (Haberly 2001; Wilson and Sullivan 2011). This allows for processes such as pattern recognition and pattern completion on subsequent exposure to those odors, and enhances discriminability of similar odors that may activate highly overlapping sets of olfactory receptors (Chapuis and Wilson 2012). This perceptual learning, therefore, is critical for odor recognition and discrimination, and as described below, sleep-dependent consolidation is involved in both the strength and acuity of odor memory.

The more commonly studied form of odor memory is associative memory. Odors often predict the occurrence of other events such as the arrival of food or the emergence of a threat. In fact, associative odor learning occurs during the perinatal period in many mammals as infants learn the odor signature of their mother (Sullivan et al. 1991, 2000). Associative odor memory involves a wider network of neural circuits, but the piriform cortex is still intimately involved as it receives monosynaptic input from regions such as the amygdala, entorhinal cortex and orbitofrontal cortex (Cleland and Linster 2003), and previous odor associations modify piriform cortical odor responses (Litaudon et al. 1997; Moriceau and Sullivan 2004; Roesch et al. 2007; Chapuis and Wilson 2012; Gire et al. 2013).

Other forms of memory include, but are not limited to, odor context (Rasch et al. 2007), odor habituation (McNamara et al. 2008), odor sequences (Staubli et al. 1995; Devito and Eichenbaum 2011), and odor spatial information (Andre and Manahan-Vaughan 2013). To our knowledge, the exploration of the role of sleep in consolidation of these forms of odor memory has been limited to odor contextual memory in both human and non-human animal models (Rasch et al. 2007; Zwaka et al. 2015).

Sleep and Olfactory System Function

Olfactory system physiology is dramatically shaped by neuromodulatory tone (Sullivan et al. 1989; Bouret and Sara 2002; Linster and Fontanini 2014) and sleep/wake state (Murakami et al. 2005; Yamaguchi et al. 2013; Barnes and Wilson 2014a). Olfactory sensitivity is also influenced by circadian rhythm (Krishnan et al. 2008; Granados-Fuentes et al. 2011; Lehmann et al. 2011; Corthell et al. 2013; Miller et al. 2014; Saleh et al. 2015; Schendzielorz et al. 2015), though this will not be explored in this review. Sleep states are associated with slow, regular respiration or immobile antenna, as opposed to active sniffing or antennal flicking often seen in alert animals. This change in sampling can influence olfactory function (Mozell et al. 1987; Courtiol et al. 2011, 2014). However, in rodents, the most extreme sleep-dependent changes are observed in central, cortical olfactory structures.

In rodents, the piriform cortex undergoes three major functional changes during slow-wave sleep. First, single-units in the piriform cortex become hyporesponsive to odor stimuli compared to waking (Murakami et al. 2005; Wilson 2010; Barnes et al. 2011; Fig. 2). That is, although olfactory bulb output neurons, mitral/tufted cells,

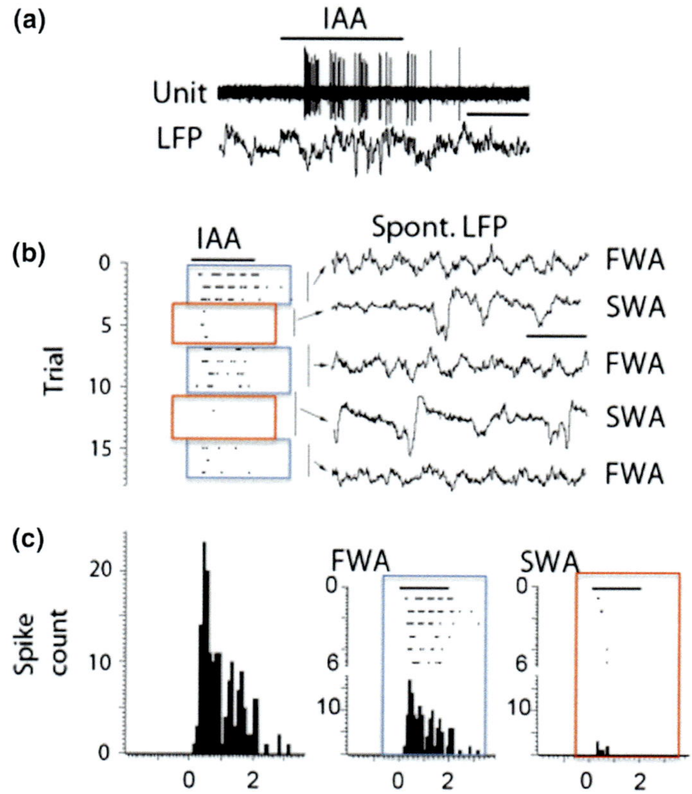

Fig. 2 The primary olfactory cortex (piriform cortex) is hyporesponsive to afferent input from the olfactory bulb during slow-wave activity (SWA) compared to during fast-wave activity (FWA). **a** Example of single-unit and local field potential (LFP) response to a 2 s presentation of isoamylacetate odor (IAA) during fast-wave activity. **b** Activity of a piriform cortex single-unit to odor over sequential fast-wave and slow-wave bouts. During neocortical fast-wave activity, LFP in the piriform cortex displays oscillations in phase with respiration and single-units are responsive to odor stimuli. During slow-wave activity, LFP in the piriform cortex displays sharp-waves and reduced coherence with respiration, while single-units show reduced responsiveness to odors. **c** Summary rasterplots and peristimulus time histograms of activity from the same single-unit during fast-wave and slow-wave states. Adapted from Wilson 2010

continue to respond to odors, odor responses in the piriform cortex, at both the single-unit and local field potential level, are greatly reduced. Piriform cortical odor responses are not reduced during REM sleep compared to waking (Barnes et al. 2011). Second, during slow-wave sleep, piriform cortical neurons switch from slow firing rates generally in phase with respiration, to a bursting mode in phase with sharp-waves recorded in the cortical local field potential (Murakami et al. 2005; Wilson 2010). These sharp-wave are similar to those observed during slow-wave sleep in the hippocampal formation, though are not coherent with them (Manabe et al. 2011), and thus have a distinct generator (Behan and Haberly 1999). Given that

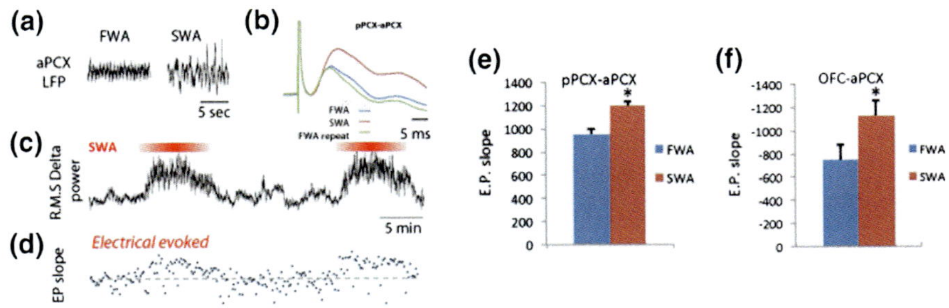

Fig. 3 While olfactory input to the piriform cortex is reduced during slow-wave activity, intra-cortical synaptic inputs are enhanced. **a** Example of LFP activity recorded in the anterior piriform cortex (aPCX) during fast-wave and slow-wave states of a urethane anesthetized rat. **b** Example monosynaptic extracellular evoked responses recorded in aPCX in response to electrical stimulation of intracortical association fibers from the posterior piriform cortex (pPCX). Stimulus intensity adjusted to evoke half maximal response. **c** Example of delta power of aPCX LFP showing spontaneous cycling between fast-wave and slow-wave activity. **d** Example of evoked potential (EP) slope measures of aPCX extracellular responses to pPCX stimulation recorded simultaneously with C, showing enhancement of the monosynaptic response during SWA. Mean changes in monosynaptic responses recorded in aPCX to either pPCX (**e**) or orbitofrontal cortex (OFC; **f**) input during FWA and SWA activity. There is a significant increase in intracortical monosynaptic responses in both pathways during SWA compared to FWA (pPCX to aPCX, $t(7) = 6.0$, $p < 0.01$; OFC to aPCX, $t(6) = 2.78$, $p < 0.05$)

piriform cortical pyramidal cells are highly inter-connected (Haberly 2001; Poo and Isaacson 2011), distributed cells fire together during these sharp-waves, potentially allowing ensembles to strengthen their interconnections. In addition, given that piriform cortical pyramidal cells project back to the olfactory bulb, slow-wave sleep is a period of strong bursts of descending input to olfactory bulb granule cells (Manabe et al. 2011). The third slow-wave sleep dependent change in piriform cortical activity is a release from pre-synaptic inhibition of intra- and inter-cortical axons. This release from inhibition is probably at least partially due to the decrease in cholinergic input to piriform cortex during slow-wave sleep, though noradrenergic inputs may also be involved (Hasselmo et al. 1997; Eschenko and Sara 2008). Acetylcholine and norepinephrine selectively suppress intra- and inter-cortical synapses in piriform cortex (Hasselmo and Bower 1992; Hasselmo et al. 1997; Linster et al. 2003)—though it should be noted that norepinephrine and locus coeruleus activity are not monotonically affected by sleep state (Eschenko and Sara 2008; Gais et al. 2011) (see also chapter by Maier and Kempter). Thus, at the same time the piriform cortex is hypo-responsive to odor input, it is maximally responsive to inputs from other cortical areas (Fig. 3). This combination of slow-wave sleep changes in olfactory system function may be ideal for interference-free memory consolidation.

Sleep and Odor Memory Consolidation

Sleep-dependent odor memory consolidation occurs in both vertebrate (rodents) and invertebrate (honey bee, *Drosophila*) systems. In all systems examined, odor memories are malleable during post-training sleep, and in general the more post-training sleep the stronger the subsequent odor memory. Given the different underlying neural substrates, vertebrate and invertebrate model systems are discussed separately.

Rodents: Sleep structure is modified following odor learning. Olfactory associative conditioning enhances the time spent in slow-wave sleep during the post-training period as recorded in the piriform cortex (Barnes et al. 2011). It also increases sharp-wave/ripple activity in the hippocampus (Eschenko et al. 2008), as well as sleep spindle density recorded in EEG (Eschenko et al. 2006). The amount of increase in post-training slow-wave sleep compared to basal sleep predicted memory strength 24 h later (Fig. 4; Barnes et al. 2011). It should also be noted that sleep spindle density was also enhanced after recall of a previously learned odor association, suggesting a possible role for sleep spindles not only in memory consolidation but also in re-consolidation (Eschenko et al. 2006; Sara 2010). Thus, similar to other forms of learning, following odor learning sleep structure is modified to include more time in slow-wave sleep and higher density of sleep spindles and hippocampal sharp-wave/ripples (see also chapter by McDevitt and colleagues). These sleep features have been hypothesized to enhance synaptic plasticity underlying memory consolidation and to promote transfer of information to neocortical storage (Stickgold 2005; Walker and Stickgold 2006; Ji and Wilson 2007).

However, in addition to displaying these general features of post-training sleep modification, the olfactory system also allows direct tests of hypotheses regarding the specific roles of sleep-dependent changes in circuit function that may underlie memory consolidation. This is facilitated by the relatively short pathway from peripheral receptors to sensory cortex, the initial spatial patterning of odor features,

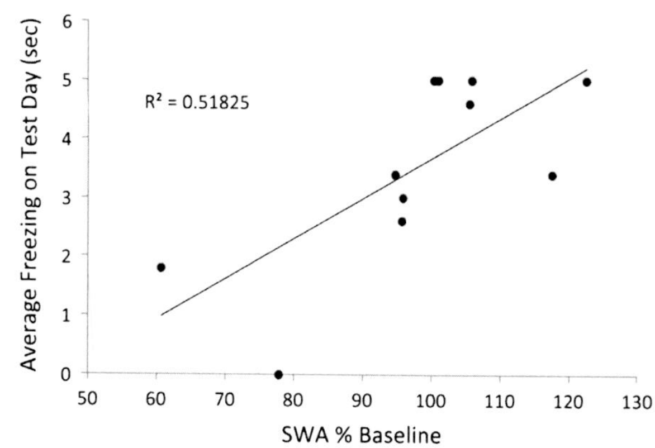

Fig. 4 In rats, increasing time in slow-wave sleep following odor-associative conditioning predicts enhanced odor memory the following day. Adapted from Barnes et al. 2011

and the lack of a thalamic relay between the peripheral stages of processing and cortex.

Does replay of learned olfactory patterns during sleep enhance odor memory consolidation? The piriform cortex displays sharp-wave activity during slow-wave sleep, similar to that observed in the hippocampal formation (Manabe et al. 2011). In hippocampus, sharp-wave/ripple activity is a period of rapid replay of recently experienced events. For example, in an animal that has recently run through a maze, activating a specific temporal sequence of hippocampal place cells, that sequence will be re-evoked during a subsequent hippocampal sharpwave (Pavlides and Winson 1989; Wilson and McNaughton 1994; Ji and Wilson 2007). This replay is hypothesized to help consolidate the representation of that spatial memory. While currently no direct evidence for similar odor replay in piriform cortex exists, there is strong indirect evidence. First, the temporal pattern of piriform cortical single-unit activity during slow-wave related piriform sharp-waves is modified if the animal has recently experienced an odor to which those cells respond (Wilson 2010). Thus, piriform cortical single-unit activity during sharp-waves reflects recent odor experience. Whether this is true "replay" or not is still to be determined. However, artificially imposed replay during post-training slow-wave sleep does enhance olfactory memory. Rats were trained in a discriminative odor-fear conditioning paradigm, however instead of using natural odors as the CS+ and CS−, different spatial patterns of olfactory bulb electrical stimulation were used (Barnes and Wilson 2014b). Rats respond to this stimulation in the same way as to a real odor and can discriminate between different patterns (Mouly et al. 1985). An advantage of using electrical odors however, is that it can overwhelm the cortical sleep-dependent "gate" and be delivered at precise times post-conditioning as a form of imposed replay. Imposing replay of the CS+ during post-training slow-wave sleep enhanced memory consolidation as evidenced by a nearly doubling of the time spent freezing to the CS+ the following day (Barnes and Wilson 2014b). If the exact same replay was instead delivered during post-training waking instead of slow-wave sleep, the animals displayed extinction (Barnes and Wilson 2014b). This suggests an important role of replay during post-training slow-wave sleep in odor memory consolidation.

State-dependent sensory gating in thalamocortical circuits has been hypothesized to reduce interference from ongoing sensory events during replay of previously learned information. As described above, a similar sleep-dependent sensory gating occurs in the piriform cortex, reducing cortical response to odors selectively during slow-wave sleep (Murakami et al. 2005; Wilson 2010). Does isolating memory circuits from ongoing sensory inputs during sleep-dependent consolidation enhance memory fidelity? Using the electrical odor paradigm just described, we tested whether presenting a novel stimulus input during presumed normal sleep-dependent replay would interfere with memory consolidation as hypothesized in other systems (Fig. 5). The electrical odors can overwhelm the normal sleep-dependent sensory gate, and potentially influence the stored information. In fact, our results demonstrated that abnormal sensory input during a period of replay severely impaired the accuracy, but not the strength, of the stored memory (Barnes and Wilson 2014b).

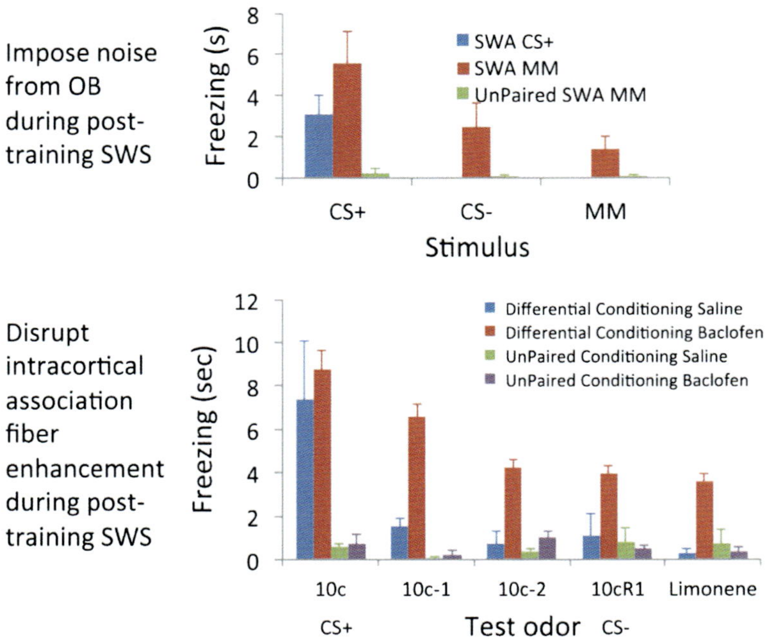

Fig. 5 The SWS-induced decrease in piriform cortical odor responsiveness and enhancement in intracortical connectivity contribute to odor memory acuity. *Top* Imposing olfactory afferent input to the PCX during SWS with patterned electrical stimulation during normal memory replay, that is a mismatch (MM) to the learned afferent pattern, interferes with accurate memory storage. When the normal hyporesponsive blockade from interference is prevented during SWS, animals subsequent show very generalized fear to stimuli. *Bottom* Pharmacological suppression of intracortical association fiber synapses during post-training SWS when they are normal enhanced, disrupts memory acuity the following day. Rats were trained with a 10 component (10c) odor mixture as the CS+ and a morphed version of this mixture (10 components with one replaced with a novel component 10cR1) as the CS-. During the post-training period baclofen or saline was bilaterally infused into the aPCX. Baclofen suppresses intracortical association fiber synapses (Tang and Hasselmo 1994; Barnes and Wilson 2014b). Disrupting the normal SWS-associated enhancement of these inputs disrupted memory accuracy as shown by generalized freezing to odors, even those never experienced before (limonene). Adapted from Barnes and Wilson (2014b)

Thus, normal sleep-dependent sensory gating is required to avoid interference during memory consolidation during slow-wave sleep. Without this gate, consolidated memory acuity is impaired.

In addition to changes in responsiveness to afferent input during slow-wave sleep, functional connectivity within the piriform cortex and between the piriform cortex and other brain regions is also modified. Functional connectivity is enhanced between piriform cortical pyramidal cells and between the piriform cortex and regions like the entorhinal cortex and amygdala during slow-wave sleep, in part due to neuromodulation of glutamatergic synapses, as described above (Fig. 3). Does this sleep-dependent enhancement in long-range intra- and inter-cortical connections influence memory consolidation? The $GABA_B$ receptor agonist baclofen selectively pre-synaptically suppresses intra- and inter-cortical glutamatergic synapses in the piriform cortex, leaving afferent synapses intact (Tang and Hasselmo 1994; Barnes and Wilson 2014b). Bilateral baclofen infusions into the piriform cortex during the post-training consolidation period (4 h) impaired memory acuity expressed the following day (Barnes and Wilson 2014b; Fig. 5). Thus, limiting the ability of cortical ensembles to link together during this post-training period, degraded the accuracy of the consolidated memory. Computational models of olfactory bulb-piriform cortex networks show similar reliance on association fiber synapse plasticity for accurate odor discrimination (Linster et al. 2009).

Another example of the importance of modified functional connectivity during slow-wave sleep in odor memory consolidation relates to the effects of piriform cortical feedback on the olfactory bulb. The piriform cortex sends a strong projection back to the olfactory bulb, primarily targeting inhibitory granule cells, which as noted above undergo adult neurogenesis. Adult-born granule cells are necessary for precise, learned, odor discrimination memory (Gheusi et al. 2000; Moreno et al. 2009; Lepousez et al. 2013). Recent work suggests that piriform cortical sharp-wave activity may play an important role in regulating selective granule cell survival. Many animals, including humans, demonstrate post-prandial sleep, i.e., are more likely to fall asleep following a meal. Mori and colleagues have demonstrated that post-prandial slow-wave sleep enhances death of adult-born granule cells (Yamaguchi et al. 2013). Their work suggests that the large volley of descending input to the granule cells during piriform cortical sharp-waves marks granule cells that have not been recently active for apoptosis (Yokoyama et al. 2011; Yamaguchi et al. 2013). This would allow those granule cells that encode information about the odors of the food recently consumed to be spared and incorporated into the olfactory bulb circuit, while those not used would be removed.

Together these results demonstrate an important role of post-training slow-wave sleep in odor memory consolidation. The features of piriform cortical physiology during sleep allow replay of recently experienced odors together with contextual information obtained from other brain regions, in a relatively interference free state. The changes in functional connectivity also allow shaping of upstream regions to more precisely define the odorant features involved in the memory.

Honey bees and Drosophila: Sleep is also involved in odor memory consolidation in invertebrates. As in rodents, sleep-wake and circadian cycles in insects are associated with changes in neuromodulatory tone and neural activity in the olfactory system (Krishnan et al. 2008; Ueno et al. 2012; Schendzielorz et al. 2015). Circadian cycles also influence odor learning (Lehmann et al. 2011). There is increasing evidence that sleep also plays a role in odor memory consolidation. For example, honey bees (*Apis mellifera carnica*) normally sleep more during the night than during the day, as determined from antennal movements as an assay of sleep-waking (Hussaini et al. 2009). Odor-sucrose associative conditioning results in reduced sleeping during the post-conditioning night, compared to groups that received either the odor or the sucrose alone (Hussaini et al. 2009). Thus, post-training sleep structure is modified by conditioning. Sleep deprivation following either initial acquisition or extinction training significantly impaired extinction but did not affect initial acquisition (Hussaini et al. 2009). This suggests that consolidation of extinction memory is at least partially sleep-dependent.

A potential role for sleep-dependent replay on memory consolidation has also been observed in honey bees (Zwaka et al. 2015). Bees were trained with a thermal stimulus CS+ that predicted sucrose reward. The conditioning took place in an odorous context. During the following nights sleep, the context odor was presented (or not), and testing of the conditioned response was performed the following day. Similar to the imposed sleep replay results described above, and work in humans (Rasch et al. 2007), exposure to the odor context during sleep in honeybees enhanced memory consolidation for the learned association (Zwaka et al. 2015).

Similar relationships between sleep and memory consolidation have been observed in Drosophila (Dissel et al. 2015a). For example, odor memory consolidation is impaired by sleep deprivation (Le Glou et al. 2012). Enhancing sleep rescued odor memory in several amnesic Drosophila mutants (Dissel et al. 2015b). Finally, sleep-dependent odor memory consolidation may be regulated by state-dependent changes in neuromodulatory tone, for example dopamine (Berry et al. 2015).

Although specific details vary, these results suggest a strong conservation of sleep-dependent odor memory consolidation across a wide swath of the evolutionary tree.

Summary

Despite dramatic differences in neuroarchitecture between thalamocortical and the mammalian olfactory system, sleep-dependent memory consolidation follows very similar rules in both olfactory and non-olfactory areas. In particular, we present evidence of odor memory replay during slow-wave sleep in piriform cortex. The strength and accuracy of this sleep-dependent memory consolidation is dependent on sensory gating of ongoing afferent activity during replay, and enhanced inter- and intra-cortical connectivity during slow-wave sleep. Sharp-waves in piriform cortex, which occur independently of those in hippocampus, help link distributed ensembles of neurons coding odor objects and their associations, as well as shape up-stream circuits in the olfactory bulb encoding odor feature information. These sleep-dependent processes are enabled by state-dependent changes in neuromodulatory tone in the olfactory structures. In addition to changes within the primary olfactory pathway, odor memory consolidation is associated with changes in global sleep architecture, including time spent in slow-wave sleep and density of sleep spindles. Importantly, odor memory consolidation in invertebrates is also sleep dependent. Together, these findings make clear that sleep-dependent memory consolidation is an evolutionarily highly conserved process, evident in invertebrates and mammalian archicortex, as well as the more widely studied hippocampal and neocortical systems. Targeting sleep dysfunction may be a fruitful path in the search for treatments of memory and other cognitive disorders.

Acknowledgements Work described here and manuscript preparation were supported by grants from NIDCD and NIAAA to D.A.W.

References

Andre MA, Manahan-Vaughan D (2013) Spatial olfactory learning facilitates long-term depression in the hippocampus. Hippocampus 23:963–968

Antonenko D, Diekelmann S, Olsen C, Born J, Molle M (2013) Napping to renew learning capacity: enhanced encoding after stimulation of sleep slow oscillations. Eur J Neurosci 37:1142–1151

Aton SJ, Seibt J, Dumoulin M, Jha SK, Steinmetz N, Coleman T, Naidoo N, Frank MG (2009) Mechanisms of sleep-dependent consolidation of cortical plasticity. Neuron 61:454–466

Barkai E, Wilson DA (eds) (2014) Odor Memory and Perception. Elsevier, Amsterdam

Barnes DC, Wilson DA (2014a) Sleep and olfactory cortical plasticity. Front Behav Neurosci 8:134

Barnes DC, Wilson DA (2014b) Slow-wave sleep imposed-replay modulates both strength and precision of memory. J Neurosci (in press)

Barnes DC, Chapuis J, Chaudhury D, Wilson DA (2011) Odor fear conditioning modifies piriform cortex local field potentials both during conditioning and during post-conditioning sleep. PLoS ONE 6:e18130–e18130

Behan M, Haberly LB (1999) Intrinsic and efferent connections of the endopiriform nucleus in rat. J Comp Neurol 408:532–548

Bellivier F, Geoffroy PA, Etain B, Scott J (2015) Sleep- and circadian rhythm-associated pathways as therapeutic targets in bipolar disorder. Expert opin Ther Targets 19:747–763

Berry JA, Cervantes-Sandoval I, Chakraborty M, Davis RL (2015) Sleep facilitates memory by blocking dopamine neuron-mediated forgetting. Cell 161:1656–1667

Binder S, Baier PC, Molle M, Inostroza M, Born J, Marshall L (2012) Sleep enhances memory consolidation in the hippocampus-dependent object-place recognition task in rats. Neurobiol Learn Mem 97:213–219

Bouret S, Sara SJ (2002) Locus coeruleus activation modulates firing rate and temporal organization of odour-induced single-cell responses in rat piriform cortex. Eur J Neurosci 16:2371–2382

Chapuis J, Wilson DA (2012) Bidirectional plasticity of cortical pattern recognition and behavioral sensory acuity. Nat Neurosci 15:155–161

Cirelli C (2009) The genetic and molecular regulation of sleep: from fruit flies to humans. Nat Rev Neurosci 10:549–560

Cirelli C (2013) Sleep and synaptic changes. Curr Opin Neurobiol 23:841–846

Cleland TA, Linster C (2003) Central olfactory structures. In: Doty RL (ed) Handbook of olfaction and gustation, 2nd edn. Marcel Dekker, New York, pp 165–180

Corthell JT, Stathopoulos AM, Watson CC, Bertram R, Trombley PQ (2013) Olfactory bulb monoamine concentrations vary with time of day. Neuroscience 247:234–241

Courtiol E, Hegoburu C, Litaudon P, Garcia S, Fourcaud-Trocme N, Buonviso N (2011) Individual and synergistic effects of sniffing frequency and flow rate on olfactory bulb activity. J Neurophysiol 106:2813–2824

Courtiol E, Lefevre L, Garcia S, Thevenet M, Messaoudi B, Buonviso N (2014) Sniff adjustment in an odor discrimination task in the rat: analytical or synthetic strategy? Front Behav Neurosci 8:145

Deuker L, Olligs J, Fell J, Ta Kranz, Mormann F, Montag C, Reuter M, Elger CE, Axmacher N (2013) Memory consolidation by replay of stimulus-specific neural activity. J Neurosci 33:19373–19383

Devito LM, Eichenbaum H (2011) Memory for the order of events in specific sequences: contributions of the hippocampus and medial prefrontal cortex. J Neurosci: Official J Soc Neurosci 31:3169–3175

Dissel S, Melnattur K, Shaw PJ (2015a) Sleep, performance, and memory in flies. Curr Sleep Med Rep 1:47–54

Dissel S, Angadi V, Kirszenblat L, Suzuki Y, Donlea J, Klose M, Koch Z, English D, Winsky-Sommerer R, van Swinderen B, Shaw PJ (2015b) Sleep restores behavioral plasticity to Drosophila mutants. Curr Biol: CB 25:1270–1281

Eschenko O, Sara SJ (2008) Learning-dependent, transient increase of activity in noradrenergic neurons of locus coeruleus during slow wave sleep in the rat: brain stem-cortex interplay for memory consolidation? Cereb Cortex 18:2596–2603

Eschenko O, Molle M, Born J, Sara SJ (2006) Elevated sleep spindle density after learning or after retrieval in rats. J Neurosci: Official J Soc Neurosci 26:12914–12920

Eschenko O, Ramadan W, Molle M, Born J, Sara SJ (2008) Sustained increase in hippocampal sharp-wave ripple activity during slow-wave sleep after learning. Learn Mem 15:222–228, Cold Spring Harbor, NY

Frank MG, Cantera R (2014) Sleep, clocks, and synaptic plasticity. Trends Neurosci 37:491–501

Gais S, Rasch B, Dahmen JC, Sara S, Born J (2011) The memory function of noradrenergic activity in non-REM sleep. J Cogn Neurosci 23:2582–2592

Gheusi G, Cremer H, McLean H, Chazal G, Vincent JD, Lledo PM (2000) Importance of newly generated neurons in the adult olfactory bulb for odor discrimination. Proc Natl Acad Sci USA 97:1823–1828

Gire DH, Whitesell JD, Doucette W, Restrepo D (2013) Information for decision-making and stimulus identification is multiplexed in sensory cortex. Nat Neurosci 16:991–993

Gottfried J (2010) Central mechanisms of odour object perception. Nat Rev Neurosci 11:628–641

Granados-Fuentes D, Ben-Josef G, Perry G, Wilson DA, Sullivan-Wilson A, Herzog ED (2011) Daily rhythms in olfactory discrimination depend on clock genes but not the suprachiasmatic nucleus. J Biol Rhythms 26:552–560

Guven-Ozkan T, Davis RL (2014) Functional neuroanatomy of Drosophila olfactory memory formation. Learn Mem 21:519–526, Cold Spring Harbor, NY

Guzman-Marin R, Ying Z, Suntsova N, Methippara M, Bashir T, Szymusiak R, Gomez-Pinilla F, McGinty D (2006) Suppression of hippocampal plasticity-related gene expression by sleep deprivation in rats. J Physiol 575:807–819

Haberly LB (2001) Parallel-distributed processing in olfactory cortex: new insights from morphological and physiological analysis of neuronal circuitry. Chem Senses 26:551–576

Hanlon EC, Faraguna U, Vyazovskiy VV, Tononi G, Cirelli C (2009) Effects of skilled training on sleep slow wave activity and cortical gene expression in the rat. Sleep 32:719–729

Hasselmo ME, Bower JM (1992) Cholinergic suppression specific to intrinsic not afferent fiber synapses in rat piriform (olfactory) cortex. J Neurophysiol 67:1222–1229

Hasselmo ME, Linster C, Patil M, Ma D, Cekic M (1997) Noradrenergic suppression of synaptic transmission may influence cortical signal-to-noise ratio. J Neurophysiol 77:3326–3339

Huber R, Ghilardi MF, Massimini M, Tononi G (2004) Local sleep and learning. Nature 430:78–81

Hussaini SA, Bogusch L, Landgraf T, Menzel R (2009) Sleep deprivation affects extinction but not acquisition memory in honeybees. Learn Mem 16:698–705, Cold Spring Harbor, NY

Ji D, Wilson MA (2007) Coordinated memory replay in the visual cortex and hippocampus during sleep. Nat Neurosci 10:100–107

Krishnan P, Chatterjee A, Tanoue S, Hardin PE (2008) Spike amplitude of single-unit responses in antennal sensillae is controlled by the Drosophila circadian clock. Curr Biol: CB 18:803–807

Lansink CS, Goltstein PM, Lankelma JV, Joosten RN, McNaughton BL, Pennartz CM (2008) Preferential reactivation of motivationally relevant information in the ventral striatum. J Neurosci: Official J Soc Neurosci 28:6372–6382

Le Glou E, Seugnet L, Shaw PJ, Preat T, Goguel V (2012) Circadian modulation of consolidated memory retrieval following sleep deprivation in Drosophila. Sleep 35:1377–1384B

Lehmann M, Gustav D, Galizia CG (2011) The early bee catches the flower—circadian rhythmicity influences learning performance in honey bees, Apis mellifera. Behav Ecol Sociobiol 65:205–215

Lepousez G, Valley MT, Lledo PM (2013) The impact of adult neurogenesis on olfactory bulb circuits and computations. Annu Rev Physiol 75:339–363

Linster C, Fontanini A (2014) Functional neuromodulation of chemosensation in vertebrates. Curr Opin Neurobiol 29:82–87

Linster C, Maloney M, Patil M, Hasselmo ME (2003) Enhanced cholinergic suppression of previously strengthened synapses enables the formation of self-organized representations in olfactory cortex. Neurobiol Learn Mem 80:302–314

Linster C, Menon AV, Singh CY, Wilson DA (2009) Odor-specific habituation arises from interaction of afferent synaptic adaptation and intrinsic synaptic potentiation in olfactory cortex. Learn Mem 16:452–459, Cold Spring Harbor, NY

Litaudon P, Mouly AM, Sullivan R, Gervais R, Cattarelli M (1997) Learning-induced changes in rat piriform cortex activity mapped using multisite recording with voltage sensitive dye. Eur J Neurosci 9:1593–1602

Luskin MB, Price JL (1983) The topographic organization of associational fibers of the olfactory system in the rat, including centrifugal fibers to the olfactory bulb. J Comp Neurol 216:264–291

Manabe H, Kusumoto-Yoshida I, Ota M, Mori K (2011) Olfactory cortex generates synchronized top-down inputs to the olfactory bulb during slow-wave sleep. J Neurosci: Official J Soc Neurosci 31:8123–8133

Marshall L, Molle M, Hallschmid M, Born J (2004) Transcranial direct current stimulation during sleep improves declarative memory. J Neurosci: Official J Soc Neurosci 24:9985–9992

McCormick DA, Bal T (1994) Sensory gating mechanisms of the thalamus. Curr Opin Neurobiol 4:550–556

McNamara AM, Magidson PD, Linster C, Wilson DA, Cleland TA (2008) Distinct neural mechanisms mediate olfactory memory formation at different timescales. Learn Mem 15:117–125, Cold Spring Harbor, NY

Miller JE, Granados-Fuentes D, Wang T, Marpegan L, Holy TE, Herzog ED (2014) Vasoactive intestinal polypeptide mediates circadian rhythms in mammalian olfactory bulb and olfaction. J Neurosci: Official J Soc Neurosci 34:6040–6046

Miller MA (2015) The role of sleep and sleep disorders in the development, diagnosis, and management of neurocognitive disorders. Front Neurol 6:224

Moreno MM, Linster C, Escanilla O, Sacquet J, Didier A, Mandairon N (2009) Olfactory perceptual learning requires adult neurogenesis. Proc Natl Acad Sci USA 106:17980–17985

Moriceau S, Sullivan RM (2004) Unique neural circuitry for neonatal olfactory learning. J Neurosci: Official J Soc Neurosci 24:1182–1189

Mouly AM, Vigouroux M, Holley A (1985) On the ability of rats to discriminate between microstimulations of the olfactory bulb in different locations. Behav Brain Res 17:45–58

Mozell MM, Sheehe PR, Hornung DE, Kent PF, Youngentob SL, Murphy SJ (1987) "Imposed" and "inherent" mucosal activity patterns. Their composite representation of olfactory stimuli. J Gen Physiol 90:625–650

Murakami M, Kashiwadani H, Kirino Y, Mori K (2005) State-dependent sensory gating in olfactory cortex. Neuron 46:285–296

Murphy MJ, Peterson MJ (2015) Sleep disturbances in depression. Sleep Med Clin 10:17–23

Pavlides C, Winson J (1989) Influences of hippocampal place cell firing in the awake state on the activity of these cells during subsequent sleep episodes. J Neurosci: Official J Soc Neurosci 9:2907–2918

Peyrache A, Khamassi M, Benchenane K, Wiener SI, Battaglia FP (2009) Replay of rule-learning related neural patterns in the prefrontal cortex during sleep. Nat Neurosci 12:919–926

Poo C, Isaacson JS (2011) A major role for intracortical circuits in the strength and tuning of odor-evoked excitation in olfactory cortex. Neuron 72:41–48

Popa D, Duvarci S, Popescu AT, Lena C, Pare D (2010) Coherent amygdalocortical theta promotes fear memory consolidation during paradoxical sleep. Proc Natl Acad Sci USA 107:6516–6519

Rasch B, Büchel C, Gais S, Born J (2007) Odor cues during slow-wave sleep prompt declarative memory consolidation. Science (New York, NY) 315:1426–1429

Roesch MR, Stalnaker TA, Schoenbaum G (2007) Associative encoding in anterior piriform cortex versus orbitofrontal cortex during odor discrimination and reversal learning. Cereb Cortex 17:643–652

Saleh M, Jurchott K, Oberland S, Neuhaus EM, Kramer A, Abraham U (2015) Genome-wide screen reveals rhythmic regulation of genes involved in odor processing in the olfactory epithelium. J Biol Rhythms 30:506–518

Sara SJ (2010) Reactivation, retrieval, replay and reconsolidation in and out of sleep: connecting the dots. Front Behav Neurosci 4:185

Schendzielorz T, Schirmer K, Stolte P, Stengl M (2015) Octopamine regulates antennal sensory neurons via daytime-dependent changes in cAMP and IP3 levels in the hawkmoth Manduca sexta. PLoS ONE 10:e0121230

Staubli U, Le TT, Lynch G (1995) Variants of olfactory memory and their dependencies on the hippocampal formation. J Neurosci: Official J Soc Neurosci 15:1162–1171

Stickgold R (2005) Sleep-dependent memory consolidation. Nature 437:1272–1278

Sullivan RM, Wilson DA, Leon M (1989) Norepinephrine and learning-induced plasticity in infant rat olfactory system. J Neurosci: Official J Soc Neurosci 9:3998–4006

Sullivan RM, Landers M, Yeaman B, Wilson DA (2000) Good memories of bad events in infancy. Nature 407:38–39

Sullivan RM, Taborsky-Barba S, Mendoza R, Itano A, Leon M, Cotman CW, Payne TF, Lott I (1991) Olfactory classical conditioning in neonates. Pediatrics 87:511–518

Tang AC, Hasselmo ME (1994) Selective suppression of intrinsic but not afferent fiber synaptic transmission by baclofen in the piriform (olfactory) cortex. Brain Res 659:75–81

Tononi G, Cirelli C (2006) Sleep function and synaptic homeostasis. Sleep Med Rev 10:49–62

Ueno T, Tomita J, Tanimoto H, Endo K, Ito K, Kume S, Kume K (2012) Identification of a dopamine pathway that regulates sleep and arousal in Drosophila. Nat Neurosci 15:1516–1523

Valdes JL, McNaughton BL, Fellous JM (2015) Offline reactivation of experience-dependent neuronal firing patterns in the rat ventral tegmental area. J Neurophysiol 114:1183–1195

Volgushev M, Chauvette S, Mukovski M, Timofeev I (2006) Precise long-range synchronization of activity and silence in neocortical neurons during slow-wave oscillations [corrected]. J Neurosci 26:5665–5672

Vorster AP, Born J (2015) Sleep and memory in mammals, birds and invertebrates. Neurosci Biobehav Rev 50:103–119

Walker MP (2009) Sleep-dependent memory processing. In: Stickgold R, Walker MP (eds) The neuroscience of sleep. Academic Press, San Diego, CA, pp 230–240

Walker MP, Stickgold R (2006) Sleep, memory, and plasticity. Annu Rev Psychol 57:139–166

Wilson DA (2010) Single-unit activity in piriform cortex during slow-wave state is shaped by recent odor experience. J Neurosci: Official J Soc Neurosci 30:1760–1765

Wilson DA, Sullivan RM (2011) Cortical processing of odor objects. Neuron 72:506–519

Wilson M, McNaughton B (1994) Reactivation of hippocampal ensemble memories during sleep. Science (New York, NY) 265:5–8

Wooden JI, Pido J, Mathews H, Kieltyka R, Montemayor BA, Ward CP (2014) Sleep deprivation impairs recall of social transmission of food preference in rats. Nature Sci Sleep 6:129–135

Xie L, Kang H, Xu Q, Chen MJ, Liao Y, Thiyagarajan M, O'Donnell J, Christensen DJ, Nicholson C, Iliff JJ, Takano T, Deane R, Nedergaard M (2013) Sleep drives metabolite clearance from the adult brain. Science (New York, NY) 342:373–377

Yamaguchi M, Manabe H, Murata K, Mori K (2013) Reorganization of neuronal circuits of the central olfactory system during postprandial sleep. Front Neural Circ 7:132

Yokoyama TK, Mochimaru D, Murata K, Manabe H, Kobayakawa K, Kobayakawa R, Sakano H, Mori K, Yamaguchi M (2011) Elimination of adult-born neurons in the olfactory bulb is promoted during the postprandial period. Neuron 71:883–897

Zwaka H, Bartels R, Gora J, Franck V, Culo A, Gotsch M, Menzel R (2015) Context odor presentation during sleep enhances memory in honeybees. Curr Biol: CB 25:2869–2874

The Effect of Sleep on Multiple Memory Systems

Monika Schönauer and Steffen Gais

Abstract Sleep is important for determining how new traces in multiple memory systems develop. Sleep helps the consolidation of both explicit declarative as well as implicit procedural memory. Apart from strengthening memory traces in distinct systems separately, sleep also favors interactions between memory systems. Sleep facilitates the dialogue between fast and slow learning systems and shifts memory representations from highly plastic to more stable networks. Sleep might also change whether memory systems compete or cooperate. In the present review, we provide evidence that memory systems interactions do not terminate after encoding but continue beyond task acquisition into off-line consolidation periods.

Keywords Memory systems interaction · Hippocampus · Striatum

The Concept of Multiple Long-Term Memory Systems

The ability to learn is a deciding factor in evolutionary fitness. It leads to lasting changes in behavior and helps to cope with future challenges. Learning and memory pathologies have a strong negative impact on life and patients with severe memory impairments often require constant care. Different forms of memories can be affected in amnesic disorders. This first became strikingly apparent when Brenda Milner tested patient H.M., who suffered from severe retrograde and anterograde amnesia after bilateral removal of his hippocampi, on an eye-hand coordination skill. Over multiple training sessions during which he had to trace the outline of figures with a pen, seeing his drawing hand only in a mirror, H.M. gained proficiency in mirror drawing without remembering ever having done the task before (Milner 1962). This finding demonstrated the existence of two separate memory systems in the human brain that support different kinds of learned behavior

M. Schönauer · S. Gais (✉)
Institute of Medical Psychology and Behavioral Neurobiology,
Eberhard-Karls-Universität Tübingen, Silcherstr. 5, 72076 Tübingen, Germany
e-mail: steffen.gais@uni-tuebingen.de

N. Axmacher and B. Rasch (eds.), *Cognitive Neuroscience of Memory Consolidation*, Studies in Neuroscience, Psychology and Behavioral Economics, DOI 10.1007/978-3-319-45066-7_7

(Squire 2004). The dichotomy between explicit declarative memory for facts and events, supported by structures in the medial temporal lobe, and implicit habit memory, supported by the striatum, has been replicated many times in different species and using diverse tasks (Poldrack and Packard 2003; Poldrack and Foerde 2008). Further evidence comes from research on neurodegenerative disorders. In Alzheimer's disease, which affects the temporal lobe early during pathogenesis, explicit memory for facts and events is impaired whereas in Parkinson's disease, degeneration of dopaminergic neurons impairs striatal functioning, which leads to a loss of procedural learning (Knowlton et al. 1996; Shohamy et al. 2004; Koenig et al. 2007; Heindel et al. 2013).

Since these first observations, a number of distinct memory systems have been postulated. The actual number of memory systems is an open question. Although Endel Tulving's discussion of 256 different kinds of memory was not meant as a definite answer, it shows that there must be many different memory systems (Tulving 2007). Because every neural system in the brain has been found to show plasticity on closer examination, it might be said that there are perhaps as many memory systems as there are functional systems in the brain. The most prominent distinct memory systems are probably episodic and semantic memory in the domain of declarative knowledge, which are supported mainly by the hippocampus and neocortical areas. Furthermore, there are habits and sensory-motor skills, which depend on the striatum, cerebellum and the respective sensory-motor regions. Finally, emotional memory and conditioning are also intensively studied, showing important contributions of the amygdala and basal ganglia (Squire and Zola-Morgan 1991; Squire and Zola 1996).

How Does Sleep Affect Distinct Memory Systems?

For a memory to last for the long term, it needs to be stabilized after encoding, a process that has been termed consolidation. Models of how new memories are processed and organized after encoding have been developed for both declarative as well as procedural memories (Squire and Alvarez 1995; Nadel and Moscovitch 1997; McGaugh 2000; Hikosaka 2002; Doyon et al. 2003; Morris 2006; Doyon et al. 2009a; Penhune and Steele 2012). It has been known for a long time that sleep benefits declarative memory consolidation. For instance, when participants are allowed to sleep after vocabulary learning, they retain this information better than when they stay awake during the same time interval (Gais et al. 2006). Navigation in a virtual environment, autobiographical recollection, temporal order in episodic memories, and prospective memory all benefit from a period of sleep after encoding (Diekelmann et al. 2013; Nguyen et al. 2013; Murre et al. 2014). Overall, most declarative types of memory, independent of the material and type of task used, seem to benefit from sleep, although the effects are of small size and therefore not always found in small samples (Schönauer et al. 2014).

Similarly, procedural memory was found to benefit from sleep, although the situation seems more ambiguous than for declarative memory. Motor adaptation skills like mirror tracing (Plihal and Born 1997; Schönauer et al. 2015) and a rotor pursuit task (Smith and MacNeill 1994; Fogel et al. 2007) as well as a visual discrimination skill (Gais et al. 2000; Stickgold et al. 2000) all are consolidated during sleep. Learning temporal rhythm (Durrant et al. 2011), gross motor movement (Kempler and Richmond 2012) and dance moves (Genzel et al. 2012) likewise profit from sleep. Consolidation of motor sequence memory has also consistently been found to gain from sleep (Fischer et al. 2002; Walker et al. 2002; Korman et al. 2007; Doyon et al. 2009b; Djonlagic et al. 2012), although there are some studies that show that other factors than sleep, e.g. circadian rhythm, have to be considered as well (Cai and Rickard 2009; Brawn et al. 2010). However, there are also some inconsistent findings. While rotation adaptation was found to benefit from sleep in one study (Huber et al. 2004), others could not replicate this finding (Donchin et al. 2002; Doyon et al. 2009b).

Finally, there are a number of studies investigating emotional memory (Wagner et al. 2005; Nishida et al. 2009; Payne and Kensinger 2011; Payne et al. 2012) (see Chapter by Cunningham and Payne) and conditioning (Graves et al. 2003; Hagewoud et al. 2011; Ognjanovski et al. 2014), indicating that these forms of memory are also influenced by sleep. Thus, a beneficial effect of sleep following learning has been observed for memory stored in a number of different systems (Rasch and Born 2013) (see also Chapter by Rauss and Born).

In the wake of the discovery of independent memory systems, sleep research has begun to ask the question whether sleep affects all types of memory equally or whether different aspects of sleep benefit different forms of memory. In particular, several theories have been developed to describe the relation between sleep and individual types of memory. Associations have been proposed between sleep stages (e.g. stage 2, SWS, REM sleep) and consolidation of distinct types of memory (Diekelmann and Born 2010). More mechanistically, specific features of sleep (e.g. sleep spindles, slow EEG waves, hippocampal ripple activity; see Chapter by Bergmann and Staresina) have all been linked to memory consolidation (Dudai et al. 2015). It is not yet clear, whether any of these sleep-related processes can be assigned to individual forms of memory. However, one mechanism that received wide support from animal and human studies and that is assumed to mediate memory consolidation is the reactivation of neuronal firing patterns during sleep (O'Neill et al. 2010) (see also Chapter by Zhang, Deuker and Axmacher).

First evidence for replay of hippocampal memory came from studies in rats, showing that the firing patterns apparent during training are reinstated in post-learning sleep (Wilson and McNaughton 1994). Subsequently, it was found that sequences of neuronal spiking are replayed in neocortical areas as well (Ji and Wilson 2007). Investigating the functional significance of this process, Rasch and colleagues linked reactivation to the improvements of performance observed after sleep in humans (Rasch et al. 2007). In their experiment, they associated an odor cue with the learning situation so that they could later reactivate the context during sleep. The odor cue was present while participants learned locations in a card pairs game. During following sleep or wakefulness, either the odor or an odorless substance was presented. Participants performed better on the following retrieval test when the odor cue was presented during sleep, but not during wakefulness. Additionally, cue presentation during sleep activated the hippocampal region in fMRI. Together, this evidence strongly supports the view that the effect of sleep on declarative memory is at least partially mediated by a reactivation of learning related neuronal activity.

Several recent studies in rodents suggest that consolidation of procedural motor memory may also depend on reactivation mechanisms, but at the neocortical level. During sleep after learning a new motor skill, neurons in the motor cortex of rats show a pattern of activity similar to that observed during task learning. This reactivation is required for new spine formation (Yang et al. 2014), and it leads to improved performance after sleep (Ramanathan et al. 2015). In humans, cueing memory reactivation during sleep with sounds that have been associated to a motor learning task can improve memory for motor sequences in a similar way as observed for odor cues and object location memory above (Antony et al. 2012; Schönauer et al. 2014) (see also Chapter by Schreiner, Lehmann and Rasch).

Whereas reactivation seems to be a mechanism common to declarative and non-declarative memory, different forms of memory also have distinct properties regarding sleep. There are indications that more than one sleep-dependent consolidation process exists and that different memory tasks are differently affected by these processes (Geyer et al. 2013; Schönauer et al. 2015). Visual discrimination learning is a form of procedural memory in which the subject learns to detect a visual stimulus that is presented only briefly. Performance in this task improves only after a period of sleep, and only if sleep follows training within a few hours. It does not show improvements after several nights of sleep when subjects are sleep deprived the first night after learning, but the benefit achieved after initial sleep remains stable even after one week (Stickgold et al. 2000). It is therefore distinct in that it requires sleep after training for improvements to occur. Similarly, Schönauer and colleagues showed that also in a mirror tracing task, benefits of sleep only emerge if sleep follows shortly after learning, but not when it is delayed by more than 12 h. Hippocampus-dependent memory, in contrast, shows these improvements even if sleep occurs up to 24 h later (Schönauer et al. 2015). Finger sequence tapping, on the other hand, behaves more similar to declarative memory. It shows improvements even if sleep is delayed. Moreover, it also benefits from memory reactivation by sound cueing during sleep (Schönauer et al. 2014).

Sleep and Memory Systems Consolidation

There is evidence that the brain areas that participate in memory retrieval shift over the course of consolidation. Declarative material is supposed to be initially encoded via long-term potentiation in the hippocampus. However, this initial dependence on the hippocampus diminishes over time. Gradients in retrograde amnesia show that older memories tend to be less affected by medial temporal lobe damage than younger memories, which indicates that independent memory traces are formed in other brain areas (Squire et al. 1989; Squire and Zola-Morgan 1991). Such a two-storage solution helps to avoid interference between newly encoded and older memory content (McClelland et al. 1995). It is therefore not surprising that behavioral effects of sleep are accompanied by changes in the neural representation of memories. Sleep after learning shifts the neural substrate of new memories from hippocampal to neocortical regions (Takashima et al. 2006; Gais et al. 2007; Takashima et al. 2009) (see Chapters by Fernandez and by Genzel and Battaglia). Altered network dynamics, in particular a release of inhibition on feedback synapses coming from the hippocampus, permit a hippocampal-neocortical dialogue during sleep that may not similarly occur during wakefulness (Buzsáki 1996; Hasselmo 1999). Moreover, during sleep, replay of learning-related neuronal activity in the hippocampus can trigger replay activity in the neocortex (Ji and Wilson 2007; Peyrache et al. 2009). Over time, this neocortical reactivation is suggested to render neocortical memory traces independent of the hippocampus (Frankland and Bontempi 2005). Thus, newly acquired memories can be integrated more tightly into existing neocortical networks and made more permanent in a less plastic but more stable substrate (Buzsáki 1996; Gais and Born 2004; Diekelmann and Born 2010).

Similar changes in brain systems contribution and additional involvement of neocortical areas after sleep have been observed for motor sequence learning (Walker et al. 2005; Albouy et al. 2013b). Also in implicit statistical learning, memory representations show changes in neural substrate after consolidation. Whereas early after learning, recognizing statistically similar tone sequences activates regions in the medial temporal lobe, sleep shifts this activation to striatal areas (Durrant et al. 2013). This shift in brain activity goes along with better behavioral performance after sleep (Durrant et al. 2011, 2013). Together, these findings support that sleep aids a transfer of memory function from the hippocampus to other cortical and subcortical areas.

The Impact of Sleep on Memory Systems Interactions

As already described above, replay of neuronal activity is not restricted to the hippocampus but can also be observed in other cortical areas. It has been shown that replay in the hippocampus can trigger replay in the neocortex (Peyrache et al. 2009) and subcortical areas, like the striatum (Pennartz et al. 2004; Lansink et al. 2008, 2009). Most interestingly, the hippocampus and the striatum are typically seen as two independent memory systems, serving different roles in declarative memory and habit formation. This raises several intriguing questions. Do memory systems cooperate and how does sleep influence this cooperation? Is there a competition between these memory systems and does this change over sleep? Does sleep allow the integration of different aspects of a memory, leading to a more unified representation?

The first indication that sleep affects the interaction between memory systems came from two studies by Brown and Robertson, who showed that learning a declarative memory task can interfere with consolidation of procedural memory and vice versa when subjects stayed awake, but sleep could abate this effect (Brown and Robertson 2007a, b; Robertson 2012). Cross-regional reactivation could impact how memory is represented in the brain by changing how multiple memory systems interact. It has been established that different memory systems, especially the hippocampus and striatum, compete during task training (Brown and Robertson 2007a, b; Keisler and Shadmehr 2010; Cohen and Robertson 2011; Robertson 2012). Data from both brain imaging as well as behavioral experiments indicate that sleep may in fact change the way in which memory systems interact later (Albouy et al. 2013a). One line of experiments indicates that the competitive interaction between brain regions might be preserved or even enhanced during memory processing in sleep. Logothetis and colleagues (Logothetis et al. 2012) found that reactivation-related hippocampal activity during sleep coincides with a deactivation of the basal ganglia. This competitive relation between multiple memory systems during memory consolidation might reduce interference between separate memory representations and thus ultimately favor successful strengthening of traces in different systems. This would enforce the initial competition between memory systems and bias processing in favor of one strategy which will then prevail over alternative strategies. So far, however, no data exist to support this idea.

During sleep, global information processing as observed during wakefulness changes to more local processing, a property that might explain the loss of consciousness during sleep (Massimini et al. 2005; Samann et al. 2011; Boly et al. 2012). Such information processing in local nodes may lead to a strengthening of traces in different memory systems, separately. In principle, this idea is compatible with the suggested competitive interaction of brain regions during sleep-dependent consolidation, as it assumes that all local nodes merely act independently of each other. However, the close temporal proximity of reactivation in the hippocampus and other cortical and subcortical areas makes it possible to induce spike-timing dependent plasticity across multiple brain regions. It has been suggested that this may bind separate facets of a memory into coherent, joint representations (Lansink et al. 2009). Supporting this proposition, Albouy and colleagues (Albouy et al. 2008) found that during motor sequence learning, an initially competitive interaction between the hippocampus and striatum turns cooperative over a 24 h delay including a night of sleep. Their results indicate that rather than enforcing competition or independence between hippocampal and striatal representations, sleep may favor cooperative interaction between the two.

To test this idea, we devised a category learning task that allows both striatal as well as hippocampal engagement during training (Schönauer et al., unpublished data). During retrieval implicit and explicit components could be readily separated. We found that the implicit and explicit memory representations compete after training and remain competitive over a 12 h period of wakefulness. If subjects got a full night of sleep after learning, however, the initial competitive interaction turned into cooperation. Performance was significantly better when retrieval allowed the cooperative use of both strategies. Furthermore, changes in brain activity reflected an integration of hippocampal and striatal aspects of memory. After sleep, the hippocampus was more active for implicit memory retrieval whereas the striatum was more engaged during an explicit memory task, and connectivity between both regions increased after sleep. These results suggest that sleep combines memory representations over multiple memory systems into one coherent representation that entails both implicit and explicit components. Finally, this systems cooperation was accompanied by greater insight into the complexity of the implicit task rules. Our findings suggest that an integration of information across systems may aid conversions of implicit skill to explicit knowledge.

Summary

Recent studies show that sleep, in particular, plays an important role in determining how traces in multiple memory systems develop. Sleep aids the consolidation of both explicit declarative as well as implicit procedural memory. Evidence points in the direction that more than one sleep-dependent consolidation mechanism exists, and that a hippocampal contribution during learning might be the determining factor regarding how memories are processed during offline periods in both sleep and wakefulness. Apart from strengthening traces in distinct systems separately, sleep has been shown to favor interactions between multiple memory systems. First, sleep facilitates a dialogue between fast and slow learning systems in the brain, aiding a shift of memory representations from highly plastic to more stable networks. Second, sleep might change whether different memory systems compete or cooperate. More work is needed to answer the questions in which way sleep changes these systems interactions. There is, however, strong indication that memory systems interactions do not terminate after encoding but continue beyond task acquisition into off-line consolidation periods.

References

Albouy G, Sterpenich V, Balteau E, Vandewalle G, Desseilles M, Dang-Vu T, Darsaud A, Ruby P, Luppi PH, Degueldre C et al (2008) Both the hippocampus and striatum are involved in consolidation of motor sequence memory. Neuron 58(2):261–272

Albouy G, King BR, Maquet P, Doyon J (2013a) Hippocampus and striatum: dynamics and interaction during acquisition and sleep-related motor sequence memory consolidation. Hippocampus 23(11):985–1004

Albouy G, Sterpenich V, Vandewalle G, Darsaud A, Gais S, Rauchs G, Desseilles M, Boly M, Dang-Vu T, Balteau E et al (2013b) Interaction between hippocampal and striatal systems predicts subsequent consolidation of motor sequence memory. PLoS ONE 8(3):e59490

Antony JW, Gobel EW, O'Hare JK, Reber PJ, Paller KA (2012) Cued memory reactivation during sleep influences skill learning. Nat Neurosci 15(8):1114–1116

Boly M, Perlbarg V, Marrelec G, Schabus M, Laureys S, Doyon J, Pelegrini-Issac M, Maquet P, Benali H (2012) Hierarchical clustering of brain activity during human nonrapid eye movement sleep. Proc Natl Acad Sci USA 109(15):5856–5861

Brawn TP, Fenn KM, Nusbaum HC, Margoliash D (2010) Consolidating the effects of waking and sleep on motor-sequence learning. J Neurosci 30(42):13977–13982

Brown RM, Robertson EM (2007a) Inducing motor skill improvements with a declarative task. Nat Neurosci 10(2):148–149

Brown RM, Robertson EM (2007b) Off-line processing: reciprocal interactions between declarative and procedural memories. J Neurosci 27(39):10468–10475

Buzsáki G (1996) The hippocampo-neocortical dialogue. Cereb Cortex 6(2):81–92

Cai DJ, Rickard TC (2009) Reconsidering the role of sleep for motor memory. Behav Neurosci 123(6):1153–1157

Cohen DA, Robertson EM (2011) Preventing interference between different memory tasks. Nat Neurosci 14(8):953–955

Diekelmann S, Born J (2010) The memory function of sleep. Nat Rev Neurosci 11(2):114–126

Diekelmann S, Wilhelm I, Wagner U, Born J (2013) Sleep to implement an intention. Sleep 36 (1):149–153

Djonlagic I, Saboisky J, Carusona A, Stickgold R, Malhotra A (2012) Increased sleep fragmentation leads to impaired off-line consolidation of motor memories in humans. PLoS ONE 7(3):e34106

Donchin O, Sawaki L, Madupu G, Cohen LG, Shadmehr R (2002) Mechanisms influencing acquisition and recall of motor memories. J Neurophysiol 88(4):2114–2123

Doyon J, Penhune V, Ungerleider LG (2003) Distinct contribution of the cortico-striatal and cortico-cerebellar systems to motor skill learning. Neuropsychologia 41(3):252–262

Doyon J, Bellec P, Amsel R, Penhune V, Monchi O, Carrier J, Lehericy S, Benali H (2009a) Contributions of the basal ganglia and functionally related brain structures to motor learning. Behav Brain Res 199(1):61–75

Doyon J, Korman M, Morin A, Dostie V, Hadj Tahar A, Benali H, Karni A, Ungerleider LG, Carrier J (2009b) Contribution of night and day sleep vs. simple passage of time to the consolidation of motor sequence and visuomotor adaptation learning. Exp Brain Res 195(1): 15–26

Dudai Y, Karni A, Born J (2015) The Consolidation and Transformation of Memory. Neuron 88 (1):20–32

Durrant SJ, Taylor C, Cairney S, Lewis PA (2011) Sleep-dependent consolidation of statistical learning. Neuropsychologia 49(5):1322–1331

Durrant SJ, Cairney SA, Lewis PA (2013) Overnight consolidation aids the transfer of statistical knowledge from the medial temporal lobe to the striatum. Cereb Cortex 23(10):2467–2478

Fischer S, Hallschmid M, Elsner AL, Born J (2002) Sleep forms memory for finger skills. Proc Natl Acad Sci USA 99(18):11987–11991

Fogel SM, Smith CT and Cote KA (2007) Dissociable learning-dependent changes in REM and non-REM sleep in declarative and procedural memory systems. Behav Brain Res 180(1): 48-61

Frankland PW, Bontempi B (2005) The organization of recent and remote memories. Nat Rev Neurosci 6(2):119–130

Gais S, Born J (2004) Declarative memory consolidation: mechanisms acting during human sleep. Learn Mem 11(6):679–685

Gais S, Plihal W, Wagner U, Born J (2000) Early sleep triggers memory for early visual discrimination skills. Nat Neurosci 3(12):1335–1339

Gais S, Lucas B, Born J (2006) Sleep after learning aids memory recall. Learn Mem 13(3): 259–262

Gais S, Albouy G, Boly M, Dang-Vu TT, Darsaud A, Desseilles M, Rauchs G, Schabus M, Sterpenich V, Vandewalle G et al (2007) Sleep transforms the cerebral trace of declarative memories. Proc Natl Acad Sci USA 104(47):18778–18783

Genzel L, Quack A, Jager E, Konrad B, Steiger A, Dresler M (2012) Complex motor sequence skills profit from sleep. Neuropsychobiology 66(4):237–243

Geyer T, Mueller HJ, Assumpcao L, Gais S (2013) Sleep-effects on implicit and explicit memory in repeated visual search. PLoS ONE 8(8):e69953

Graves LA, Heller EA, Pack AI, Abel T (2003) Sleep deprivation selectively impairs memory consolidation for contextual fear conditioning. Learn Mem 10(3):168–176

Hagewoud R, Bultsma LJ, Barf RP, Koolhaas JM, Meerlo P (2011) Sleep deprivation impairs contextual fear conditioning and attenuates subsequent behavioural, endocrine and neuronal responses. J Sleep Res 20(2):259–266

Hasselmo ME (1999) Neuromodulation: acetylcholine and memory consolidation. Trends Cog Sci 3(9):351–359

Heindel WC, Festa EK, Ott BR, Landy KM, Salmon DP (2013) Prototype learning and dissociable categorization systems in Alzheimer's disease. Neuropsychologia 51(9):1699–1708

Hikosaka O (2002) A new approach to the functional systems of the brain. Epilepsia 43(Suppl 9): 9–15

Huber R, Ghilardi MF, Massimini M, Tononi G (2004) Local sleep and learning. Nature 430 (6995):78–81

Ji D, Wilson MA (2007) Coordinated memory replay in the visual cortex and hippocampus during sleep. Nat Neurosci 10(1):100–107

Keisler A, Shadmehr R (2010) A shared resource between declarative memory and motor memory. J Neurosci 30(44):14817–14823

Kempler L, Richmond JL (2012) Effect of sleep on gross motor memory. Memory 20(8):907–914

Knowlton BJ, Mangels JA, Squire LR (1996) A neostriatal habit learning system in humans. Science 273(5280):1399–1402

Koenig P, Smith EE, Moore P, Glosser G, Grossman M (2007) Categorization of novel animals by patients with Alzheimer's disease and corticobasal degeneration. Neuropsychology 21(2): 193–206

Korman M, Doyon J, Doljansky J, Carrier J, Dagan Y, Karni A (2007) Daytime sleep condenses the time course of motor memory consolidation. Nat Neurosci 10(9):1206–1213

Lansink CS, Goltstein PM, Lankelma JV, Joosten RN, McNaughton BL, Pennartz CM (2008) Preferential reactivation of motivationally relevant information in the ventral striatum. J Neurosci 28(25):6372–6382

Lansink CS, Goltstein PM, Lankelma JV, McNaughton BL, Pennartz CM (2009) Hippocampus leads ventral striatum in replay of place-reward information. PLoS Biol 7(8):e1000173

Logothetis NK, Eschenko O, Murayama Y, Augath M, Steudel T, Evrard HC, Besserve M, Oeltermann A (2012) Hippocampal-cortical interaction during periods of subcortical silence. Nature 491(7425):547–553

Massimini M, Ferrarelli F, Huber R, Esser SK, Singh H, Tononi G (2005) Breakdown of cortical effective connectivity during sleep. Science 309(5744):2228–2232

McClelland JL, McNaughton BL, O'Reilly RC (1995) Why there are complementary learning systems in the hippocampus and neocortex: insights from the successes and failures of connectionist models of learning and memory. Psychol Rev 102(3):419–457

McGaugh JL (2000) Memory—a century of consolidation. Science 287(5451):248–251

Milner B (1962) Les troubles de la mémoire accompagnant des lésions hippocampiques bilatérales. Physiologie de l'hippocampe. Paris, Centre National de la Recherche Scientifique, pp 257–272

Morris RG (2006) Elements of a neurobiological theory of hippocampal function: the role of synaptic plasticity, synaptic tagging and schemas. Eur J Neurosci 23(11):2829–2846

Murre JM, Kristo G, Janssen SM (2014) The effect of self-reported habitual sleep quality and sleep length on autobiographical memory. Memory 22(6):633–645

Nadel L, Moscovitch M (1997) Memory consolidation, retrograde amnesia and the hippocampal complex. Curr Opin Neurobiol 7(2):217–227

Nguyen ND, Tucker MA, Stickgold R, Wamsley EJ (2013) Overnight sleep enhances hippocampus-dependent aspects of spatial memory. Sleep 36(7):1051–1057

Nishida M, Pearsall J, Buckner RL, Walker MP (2009) REM sleep, prefrontal theta, and the consolidation of human emotional memory. Cereb Cortex 19(5):1158–1166

Ognjanovski N, Maruyama D, Lashner N, Zochowski M, Aton SJ (2014) CA1 hippocampal network activity changes during sleep-dependent memory consolidation. Front Syst Neurosci 8:61

O'Neill J, Pleydell-Bouverie B, Dupret D, Csicsvari J (2010) Play it again: reactivation of waking experience and memory. Trends Neurosci 33(5):220–229

Payne JD, Kensinger EA (2011) Sleep leads to changes in the emotional memory trace: evidence from FMRI. J Cogn Neurosci 23(6):1285–1297

Payne JD, Chambers AM, Kensinger EA (2012) Sleep promotes lasting changes in selective memory for emotional scenes. Front Integr Neurosci 6:108

Penhune VB, Steele CJ (2012) Parallel contributions of cerebellar, striatal and M1 mechanisms to motor sequence learning. Behav Brain Res 226(2):579–591

Pennartz CM, Lee E, Verheul J, Lipa P, Barnes CA, McNaughton BL (2004) The ventral striatum in off-line processing: ensemble reactivation during sleep and modulation by hippocampal ripples. J Neurosci 24(29):6446–6456

Peyrache A, Khamassi M, Benchenane K, Wiener SI, Battaglia FP (2009) Replay of rule-learning related neural patterns in the prefrontal cortex during sleep. Nat Neurosci 12(7):919–926

Plihal W, Born J (1997) Effects of early and late nocturnal sleep on declarative and procedural memory. J Cogn Neurosci 9(4):534–547

Poldrack RA, Foerde K (2008) Category learning and the memory systems debate. Neurosci Biobehav Rev 32(2):197–205

Poldrack RA, Packard MG (2003) Competition among multiple memory systems: converging evidence from animal and human brain studies. Neuropsychologia 41(3):245–251

Ramanathan DS, Gulati T, Ganguly K (2015) Sleep-dependent reactivation of ensembles in motor cortex promotes skill consolidation. PLoS Biol 13(9):e1002263

Rasch B, Born J (2013) About sleep's role in memory. Physiol Rev 93(2):681–766

Rasch B, Buchel C, Gais S, Born J (2007) Odor cues during slow-wave sleep prompt declarative memory consolidation. Science 315(5817):1426–1429

Robertson EM (2012) New insights in human memory interference and consolidation. Curr Biol 22(2):R66–R71

Samann PG, Wehrle R, Hoehn D, Spoormaker VI, Peters H, Tully C, Holsboer F, Czisch M (2011) Development of the brain's default mode network from wakefulness to slow wave sleep. Cereb Cortex 21(9):2082–2093

Schönauer M, Pawlizki A, Kock C, Gais S (2014) Exploring the effect of sleep and reduced interference on different forms of declarative memory. Sleep 37(12):1995–2007

Schönauer M, Geisler T, Gais S (2014) Strengthening procedural memories by reactivation in sleep. J Cogn Neurosci 26(1):143–153

Schönauer M, Grätsch M, Gais S (2015) Evidence for two distinct sleep-related long-term memory consolidation processes. Cortex 63:68–78

Shohamy D, Myers CE, Grossman S, Sage J, Gluck MA, Poldrack RA (2004) Cortico-striatal contributions to feedback-based learning: converging data from neuroimaging and neuropsychology. Brain 127(Pt 4):851–859

Smith C, MacNeill C (1994) Impaired motor memory for a pursuit rotor task following stage 2 sleep loss in college students. J Sleep Res 3:206–213

Squire LR (2004) Memory systems of the brain: a brief history and current perspective. Neurobiol Learn Mem 82(3):171–177

Squire LR, Alvarez P (1995) Retrograde amnesia and memory consolidation: a neurobiological perspective. Curr Opin Neurobiol 5(2):169–177

Squire LR, Haist F, Shimamura AP (1989) The neurology of memory: quantitative assessment of retrograde amnesia in two groups of amnesic patients. J Neurosci 9(3):828–839

Squire LR, Zola SM (1996) Structure and function of declarative and nondeclarative memory systems. Proc Natl Acad Sci USA 93(24):13515–13522

Squire LR, Zola-Morgan S (1991) The medial temporal lobe memory system. Science 253(5026):1380–1386

Stickgold R, James L, Hobson JA (2000) Visual discrimination learning requires sleep after training. Nat Neurosci 3(12):1237–1238

Takashima A, Petersson KM, Rutters F, Tendolkar I, Jensen O, Zwarts MJ, McNaughton BL, Fernandez G (2006) Declarative memory consolidation in humans: a prospective functional magnetic resonance imaging study. Proc Natl Acad Sci USA 103(3):756–761

Takashima A, Nieuwenhuis IL, Jensen O, Talamini LM, Rijpkema M, Fernandez G (2009) Shift from hippocampal to neocortical centered retrieval network with consolidation. J Neurosci 29(32):10087–10093

Tulving E (2007) Are there 256 different kinds of memory? In: Nairne JS The foundations of remembering: essays in honor of Henry L. Roediger, III. New York, Psychology Press, p 39–52

Wagner U, Degirmenci M, Drosopoulos S, Perras B, Born J (2005) Effects of cortisol suppression on sleep-associated consolidation of neutral and emotional memory. Biol Psychiatry 58 (11):885–893

Walker MP, Brakefield T, Morgan A, Hobson JA, Stickgold R (2002) Practice with sleep makes perfect: sleep-dependent motor skill learning. Neuron 35(1):205–211

Walker MP, Stickgold R, Alsop D, Gaab N, Schlaug G (2005) Sleep-dependent motor memory plasticity in the human brain. Neuroscience 133(4):911–917

Wilson MA, McNaughton BL (1994) Reactivation of hippocampal ensemble memories during sleep. Science 265(5172):676–679

Yang G, Lai CS, Cichon J, Ma L, Li W, Gan WB (2014) Sleep promotes branch-specific formation of dendritic spines after learning. Science 344(6188):1173–1178

A Role of Sleep in Forming Predictive Codes

Karsten Rauss and Jan Born

Introduction

Memory consolidation is traditionally described and investigated in terms of the strengthening of memory traces, particularly those pertaining to recently encoded information. Over the last decade, this traditional view has been complemented by conceptual and empirical work indicating that memory consolidation is both selective and transformative: only a subset of new experiences undergoes consolidation, and in this subset, representations are not merely strengthened, but also restructured. In this chapter, we briefly review evidence for selection and transformation of memory traces during consolidation, and the pivotal role of sleep in these processes. We will then argue that these findings might be best understood in the context of recent Bayesian theories of brain function, and we will show how this perspective helps generate new hypotheses concerning the mechanisms of memory consolidation during sleep.

Memory Consolidation During Sleep

Sleep is known to play a central role in memory formation, and the overwhelming evidence for this assertion has been reviewed elsewhere (Diekelmann and Born 2010; Maquet 2001; Rasch and Born 2013; Walker et al. 2005) (see also chapter by Schönauer and Gais and by Kreutzmann and colleagues). Current research indicates

K. Rauss · J. Born (✉)
Institute of Medical Psychology and Behavioral Neurobiology,
University of Tübingen, Otfried-Müller-Str. 25, 72076 Tübingen, Germany
e-mail: jan.born@uni-tuebingen.de

K. Rauss
e-mail: karsten.rauss@uni-tuebingen.de

N. Axmacher and B. Rasch (eds.), *Cognitive Neuroscience of Memory Consolidation*, Studies in Neuroscience, Psychology and Behavioral Economics, DOI 10.1007/978-3-319-45066-7_8

that this effect relies on an active role of sleep in consolidating newly acquired information via processes of reactivation (Diekelmann and Born 2010; Ellenbogen et al. 2006; Lewis and Durrant 2011; Peigneux et al. 2004). Much of this research in humans goes back to the demonstration by Wilson and McNaughton (1994) that sequences of neuronal activity recorded in the rat hippocampus during learning are replayed in a very similar way offline during sleep. In addition, human neuroimaging studies have shown that sleep as compared to wakefulness leads to a reorganization of brain regions involved in memory processing, with a redistribution of memory representations from hippocampal sites serving as a temporary store to neocortical sites for long-term storage (Gais et al. 2007; Takashima et al. 2006) (see also chapters by Fernandez, by Genzel and Battaglia, and by Zhang, Deuker and Axmacher).

Recent evidence suggests that a main function of sleep is to select and preferentially consolidate memory representations that are deemed relevant for future behavior. For example, if participants learn two sets of picture-location associations and are then, after learning, instructed that only one of the sets will be relevant at a later retrieval test, those who are allowed to sleep during the retention interval show superior memory consolidation compared to a wake control group, but only for the designated test set and not for the irrelevant set (van Dongen et al. 2012) (see chapter by Fernandez). Employing paradigms of "directed forgetting", several studies found that material that was instructed to be remembered showed superior consolidation over sleep compared to wake intervals, whereas no or lower benefits from sleep were observed for material that was instructed to be forgotten (Fischer et al. 2011; Rauchs et al. 2011; Saletin et al. 2011).

We directly tested the hypothesis that sleep-dependent consolidation selectively facilitates information that is relevant for the future. Subjects learned a set of word pairs and after learning, they either were instructed or kept misinformed about a recall test that would take place on the next day (Wilhelm et al. 2011). Only subjects who expected the retrieval test showed a pronounced sleep benefit for word pair memories. These subjects also displayed significantly increased EEG power in the slow-oscillation frequency band (0.7–1.2 Hz) during the post-learning non-rapid eye movement (NonREM) sleep, and slow oscillation power was correlated with recall performance.

The selection of memory for consolidation during sleep might be linked to the selective reactivation of the respective representations during sleep. We used odors to cue, and thereby experimentally boost, the reactivation of memories during sleep in humans (Diekelmann et al. 2011; Rasch et al. 2007) (see also chapters by Wilson, Kondrakiewicz and Barnes and by Shanahan and Gottfried). Participants learned a 2D object location task in the presence of an odor that became associated with the learning material. During subsequent sleep, the odor was re-exposed to the participant, to reactivate the associated memories. At a recall test after sleep, subjects displayed better performance if they received the odor during slow wave sleep (SWS), as compared to an odorless vehicle control condition. Functional magnetic resonance imaging (fMRI) indicated that presentation of the odor during SWS activated hippocampal and selected neocortical regions possibly involved in the transmission of reactivated memory information from temporary hippocampal stores to long-term memory storage in the neocortex. Thus, externally reinstating the context (odor) in which information was encoded benefits subsequent consolidation and recall of this information.

Sleep-dependent consolidation does not only comprise a selection process, but once representations are selected, sleep can also qualitatively transform these memories (see also chapter by Schönauer and Gais). For instance, sleep supports the transition from implicit to explicit memory in a serial reaction time task (SRTT), in which the response sequence on each experimental trial follows a hidden structure. Compared with a wake control condition, subjects after a period of sleep were better at explicitly generating the sequence underlying the SRTT which they had implicitly learned before sleep (Drosopoulos et al. 2011; Fischer et al. 2006). Sleep, and especially SWS, after practicing a more complex cognitive problem solving task likewise promotes explicit insight into the hidden structure embedded in the task, which was not seen before sleep or after corresponding retention periods of wakefulness (Wagner et al. 2004; Yordanova et al. 2008; Yordanova et al. 2011). Such findings suggest that sleep-dependent memory consolidation not only acts to stabilize information storage, but also supports the explicit detection of contingencies in previously encoded material. Crucially, once contingencies have been detected, this information can be used to generate predictions for similar situations encountered in the future. Thus, the studies discussed here provide initial evidence for the idea that sleep could be involved in transforming memories into a format that allows for better predictions of the future (see also chapter by Nissen and colleagues).

Memory for the Future

The idea that memory is not so much an exact record of past experience, but a constructive process that guides our actions in the present and helps us plan for the future was first articulated by Bartlett (1932).

Prospective Memory

This idea is perhaps most clearly encapsulated in the concept of prospective memory (Kliegel et al. 2007). Prospective memory refers to the process of retrieving previously encoded information about actions to be performed in the future, and to do so at the right time or within the appropriate context. A long history of clinical research has demonstrated that deficits in prospective memory are linked to poor functioning in diseases such as Parkinson's disease (Foster et al. 2009) and schizophrenia

(Ordemann et al. 2014), as well as during healthy aging (Einstein and McDaniel 1990).

Prospective memory is not a unitary process, but is composed of at least two different components, namely the prospective component (to remember *that* something has to be done) and the retrospective component (to remember *what* has to be done). These two components can be independently manipulated (Cohen et al. 2003). A human lesion study has suggested that the prospective component essentially recruits prefrontal regions, whereas medial-temporal structures including the hippocampus are involved in the retrospective component (Umeda et al. 2006). Prospective memory can be successfully accomplished using different strategies (McDaniel and Einstein 2000): either via a capacity-dependent, attention-based strategy relying on voluntary monitoring of the environment to detect relevant cues, or using a capacity-independent, memory-based strategy. When relying on the memory-based strategy, prospective remembering takes place spontaneously if the prospective memory cue is sufficiently connected to the intended action (McDaniel et al. 2004). Whereas the attention-based strategy relies on prefrontal executive functions, the memory-based strategy requires the hippocampal system (Einstein et al. 2005).

Importantly, sleep exerts a robust enhancing effect on prospective memory (Scullin and McDaniel 2010). In one of our own studies, all of the subjects trained on a prospective memory task in the evening before a night of sleep consistently remembered to execute the designated task when tested two days later, whereas half of the subjects who stayed awake on the first night following encoding of the prospective-memory task forgot about their intention (Diekelmann et al. 2013b). The beneficial effect of sleep critically depended on SWS, rather than on REM sleep. Another study (Diekelmann et al. 2013a) showed an effect of sleep specifically on the prospective component, in addition to the benefits on the retrospective component, and also demonstrated that the effect of sleep eventually expresses itself in the preferential use of the memory-based strategy over the attention-based strategy at the time of execution of the intention. In this study, subjects learned a longer list of word-pairs and afterwards were instructed to respond to certain cue words in the list by typing in the associated word (instructed intention) when encountering them at a test session two days later. During the test session, subjects who had slept after the instructed intention detected more cue words (indicating the prospective memory component) and also correctly recalled more of the associated words (retrospective memory component) than subjects who had stayed awake after the instructed intention. The effect of sleep on prospective memory was fully developed only under divided attention conditions, where subjects performed a vigilance task in parallel to the lexical decision task. The latter finding indicates that after sleep, subjects prefer a memory-based strategy during execution of the intention, corroborating the notion of an immediate effect of sleep on the prospective component of memory.

False Memories and Gist Abstraction

One of the main advantages of regarding memory as something that is for the future rather than about the past is that it allows for functional interpretations of many of the putative failures of memory. There is a rich literature on such failures, particularly in relation to the trustworthiness of eyewitness accounts (Loftus 2003). False memories are among the best-investigated examples of such failures. At the same time, they nicely illustrate the constructive and transformative features of memory formation.

False memories designate an individual's remembering of stimuli or episodes that never occurred. In the laboratory, this has mostly been investigated using versions of the Deese-Roediger-McDermott (DRM) paradigm (Deese 1959; Roediger and McDermott 1995). In these tasks, participants learn lists of semantically associated words (e.g. dark, bed, pillow, pyjama) with one common theme-word (e.g. sleep) missing. When later asked to recall words on the lists, subjects typically produce the theme-word as well as the actually encoded ones. In addition, they are subjectively very confident of having encoded the theme-word when asked to give confidence ratings. While obviously false in an objective sense, these memories likely reflect the successful adaptation of biological memory systems to ever-changing environments.

In this view, many cases of false memories may be better interpreted as instances of successful abstraction and generalization, optimizing memories for future use. For instance, both verbal and non-verbal versions of the DRM paradigm nudge subjects towards falsely remembering omitted theme-words or prototype stimuli by the substantial semantic or perceptual overlap between the encoded items. Indeed, it has been argued that extracting commonalities from individual pieces of information is one of the central functions of systems consolidation (Inostroza and Born 2013; Lewis and Durrant 2011) (see chapters by Cheng and by Fernandez). Such abstracted representations would offer substantial benefits both in terms of efficient information storage and in order to balance the complementary needs for specificity and generalizability of long-term memories. They are most frequently described as schemas, but we prefer to label them as gist memories, to emphasize the idea that information is extracted and abstracted from newly encoded material in order to build and update these memories.

If gist abstraction represents one of the main functions of memory, one would expect it to be affected by sleep. However, the available evidence is equivocal: work from our group has demonstrated that sleep can induce qualitative changes of newly acquired memory representations that increase the number of false memories, (Diekelmann et al. 2010a, b), and similar findings were reported by Payne et al. (2009). In contrast, others found no such effects or even a reduction of false memories following sleep (Fenn et al. 2009; Lo et al. 2014). A potential problem of most previous studies is that they are designed like traditional studies of memory consolidation, with relatively short intervals of only a few days between encoding and recall. However, theoretical considerations (Dudai 2012; Lewis and

Durrant 2011) and empirical evidence (Cox et al. 2014; Jurewicz et al. 2016) suggest that gist abstraction may evolve across extended intervals, spanning multiple nights of sleep and perhaps time periods of months and years. Unpublished results from our group using a visual analogue of the DRM paradigm (Diekelmann et al. 2010a, b; Slotnick and Schacter 2004) indeed suggest that beneficial effects of sleep on gist abstraction may emerge only after prolonged periods of several months.

In summary, the notion that memory is less about the past than for the future is intuitively appealing, clinically relevant, and supported by substantial empirical evidence. This notion implies the selective and transformative nature of the process of memory formation and consolidation, which could particularly benefit from sleep. What seems to be lacking, however, is a framework that would unite the different strands of evidence for the prospective component inherent to memory, and the idea that such memories for the future are formed and transformed in a consolidation process which benefits from sleep. In the following, we propose that hierarchical predictive coding offers a framework that might enable such integration.

Predictive Codes

The conceptual roots of predictive coding are usually traced back to Helmholtz' (1867) assertion that perception equals unconscious inference. The underlying assumption is that noise both in the environment and inherent in the numerous transformations performed by perceptual systems renders perception probabilistic. One way to deal with the probabilistic nature of perception is to use the context in which a particular piece of information occurs to predict that information. For example, in DRM-like studies, the recall of a "false" memory would be considered an outcome based on the comparison of a candidate word against other remembered words, essentially yielding an estimate of the probability that the candidate word is part of this particular list of related words. Such reliance on probabilistic predictions can be seen as a viable strategy to reconstruct a maximum of information from intrinsically noisy inputs that have been converted into similarly noisy memory representations. Such a strategy confers important advantages, both in terms of information compression and in terms of redundancy reduction (Barlow 1985).

Indeed, predictive coding was first described as a technical means of compressing image and video data in telecommunications (Oliver 1952). Later on, the approach was adopted in the neurosciences, mainly as a model of basic visual perception (e.g., Srinivasan et al. 1982; Hosoya et al. 2005). Rao and Ballard (1999) reported computational work showing that the cortex may also employ predictive coding. Training a neural-network model of predictive coding on natural images, they demonstrated that different levels of the network developed response profiles that matched those observed in visual cortex. In essence, their model suggests that response patterns of visual cortex neurons can be explained by the increased predictability of what a particular neuron "sees" when a given stimulus can also be captured by neighboring cells. Importantly, the model is hierarchical, with all but the lowest level of the network predicting, not the input itself, but activity patterns at the next lower level. Responses at lower network levels are disinhibited when predictive feedback from higher levels is disabled, similar to what is observed electrophysiologically (Hupe et al. 1998, 2001). More recently, predictive coding has been proposed as a general mechanism underlying the nervous system's ability to generate predictions about the future (for a comprehensive review, see Clark 2013). The most general model has been outlined by Friston (Friston 2005; Friston and Kiebel 2009), who describes predictive coding as one way of minimizing free-energy (and thus surprise) through perception and action.

Neurophysiologically, hierarchical predictive coding should yield two principal types of signals: first, an activation of representations that refer to the predicted stimulus before it actually occurs or even in its complete absence (i.e., an "active prediction" signal); and second, mismatch responses when predicted stimuli do not occur in the expected manner (i.e. prediction error signals). Both types of signals have been observed in numerous studies (Bannert and Bartels 2013; Schultz et al. 1997; Summerfield et al. 2006). However, as outlined below, their evolution over the course of learning is much less investigated.

Hierarchical Predictive Coding, Memory, and Sleep

A central assumption in hierarchical predictive coding is that predictions are generated on the basis of statistical representations of previous experience, which are used to construct an internal model, i.e., a memory representation, of how perceptions are generated. Learning thus amounts to adapting estimates of prior probabilities based on observed likelihoods, in order to optimize estimates of posterior probabilities and minimize prediction errors.

Numerous studies investigating predictive coding or the more general notion of Bayesian coding (Knill and Pouget 2004) have demonstrated that the brain is surprisingly adept at detecting arbitrary statistical regularities in both space (Fiser and Aslin 2002b) and time (Fiser and Aslin 2002a), and subsequently adapts its prior estimates. A typical example is the learning of probabilistic sequences in a serial reaction time task (SRTT, Nissen and Bullemer 1987). Effective training results in speeded response times to the probabilistic sequence in comparisons to stimuli occurring in random order. These studies usually assume that the probabilities underlying the task are acquired from scratch. However, any general notion of hierarchical predictive coding implies that similar processes operate in natural environments. Accordingly, from a predictive-coding perspective, memory should be regarded not so much as a record of past experience, but as information stored for the purpose of predicting future perceptions and acting upon them.

To date, only a small number of studies have explicitly addressed the links between predictive coding and memory; and none has investigated the contribution of sleep-dependent memory consolidation to predictive coding. An attractive hypothesis is that sleep-dependent memory consolidation, including gist extraction, underlies the emergence and fine-tuning of predictive coding over the course of learning.

A harbinger of this idea is the "wake-sleep algorithm" for unsupervised learning proposed by Hinton et al. (1995). In their artificial network model, bottom-up information transmission along recognition connections is limited to a "wake" phase. At the same time, top-down activity carried by "generative" connections attempts to match representations at a given level to those of the next higher level. During a subsequent "sleep" phase, the system is driven exclusively via top-down connections, producing what Hinton et al. (1995) term "fantasies", which percolate down to the level of the input units. These fantasies reflect, not a simple replay of previously encountered activation patterns, but a top-down-initiated replay of the generative model acquired by the network during the preceding wake phase. Over the course of the sleep phase, feedforward connections are fine-tuned so as to reproduce the fantasy replay patterns at the next higher level, which essentially adapts bottom-up information flow to the network's current generative model. This model integrates central conceptual elements of both hierarchical predictive coding and active systems consolidation theories, such as the idea that feedback connections essentially implement predictions of activity patterns at lower levels (Rao and Ballard 1999), or the notion of an active consolidation process during sleep where cortical feedback to hippocampal networks initiates the reactivation and redistribution of memory information towards networks storing representations at a less integrative level (Diekelmann and Born 2010). These links are schematically presented in Fig. 1.

More recently, Hobson and Friston (2012) proposed that sleep, and particularly REM sleep, is crucial for tuning the brain's generative model towards Bayes-optimal inference. Based on Friston's free-energy formulation of brain function, they argue that the essential function of sleep is to allow the brain to optimize itself in the absence of sensory prediction errors. The goal of this optimization is to reduce model complexity by pruning synaptic connections - in the words of Hobson and Friston (2012), sleep "allows the brain to concentrate on statistical housekeeping" (p. 97). Beyond shared aspects with the model of Hinton et al. (1995), there are clear links between the theory laid out by Hobson and Friston (2012) and theories of synaptic homeostasis (Tononi and Cirelli, 2014), although the latter assume a central role of SWS rather than REM sleep for processes of synaptic pruning and down-scaling.

Interestingly, a study by Strauss et al. (2015) found that spontaneous predictive coding of auditory sequences seems to be abolished during sleep, which nicely fits with the models of Hinton et al. (1995) and Hobson and Friston (2012), as both assume that insulation from sensory prediction errors is one of the central requirements for model optimization during sleep. However, the study by Strauss et al. (2015) was not designed to address the question of how sleep translates wake

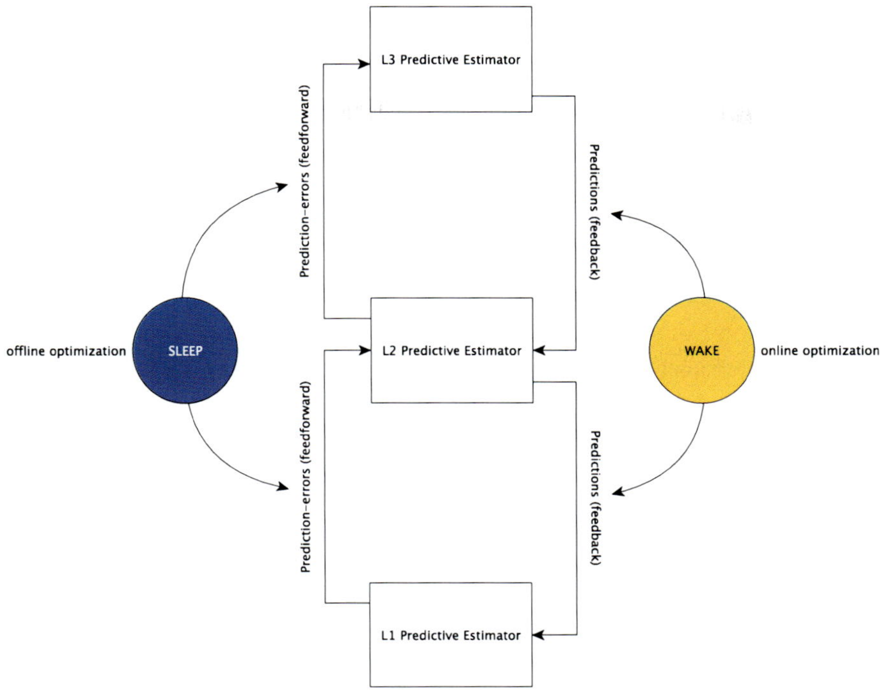

Fig. 1 Hierarchical predictive coding during sleep and wakefulness. Feedforward prediction-errors are compared against feedback predictions across several layers of a hierarchically organized system. Following Hinton et al. (1995), it is assumed that prediction-error signals during wakefulness are used to adapt predictions online, so as to explain as much of the input signal as possible at each level. Conversely, during sleep, the system is driven in a top-down fashion by internally generated signals that reflect what has been learned during wakefulness. Feedforward connections are then optimized so as to recover activity at higher levels. Thus, optimization always happens on the branch of the system along which information flow is less involved in each state (feedback during wakefulness, feedforward during sleep). This scheme implies that, over the course of learning, maximally efficient predictions are implemented at the lowest possible level in the hierarchy. It should thus be possible to track successful memory consolidation by a shift of prediction-error signals to successively lower hierarchical levels

experience into predictive memory codes. Nevertheless, it seems interesting that sleep, which we believe is involved in forming enduring predictive codes, reflects a brain state in which active predictions of external stimulation are temporarily suspended.

In the absence of direct empirical evidence for a link between sleep-dependent memory consolidation and hierarchical predictive coding, it is important to highlight that many of the classical findings on sleep and memory consolidation can be readily interpreted in terms of predictive coding. Even for simple tasks such as word-pair learning, one can argue that the brain does not passively associate two stimuli, but actively predicts the second one (Schultz et al. 1997). This interpretation is supported by the observation that after encoding of word triplets, sleep specifically consolidates associations in the temporal order in which they were acquired, but not the reverse order (Drosopoulos et al. 2007). In a similar vein, previous findings of sleep-dependent consolidation of sequence learning (Albouy et al. 2013), including the emergence of explicit sequence knowledge after sleep (Fischer et al. 2006), could reflect fine-tuning of predictive mechanisms during sleep. Here, active predictions of upcoming stimuli allow for preparing appropriate motor movements ahead of time, which in turn supports behavioral performance both in terms of accuracy as well as reaction times. Whether neural activity patterns indeed predict individual stimuli in visually presented sequences, and whether unexpected deviants elicit corresponding prediction-error signals, remains to be tested.

Sequence learning is particularly important in the context of language acquisition. Studies using language-like stimuli have consistently shown that even infants can use statistical regularities in auditory sequences to segment words from continuous speech (Saffran et al. 1996), and that previously segmented information is later available for labeling classes of objects (Estes et al. 2007). Given the obvious importance of such knowledge for language acquisition, one would expect it to be consolidated for the long term. Indeed, a growing number of studies suggest that artificial grammar learning (Gomez et al. 2006) as well as category learning are boosted by sleep (Djonlagic et al. 2009; Friedrich et al. 2015). The speed of verbal communication suggests that mechanisms of predictive coding at very low levels play an important role, both in language acquisition and adult communication. In this context, it is interesting that Dambacher et al. (2009) observed mismatch responses in the EEG as early as 90 ms after adults read unexpected words. Such low-level predictive codes could be fostered by sleep.

Going beyond mismatch or prediction-error responses, predictive neural signals can be examined using stimulus-evoked responses reflecting very early stages of stimulus processing that are unlikely to be affected by re-entrant feedback linked to the processing of the current stimulus. Pourtois et al. (2008) investigated modulations of the earliest cortical component of the visual evoked potential, the so-called C1, thought to be generated in primary visual cortex (Jeffreys and Axford 1972). Following a procedure developed by Karni and Sagi (1991), participants were trained on a texture-discrimination task in which learning is known to depend on sleep (Gais et al. 2000; Karni et al. 1994). Results indicated a modulation of the C1, selectively at previously trained locations. In light of the early onset and peak latencies of the C1, such modulations (Rauss et al. 2009, 2012) likely reflect mechanisms of predictive coding (Rauss et al. 2011). The underlying assumption is that the visual system learns to suppress or enhance input even before it occurs. Particularly in the case of highly demanding perceptual tasks, such initial gating effects seem to emerge only after intervening periods of sleep (Bao et al. 2010; Pourtois et al. 2008).

In summary, theoretical and modeling work suggests close ties between predictive coding and memory formation during sleep. Empirical studies directly addressing this issue are almost entirely missing. However, there is first promising, albeit indirect evidence for the idea that sleep-dependent memory consolidation specifically supports the emergence of predictive codes.

Future Directions

Hierarchical predictive coding implies that based on statistical regularities in its inputs, the brain forms persisting representations of predictive codes to optimize future perception and action. In computational terms, this refers to the optimization of prior beliefs based on new information. An important reason to be interested in the long-term consolidation and optimization of prior beliefs lies in the potentially divergent evaluations of incoming information at different hierarchical levels. For example, passing John's house in the evening, one may not be surprised to see John unlocking his door. However, knowing that he left yesterday for a three-week holiday would make the same percept highly surprising. In other words, suppression of predictable information, as advocated by predictive-coding theories and their predecessors (Barlow 1985, 2001), cannot occur in a purely local fashion, and even relatively stable parameters ("This is John's house") should remain temporarily malleable ("But he's on vacation for three weeks").

Previous studies of hierarchical predictive coding have mostly relied on one-shot designs where the relevant probability distributions are encoded and employed on the fly. Consequently, the process of gradually forming such predictive representations and how this process is specifically influenced by sleep has so far been largely ignored. Based on indirect evidence, it is tempting to suggest that sleep does not only passively stabilize newly formed predictive codes. Active processes of systems consolidation during sleep, and specifically neural replay of representations that occurs during SWS might additionally help to refine predictive codes and to advance them towards lower levels of perceptual and motor processing. A central question that hierarchical predictive coding will have to answer is how the statistical regularities encountered during wakefulness are refined during sleep in order to retain only a transformed subset of these experiences which in turn should be optimal in a Bayesian sense for dealing with future events.

In the field of sleep and memory research, existing theories of active systems consolidation (Diekelmann and Born 2010) and synaptic homeostasis (Tononi and Cirelli 2014) might benefit from integrating principles of hierarchical predictive coding. Predictive-coding theory offers a conceptual and computational handle on the transformations that newly encoded information should undergo during sleep-dependent memory consolidation. Specifically, cortical representations optimized during sleep should subsequently exhibit Bayes-optimal patterns of baseline activity and equally optimal responses to prediction errors; which means a minimization of prediction errors and a delegation of perception and action to the lowest possible levels in the respective hierarchies (for a similar argument, cf. Ahissar and Hochstein 2004). Importantly, these effects can be traced down to changes in activity profiles across cortical layers (Bastos et al. 2012; Kok et al. 2016). In this context, questions of consolidation and re-consolidation of predictive codes and how sleep contributes to modifying pre-existing codes are of utmost interest (Dudai et al. 2015).

In terms of experimental approaches, research needs to move from examining the acquisition and prediction of fleeting regularities into investigations of how the brain builds and maintains generative models in natural environments. Here, the study of artificial category learning might serve as a promising intermediate step, because it is inherently probabilistic, and uncertainty associated with object boundaries can be systematically manipulated (e.g. Friedrich et al. 2015). Indeed, highlighting the role of uncertainty may turn out to be one of the most important contributions of predictive coding to research on sleep and memory. Predictive coding and other Bayesian theories of brain function require beliefs about states, parameters, and precisions to be represented in terms of probability distributions. However, uncertainty is rarely systematically considered or manipulated in studies of memory consolidation (but see Barsky et al. 2015). By highlighting the pivotal role of probabilistic representations for perception and action, hierarchical predictive coding may offer a unifying perspective on previous studies and suggest new experimental manipulations.

In summary, we have reviewed theoretical and empirical evidence that argues for an involvement of sleep in the extraction and optimization of predictive codes. However, given the paucity of currently available data, studies are needed which directly apply and scrutinize the concept of hierarchical predictive coding in research on sleep-dependent memory consolidation.

Acknowledgements We would like to thank Susanne Diekelmann for helpful suggestions on the manuscript.

References

Ahissar M, Hochstein S (2004) The reverse hierarchy theory of visual perceptual learning. Trends Cogn sci 8:457–464. doi:10.1016/j.tics.2004.08.011

Albouy G, Fogel S, Pottiez H, Nguyen VA, Ray L, Lungu O, Doyon J (2013) Daytime sleep enhances consolidation of the spatial but not motoric representation of motor sequence memory. PloS ONE, 8. doi:10.1371/journal.pone.0052805

Bannert MM, Bartels A (2013) Decoding the yellow of a gray banana. Curr Biol 23:2268–2272. doi:10.1016/j.cub.2013.09.016

Bao M, Yang L, Rios C, He B, Engel SA (2010) Perceptual learning increases the strength of the earliest signals in visual cortex. J Neurosci 30:15080–15084. doi:10.1523/JNEUROSCI.5703-09.2010

Barlow HB (1985) Cerebral cortex as model builder. In: Rose D, Dobson VG (eds) Models of the visual cortex, Wiley, Chichester, pp 37–46

Barlow HB (2001) Redundancy reduction revisited. Network, 12: 241–253

Barsky MM, Tucker MA, Stickgold R (2015) REM sleep enhancement of probabilistic classification learning is sensitive to subsequent interference. Neurobiol Learn Mem 122: 63–68. doi:10.1016/j.nlm.2015.02.015

Bartlett FC (1932) Remembering: a study in experimental and social psychology. Cambridge University Press, Cambridge

Bastos AM, Usrey WM, Adams RA, Mangun GR, Fries P, Friston KJ (2012) Canonical microcircuits for predictive coding. Neuron 76:695–711. doi:10.1016/j.neuron.2012.10.038

Clark A (2013) Whatever next? predictive brains, situated agents, and the future of cognitive science. Behav Brain Sci 181–253. doi:10.1017/S0140525X12000477

Cohen AL, Dixon RA, Lindsay DS, Masson ME (2003) The effect of perceptual distinctiveness on the prospective and retrospective components of prospective memory in young and old adults. Can J Exp Psychol 57(4):274–289

Cox R, Tijdens RR, Meeter MM, Sweegers CC, Talamini LM (2014) Time, not sleep, unbinds contexts from item memory. PLoS ONE 9(2):e88307. doi:10.1371/journal.pone.0088307

Dambacher M, Rolfs M, Göllner K, Kliegl R, Jacobs AM (2009) Event-related potentials reveal rapid verification of predicted visual input. PLoS ONE 4:e5047. doi:10.1371/journal.pone.0005047

Deese J (1959) On the prediction of occurrence of particular verbal intrusions in immediate recall. J Exp Psychol 58(1):17–22

Diekelmann S, Born J (2010) The memory function of sleep. Nat Rev Neurosci 11:114–126. doi:10.1038/nrn2762

Diekelmann S, Born J, Wagner U (2010a) Sleep enhances false memories depending on general memory performance. Behav Brain Res 208:425–429. doi:10.1016/j.bbr.2009.12.021

Diekelmann S, Wilhelm I, Wagner U, Born J (2010b) Elevated cortisol at retrieval suppresses false memories in parallel with correct memories. J Cogn Neurosci 23:772–781. doi:10.1162/jocn.2010.21493

Diekelmann S, Buchel C, Born J, Rasch B (2011) Labile or stable: opposing consequences for memory when reactivated during waking and sleep. Nat Neurosci 14:381–386. doi:10.1038/nn.2744

Diekelmann S, Wilhelm I, Wagner U, Born J (2013a) Sleep improves prospective remembering by facilitating spontaneous-associative retrieval processes. PLoS ONE 8(10):e77621. doi:10.1371/journal.pone.0077621

Diekelmann S, Wilhelm I, Wagner U, Born J (2013b) Sleep to implement an intention. Sleep 36 (1):149–153. doi:10.5665/sleep.2322

Djonlagic I, Rosenfeld A, Shohamy D, Myers C, Gluck M, Stickgold R (2009) Sleep enhances category learning. Learn Mem 16(12):751–755. doi:10.1101/lm.1634509

Drosopoulos S, Windau E, Wagner U, Born J (2007) Sleep enforces the temporal order in memory. PLoS ONE 2:e376. doi:10.1371/journal.pone.0000376

Drosopoulos S, Harrer D, Born J (2011) Sleep and awareness about presence of regularity speed the transition from implicit to explicit knowledge. Biol Psychol 86:168–173. doi:10.1016/j.biopsycho.2010.11.005

Dudai Y (2012) The restless engram: consolidations never end. Annu Rev Neurosci 35:227–247. doi:10.1146/annurev-neuro-062111-150500

Dudai Y, Karni A, Born J (2015) The consolidation and transformation of memory. Neuron 88 (1):20–32. doi:10.1016/j.neuron.2015.09.004

Einstein GO, McDaniel MA (1990) Normal aging and prospective memory. J Exp Psychol Learn Mem Cogn 16(4):717–726

Einstein GO, McDaniel MA, Thomas R, Mayfield S, Shank H, Morrisette N, Breneiser J (2005) Multiple processes in prospective memory retrieval: factors determining monitoring versus spontaneous retrieval. J Exp Psychol Gen 134(3):327–342. doi:10.1037/0096-3445.134.3.327

Ellenbogen JM, Hulbert JC, Stickgold R, Dinges DF, Thompson-Schill SL (2006) Interfering with theories of sleep and memory: sleep, declarative memory, and associative interference. Curr Biol: CB 16:1290–1294. doi:10.1016/j.cub.2006.05.024

Estes KG, Evans JL, Alibali MW, Saffran JR (2007) Can infants map meaning to newly segmented words? Statistical segmentation and word learning. Psychol Sci 18:254–260. doi:10.1111/j.1467-9280.2007.01885.x

Fenn KM, Gallo DA, Margoliash D, Roediger HL 3rd, Nusbaum HC (2009) Reduced false memory after sleep. Learn Mem 16(9):509–513. doi:10.1101/lm.1500808

Fischer S, Drosopoulos S, Tsen J, Born J (2006) Implicit learning–explicit knowing: a role for sleep in memory system interaction. J Cogn, 18, 311–319. doi:10.1162/jocn.2006.18.3.311

Fischer S, Diekelmann S, Born J (2011) Sleep's role in the processing of unwanted memories. J Sleep Res 20(2):267–274. doi:10.1111/j.1365-2869.2010.00881.x

Fiser J, Aslin RN (2002a) Statistical learning of higher-order temporal structure from visual shape sequences. J Exp Psychol Learn Mem Cogn 28:458–467. doi:10.1037//0278-7393.28.3.458

Fiser J, Aslin RN (2002b) Statistical learning of new visual feature combinations by infants. Proc Natl Acad Sci USA 99:15822–15826. doi:10.1073/pnas.232472899

Foster ER, McDaniel MA, Repovs G, Hershey T (2009) Prospective memory in Parkinson disease across laboratory and self-reported everyday performance. Neuropsychology 23(3):347–358. doi:10.1037/a0014692

Friedrich M, Wilhelm I, Born J, Friederici AD (2015) Generalization of word meanings during infant sleep. Nat Commun 6:6004. doi:10.1038/ncomms7004

Friston K (2005) A theory of cortical responses. Philos Trans R Soc B Biol Sci 360:815–836. doi:10.1098/rstb.2005.1622

Friston K, Kiebel S (2009) Predictive coding under the free-energy principle. Philos Trans R Soc B Biol Sci 364:1211–1221. doi:10.1098/rstb.2008.0300

Gais S, Plihal W, Wagner U, Born J (2000) Early sleep triggers memory for early visual discrimination skills. Nat Neurosci 3:1335–1339

Gais S, Albouy G, Boly M, Dang-Vu TT, Darsaud A, Desseilles M, Peigneux P (2007) Sleep transforms the cerebral trace of declarative memories. Proc Natl Acad Sci 104:18778–18783. doi:10.1073/pnas.0705454104

Gomez RL, Bootzin RR, Nadel L (2006) Naps promote abstraction in language-learning infants. Psychol Sci 17(8):670–674. doi:10.1111/j.1467-9280.2006.01764.x

Helmholtz H (1867) Handbuch der Physiologischen Optik. Leopold Voss, Leipzig

Hinton GE, Dayan P, Frey BJ, Neal RM (1995) The "wake-sleep" algorithm for unsupervised neural networks. Science (New York, N.Y.) 268:1158–1161. doi:10.1126/science.7761831

Hobson JA, Friston KJ (2012) Waking and dreaming consciousness: neurobiological and functional considerations. Prog Neurobiol 98:82–98. doi:10.1016/j.pneurobio.2012.05.003

Hosoya T, Baccus SA, Meister M (2005) Dynamic predictive coding by the retina. Nature 436: 71–77. doi:10.1038/nature03689

Hupe JM, James AC, Payne BR, Lomber SG, Girard P, Bullier J (1998) Cortical feedback improves discrimination between figure and background by V1, V2 and V3 neurons. Nature 394:784–787

Hupe JM, James AC, Girard P, Lomber SG, Payne BR, Bullier J (2001) Feedback connections act on the early part of the responses in monkey visual cortex. J Neurophysiol 85:134–145

Inostroza M, Born J (2013) Sleep for preserving and transforming episodic memory. Annu Rev Neurosci 36:79–102. doi:10.1146/annurev-neuro-062012-170429

Jeffreys DA, Axford JG (1972) Source locations of pattern-specific components of human visual evoked potentials. I. Component of striate cortical origin. Exp Brain Res 16:1–21

Jurewicz K, Cordi MJ, Staudigl T, Rasch B (2016) No evidence for memory decontextualization across one night of sleep. Front Human Neurosci 10. doi:10.3389/fnhum.2016.00007

Karni A, Sagi D (1991) Where practice makes perfect in texture discrimination: evidence for primary visual cortex plasticity. Proc Natl Acad Sci U S A 88:4966–4970

Karni A, Tanne D, Rubenstein BS, Askenasy JJ, Sagi D (1994) Dependence on REM sleep of overnight improvement of a perceptual skill. Science 265:679–682

Kliegel M, McDaniel MA, Einstein GO (2007) Prospective memory: cognitive, neuroscience, developmental, and applied perspectives. Lawrence Earlbaum Assoc Inc, Mahwah, NJ

Knill DC, Pouget A (2004) The Bayesian brain: the role of uncertainty in neural coding and computation. Trends Neurosci 27:712–719. doi:10.1016/j.tins.2004.10.007

Kok P, Bains LJ, van Mourik T, Norris DG, de Lange FP (2016) Selective activation of the deep layers of the human primary visual cortex by top-down feedback. Curr Biol 26(3):371–376. doi:10.1016/j.cub.2015.12.038

Lewis PA, Durrant SJ (2011) Overlapping memory replay during sleep builds cognitive schemata. Trends Cogn Sci 15:343–351. doi:10.1016/j.tics.2011.06.004

Lo JC, Sim SK, Chee MW (2014) Sleep reduces false memory in healthy older adults. Sleep 37(4): s 665–671, 671A. doi:10.5665/sleep.3564

Loftus EF (2003) Make-believe memories. Am Psychol 58:867–873. doi:10.1037/0003-066X.58.11.867

Maquet P (2001) The role of sleep in learning and memory. Science 294:1048–1052

McDaniel MA, Einstein GO (2000) Strategic and automatic processes in prospective memory retrieval: a multiprocess framework. Appl Cogn Psychol 14:S127–S144. doi:10.1002/acp.775

McDaniel MA, Guynn MJ, Einstein GO, Breneiser J (2004) Cue-focused and reflexive-associative processes in prospective memory retrieval. J Exp Psychol Learn Mem Cogn 30(3):605–614. doi:10.1037/0278-7393.30.3.605

Nissen MJ, Bullemer P (1987) Attentional requirements of learning—evidence from performance-measures. Cogn Psychol 19(1):1–32. doi:10.1016/0010-0285(87)90002-8

Oliver BM (1952) Efficient Coding. Bell Syst Tech J 31(4):724–750. doi:10.1002/j.1538-7305.1952.tb01403.x

Ordemann GJ, Opper J, Davalos D (2014) Prospective memory in schizophrenia: a review. Schizophr Res 155(1–3):77–89. doi:10.1016/j.schres.2014.03.008

Payne JD, Schacter DL, Propper RE, Huang L-W, Wamsley EJ, Tucker MA, Stickgold R (2009) The role of sleep in false memory formation. Neurobiol Learn Mem 92:327–334. doi:10.1016/j.nlm.2009.03.007

Peigneux P, Laureys S, Fuchs S, Collette F, Perrin F, Reggers J, Maquet P (2004) Are spatial memories strengthened in the human hippocampus during slow wave sleep? Neuron 44:535–545. doi:10.1016/j.neuron.2004.10.007

Pourtois G, Rauss KS, Vuilleumier P, Schwartz S (2008) Effects of perceptual learning on primary visual cortex activity in humans. Vision Res 48:55–62

Rao RPN, Ballard DH (1999) Predictive coding in the visual cortex: a functional interpretation of some extra-classical receptive-field effects. Nat Neurosci 2:79–87. doi:10.1038/4580

Rasch B, Born J (2013) About sleep's role in memory. Physiol Rev 93:681–766. doi:10.1152/physrev.00032.2012

Rasch B, Büchel C, Gais S, Born J (2007) Odor cues during slow-wave sleep prompt declarative memory consolidation. Science 315:1426–1429. doi:10.1126/science.1138581

Rauchs G, Feyers D, Landeau B, Bastin C, Luxen A, Maquet P, Collette F (2011) Sleep contributes to the strengthening of some memories over others, depending on hippocampal activity at learning. J Neurosci 31(7):2563–2568. doi:10.1523/JNEUROSCI.3972-10.2011

Rauss K, Pourtois G, Vuilleumier P, Schwartz S (2009) Attentional load modifies early activity in human primary visual cortex. Hum Brain Mapp 30:1723–1733. doi:10.1002/hbm.20636

Rauss K, Schwartz S, Pourtois G (2011) Top-down effects on early visual processing in humans: a predictive coding framework. Neurosci Biobehav Rev 35:1237–1253. doi:10.1016/j.neubiorev.2010.12.011

Rauss K, Pourtois G, Vuilleumier P, Schwartz S (2012) Effects of attentional load on early visual processing depend on stimulus timing. Hum Brain Mapp 33:63–74. doi:10.1002/hbm.21193

Roediger H, McDermott K (1995) Creating false memories—remembering words not presented in lists. J Exp Psychol Learn Mem Cogn 21:803–814. doi:10.1037/0278-7393.21.4.803

Saffran JR, Aslin RN, Newport EL (1996) Statistical learning by 8-month-old infants. Science 274:1926–1928. doi:10.1126/science.274.5294.1926

Saletin JM, Goldstein AN, Walker MP (2011) The role of sleep in directed forgetting and remembering of human memories. Cereb Cortex 21(11):2534–2541. doi:10.1093/cercor/bhr034

Schultz W, Dayan P, Montague PR (1997) A neural substrate of prediction and reward. Science 275:1593–1599. doi:10.1126/science.275.5306.1593

Scullin MK, McDaniel MA (2010) Remembering to execute a goal: sleep on it! Psychol Sci 21 (7):1028–1035. doi:10.1177/0956797610373373

Slotnick SD, Schacter DL (2004) A sensory signature that distinguishes true from false memories. Nat Neurosci 7:664–672. doi:10.1038/nn1252

Srinivasan MV, Laughlin SB, Dubs A (1982) Predictive coding: a fresh view of inhibition in the retina. Proc R Soc Lond Ser B, Containing Papers of a Biological Character. Royal Society (Great Britain), 216, 427–459

Strauss M, Sitt JD, King J-R, Elbaz M, Azizi L, Buiatti M, Dehaene S (2015) Disruption of hierarchical predictive coding during sleep. Proc Natl Acad Sci U S A 112:E1353–E1362. doi:10.1073/pnas.1501026112

Summerfield C, Egner T, Greene M, Koechlin E, Mangels J, Hirsch J (2006) Predictive codes for forthcoming perception in the frontal cortex. Science 314:1311–1314. doi:10.1126/science.1132028

Takashima A, Petersson KM, Rutters F, Tendolkar I, Jensen O, Zwarts MJ, Fernandez G (2006) Declarative memory consolidation in humans: a prospective functional magnetic resonance imaging study. Proc Natl Acad Sci U S A 103(3):756–761. doi:10.1073/pnas.0507774103

Tononi G, Cirelli C (2014) Sleep and the price of plasticity: from synaptic and cellular homeostasis to memory consolidation and integration. Neuron 81:12–34

Umeda S, Nagumo Y, Kato M (2006) Dissociative contributions of medial temporal and frontal regions to prospective remembering. Rev Neurosci 17(1–2):267–278

van Dongen EV, Thielen J-W, Takashima A, Barth M, Fernández G (2012) Sleep supports selective retention of associative memories based on relevance for future utilization. PLoS ONE 7:e43426. doi:10.1371/journal.pone.0043426

Wagner U, Gais S, Haider H, Verleger R, Born J (2004) Sleep inspires insight. Nature 427: 352–355. doi:10.1038/nature02223

Walker MP, Stickgold R, Alsop D, Gaab N, Schlaug G (2005) Sleep-dependent motor memory plasticity in the human brain. Neuroscience 133:911–917. doi:10.1016/j.neuroscience.2005.04.007

Wilhelm I, Diekelmann S, Molzow I, Ayoub A, Mölle M, Born J (2011) Sleep Selectively Enhances Memory Expected to Be of Future Relevance. J Neurosci 31:1563–1569. doi:10.1523/JNEUROSCI.3575-10.2011

Wilson MA, McNaughton BL (1994) Reactivation of hippocampal ensemble memories during sleep. Science 265:676–679

Yordanova J, Kolev V, Verleger R, Bataghva Z, Born J, Wagner U (2008) Shifting from implicit to explicit knowledge: different roles of early—and late-night sleep. Learn Mem 15:508–515. doi:10.1101/lm.897908

Yordanova J, Kolev V, Wagner U, Born J, Verleger R (2011) Increased alpha (8–12 Hz) activity during slow wave sleep as a marker for the transition from implicit knowledge to explicit insight. J Cogn Neurosci 24:119–132. doi:10.1162/jocn_a_00097

Emotional Memory Consolidation During Sleep

Tony J. Cunningham and Jessica D. Payne

Abstract Emotional experiences have a privileged place in our memories. As humans are constantly inundated with information, this adaptive process allows for the preservation of more important emotionally salient events, while allowing less important information to be forgotten. Although the experience of emotion itself releases a bevy of neurochemical and neurophysiological reactions that can benefit memory, recent research indicates that sleep plays a vital role in the preferential processing that leads to long-term consolidation of emotional information. Behavioral, neuroimaging, and neuropsychological evidence suggests that sleep benefits the long-term storage of emotional memories by increasing activation and connectivity among the same brain regions responsible for the initial encoding of emotional experience, and by facilitating the integration of the various elements of the experience into higher level cortical networks. This chapter will review the behavioral and neurobiological evidence supporting enhanced memory for emotionally salient information and the role that sleep plays in this preferential processing. We will conclude with questions that remain in this field, highlighting areas that may be fruitful for future sleep and memory researchers to explore.

Keywords Emotional memory · Memory consolidation · Sleep · Amygdala · Hippocampus · REM sleep

T.J. Cunningham · J.D. Payne (✉)
Department of Psychology, University of Notre Dame,
Haggar Hall, Room 122B, Notre Dame, IN 46556, USA
e-mail: jpayne7@nd.edu

T.J. Cunningham
e-mail: acunnin1@nd.edu

N. Axmacher and B. Rasch (eds.), *Cognitive Neuroscience of Memory Consolidation*, Studies in Neuroscience, Psychology and Behavioral Economics, DOI 10.1007/978-3-319-45066-7_9

Introduction

Emotional experiences shape the critical components of our life narratives. The integration of these experiences into our personal stories provides us with the scaffolding necessary to learn from our greatest successes as well as our most painful failures. These emotionally salient events enjoy a privileged place in our memories, such that the central components of our experience can be recalled later with ease and in vivid detail. Evolutionarily, the development of higher cognitive capacity and memory capabilities was vital for remembering positive information, such as the location of a regular food or water source, as well as experiences that induced fear or negative emotions, like remembering to avoid a cave that was previously found to be filled with snakes. As humans have become more reliant on socialization for survival, the role of the brain's memory structures have evolved to help us navigate the social world and be successful in forming relationships and avoid social gaffes. Through millennia of refinement, these systems have evolved to allow the formation of long-term emotional episodic memories, enabling us to recall and utilize the most critical information to survive and predict the future without being consumed by less relevant details (Buchanan and Adolphs 2002; Schacter and Addis 2007a, b).

Although the history of long-term memory and emotion research is extensive, new light continues to be shed on a variety of factors that influence their relationship (e.g. attention, stress, mental health, etc.; see also chapter by Meir Drexler and Wolf). In recent years, sleep has received growing attention as a critical state for the consolidation and storage of long term emotional memories. When we go to sleep, the brain does not simply shut down for the night or remain in a single state until morning. Instead, the sleeping brain is incredibly active, at times even more active than the awake brain (Nir and Tononi 2010), and research in the last 20 years suggests that this activity may be crucial (among other things) for the selective processing and consolidation of our emotional experiences, preferentially storing them over less pertinent information (Wagner et al. 2001; Hu et al. 2006). Furthermore, some of the latest research suggests that sleep may also be important for stripping the emotional tone from these memories so that we are able to recall the content of these experiences without lasting negative emotional impact (van der Helm et al. 2011; Cunningham et al. 2014b).

This chapter aims to lay the foundation for an understanding of sleep's role in emotional memory processing, beginning with a description of some of the evidence supporting enhanced long-term memory for emotional events and the neurobiological mechanisms that are currently theorized to underlie this process. We will then highlight evidence supporting the importance of sleep for successful emotional memory consolidation, and discuss how the physiology of sleep reactivates the neural systems that were engaged when these experiences were initially encountered. We will conclude with a brief review of the evidence suggesting that sleep plays a part in processing the affective reactivity initially tied to an emotionally negative and arousing experience, and a look ahead at the future of sleep and emotional memory research.

The Impact of Emotion on Memory

Since the earliest periods of recorded history, it is clear that humans had awareness that events in our past shape our future experiences and actions. The Roman Poet Ovid (43 BC–AD 17) wrote, "Pain makes you stronger, tears make you braver, and heartbreak makes you wiser. So thank the past for a better future." Since then, human fascination for how our past experiences are integrated into our memories for later recall has continued to flourish. In the modern era of research, exploration into the impact of emotionally salient experiences on long-term memory and behavior was largely restricted to animal behaviorists until the seminal paper by Kleinsmith and Kaplan (1963) broke into human research by reporting that high arousal during paired associates learning leads to enhanced long-term memory after a week delay compared to pairs learned under a low arousal condition. Research in this area in the subsequent five decades has been extensive and thorough, integrating a number of different types of stimuli and experimental designs to demonstrate the impact of emotion on memory (see Buchanan and Adolphs 2002; Hamann 2001; LaBar and Cabeza 2006 for reviews).

Emotion can influence memory processing in a variety of ways. Two of the most prominent methods include (a) testing memory for any kind of information following an induction of a certain mood, often through music or videos (similar to the above work by Kleinsmith and Kaplan), and (b) testing memory for emotionally salient information itself, regardless of mood at encoding. For the purpose of this review, we will be focusing primarily on the latter and explore how memory is impacted when the information to be tested carries some degree of emotional salience (e.g. an image of a vicious looking snake). Evidence suggests that even when in a euthymic mood state, emotionally salient information tends to be prioritized in memory. For instance, Cahill and McGaugh (1995) had participants watch a slide show while listening to a narrative about what was happening in the scenes. In this study, all participants viewed the exact same pictures, but participants were randomly assigned to listen to either a neutral or emotionally arousing version of the narrative. Both stories portrayed a mother taking her son to watch his father work in the hospital. In the negative version, the boy is badly injured on the way to the hospital and the surgeons must scramble to save his life. In the neutral version, the narration explains that hospital staff is conducting a practice disaster drill. After a 2 week delay, the subjects completed an unexpected memory test based on the content of the pictures and narration. Both free recall and recognition tests revealed that the group who listened to the arousing narrative had significantly better memory for the arousing portion of the slide show, while memory for the neutral portions were similar between groups. In another study exploring the impact of emotion on memory for words, Kensinger and Corkin (2003) presented 70 negative and 70 neutral words to participants and found that after a 15 min delay occupied by a distractor task, recognition memory was better for the negative over the neutral words. In a second experiment using the same design, the authors presented the words in different colors and asked the participants to recall what color each word was when originally viewed. Memory for the color of the word was also better for the emotionally negative word, indicating that emotional salience benefits both source and detail memory (Kensinger and Corkin 2003). In a study from Kensinger et al. (2006), participants viewed a series of negative and neutral images presented for various lengths of time (250, 500, and 1000 ms). After a two day delay, participants performed a surprise recognition test, and were asked to determine if the objects presented were the *same*, *similar*, or *new* compared to the objects presented during the initial presentation. The authors found that after this extended delay, memory for negatively arousing content was enhanced over memory for neutral objects, particularly when the items were shown for longer durations (500 or 1000 ms). Studies such as these indicate that emotional arousal benefits memory in part by facilitating the processes of encoding and consolidation across both short and long delays.

Not only are emotional memories more easily retained, but they are also harder to forget. In one study, participants were shown 60 neutral and 60 emotionally negative images, and the participants were instructed to remember half of each valence while forgetting the other half (Nowicka et al. 2010). They found that the recognition rate of emotionally negative to-be-forgotten images was higher compared to the neutral ones. Imaging data also showed that the intention to forget emotionally negative pictures elicited strong activation from a distributed neural network, while attempting to forget neutral image generated a much narrower neural response. From this, the authors concluded that intentional forgetting of emotional information is more difficult and requires more neural effort than forgetting of neutral information.

Neurobiology of Emotional Memory Consolidation

The results of the studies highlighted above motivate examination of the neurobiological underpinnings of emotional memory. The limbic system is considered to be the hub of memory activity as it is initially encoded and consolidated before being integrated into neocortical networks for long term storage. The limbic system can be further divided into several separate structures and systems, the two considered most important for emotional memory being the medial temporal lobe (MTL) memory system (which includes the hippocampus) and the amygdala (LaBar and Cabeza 2006). The MTL memory system is a critical structure for the formation of most types of long-term declarative memories, regardless of the valence or arousal (Huijgen and Samson 2015; Squire 1992). The most infamous case demonstrating the role of the MTL system in memory is Henry Molaison (patient H.M., 1926–2008). In 1953, Mr. Molaison had bilateral medial temporal lobectomy in an attempt to treat a severe epileptic disorder. While the seizures subsided substantially, the surgery left him with severe anterograde amnesia, or the inability to create new declarative memories (Scoville and Milner 1957; Corkin 2002). The case and sacrifice of H.M. inspired decades of research confirming importance of the MTL region for memory (see Nadel and Moscovitch 1997; Burgess et al. 2002; and Bird and Burgess 2008 for review).

While the hippocampus and the rest of the MTL memory system are critical for the consolidation of all forms of declarative memory, regardless of valence or arousal, the experience of an emotionally arousing event unleashes a bevy of additional neurological and physiological reactions in a human observer. Among them are stress hormones such as cortisol and norepinephrine (see the chapter by Meir Drexler and Wolf). When an emotional event is experienced, these stress-related neuromodulators flood the central nervous system, which in turn lead to increased activation in the amygdala, an almond-sized structure set directly adjacent to the MTL (Payne and Kensinger 2010). Evidence suggests that this amplified amygdala activation and its strong connectivity to the hippocampus (an area that has been shown to be important for memory consolidation) and other MTL and neocortical regions plays a particularly critical role in the enhanced consolidation of memory for emotionally salient information and experiences. In particular, during an emotional experience, activation of the amygdala is thought to modulate hippocampal activity, leading to enhanced memory for the event.

Similar to the initial determination of the structure-function relationships of the MTL and many other neural regions, studies of brain lesion patients create the foundation in understanding the role of the amygdala in cognition and memory. Bilateral amygdala lesions have been shown to reduce attentional focusing on the emotionally salient features of a stimulus or event (Adolphs et al. 2005). These bilateral (but not unilateral) lesions reduce patients' ability to abstract and remember the gist of aversive stimuli, while leaving memory for the details intact (Adolphs et al. 2001). In regards to evolutionary fitness, this would suggest that if someone with bilateral amygdala damage was attacked at a watering hole, he or she may remember to avoid that particular watering hole, but they may not be able to integrate this experience in a way that would remind them to be more wary at all watering holes in the future. Individuals with Urbach-Wiethe syndrome, a rare genetic disorder that leads to the bilateral calcification of the amygdala, also have shown impaired abilities in long-term memory of emotional stories, pictures, and words (Markowitsch et al. 1994; Cahill et al. 1995; Adolphs et al. 1997).

Imaging studies have also influenced our understanding of the role that the amygdala plays in the consolidation of emotional memories (see Strange and Dolan 2006; Dolcos et al. 2013 for review). For instance, Cahill et al. (1996) showed participants emotional and neutral films while brain activity was assessed using PET imaging, and found a correlation between increased right amygdala activation and enhanced memory for the emotionally arousing film. This correlation was not found in the neutral condition. In a replication and extension of this study, Hamann et al. (1999) found increased bilateral amygdala activity at encoding correlated with enhanced memory for both aversive and pleasant emotional stimuli, while again having no relationship with the successful consolidation of neutral information. Critically, for both the positive and negative stimuli, they found that the degree of amygdala response during encoding also significantly correlated with modulated activity in the corresponding hippocampal or parahippocampal region. This study

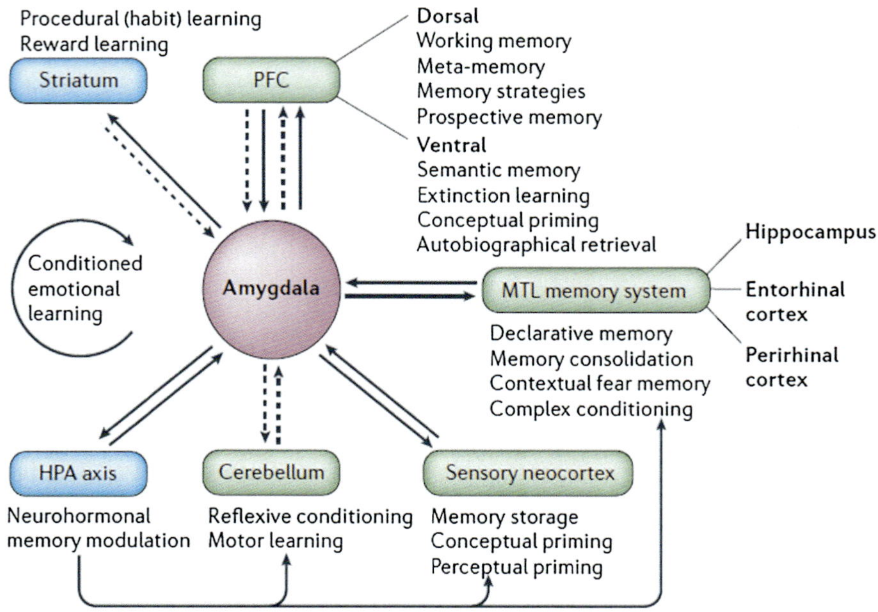

Fig. 1 Potential mechanisms in which the amygdala may impact emotional memory consolidation. In addition to impacting the MTL, neural projections from the amygdala may directly and indirectly influence several memory systems in the brain, including working memory, declarative memory and some forms of non-declarative memory. Blue labels indicate connections with subcortical structures. Dashed arrows indicate indirect connections. Solid arrows indicate direct connections (from LaBar and Cabeza 2006)

provided the first strong support for the memory-modulation theory, which suggests that the amygdala's influence on memory occurs by modulating activity in the areas of the brain (particularly the MTL) that are important for long-term memory consolidation (see Phelps 2004; Talmi 2013 for review; see Fig. 1). More recently, Ritchey et al. (2008) used functional magnetic resonance imaging (fMRI) to explore the contribution of the connectivity between the amygdala and MTL to the subsequent memory for emotional information various delays of various length. They again found that amygdala activity correlated with enhanced emotional, but not neutral memories, and that this pattern was similar across both short (20 min) and long (1 week delays). Interestingly, however, they found that the greater *connectivity* established between the amygdala and MTL regions when encoding emotional stimuli at encoding predicted better memory for the emotionally salient information after a long delay, but had less of an impact after the shorter delay. This suggests that the longer the delay is between encoding and retrieval, the more vital this connection between the amygdala and MTL becomes, lending further credence to the consolidation hypothesis. The interaction between the amygdala and MTL may be valence-dependent, with amygdala-MTL interaction being stronger for negative stimuli and MTL and prefrontal cortex (PFC) connectivity being stronger for positive stimuli (Ritchey et al. 2011).

Such studies have garnered considerable support for several key assumptions regarding the amygdala's role in the processing of emotional memory (see Box 1; Hamann 2001). The boost in consolidation processing for emotional experiences by the amygdala is so potent that altered neural traces can still be found a year following initial exposure (e.g. Mueller and Pizzagalli 2015). Importantly, many of the studies supporting these claims range in the timing of the delay between encoding and retrieval from a matter of minutes to several months or even years. As noted earlier, one critical component that may further influence the consolidation of emotional memory over time is sleep. The rest of this chapter will explore the latest research on the contribution of sleep to the consolidation of emotionally salient information.

Sleep and Emotional Memory Consolidation

While sleep has been shown to benefit most forms of declarative (Plihal and Born 1997; Peigneux et al. 2004) and non-declarative memories (Plihal and Born 1997; Walker et al. 2002; see chapters by Rauss and Born and by Schönauer and Gais), recent research has begun to highlight the particularly important role that sleep plays in the consolidation of emotionally salient memories. For example, Hu et al. (2006) created two matched sets of 90 emotionally arousing and 90 neutral scenes taken from the International Affective Picture System (IAPS), a standardized series of picture slides with previously normed emotional ratings (Lang et al. 1997). During the study phase, participants individually rated each of the slides on valence and arousal either in the evening or the morning, followed by a 12 h delay of wakefulness (following morning encoding) or a 12 h delay including a night of sleep (following evening encoding). Importantly, the authors used a within-subjects design such that each participant had a turn to be in both the morning and evening groups with the phases and lists counterbalanced. The two phases were also always at least one week apart. At the testing phase, all 180 pictures from the initial study session were presented along with 120 new pictures (50% arousing, 50% neutral), and the participants were asked to report if they had a conscious recollection of seeing the specific picture from the study session (a "Remember" judgment), knew that the picture was presented but could not recall any contextual information about its initial occurrence (a "Know" judgment), or thought the picture was entirely new (a "New" judgment). Across both conditions, more emotional images were more likely to be correctly identified as *remembered* or *known* than neutral images. When separated by condition however, participants that were given a chance to sleep were shown to have better memory for emotional images compared to those that remained awake, particularly for accurate *know* judgments, which increased by more than 40%. Additionally, after sleep, participants demonstrated a more *conservative* use of remember judgments compared to wakefulness, suggesting that sleep produces less indiscriminate responding, possibly through an enhancement in confidence for remembered emotional stimuli due to deepened memory traces (Hu et al. 2006).

The benefit of sleep for emotional memory consolidation has been shown to begin at a young age as well. In a similar study, negative and neutral IAPS images were shown to adolescents aged 10–13 in a within-subjects design such that each child had an opportunity to participate in both the sleep and wake groups (Prehn-Kristensen et al. 2009). Again, sleeping between sessions had a clear benefit specifically for emotional memory as shown by significant improvement in memory performance for emotional scenes after sleep, but no difference in memory for the neutral scenes between conditions. Thus, sleep's influence on emotional memory seems to be a critical feature that begins early in development.

One important consideration, however, is that emotional memories are often complex and comprised of both emotionally arousing and less important, neutral details. This provokes the question of whether sleep serves to bolster memory for all aspects of an emotional experience or if the sleeping brain is somehow able to selectively enhance memory just for the parts that it deems to be particularly important to remember. One ecologically-relevant example of this can be seen in the weapons-focus effect (Loftus et al. 1987; see Fig. 2), in which victims of a robbery frequently have excellent memory for the brandished weapon, but poor memory for the surrounding details, even including the face of their assailant. In a 2008 study, we explored this phenomenon by investigating how different components of complex negatively arousing and neutral scenes would change across delays of sleep and wakefulness (Payne et al. 2008). The goal of this study was to systematically determine if sleep would differently affect the consolidation of the emotional scenes compared to neutral and to determine if sleep would benefit the emotional scenes as single, intact units or if it would lead to a selective emphasis on the most salient information within the scene. To do this, participants were randomly assigned to come in for their initial session at 9 am or 9 pm. Participants were further separated into short delay (30 min) and long delay [12 h; wake (9 am) vs. sleep (9 pm)] groups. During the first session, all participants rated 64 complex scenes, 32 with a negative object placed on a neutral background and 32 with a neutral object on a neutral background. After their respective delays, each group completed an unexpected memory test in which the objects and backgrounds initially shown together were presented separately and one at a time (see Fig. 3). During this recognition task, participants saw a mix of *same*, *similar* and *new* negative and neutral objects and *same*, *similar*, and *new* backgrounds that had previously been paired with negative or neutral objects (unless new) and they were asked to determine which ones they had seen before. Compared to a delay of just 30 min,[1] daytime wakefulness led to forgetting of all components of the negative and neutral scenes in their entirety, with memory for the negative objects, neutral objects, and all backgrounds degrading similarly over time. While the sleep group

[1]Importantly, there we no performance differences between the 9 am and 9 pm 30 min delay groups, and thus they were collapsed.

Fig. 2 Pictorial representation of the weapon-focus effect. Eye-witnesses of a robbery frequently have excellent memory for the weapon but poor memory of other details surrounding the event, even about the assailant themselves

showed a similar pattern for neutral objects and both backgrounds, sleep led to a selective *increase* in negative object memory, even above and beyond performance just 30 min after encoding (Payne and Kensinger 2010; see Fig. 4). This "emotional memory tradeoff effect" suggests that rather than preserving all components of emotionally salient experiences, the sleeping brain is able to somehow determine what components of the scenes are most vital for consolidation and then "unbinds" the scenes to allow for additional processing of the emotionally salient information, deepening their memory traces, while allowing neutral background information to decay. In a 2014 follow up study, we again recruited individual sleep and wake groups to complete the same initial encoding session, but this time immediately following encoding we informed participants that their memory would be tested at the next session, thereby removing the unexpected nature of the test and creating a new level of expectation during the consolidation phase in both the sleep and wake groups (Cunningham et al. 2014a). Interestingly, when wake participants were aware of the pending memory test, they showed a dramatic *increase* in this trade-off memory pattern, such that their performance was now similar to the sleep group with superior memory for the emotional objects over their paired neutral backgrounds. In the sleep group, however, there was no significant difference in memory performance whether the memory test was expected or unexpected. This suggests that the sleeping brain is so proficient at identifying and "tagging" emotional memories for enhanced consolidation via memory triage processing (see Stickgold and Walker 2013 for review), that even the expectation of a memory test offers no further benefit (Cunningham et al. 2014a). The sleeping brain may be able to do it all on its own.

Fig. 3 Example encoding and recognition materials used in emotional memory tradeoff studies. Complex scenes are presented at encoding with negative or neutral objects placed on plausible neutral backgrounds. At recognition, objects and backgrounds are presented separately and one at a time along with similar lures and completely new foils to test for specificity of memory for the different scene components

This "tagging" may occur through a cascade of physiological and neurochemical processes. For instance, in another replication of the 'emotional memory tradeoff' effect, we collected measures of heart rate and skin conductance response during the encoding phase of the complex negative and neutral scenes (Cunningham et al. 2014b). We found correlations between increased measures of heart rate and skin conductance reactivity during encoding and subsequent memory for the negative objects at recognition. Critically, however, this relationship was found *only* if the delay period included sleep. Similarly, Bennion et al. (2013) measured baseline cortisol concentrations at encoding of the tradeoff task images. They found that

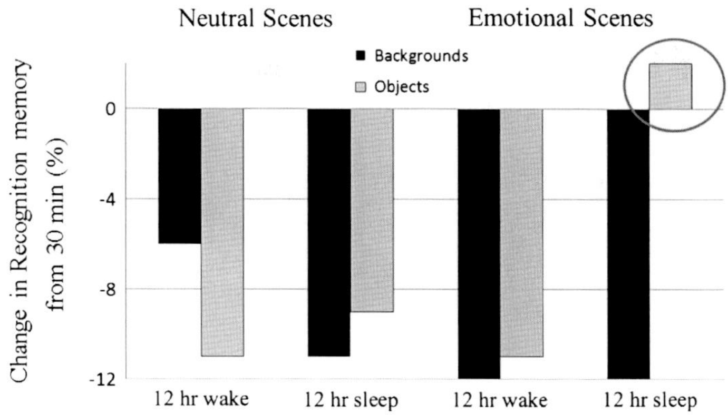

Fig. 4 Compared to a 30 min delay, Payne et al. (2008) found that memory for neutral objects and their corresponding backgrounds deteriorated at similar rates across both sleep and wake. Critically, however, while the wake group also showed deterioration for negative objects and their backgrounds, the sleep group showed *increased* memory for negative objects compared to a test just 30 min after encoding, yet typical deterioration of the neutral backgrounds on which the negative objects were originally presented. This suggests that sleep actively processes information during sleep, leading to enhanced memory for emotionally salient information compared to even a shorter delay without sleep

higher levels of baseline cortisol correlated with enhanced memory for the negative content, but again, only if the delay included sleep.

Together, these studies suggest that sleep interacts with the neurohormonal environment at encoding to prioritize emotional content for additional processing. These findings, however, raise another pertinent question: what are the processes that occur *during* sleep that benefits consolidation of emotionally salient information?

The Neurobiology of Sleep and Emotional Memory

The systems consolidation theory of declarative memories (both emotional and neutral) suggests that long-term memory storage involves a reorganization of memory traces such that neural representations of the event are transferred from the hippocampus to the neocortex (Takashima et al. 2006; Diekelmann and Born 2010; see chapters by Fernandez, by Genzel and Battaglia and by Nissen and colleagues). Recent evidence suggests that sleep may play an important role in modulating these cortical storage sites, particularly for memories with an emotional tone associated with them. In one seminal study, participants encoded emotionally arousing negative and positive pictures along with neutral pictures, and were then allowed to sleep normally or were sleep-deprived (Sterpenich et al. 2007). Three days later they were asked to return and complete a memory test in a functional MRI scanner.

They found that those who were allowed to sleep following encoding had better memory for negative, neutral, and positive scenes compared to the individuals who had been sleep deprived. Critically, compared to the sleep deprived group, participants who were allowed a normal night of sleep after encoding also had greater activation and connectivity between the hippocampus and cortical areas (including medial prefrontal cortex) during successful recollection of the emotional stimuli (Sterpenich et al. 2007). This distilment of neural correlates from diffuse neocortical networks to a refined network of regions with enhanced limbic-cortical connectivity (including the hippocampus, amygdala, and ventromedial prefrontal cortex; vmPFC) can occur in as little as 12 h if sleep is allowed during the delay (Payne and Kensinger 2011). After 6 months, Sterpenich et al. (2009) retested the participants in their earlier study and discovered another shift in neural traces such that successful recognition of the emotional scenes no longer required hippocampal activation, but instead relied largely on increased connectivity between the vmPFC and precuneus (areas important for retrieval: Henson et al. 2005; Cavanna and Trimble 2006) and the amygdala and occipital cortex (Sterpenich et al. 2009). This suggests that not only are the changes to neural representations of emotional memory generated by sleep lasting, but sleep may also be paramount in getting the amygdala involved in the modulation of the emotional memory network.

While these studies have begun to highlight the effect that a total night of sleep may have on the neurobiology of emotional memory, as noted in the *Introduction*, when we go to sleep our brains do not remain in a single, constant state. Instead, the brain moves through a cyclical pattern of states, each with its own host of neurological and neurochemical markers. Recent research has attempted to determine how these individual sleep stages may be related to behavioral outcomes (e.g. Wagner et al. 2001; Nishida et al. 2009). To understand this, we will take a brief look at what we know about these sleep stages.

Sleep Stages and Emotional Memory

As we drift off to sleep at night, we quickly move through a transitory stage of very light sleep (Stage 1) and enter into a slightly deeper stage known as Stage 2, marked by high frequency bursts in the sigma band, called sleep spindles, and K-complexes, or brief, high-amplitude negative peaks followed by a slower positive complex (see Smith et al. 2004 for review). From there, the first half of the night is dominated by a stage known as slow-wave sleep (SWS; see Fig. 5), characterized by high amplitude, low frequency brain wave oscillations, or delta waves (0.5–4 Hz), as well as a distinct change in neurochemical makeup (together Stages 1, 2, and SWS are referred to as NREM sleep; Steriade et al. 1993). In particular, cholinergic firing reduces so dramatically that acetylcholine levels become almost negligible (Hasselmo and McGaughy 2004). Noradrenergic and serotonergic firing also slow compared to waking levels, but they remain relatively active compared to the cholinergic systems (see Fig. 6). In the second half of the night, there is a shift

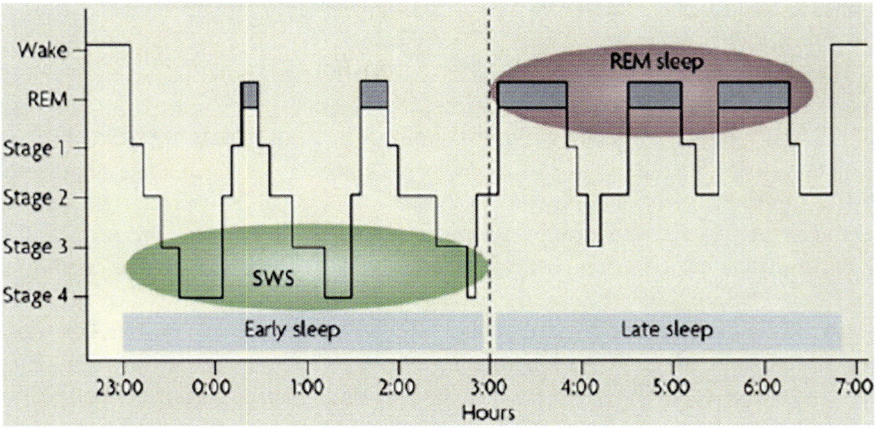

Fig. 5 The first half of the night is dominated by slow wave sleep (SWS), while the latter half of the night is REM sleep-rich (from Payne 2010)

Fig. 6 Graphical representation of varying concentrations of neuromodulators across a variety of brain states. 5-HT = Serotonin, ACh = Acetylcholine, NE = Norepinephrine

	Active Wake	Quiet Wake	SWS	REM
ACh	++	+	–	+++
NE 5-HT	++	+	+	–

away from the deeper SWS stage and an increase in the amount of the more active rapid-eye movement (REM) sleep stage (see Fig. 5). During REM sleep, the electroencephalographic (EEG) recordings show low-amplitude, high-frequency beta and theta waves (15–30 Hz), along with a dramatic decrease in muscle tone and periodic rapid eye movements (REMs; Carskadon and Dement 2000). Critically, during REM sleep the neurochemical milieu also changes such that cholinergic firing increases to levels even higher than waking, while noradrenergic and serotonergic systems take a turn being silent (Hasselmo 1999; see Fig. 6). Activity during REM sleep is also marked by substantial increase in activity in the amygdala and MTL regions of the brain (Braun et al. 1997; Maquet et al. 1996; Nofzinger et al. 1997).

The earliest studies attempting to parse out the differentiating roles of SWS and REM frequently relied on attempts to deprive participants of particular stages of sleep across the night (see Cunningham et al. 2014c). One study, for instance, found that REM-deprived (REM-D) participants were less able to recall words from a more threatening task compared to NREM-deprived individuals (NREM-D; Grieser et al. 1972). In a similar study, REM-D participants were less able to retrieve emotionally meaningful memories or stories from childhood compared to NREM-D and normal sleep controls (Greenberg et al. 1983), leading these early sleep research pioneers to theorize that REM sleep may have a role in the consolidation of emotionally-charged information.

Sleep stage deprivation studies quickly fell out of favor, however, due to the high levels of stress caused by repetitive arousals throughout the night (Born and Gais 2000). In an effort to reduce the stress induced by deprivation, the next wave of studies instead took advantage of the natural dichotomy of sleep stage dominance across the night mentioned above (Wagner et al. 2001). To do this, the authors randomly separated two groups of men into sleep and wake groups. In this within-subjects design, each participant was tested over two nights separated by at least one week. Participants in the sleep group completed an early and late sleep condition, with the expectation that early sleep would allow for more SWS while sleep later in the night would be dominated by REM. On the night they completed the early sleep condition, they were trained on emotional and neutral texts before going to sleep at approximately 11 pm, and then were aroused 3 h later for testing. For the late sleep condition, they were put to bed first, and then were aroused 3 h later (around 2 am) for initial learning of the texts. They were then allowed to go back to sleep for another 3 h before they were aroused again for testing (see Fig. 7). Wake participants followed the same design, except in between the learning and testing they were kept awake. Early and late conditions and the texts learned were counterbalanced to control for order effects. As the authors predicted, when the sleep participants slept later in the night they experienced nearly three times as much REM sleep as when they did the early sleep condition, and on the nights they slept earlier they had significantly more SWS. Behavioral results demonstrated that memory for the emotional texts was significantly better after the late, REM-rich sleep condition compared to both the early sleep condition and performance by the wake group, leading the authors to conclude that REM sleep is particularly important for the consolidation of emotional memory (Wagner et al. 2001). Interestingly, the authors interviewed the same participants four years later and tested their memory for the texts again (Wagner et al. 2006). The sleep group as a whole continued to outperform the wake group on recognition memory for the emotional texts, but this time there was no difference in memory performance for

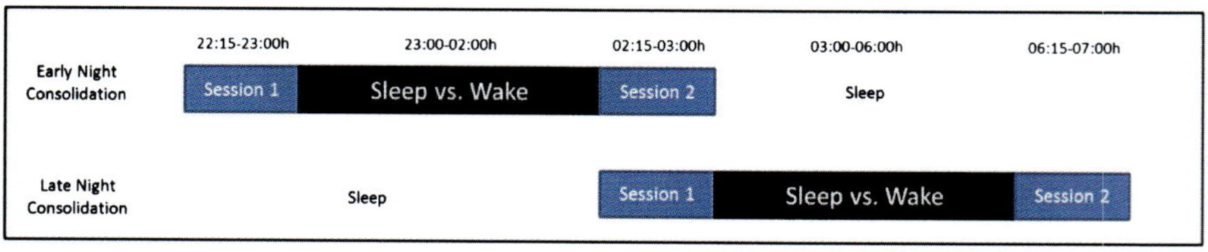

Fig. 7 Pictoral representation of the early sleep versus late sleep split night design used in Wagner et al. (2001). Early sleep participants complete the initial encoding task early in the evening and are then allowed 3 h of undisturbed, SWS-heavy early night sleep before being awoken and asked to complete the testing session at approximately 2:15 am. Late night sleep participants go to sleep first and are then awoken around 2:00 am to complete the initial encoding session. They are then given 3 h of late night sleep, typically dominated by REM sleep, before being awoken at approximately 6:00 am for testing

the emotional texts between the early and late sleep conditions. The authors suggest that this is likely because after testing was complete in the early condition, the participants were then allowed to go back to sleep, thereby receiving the second, REM-heavy period of sleep. Thus, they were still able to benefit from the selective consolidation during REM allowing them to perform equally well 4 years later (Wagner et al. 2006).

Since these initial studies by Wagner and colleagues, several studies have corroborated the theory that REM sleep is particularly important for emotional memory processing. For instance, Nishida et al. (2009) randomly separated participants into Nap and No-nap groups and then presented them with emotional and neutral picture stimuli. Following learning, those in the Nap group took a 90 min polysomnograph-recorded nap, while those in the No-nap group were asked to stay awake. Four hours after the initial learning, the participants were exposed to another set of negative and neutral pictures, and after a 15 min delay, their memory was tested for pictures shown during both the initial and secondary learning phases. Similar to previous findings, participants who were allowed to sleep between encoding and recognition demonstrated a selective consolidation enhancement for the emotional images over the neutral images. Critically, however, the extent of emotional memory facilitation negatively correlated with REM sleep latency (i.e. the faster participants entered REM, the greater the memory benefit) and positively correlated with the amount of REM sleep received during the nap and with right dominant prefrontal theta power during the REM sleep periods (Nishida et al. 2009). This increased theta band activity is of particular interest as previous studies have found a relationship between emotional processing and theta oscillations in limbic and prefrontal regions (Jones and Wilson 2005; Paré et al. 2002). In addition to measures of latency, duration, and theta power, other measures, such as REM sleep density, have been shown to be related to emotional memory performance as well (Gilson et al. 2015).

A recent replication of the "emotional memory tradeoff" effect discussed above also reported findings suggesting that REM sleep may have a particularly important role in facilitating selective memory processing for emotional components of scenes. In this study, we found that memory for the foreground emotional objects positively correlated with measures of overnight REM sleep duration (Payne et al. 2012). Thus, out of all of the various scene components tested, REM sleep obtained across a full night of sleep only predicted memory for the emotionally salient stimuli. Interestingly, when sleep occurred soon after encoding, this memory boost was stable even across longer delays (12–24 h), but participants that encoded in the morning and waited until evening to sleep showed an overall deterioration of this trade-off effect. This finding also combats the idea that sleep merely prevents interference as both the sleep-first and wake-first groups experienced roughly the same amount of waking interference.

Together, the studies highlighted above make a strong case for a vital role of REM sleep in deepening memory traces for emotional information over time. What then makes this stage of sleep so important for the processing of emotional memories?

Why might REM Sleep Benefit Emotional Memory Processing?

As previously mentioned, each stage of sleep has a unique neurochemical and electrophysiological fingerprint, and many theorize that the conditions during REM sleep create the perfect storm for emotional memory processing and consolidation. For instance, during REM sleep the limbic (including the amygdala) and medial temporal (including the hippocampus and surrounding cortex) areas of the brain are incredibly active, sometimes reaching levels of activity even greater than wakefulness (Nir and Tononi 2010). Behaviorally, this enhanced activation has been shown to lead to a broader emotional semantic network when awakened from REM sleep compared to NREM sleep (Carr and Nielsen 2015). Along with increased limbic activation, REM sleep is also characterized by a surge in synchronized oscillatory activity between these regions and neocortical areas, which are believed to be the regions in which these emotional memories become integrated for long-term memory storage (Paré et al. 2002; Jones and Wilson 2005). Not only is communication between these structures important for the modulation of affective experiences (Paré et al. 2002), but as previously discussed, sleep as a whole has been shown to modulate the neural traces of emotional memory leading to enhanced activation and connectivity between these regions during successful recognition of emotionally salient information (Sterpenich et al. 2007, 2009; Payne and Kensinger 2011). The enhanced connectivity between these structures during REM sleep has led many to theorize that REM sleep in particular may modulate plastic changes important for emotional memory consolidation (Maquet 1997; Nishida et al. 2009; Payne and Kensinger 2011).

Interestingly, some recent evidence supports the theory that REM sleep plays a role in the shift to limbic-neocortical networks for long-term storage of emotional memory traces. In another replication of the study by Wagner et al. (2001), participants were again shown negative and neutral pictures before 3 h of early, SWS-heavy sleep or 3 h of late, REM-rich sleep (Groch et al. 2013). This time, however, during recognition they used EEG to explore how event related potentials (ERPs) to correct responses would be altered across the two different sleep states. Participants in the late sleep group once again had enhanced memory for emotional pictures. Moreover, after the REM dominant period of sleep, correct identification of old, emotionally salient information was associated with increased late positive potential (LPP) of ERPs over regions of the frontal cortex, while there were no notable ERP markers to new emotional or any kind of neutral pictures. The early, SWS-rich group also did not demonstrate any changes to ERP patterns during recognition (Groch et al. 2013). While ERP research lacks the ability to determine specific neural structures involved, this study provides preliminary evidence that a period of REM-heavy sleep may lead to a shift in activation to include frontal regions during accurate recall of emotional information.

In addition to substantial neurophysiological differences, REM sleep is also characterized as having markedly elevated ACh levels in the limbic and forebrain regions of the brain, with some reporting approximately double the concentrations found during quiet wake and nearly 4 times greater than those found during SWS (Vazquez and Baghdoyan 2001; Marrosu et al. 1995). Importantly, ACh is known to be critical for the encoding and long-term consolidation of amygdala-dependent emotional memories (McGaugh 2004; Walker and van der Helm 2009), as made evident through previous experimental manipulations of ACh (see Power 2004 for review). This suggests that the increased cholinergic state found in REM sleep may be particularly beneficial for the selective enhancement in memory for emotionally salient experiences. Furthermore, REM sleep is also differentiated by the absence of aminergic (particularly noradrenergic) input (Pace-Schott and Hobson 2002). Thus, the enhanced processing of emotional stimuli without the presence of stress related neuromodulators may not only permit increased consolidation, but may help to reduce the autonomic charge that was initially associated with the emotional experience (Walker and van der Helm 2009). Together these unique attributes of REM sleep may allow for the integration of these emotional experiences into our life narratives, without creating lasting anxieties or negative impact.

In summation, a robust line of research has provided strong evidence suggesting that sleep, and possibly REM sleep in particular, may be important for the deepening of emotional memory traces through systems consolidation involving a shift from storage across a diffuse neural network to a more refined limbic-cortical network. Strengthening memory, however, may not be the only job that sleep has when processing emotionally salient information. Some recent research has begun to suggest that sleep may also play a role in reducing the affective tone initially associated with the emotional experience as well.

Sleep and Emotional Reactivity

Emotional experiences are necessarily shrouded in a layer of emotional reactivity, consistently eliciting increased subjective ratings of arousal and greater physiological responses compared to their neutral counterparts (Lang et al. 1993; Lang 1995). This increased arousal also leads to the activation of different neural substrates, such as the amygdala, which, as discussed, is also vital for the privileged memory processing that these experiences receive (Hamann 2001; and LaBar and Cabeza 2006). While the enhanced reactivity in response to an emotional event may be initially adaptive in preparing fight or flight mechanisms and tagging the information as important for preferential memory consolidation, having the same degree of affective response each time the event is called to mind would be maladaptive, and could potentially lead to increased anxiety and negative affect. Interestingly, a small body of literature supports the idea that sleep may play a dual role by enhancing memory consolidation of emotionally salient information while simultaneously reducing the affectivity initially associated with it (Greenberg et al. 1972; Gujar et al. 2011; Pace-Schott et al. 2011; van der Helm et al. 2011; Rosales-Lagarde et al. 2012; Cunningham et al. 2014b).

For instance, in a recent replication and extension of the emotional memory tradeoff effect, we showed participants the same complex scenes with negative and neutral foreground objects on neutral backgrounds, and tested their memories across delays of sleep and wake (Cunningham et al. 2014b). This time, however, we also collected measures of skin conductance response (SCR) and heart rate deceleration (HRD; a phasic response that maps onto the affective arousal of a stimulus) at both encoding and recognition. Those who slept showed a significant decrease in SCR and HRD activation from encoding to recognition, while those who remained awake stayed equally reactive (Cunningham et al. 2014b). Similarly, another study measured SCR, HRD, and facial electromyogram (EMG) during repeated intra-session viewing of negative scenes and explored how these responses changed across nap- or wake-filled delays (Pace-Schott et al. 2011). Participants in the nap group showed greater SCR and EMG habituation to the previously viewed emotional stimuli compared to those who stayed awake, but this time the authors found no difference in HRD measures between the groups. Interestingly, despite the decrease in physiological reactivity measures, subjective ratings of arousal remained unchanged in the nap group (Pace-Schott et al. 2011).

Similar to emotional memory consolidation, REM sleep may also aid the processing of emotional reactivity. REM sleep deprivation studies were again the first line of inquiry, with multiple studies showing that REM-D groups showed enhanced reactivity compared to NREM-D or control groups, leading the experimenters to conclude that REM must be important to behaviorally and neurally regulate our emotional responsiveness to stored memories (Greenberg et al. 1972; Rosales-Lagarde et al. 2012). In a nap study, participants who slept after viewing affective faces showed a decrease in ratings of fear expressions, and sleep stage analysis revealed that those who obtained REM sleep during the nap had the most pronounced reductions toward the fearful faces.[2] In perhaps the most seminal study on this topic, emotional pictures were shown to participants during fMRI scanning before and after a night of polysomnograph-recorded nocturnal sleep or a matched period of wakefulness (van der Helm et al. 2011). The authors found that both amygdala activation and subjective ratings of emotional intensity significantly depotentiated across a night of sleep, while the wake group experienced no change. They also found increased connectivity between the ventromedial PFC (an area known to have top-down inhibitory effects on the amygdala; Sotres-Bayon et al. 2004) only after sleep. Critically, reduced REM gamma EEG activity (a marker of reduced central norepinephrine activity) correlated with decreases in both subjective and amygdala reactivity, leading the authors to conclude that REM sleep physiology is linked to the dissipation of brain activation and behavioral responses across sleep (van der Helm et al. 2011).

[2]Interestingly, participants who achieved REM during their naps also showed the most significant enhancements toward happy face stimuli, perhaps indicating a valence-dependent system. Few other studies have been able to replicate this effect, however.

These studies have given rise to the *Sleep to Forget, Sleep to Remember* (SFSR) theory of emotional memory processing (see Walker and van der Helm 2009; van der Helm and Walker 2012; Cunningham et al. 2014c for review). According to this idea, an emotional memory includes both a memory and an affective tone. When the experience is first integrated into our experience, it is wrapped in a "blanket" of emotional salience that initially helps consolidate the memory. Over time, however, sleep chips away at these layers of emotion until only the content for the memory remains (van der Helm and Walker 2012). Van der Helm and Walker (2012) theorize that the same neurobiological milieu of REM sleep discussed above offers the prime state for the decoupling of emotion from memory. Thus, during REM sleep we sleep to "forget" the emotional tone while simultaneously sleeping to "remember" the content of the experience itself. Without this dual processing during sleep, emotional experiences may lead to longer lasting, negative impacts, perhaps impacting our mental health (Walker and van der Helm 2009).

Sleep and Emotional Memory: An Incomplete Story

While the story of sleep and emotional memory discussed above is both empirically supported and currently the most popular theory, the final conclusion is far from complete. Few would argue against the assertion that sleep as a whole benefits memory. Numerous studies have convincingly shown that a consolidation period filled with sleep leads to enhanced memory performance compared to periods of wakefulness (see Rasch and Born 2013), particularly when the material has some degree of emotional salience (see Walker and van der Helm 2009; Walker 2009). The active system consolidation hypothesis assumes that sleep is an active process that creates optimal conditions to facilitate the consolidation and stabilization of memories into long-term memory stores (Diekelmann and Born 2010; Born and Wilhelm 2012). However, many of the more intricate details of how this processing occurs remain hotly contested, especially when considering the preferential consolidation of emotional stimuli.

For instance, despite the enhanced activation of limbic networks during REM and the impressive list of studies above that have reported a relationship between REM sleep and emotional memory performance, several studies have been unable to find a specific REM sleep effect, even despite the usual benefit to emotional memory across sleep (e.g. Baran et al. 2012; Bennion et al. 2013; Cunningham et al. 2014b; Payne et al. 2015). Moreover, the last 5 years has seen a rise in studies reporting a positive relationship between emotional memory and SWS (Groch et al. 2011; Payne et al. 2015; Alger et al. 2014). The low concentration of ACh during SWS has been shown to be critical for the enhancement of hippocampus-dependent declarative memory across sleep (Gais and Born 2004), while the presence of norepinephrine during SWS has been shown to be important for preserving some aspects of emotional events (Groch et al. 2011). Evidence has even been reported suggesting that sleep spindles during Stage 2 sleep may indirectly support the consolidation of emotionally salient contextual information by suppressing neutral contexts (Cairney et al. 2014). Additionally, when considering the role that sleep plays in modulating the affective tone associated with these memories the picture becomes even murkier, with some reporting findings that REM sleep may actually serve to protect (Baran et al. 2012; Groch et al. 2013) or even potentiate (Wagner et al. 2002; Lara-Carrasco et al. 2009) emotional reactivity over time.

While these results may appear to be in conflict, many have instead integrated these findings into new theories of sleep and emotional memory processing, such as the "dual process" (see Ackermann and Rasch 2014) and the "sequential" (see Walker and Stickgold 2010) hypotheses of sleep-dependent memory consolidation. The shared concept between these theories is that instead of working in isolation, the individual stages of sleep are in fact working in *collaboration* to deepen the memory traces for emotionally salient information. Thus, instead of being completely reliant on REM sleep to consolidate emotional information, it may actually be a more complex dialogue between sleep stages that allows for this preferential processing to occur.

Some evidence supporting these theories of collaboration between sleep stages have already begun to arise. For instance, Cairney et al. (2014) had participants encode positive, negative, and neutral images 24 h before a recognition test in an fMRI machine. They found that SWS during the intervening night was correlated with enhanced memory for remote negative images as well as reduced activation in the right hippocampus during the recollection of these items. REM sleep, meanwhile, predicted an overnight increase in connectivity between the hippocampus and neocortex, associated with negative memory performance. Together the authors conclude that SWS and REM sleep serve distinct but complementary functions to facilitate emotional memory consolidation (Cairney et al. 2014). Another study using the early SWS-rich vs. late REM-heavy paradigm found that REM sleep may facilitate the consolidation of emotional *item* memory, while SWS may play a role in the consolidation of *context* information associated with the item (Groch et al. 2015).

Neurophysiological explanations of how each of the stages may contribute to emotional memory processing have only begun to be explored. The high levels of limbic activity, neocortical connectivity, and cholinergic conditions during REM sleep certainly continue to be germane when considering the partial role that REM sleep plays in the processing of emotional events. In regards to SWS, the presence of norepinephrine during SWS has been shown to be important for preserving the temporal order of emotional events (Groch et al. 2011). While research on the role of dopamine (DA) in emotional memory consolidation is relatively new, early studies suggest that DA may also play a role in enhancing emotional memory (Mehta et al. 2005). Interestingly, DA levels have been shown to be higher during REM sleep than other stages of sleep (Maloney et al. 2002; Lena et al. 2005), warranting further investigation into how DA and other neurohormones may be involved in the selective enhancement of emotional content across sleep.

These studies offer an excellent framework for future research to be built upon. The conflicting sleep stage data suggests that we cannot simply place behavioral functions into certain boxes in which only a single stage of sleep is responsible for them, but instead we need to take a big picture approach and explore how these stages may actually be working in cooperation to enhance both emotional and other forms of memory across periods of wakefulness. It will be critical for the next generation of sleep and memory researchers to continue to develop creative ways to separate the individual contributions of each stage of sleep while simultaneously developing a model for how these processes may work in conjunction with the other stages of sleep. Additionally, simply calculating the amount of time spent in each stage of sleep may not be enough to fully appreciate sleep's role in the processing of emotional memory over time. Understanding deeper measures, such as power spectra, spindle/theta density, and hippocampal sharp-wave ripple complexes, will be helpful in unlocking the true nature behind the role of sleep in emotional memory processing (e.g. Louie and Wilson 2001; Mölle et al. 2006; Nishida et al. 2009; Payne et al. 2015; Siapas and Wilson 1998; Sirota et al. 2003; Steriade et al. 1993; see also chapters by Bergmann and Staresina and by Maier and Kempter).

Summary

For millennia humans believed that emotional experiences played a formative role in the development of our character and life narratives. This belief has garnered significant empirical support as emotion has been shown to generate a host of neurophysiological and neurochemical changes that enhance activation and connectivity among regions of the brain important for emotional memory consolidation and storage. More recently, sleep has been shown to be vital in facilitating this preferential processing of emotionally salient events over more neutral ones. The different stages of sleep have different neurochemical milieus and are associated with different brain activity patterns, involving regions such as the amygdala and MTL, which may create the optimal conditions for integrating initially labile memories into higher level neocortical networks for long term memory storage. While there is a large body of evidence suggesting that sleep as a whole benefits memory, it remains unclear precisely which sleep cycle attributes and processes are important for emotional memory processing and storage. In order to fully understand the impact of sleep on emotional experiences, it will be critical for future sleep and memory researchers to develop new and creative techniques to distinguish the individual and interactive contributions that sleep stages, and their associated plastic events (e.g. sleep spindles, hippocampal sharp-wave complexes, the theta rhythm) play in integrating these life-defining events into our memories and personal stories.

Box 1 Key assumptions about the role of the amygdala in emotional memory processing:

1. The amygdala is the primary orchestrator of processes of emotional memory, without which emotional effects on memory cannot occur.
2. The amygdala can affect explicit memory by modulating or enhancing the activity of other brain regions involved in memory.
3. Emotional arousal can affect explicit memory through the release of stress hormones that interact with the amygdala
4. The modulatory influence of emotional arousal via the amygdala acts specifically on consolidation processes in memory regions such as the hippocampus.

From Hamann (2001)

Acknowledgements National Science Foundation (*BCS-0963581*) *and* (*BCS-1539361*).
Conflicts of Interest: The authors declare no competing financial interests.

References

Ackermann S, Rasch B (2014) Differential effects of non-REM and REM sleep on memory consolidation? Curr Neurol Neurosci Rep 14(2):1–10

Adolphs R, Cahill L, Schul R, Babinsky R (1997) Impaired declarative memory for emotional material following bilateral amygdala damage in humans. Learn Mem 4:291–300

Adolphs R, Denburg NL, Tranel D (2001) The amygdala's role in long-term declarative memory for gist and detail. Behav Neurosci 115(5):983

Adolphs R, Tranel D, Buchanan T (2005) Amygdala damage impairs emotional memory for gist but not details of complex stimuli. Nat Neurosci 8:512–518

Alger S, Chambers A, Payne JD (2014). A nap rich in slow wave sleep selectively preserves emotional scene components. Poster session presented at the meeting of the Associated Professional Sleep Societies, LLC, Minneapolis, MN

Baran B, Pace-Schott EF, Ericson C, Spencer RM (2012) Processing of emotional reactivity and emotional memory over sleep. J Neurosci 32(3):1035–1042

Bennion KA, Steinmetz KRM, Kensinger EA, Payne JD (2013) Sleep and cortisol interact to support memory consolidation. Cereb Cortex bht255

Bird CM, Burgess N (2008) The hippocampus and memory: insights from spatial processing. Nat Rev Neurosci 9(3):182–194

Born J, Gais S (2000) REM sleep deprivation: the wrong paradigm leading to wrong conclusions. Behav Brain Sci 23(06):912–913

Born J, Wilhelm I (2012) System consolidation of memory during sleep. Psychol Res 76(2):192–203

Braun AR, Balkin TJ, Wesenten NJ, Carson RE, Varga M, Baldwin P, Selbie S, Belenky G, Herscovitch P (1997) Regional cerebral blood flow throughout the sleep-wake cycle. An H2 (15) O PET study. Brain 120(7):1173–1197

Buchanan TW, Adolphs R (2002) The role of the human amygdala in emotional modulation of long-term declarative memory. Adv Conscious Res 44:9–34

Burgess N, Maguire EA, O'Keefe J (2002) The human hippocampus and spatial and episodic memory. Neuron 35(4):625–641

Cahill L, Babinsky R, Markowitsch HJ, McGaugh JL (1995) The amygdala and emotional memory. Nature 377:295–296

Cahill L, Haier RJ, Fallon J, Alkire MT, Tang C, Keator D, Wull J, Mcgaugh JL (1996) Amygdala activity at encoding correlated with long-term, free recall of emotional information. Proc Natl Acad Sci 93(15):8016–8021

Cahill L, McGaugh JL (1995) A novel demonstration of enhanced memory associated with emotional arousal. Conscious Cogn 4(4):410–421

Cairney SA, Durrant SJ, Power R, Lewis PA (2014) Complimentary roles of slow-wave sleep and rapid eye movement sleep in emotional memory consolidation. Cereb Cortex bht349

Carr M, Nielsen T (2015) Morning rapid eye movement sleep naps facilitate broad access to emotional semantic networks. Sleep 38(3):433–443

Carskadon MA, Dement WC (2000) Normal human sleep: an overview. Principles Pract Sleep Med 4:13–23

Cavanna AE, Trimble MR (2006) The precuneus: a review of its functional anatomy and behavioural correlates. Brain 129(3):564–583

Corkin S (2002) What's new with the amnesic patient HM? Nat Rev Neurosci 3(2):153–160

Cunningham TJ, Chambers AM, Payne JD (2014a) Prospection and emotional memory: how expectation affects emotional memory formation following sleep and wake. Front Psychol 5 (862)

Cunningham TJ, Crowell CR, Alger SE, Kensinger EA, Villano MA, Mattingly SM, Payne JD (2014b) Psychophysiological arousal at encoding leads to reduced reactivity but enhanced emotional memory following sleep. Neurobiol Learn Mem 114:155–164

Cunningham TJ, Pardilla-Delgado E, Alger S, Payne J (2014c) The role of REM sleep in emotional memory and affective reactivity in humans. In: Saylor C (ed) REM sleep: characteristics, disorders and physiological effects. Nova Science Publishers, New York

Diekelmann S, Born J (2010) The memory function of sleep. Nat Rev Neurosci 11(2):114–126

Dolcos F, Iordan AD, Kragel J, Stokes J, Campbell R, McCarthy G, Cabeza R (2013) Neural correlates of opposing effects of emotional distraction on working memory and episodic memory: an event-related FMRI investigation. Front Psychol 4:293. doi:10.3389/fpsyg.2013.00293

Gais S, Born J (2004) Low acetylcholine during slow-wave sleep is critical for declarative memory consolidation. Proc Natl Acad Sci U S A 101(7):2140–2144

Gilson M, Deliens G, Leproult R, Bodart A, Nonclercq A, Ercek R, Peigneux P (2015) REM-enriched naps are associated with memory consolidation for sad stories and enhance mood-related reactivity. Brain Sci 6(1):1

Greenberg R, Pearlman C, Schwartz WR, Grossman HY (1983) Memory, emotion, and REM sleep. J Abnorm Psychol 92(3):378

Greenberg R, Pillard R, Pearlman C (1972) The effect of dream (stage REM) deprivation on adaptation to stress. Psychosom Med 34(3):257–262

Grieser C, Greenberg R, Harrison RH (1972) The adaptive function of sleep: the differential effects of sleep and dreaming on recall. J Abnorm Psychol 80(3):280

Groch S, Wilhelm I, Diekelmann S, Born J (2013) The role of REM sleep in the processing of emotional memories: evidence from behavior and event-related potentials. Neurobiol Learn Mem 99:1–9

Groch S, Wilhelm I, Diekelmann S, Sayk F, Gais S, Born J (2011) Contribution of norepinephrine to emotional memory consolidation during sleep. Psychoneuroendocrinology 36(9):1342–1350

Groch S, Zinke K, Wilhelm I, Born J (2015) Dissociating the contributions of slow-wave sleep and rapid eye movement sleep to emotional item and source memory. Neurobiol Learn Mem 122:122–130

Gujar N, McDonald SA, Nishida M, Walker MP (2011) A role for REM sleep in recalibrating the sensitivity of the human brain to specific emotions. Cereb Cortex 21(1):115–123

Hamann S (2001) Cognitive and neural mechanisms of emotional memory. Trends Cogn Sci 5 (9):394–400

Hamann SB, Ely TD, Grafton ST, Kilts CD (1999) Amygdala activity related to enhanced memory for pleasant and aversive stimuli. Nat Neurosci 2(3):289–293

Hasselmo ME (1999) Neuromodulation: acetylcholine and memory consolidation. Trends Cogn Sci 3(9):351–359

Hasselmo ME, McGaughy J (2004) High acetylcholine levels set circuit dynamics for attention and encoding and low acetylcholine. Acetylcholine Cereb Cortex 24:207

Henson RN, Hornberger M, Rugg MD (2005) Further dissociating the processes involved in recognition memory: an FMRI study. J Cogn Neurosci 17(7):1058–1073

Hu P, Stylos-Allan M, Walker MP (2006) Sleep facilitates consolidation of emotional declarative memory. Psychol Sci 17(10):891–898

Huijgen J, Samson S (2015) The hippocampus: a central node in a large-scale brain network for memory. Revue Neurologique 171(3):204–216

Jones MW, Wilson MA (2005) Phase precession of medial prefrontal cortical activity relative to the hippocampal theta rhythm. Hippocampus 15(7):867

Kensinger EA, Corkin S (2003) Memory enhancement for emotional words: are emotional words more vividly remembered than neutral words? Mem Cogn 31(8):1169–1180

Kensinger EA, Garoff-Eaton RJ, Schacter DL (2006) Memory for specific visual details can be enhanced by negative arousing content. J Mem Lang 54(1):99–112

Kleinsmith LJ, Kaplan S (1963) Paired-associate learning as a function of arousal and interpolated interval. J Exp Psychol 65(2):190

LaBar KS, Cabeza R (2006) Cognitive neuroscience of emotional memory. Nat Rev Neurosci 7 (1):54–64

Lang PJ (1995) The emotion probe: studies of motivation and attention. Am Psychol 50(5):372

Lang PJ, Bradley MM, Cuthbert BN (1997) International affective picture system (IAPS): technical manual and affective ratings. NIMH Cent Study Emot Attention 39–58

Lang PJ, Greenwald MK, Bradley MM, Hamm AO (1993) Looking at pictures: Affective, facial, visceral, and behavioral reactions. Psychophysiology 30(3):261–273

Lara-Carrasco J, Nielsen TA, Solomonova E, Levrier K, Popova A (2009) Overnight emotional adaptation to negative stimuli is altered by REM sleep deprivation and is correlated with intervening dream emotions. J Sleep Res 18(2):178–187

Lena I, Parrot S, Deschaux O, Muffat-Joly S, Sauvinet V, Renaud BEEA, Gottesmann C (2005) Variations in extracellular levels of dopamine, noradrenaline, glutamate, and aspartate across the sleep–wake cycle in the medial prefrontal cortex and nucleus accumbens of freely moving rats. J Neurosci Res 81(6):891–899

Loftus EF, Loftus GR, Messo J (1987) Some facts about "weapon focus". Law Hum Behav 11 (1):55

Louie K, Wilson MA (2001) Temporally structured replay of awake hippocampal ensemble activity during rapid eye movement sleep. Neuron 29(1):145–156

Maloney KJ, Mainville L, Jones BE (2002) c-Fos expression in dopaminergic and GABAergic neurons of the ventral mesencephalic tegmentum after paradoxical sleep deprivation and recovery. Eur J Neurosci 15(4):774–778

Maquet P (1997) REM sleep and amygdala. Mol Psychiatry 2(3)

Maquet P, Péters JM, Aerts J, Delfiore G, Degueldre C, Luxen A, Franck G (1996) Functional neuroanatomy of human rapid-eye-movement sleep and dreaming. Nature 383(6596):163–166

Markowitsch HJ, Calabrese P, Würker M, Durwen HF, Kessler J, Babinsky R, Brechtelsbauer D, Heuser L, Gehlen W (1994) The amygdala's contribution to memory-a study on two patients with Urbach-Wiethe disease. Neuroreport 5(11):1349–1352

Marrosu F, Portas C, Mascia MS, Casu MA, Fà M, Giagheddu M, Imperato A, Gessa GL (1995) Microdialysis measurement of cortical and hippocampal acetylcholine release during sleep-wake cycle in freely moving cats. Brain Res 671(2):329–332

McGaugh JL (2004) The amygdala modulates the consolidation of memories of emotionally arousing experiences. Annu Rev Neurosci 27:1–28

Mehta MA, Hinton EC, Montgomery AJ, Bantick RA, Grasby PM (2005) Sulpiride and mnemonic function: effects of a dopamine D2 receptor antagonist on working memory, emotional memory and long-term memory in healthy volunteers. J Psychopharmacol 19(1):29–38
Mölle M, Yeshenko O, Marshall L, Sara SJ, Born J (2006) Hippocampal sharp wave-ripples linked to slow oscillations in rat slow-wave sleep. J Neurophysiol 96(1):62–70
Mueller EM, Pizzagalli DA (2015) One-year-old fear memories rapidly activate human fusiform gyrus. Soc Cogn Affect Neurosci nsv122
Nadel L, Moscovitch M (1997) Memory consolidation, retrograde amnesia and the hippocampal complex. Curr Opin Neurobiol 7(2):217–227
Nir Y, Tononi G (2010) Dreaming and the brain: from phenomenology to neurophysiology. Trends Cogn Sci 14(2):88–100
Nishida M, Pearsall J, Buckner RL, Walker MP (2009) REM sleep, prefrontal theta, and the consolidation of human emotional memory. Cereb Cortex 19(5):1158–1166
Nofzinger EA, Mintun MA, Wiseman M, Kupfer DJ, Moore RY (1997) Forebrain activation in REM sleep: an FDG PET study. Brain Res 770(1):192–201
Nowicka A, Marchewka A, Jednoróg K, Tacikowski P, Brechmann A (2010) Forgetting of emotional information is hard: an fMRI study of directed forgetting. Cereb Cortex bhq117
Pace-Schott EF, Hobson JA (2002) The neurobiology of sleep: genetics, cellular physiology and subcortical networks. Nat Rev Neurosci 3(8):591–605
Pace-Schott EF, Shepherd E, Spencer RM, Marcello M, Tucker M, Propper RE, Stickgold R (2011) Napping promotes inter-session habituation to emotional stimuli. Neurobiol Learn Mem 95(1):24–36
Paré D, Collins DR, Pelletier JG (2002) Amygdala oscillations and the consolidation of emotional memories. Trends Cogn Sci 6(7):306–314
Payne JD (2010) Memory consolidation, the diurnal rhythm of cortisol, and the nature of dreams: a new hypothesis. Int rev neurobiol 92:101–134
Payne JD, Chambers AM, Kensinger EA (2012) Sleep promotes lasting changes in selective memory for emotional scenes. Front Integr Neurosci 6(108):1–11
Payne JD, Kensinger EA (2010) Sleep's role in the consolidation of emotional episodic memories. Curr Dir Psychol Sci 19(5):290–295
Payne JD, Kensinger EA (2011) Sleep leads to changes in the emotional memory trace: evidence from FMRI. J Cogn Neurosci 23(6):1285–1297
Payne JD, Kensinger EA, Wamsley EJ, Spreng RN, Alger SE, Gibler K, Stickgold R (2015) Napping and the selective consolidation of negative aspects of scenes. Emotion 15(2):176
Payne JD, Swanberg K, Stickgold R, Kensinger EA (2008) Sleep preferentially enhances memory for emotional components of scenes. Psychol Sci 19:781–788
Peigneux P, Laureys S, Fuchs S, Collette F, Perrin F, Reggers J, Phillips C, Degueldre C, Del Fiore G, Aerts J, Luxen A, Maquet P (2004) Are spatial memories strengthened in the human hippocampus during slow wave sleep? Neuron 44(3):535–545
Phelps EA (2004) Human emotion and memory: interactions of the amygdala and hippocampal complex. Curr Opin Neurobiol 14(2):198–202
Plihal W, Born J (1997) Effects of early and late nocturnal sleep on declarative and procedural memory. J Cogn Neurosci 9(4):534–547
Power AE (2004) Muscarinic cholinergic contribution to memory consolidation: with attention to involvement of the basolateral amygdala. Curr Med Chem 11(8):987–996
Prehn-Kristensen A, Göder R, Chirobeja S, Breßmann I, Ferstl R, Baving L (2009) Sleep in children enhances preferentially emotional declarative but not procedural memories. J Exp Child Psychol 104(1):132–139
Rasch B, Born J (2013) About sleep's role in memory. Physiol Rev 93(2):681–766
Ritchey M, Dolcos F, Cabeza R (2008) Role of amygdala connectivity in the persistence of emotional memories over time: an event-related fMRI investigation. Cereb Cortex 18 (11):2494–2504
Ritchey M, LaBar KS, Cabeza R (2011) Level of processing modulates the neural correlates of emotional memory formation. J Cogn Neurosci 23(4):757–771

Rosales-Lagarde A, Armony JL, del Río-Portilla Y, Trejo-Martínez D, Conde R, Corsi-Cabrera M (2012) Enhanced emotional reactivity after selective REM sleep deprivation in humans: an fMRI study. Mem Motivational/Emotional Processes 182

Scoville WB, Milner B (1957) Loss of recent memory after bilateral hippocampal lesions. J Neurol Neurosurg Psychiatry 20(1):11

Schacter DL, Addis DR (2007a) Constructive memory: the ghosts of past and future. Nature 445 (7123):27–27

Schacter DL, Addis DR (2007b) The cognitive neuroscience of constructive memory: remembering the past and imagining the future. Philos Trans R Soc Lond B Biol Sci 362 (1481):773–786

Siapas AG, Wilson MA (1998) Coordinated interactions between hippocampal ripples and cortical spindles during slow-wave sleep. Neuron 21:1123–1128

Sirota A, Csicsvari J, Buhl D, Buzsáki G (2003) Communication between neocortex and hippocampus during sleep in rodents. Proc Natl Acad Sci U S A 100(4):2065–2069

Smith CT, Aubrey JB, Peters KR (2004) Different roles for REM and stage 2 sleep in motor learning: a proposed model. Psychol Belg 44:81–104

Sotres-Bayon F, Bush DE, LeDoux JE (2004) Emotional perseveration: an update on prefrontal-amygdala interactions in fear extinction. Learn Mem 11(5):525–535

Squire LR (1992) Memory and the hippocampus: a synthesis from findings with rats, monkeys, and humans. Psychol Rev 99(2):195

Steriade M, Nuñez A, Amzica F (1993) Intracellular analysis of relations between the slow (<1 Hz) neocortical oscillation and other sleep rhythms of the electroencephalogram. J Neurosci 13 (8):3266–3283

Sterpenich V, Albouy G, Boly M, Vandewalle G, Darsaud A, Balteau E, Vu TTD, Desseilles M, D'Argembeau A, Gais S, Rauchs G, Schabus CD, Luxen A, Collette F, Maquet P (2007) Sleep-related hippocampo-cortical interplay during emotional memory recollection. PLoS Biol 5(11):e282

Sterpenich V, Albouy G, Darsaud A, Schmidt C, Vandewalle G, Vu TTD, Desseilles M, Phillips C, Degueldre C, Balteau E, Collette F, Luxen A, Maquet P (2009) Sleep promotes the neural reorganization of remote emotional memory. J Neurosci 29(16):5143–5152

Strange BA, Dolan RJ (2006) Anterior medial temporal lobe in human cognition: memory for fear and the unexpected. Cogn Neuropsychiatry 11(3):198–218

Stickgold R, Walker MP (2013) Sleep-dependent memory triage: evolving generalization through selective processing. Nat Neurosci 16(2):139–145

Takashima A, Petersson KM, Rutters F, Tendolkar I, Jensen O, Zwarts MJ, McNaughton B, Fernandez G (2006) Declarative memory consolidation in humans: a prospective functional magnetic resonance imaging study. Proc Natl Acad Sci U S A 103(3):756–761

Talmi D (2013) Enhanced emotional memory cognitive and neural mechanisms. Curr Dir Psychol Sci 22(6):430–436

Wagner U, Fischer S, Born J (2002) Changes in emotional responses to aversive pictures across periods rich in slow-wave sleep versus rapid eye movement sleep. Psychosom Med 64(4):627–634

Wagner U, Gais S, Born J (2001) Emotional memory formation is enhanced across sleep intervals with high amounts of rapid eye movement sleep. Learn Mem 8(2):112–119

Wagner U, Hallschmid M, Rasch B, Born J (2006) Brief sleep after learning keeps emotional memories alive for years. Biol Psychiatry 60(7):788–790

Walker MP (2009) Sleep, memory and emotion. Prog Brain Res 185:49–68

Walker MP, Brakefield T, Morgan A, Hobson JA, Stickgold R (2002) Practice with sleep makes perfect: sleep-dependent motor skill learning. Neuron 35(1):205–211

Walker MP, Stickgold R (2010) Overnight alchemy: sleep-dependent memory evolution. Nat Rev Neurosci 11(3):218–218

Walker MP, van Der Helm E (2009) Overnight therapy? The role of sleep in emotional brain processing. Psychol Bull 135(5):731

van der Helm E, Walker MP (2012) Sleep and affective brain regulation. Soc Pers Psychol Compass 6(11):773–791

van der Helm E, Yao J, Dutt S, Rao V, Saletin JM, Walker MP (2011) REM sleep depotentiates amygdala activity to previous emotional experiences. Curr Biol 21(23):2029–2032

Vazquez J, Baghdoyan HA (2001) Basal forebrain acetylcholine release during REM sleep is significantly greater than during waking. Am J Physiol-Regul Integr Comp Physiol 280(2): R598–R601

Daily Life Experiences in Dreams and Sleep-Dependent Memory Consolidation

Jean-Baptiste Eichenlaub, Sydney S. Cash and Mark Blagrove

Abstract Memories constitute much of the source material for our dreams. Although waking life events are not faithfully replayed in dreams, dream content arises from recent daily experiences. Numerous empirical studies and theoretical accounts highlight the key function of sleep in the consolidation of newly learned memories, raising the question how reference to waking memories in dreams relates to ongoing memory-related processes that take place during sleep. This review attempts to present first the current knowledge of the incorporation of waking memories in dreams by highlighting three main features of this phenomenon i.e. (1) dreaming contains abundant references from recent dreamer's own life, (2) the wake-dream relation can follow a surprising 7 day U-shaped timescale and (3) salient/intense waking events are more easily incorporated than indistinct/less-intense waking events. Second, this review attempts to discuss the relationship between this phenomenon and the memory-related processes that take place during sleep. The features of the incorporation of waking memories in dreams are in line with some characteristics of the memory processing hypothesized to take place during sleep, suggesting that dreaming might reflect this memory processing. However, substantial limitations and alternative hypotheses must be regarded and addressed in future studies to clarify the link between dream content and sleep-dependent memory consolidation.

Keywords Dream content · Sleep · Day-residue · Dream-lag · Memory consolidation · Emotion

J.-B. Eichenlaub (✉) · S.S. Cash
Department of Neurology, Massachusetts General Hospital,
Harvard Medical School, Boston, USA
e-mail: jb.eichenlaub@gmail.com

M. Blagrove
Department of Psychology, Swansea University, Swansea, UK

N. Axmacher and B. Rasch (eds.), *Cognitive Neuroscience of Memory Consolidation*, Studies in Neuroscience, Psychology and Behavioral Economics, DOI 10.1007/978-3-319-45066-7_10

Recent Daily Life Experiences Are Incorporated into Dream Content

The ties of dreaming with concrete experiences from recent waking life were already theorized in the 19th century by Freud using the term "day-residues". In his book *the Interpretation of dreams* (1899), Freud asserted that: "in every dream we may find some reference to the experiences of the preceding day, (…) that *day-residues* are left over from the waking activity" (Freud 1899). Since Freud's seminal work, many studies have confirmed and refined this phenomenon by experimentally analyzing the memory sources of dreams (e.g. Cavallero et al. 1990; Nielsen et al. 2004; Blagrove et al. 2011b; van Rijn et al. 2015). An example of incorporation of daily events in dream content is shown in Fig. 1.

A fruitful set of studies analyzed the memory sources of dreams by using a "free association" approach (Cavallero 1993). The rationale for this procedure was that, as initially proposed by Freud, associations to the elements of dream content could lead back to memories that acted as sources for the dream. Immediately after collecting dream reports, dreamers were requested to formulate, for each segment of the dream report, anything that came to mind in conjunction with that given segment. Finally, external judges classified the source of each segment as either episodic memory (autobiographical events), semantic memory (general knowledges) or semantic memory concerning the self (category called abstract self-references). Using this approach, it has been reported that up to ~50% of dream reports have episodic memory sources (Cavallero et al. 1990, 1992; Cicogna et al. 1991, 1998, 2000).

More recently, Fosse et al. (2003) applied another approach by asking participants to match their own dream experiences to their daytime activities recorded using a daily log. In accordance with the results above, the authors found that ~65% of the dream reports were scored as reflecting aspects of recent waking life, and ~51% of them contained at least one dream element exhibiting a strong similarity to a waking experience. However, the authors estimated that only 1–2% of dream reports contained a direct, complete, and transparent replay of waking episodes (Fosse et al. 2003). This was confirmed by Malinowski and Horton

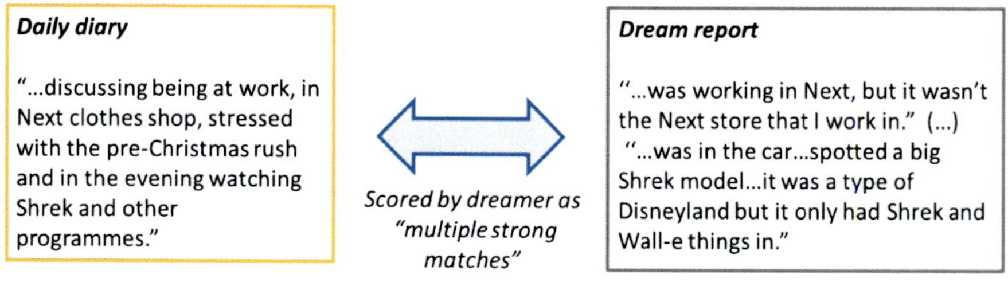

Fig. 1 Example of incorporation of recent daily events into dream content (Blagrove et al. 2011a). In this example, the link between the daily activity and the dream report is obvious, and includes the workplace (Next store) as well as the main character of a cartoon (Shrek)

(2014a), who by using an "episodic richness in dream" scale, showed that only 0.5% of dreams (1/186) exhibited a "highly similar replaying of the original experience(s)".

Following early efforts in the'60s and'70s to influence dream content by manipulating sleepers' prior experience, recent studies explored the incorporation of a predefined/controlled task. Using as "control event" interactive video games such as Tetris (Stickgold et al. 2000; Kusse et al. 2012), a skiing arcade game (Alpine Racer II; Wamsley et al. 2010a), or a virtual navigation task (Wamsley et al. 2010b), these studies reported a robust influence of the task on subsequent dreams. For instance, Stickgold et al. (2000) showed that 63% (17/27) of the participants reported in their dreams stereotypical images of Tetris (e.g. "[I'm] seeing in my mind how the game pieces kind of float down and fit into the other pieces, and am also rotating them."). Accordingly, Wamsley et al. (2010a) reported that 30% of the participants' dreams contained direct (e.g. "I'm thinking about the game, (...) that I should have used my knees more") or indirect (e.g. "I was in a race, like a running race") incorporation of the skiing game.

Together, these data teach us that dreaming incorporates recent autobiographical memories in its content. While true (exact) replays of the waking events are extremely rare, numerous elements/fragments from recent waking life experiences emerge in dreams.

Dreaming Incorporates Memories on a 7-Day U-Shaped Timescale

The reference in dreams of elements from the immediate previous day raises the question of the extent to which events from older days/time periods are incorporated as well. In this regard, several studies have identified a 7-day U-shaped timescale in the periodicity of the memories incorporated in dreams. In these studies memories in dreams are found to originate more often from the two days prior to the dream (the day-residue effect), or from the 5–7 days prior to the dream (the dream-lag effect), than from the intermediate 3–4 days period before the dream (e.g. Nielsen and Powell 1989; Powell et al. 1995; Blagrove et al. 2011a; van Rijn et al. 2015).

These results were reported from both between- and within-subjects designs. In 2004, Nielsen et al. used a between-subjects design in which each participant was instructed to report a dream and then to retrieve past events related to it that arose from one of seven randomly determined days prior to the dream. Each participant then rated the correspondence or level of matching between the dream report and the experiences from that randomly allocated day. This matching could involve, for example, the same person or emotion or setting. Finally, the mean matching scores for the 7 time period groups were calculated. Level of matching ratings were found to be significantly greater for comparing a dream report to a diary record of the day before the dream (day-residue), and for diary records of the 5th–7th days before the dream (dream-lag effect), than for diary records of the 2nd–4th days before the dream (Nielsen et al. 2004).

More recently, Blagrove et al. (2011a) replicated the day-residue and dream-lag effects using a within-subjects design. Participants kept a daily diary and a dream diary for 14 days, before comparing all the daily diaries with all the dream diaries during a "matching task". This involved scoring the level of similarities/correspondences between each of the daily diary/dream report pairs. The authors showed that matching ratings were significantly greater for comparing a dream report to a diary record of the day before the dream, and for diary records of the 5th–7th days before the dream, than for comparisons with the diary records of the 2nd–4th days before the dream (Blagrove et al. 2011a). In addition, this study introduced a "baseline" by comparing dream reports to diary records for days that occurred after the dreams. This baseline matching controlled for general commonalities that were present in the participant's life and their dream content. Importantly, the authors reported that levels of matching for both the day-residue and dream-lag effects were significantly higher than during this baseline period (Blagrove et al. 2011a).

Instead of scoring a general level of similarity, van Rijn et al. (2015) assessed the actual number of correspondences between daily logs and dream reports. Applying a within-subjects design combined with a "correspondence identification" task, participants were instructed to spot similar aspects between the log items and dream reports, such as of characters, objects, actions, locations, or themes. The mean number of incorporations per daily log-dream report pair was then computed. The authors reported a significantly higher number of waking references from 1 to 2 and 5 to 7 days before the dream, than from 3 to 4 days before the dream.

Interestingly, Blagrove's group recently highlighted individual differences in overall number of correspondences identified between diary records and dream reports (Blagrove et al. 2014; Henley-Einion and Blagrove 2014; van Rijn et al. 2015). This individual difference in tendency to find connections between daily life records and dreams reports was found to result in a dilution or eradication of time-course relationships for individuals who identify high numbers of such incorporations. Indeed, it appears that individuals who identify large numbers of correspondences did not exhibit the day-residue (Henley-Einion and Blagrove 2014) or both the day-residue and dream-lag effects (Blagrove et al. 2014; van Rijn et al. 2015).

In summary, data from different laboratories and using various approaches argue for a 7 days U-shaped timescale in the incorporation of content into dreams. Nevertheless, it is important to mention here that several studies did not report a dream-lag effect (e.g. Schredl 2006; Henley-Einion and Blagrove 2014), or in delimited conditions such as in rapid eye movement (REM) dreams (Blagrove et al. 2011b) or for salient waking events in low-incorporators who spontaneously recalled home dreams (van Rijn et al. 2015), suggesting limiting conditions for the observation of this effect.

The Emotional Component of Memories Incorporated in Dreams

A large number of dreams contain a high level of emotional involvement, including both positive and negative emotions (for reviews, see Schwartz and Maquet 2002; Blagrove and Pace-Schott 2010; Nir and Tononi 2010; Schredl 2010; Ruby 2011; see also chapter by Cunningham and Payne on emotional memory consolidation).

Accordingly, the emotion associated with waking experiences is thought to be determinant in the incorporation of these events into dreams. Intense negative waking experiences clearly affect dreaming. The degree of concern about the ex-spouse in individuals going through divorce correlates with the number of dreams in which the former partner appears as a dream character (Cartwright et al. 2006). Individuals who lived in a violent environment report more threatening events in their dream content that those who were less exposed, or not exposed to threatening real-life events (Valli et al. 2005). The number of threatening dreams (Propper et al. 2007) and the intensity of the main/most salient elements in dream content (i.e. conceptualized as "central image", Hartmann and Brezler 2008) was higher following the 09.11.01 terrorist attack than in the preceding weeks. Similarly, the viewing of September 11 media coverage was followed by more negative emotions and literal and non-literal references to 9/11 in comparison with the viewing of an educational course (Davidson and Lynch 2012).

However, this effect is not exclusive to negative waking experiences, and recent studies suggest instead that the salience (or intensity) of the experiences determines incorporations into dreams, for both positive and negative emotions. Schredl (2006) instructed participants to list and score their waking life events for both emotional intensity and tone every evening during a 2-week period (see also chapter by Schredl). Every morning, participants were also asked to report their dreams and to identify whether events from previous days occurred in the dream content. By comparing the tone and intensity of the incorporated *vs* not-incorporated daily events, the author reported no difference in emotional tone, but a higher emotional intensity for the incorporated events (Schredl 2006). Malinowski and Horton (2014b) recently confirmed, in a larger sample size study, that the waking-life experiences that are incorporated are more emotionally intense than those that are not incorporated into dreams. Accordingly, the authors reported that waking events categorized as personally significant experiences, major concerns, or novel experiences tend to be more incorporated than less-salient, everyday experiences that take up most of the dreamers' time during the day (i.e. events classified as major daily activities; Malinowski and Horton 2014b).

These studies highlight the salience of daily experiences as a determinant of the subsequent incorporation of these events into dreams. As the perceived intensity of waking experiences is subjective, it is essential here to consider participants' own perception of these events, even for pre-defined, controlled events. In this regard, individuals recruited in a sleep laboratory experience and who reported the experimental night as being a major concern, exhibited both the day-residue and dream-lag effects for the incorporations of this experimental night on subsequent dreams, while no-effect was observed for the unconcerned individuals from the same group (van Rijn et al. 2015).

Is the Incorporation of Recent Salient Waking Events in Dreams Related to the Sleep-Dependent Consolidation of New Memories?

As multiple empirical studies (e.g. Plihal and Born 1997; Walker et al. 2002) and theoretical models (Diekelmann and Born 2010; Stickgold and Walker 2013) highlight the role of sleep in memory consolidation, the question has been raised of whether the incorporation of recent salient memories in dreams, as reviewed above, is related to these ongoing memory-related processes that take place during sleep.

First, the abundant occurrence of recent memories in dreams may reflect the prioritized processing of newly formed memories by the sleeping brain. An argument in favor of this hypothesis comes from Wamsley et al. (2010b), who demonstrated that improved performance on a virtual navigation task was associated with task-related dreams during an intervening nap, but not with task-related daydreams during an intervening period of wake. Similarly, De Koninck et al. (1988) showed that among students enrolled in a 6-week French language course, those who exhibited better acquisition of the new language tended to incorporate more French into dreams than students who were less successful in the class.

Second, the 7 day U-shaped timescale is in line with the time-limited dependence of memory in the hippocampus and its gradual transfer into the neocortex (Nielsen and Stenstrom 2005; see, e.g., the chapter by Fernandez), relocation that takes place over a period of about one week according to animal studies. By tracking the excitability of hippocampal neurons during several days following a new memory acquisition, learning-specific changes in excitability gradually decreased and returned to baseline within 7 days after training, suggesting a long-lasting 7 days window critical for the storage of memory elsewhere in the brain (Moyer et al. 1996; Thompson et al. 1996; Oh and Disterhoft 2015). At the end of this critical 7-day time window, the achieved cortical form of the memory may trigger late neocortical-dependent processing that is compatible with the dream-lag effect. In the same line, by exploring the time-course of learning-related hippocampal protein synthesis (e.g. Early Growth Response-1, EGR-1), waves of protein expression were not involved in the mechanism of memory formation within the first 24 h, but were critical for long-term memory consolidation as assessed 7 days after learning (Katche et al. 2012), highlighting long-lasting memory mechanisms that may be reflected in the dream-lag effect. However, the link between the temporal attributes of such hippocampal-neocortical transfer and delayed memory sources in dreams is hypothetical (Nielsen and Stenstrom 2005), and needs to be confirmed experimentally.

Third, the preferential incorporation of salient/intense waking events in dreams matches the selective consolidation of affective/emotional experiences by the sleeping brain, especially in REM sleep (for reviews, see Walker 2009; Walker and van der Helm 2009; Genzel et al. 2015; see also chapter by Cunningham and Payne). By comparing the sleep-dependent consolidation of negative *vs* neutral narratives (Wagner et al. 2001, 2006), pictures (Hu et al. 2006; Nishida et al. 2009) or complex scenes (Payne et al. 2008, 2012, 2015), sleep has been repeatedly shown to preferentially enhance negative in comparison to neutral items. This selection extends to positive/rewarded memories as well (Perogamvros and Schwartz 2012; Igloi et al. 2015). It is thought that the salience tags attached to memories govern the sleep-dependent memory triage that discriminates between relevant and irrelevant memories and "allows the organism to adapt to environmental change rapidly and effectively, guided by the most relevant information from its own autobiographical history" (Stickgold and Walker 2013). Accordingly, the preferential incorporation of salient/intense waking events in dreams (including both positive and negative events) may reflect this preferential processing of memories that have an adaptive advantage (see also chapter by Rauss and Born).

More broadly, and in accordance with literature from rodent experiments showing that neural sequences that underlie new memories are replayed during subsequent sleep (e.g. Skaggs and McNaughton 1996; Lee and Wilson 2002; Ji and Wilson 2007), dreaming might be the mental experience of this replay (for reviews, see Wamsley and Stickgold 2011; Wamsley 2014). Alternatively, as cueing during sleep with stimuli paired with the learned material leads to enhanced consolidation (e.g. Rasch et al. 2007; Rudoy et al. 2009; Cousins et al. 2014; see chapters by Schreiner, Lehmann and Rasch and by Talamini), it may be that the dreaming process acts as an endogenous cueing, either for the memories themselves or for the environmental context in which the memories were learned.

Link Between Incorporated Memories in Dreams and Sleep-Dependent Memory Consolidation: Limitations and Alternative Explanations

Dream recall does not occur every night, and, although this may be solely a recall phenomenon, it is possible that it indicates differences between individuals in dream production. Based on a representative sample of 937 persons interviewed retrospectively about their dreams, the mean dream recall frequency (DRF) was estimated at about 1 dream per week (Schredl 2008; see also Nielsen 2012). In addition, by asking 196 participants to keep a dream diary for at least 2 weeks and by expressing DRF on a scale from 0 to 1 (i.e. proportion of diaries with a recalled dream), Schredl and Fulda (2005) showed that DRF values can spread far from the mean (mean = 0.56, standard deviation = 0.26) i.e. $\sim$11% (21/196) of the participants exhibited a DRF below 0.2 while $\sim$21% (41/196) above 0.8. That dreaming and by extension the incorporation of waking events in its content are somehow irregular events in a large part of the population challenge the hypothesis that dream content "*transparently reflects* recently encoded memory" (Wamsley 2014) or is "*a direct reflection* of concomitant memory processes in the brain" (Wamsley and Stickgold 2011). Furthermore, and since dream studies recruit (almost) systematically frequent dream-recallers to reduce the occurrence of missing observations, the generalizability of the findings is limited. The latter is especially true since the two groups (infrequent *vs* frequent dream-recallers) exhibit behavioral/psychological differences (for reviews, see Blagrove and Pace-Schott 2010; Ruby 2011), and functional brain differences (Ruby et al. 2013; Eichenlaub et al. 2014a, b) that are compatible with differences in both mental imagery and memory encoding of dreams.

As we discussed previously, a strong support for dreaming as a reflection of the sleep-dependent memory processing comes from studies showing a direct link between improved performance and task-related dreams (De Koninck et al. 1988, 1990; Wamsley et al. 2010b). By showing that the occurrence of task-related content in dreams predicted enhanced memory performance, these data suggest interplay between dreaming and sleep-dependent memory consolidation, in a way that dream content is influenced by/reflects the nature of the ongoing memory processing during sleep. However, an alternative interpretation of the data can be formulated by taking into account the difficulty of, or interest in performing the task. This hypothesis has been recently supported by De Koninck et al. (2012), showing that incorporations in dreams of content related to the learning of a new language were often expressions of frustration/anxiety encounter during the acquisition process. Instead of being the reflection of the sleep-dependent consolidation of the language itself, the authors suggested that dream content may be a reflection of obstacles encountered to acquire the language (De Koninck et al. 2012). Accordingly, it appeared in Wamsley et al.'s navigation study that those subjects who reported dreaming about the task also experienced greater difficulty in learning the maze, that is, they were among those with the poorest baseline performance before the nap (Wamsley et al. 2010b). Poor initial performance might lead to frustration/anxiety that determined the occurrence of task-related content in dreams, while making easier a post-sleep performance improvement. The link between dream content and post-sleep performance improvement might therefore be explained by a third variable i.e. the emotional response to the difficulty of acquiring the new material, and the anxiety/frustration related to this.

The latter objection can lead to a variant of the dreaming indexing memory consolidation proposal, in that it may be that dream content reflects the social and interpersonal learning involved in the emotional response to being tested, and may indeed index social and interpersonal learning more widely (Blagrove et al. 2013). This speculation would be supported by the high frequency of interactions with other characters in dreams (Nielsen and Lara-Carrasco 2007). However, although dreaming might be proposed to reflect such higher level cognition, there is a problem here that it may be in Slow Wave Sleep (SWS), rather than REM sleep, that the consolidation of episodic (Diekelmann and Born 2010) and complex schemata (Lewis and Durrant 2011) occur, while dream mentation rate is on average higher in REM sleep than in SWS (for a review, see Nielsen 2000). A response to this downplaying of REM sleep might, in time, occur when more is known of the relevance of the default network to memory consolidation and the creative reorganization of memories, and any possible link to dreaming, as reviewed and speculated upon by Domhoff (2011).

More broadly, however, the main flaw in the proposal that dream content indexes memory consolidation is that all studies to date that may have a bearing here are correlational rather than experimental (Blagrove 1992, 2011). Only when dream content can be altered such that participants can be randomly allocated to conditions with different dream content will it be possible to test the hypothesis that particular dream content can affect memory consolidation during sleep and memory performance after waking. And, even if that is achieved, there will still be a generalizability issue as studies here will of necessity use participants with a high level of dream recall, and the role of dreaming in participants who rarely or never report dreams would require further investigation.

In summary, current dream studies face substantial limitations in their generalizability as findings are drawn almost exclusively from participants with high levels of dream recall, or from experimental designs that are necessarily correlational rather than a random assignment to dream content conditions. Furthermore, alternative hypotheses highlight the difficulty of acquiring new material or incorporating and accounting for the emotional valence which could explain the apparent link between dream content and sleep-dependent memory consolidation. None-the-less, we are entering an era in which the nature of dreaming and sleep are open to detailed investigation, allowing clarifying the true link between dream content and the memory processing that take place during sleep.

Acknowledgements This work was supported by ONR N00014-13-1-0672, UK Economic and Social Research Council ES/I037555/1 and the Fyssen Foundation.

References

Blagrove M (1992) Dreams as the reflection of our waking concerns and abilities: a critique of the problem solving paradigm in dream research. Dreaming 2:205–220

Blagrove M (2011) Distinguishing continuity/discontinuity, function, and insight when investigating dream content. Int J of Dream Res 4:45–47

Blagrove M, Pace-Schott EF (2010) Trait and neurobiological correlates of individual differences in dream recall and dream content. Int Rev Neurobiol 92:155–180

Blagrove M, Ruby P, Eichenlaub JB (2013) Dreams are made of memories, but maybe not for memory. Behav Brain Sci 36:609–610

Blagrove M, Henley-Einion J, Barnett A, Edwards D, Heidi Seage C (2011a) A replication of the 5–7 day dream-lag effect with comparison of dreams to future events as control for baseline matching. Conscious Cogn 20:384–391

Blagrove M, Fouquet NC, Henley-Einion JA, Pace-Schott EF, Davies AC, Neuschaffer JL, Turnbull OH (2011b) Assessing the dream-lag effect for REM and NREM stage 2 dreams. PLoS ONE 6:e26708

Blagrove M, Eichenlaub JB, Van Rijn E, Phelan M, Ryder L, Gaskell MG, Lewis PA, Walker M (2014) The nature of delayed dream incorporation: personally significant events persist, but not major daily activities or concerns. J Sleep Res 23:48–49
Cartwright R, Agargun MY, Kirkby J, Friedman JK (2006) Relation of dreams to waking concerns. Psychiatry Res 141:261–270
Cavallero C (1993) Symposium: dream research methodology: the quest for dream sources. J Sleep Res 2:13–16
Cavallero C, Foulkes D, Hollifield M, Terry R (1990) Memory sources of REM and NREM dreams. Sleep 13:449–455
Cavallero C, Cicogna P, Natale V, Occhionero M, Zito A (1992) Slow wave sleep dreaming. Sleep 15:562–566
Cicogna P, Cavallero C, Bosinelli M (1991) Cognitive aspects of mental activity during sleep. Am J Psychol 104:413–425
Cicogna P, Natale V, Occhionero M, Bosinelli M (2000) Slow wave and REM sleep mentation. Sleep Res Online 3:67–72
Cicogna PC, Natale V, Occhionero M, Bosinelli M (1998) A comparison of mental activity during sleep onset and morning awakening. Sleep 21:462–470
Cousins JN, El-Deredy W, Parkes LM, Hennies N, Lewis PA (2014) Cued memory reactivation during slow-wave sleep promotes explicit knowledge of a motor sequence. J Neurosci 34:15870–15876
Davidson J, Lynch S (2012) Thematic, literal and associative dream imagery following a high-impact event. Dreaming 22:58–69
De Koninck J, Wong C, Hebert G (2012) Types of dream incorporations of language learning and learning efficiency. J Sleep Res 21:190
De Koninck J, Christ G, Rinfret N, Proulx G (1988) Dreams during language learning: when and how is the new language integrated. Psychiatr J Univ Ott 13:72–74
De Koninck J, Christ G, Hebert G, Rinfret N (1990) Language learning efficiency, dreams and REM sleep. Psychiatr J Univ Ott 15:91–92
Diekelmann S, Born J (2010) The memory function of sleep. Nat Rev Neurosci 11:114–126
Domhoff GW (2011) The neural substrate for dreaming: is it a subsystem of the default network? Conscious Cogn 20:1163–1174
Eichenlaub JB, Bertrand O, Morlet D, Ruby P (2014a) Brain reactivity differentiates subjects with high and low dream recall frequencies during both sleep and wakefulness. Cereb Cortex 24:1206–1215
Eichenlaub JB, Nicolas A, Daltrozzo J, Redoute J, Costes N, Ruby P (2014b) Resting brain activity varies with dream recall frequency between subjects. Neuropsychopharmacology 39:1594–1602
Fosse MJ, Fosse R, Hobson JA, Stickgold RJ (2003) Dreaming and episodic memory: a functional dissociation? J Cogn Neurosci 15:1–9
Freud S (1899) Die Traumdeutung. Fischer-Taschenbuch (L'interprétation des rêves, PUF, 1926 & 1967)
Genzel L, Spoormaker VI, Konrad BN, Dresler M (2015) The role of rapid eye movement sleep for amygdala-related memory processing. Neurobiol Learn Mem
Hartmann E, Brezler T (2008) A systematic change in dreams after 9/11/01. Sleep 31:213–218
Henley-Einion J, Blagrove M (2014) Assessing the day-residue and dream-lag effects using the identification of multiple correspondences between dream reports and waking life diaries. Dreaming 24:71–88
Hu P, Stylos-Allan M, Walker MP (2006) Sleep facilitates consolidation of emotional declarative memory. Psychol Sci 17:891–898
Igloi K, Gaggioni G, Sterpenich V, Schwartz S (2015) A nap to recap or how reward regulates hippocampal-prefrontal memory networks during daytime sleep in humans. Elife 4
Ji D, Wilson MA (2007) Coordinated memory replay in the visual cortex and hippocampus during sleep. Nat Neurosci 10:100–107

Katche C, Goldin A, Gonzalez C, Bekinschtein P, Medina JH (2012) Maintenance of long-term memory storage is dependent on late posttraining Egr-1 expression. Neurobiol Learn Mem 98:220–227
Kusse C, Shaffii LEBA, Schrouff J, Matarazzo L, Maquet P (2012) Experience-dependent induction of hypnagogic images during daytime naps: a combined behavioural and EEG study. J Sleep Res 21:10–20
Lee AK, Wilson MA (2002) Memory of sequential experience in the hippocampus during slow wave sleep. Neuron 36:1183–1194
Lewis PA, Durrant SJ (2011) Overlapping memory replay during sleep builds cognitive schemata. Trends Cogn Sci 15:343–351
Malinowski JE, Horton CL (2014a) Memory sources of dreams: the incorporation of autobiographical rather than episodic experiences. J Sleep Res
Malinowski JE, Horton CL (2014b) Evidence for the preferential incorporation of emotional waking-life experiences into dreams. Dreaming 24:18–31
Moyer JR Jr, Thompson LT, Disterhoft JF (1996) Trace eyeblink conditioning increases CA1 excitability in a transient and learning-specific manner. J Neurosci 16:5536–5546
Nielsen T (2012) Variations in dream recall frequency and dream theme diversity by age and sex. Front Neurol 3:106
Nielsen TA (2000) A review of mentation in REM and NREM sleep: "covert" REM sleep as a possible reconciliation of two opposing models. Behav Brain Sci 23:851–866; discussion 904–1121
Nielsen TA, Powell RA (1989) The 'dream-lag' effect: a 6-day temporal delay in dream content incorporation. Psychiatr J Univ Ott: Revue de psychiatrie de l'Universite d'Ottawa 14:561–565
Nielsen TA, Stenstrom P (2005) What are the memory sources of dreaming? Nature 437:1286–1289
Nielsen TA, Lara-Carrasco J (2007) Nightmares, dreaming and emotion regulation: a review. New Sci Dreaming 2:253–284
Nielsen TA, Kuiken D, Alain G, Stenstrom P, Powell RA (2004) Immediate and delayed incorporations of events into dreams: further replication and implications for dream function. J Sleep Res 13:327–336
Nir Y, Tononi G (2010) Dreaming and the brain: from phenomenology to neurophysiology. Trends Cogn Sci 14:88–100
Nishida M, Pearsall J, Buckner RL, Walker MP (2009) REM sleep, prefrontal theta, and the consolidation of human emotional memory. Cereb Cortex 19:1158–1166
Oh MM, Disterhoft JF (2015) Increased excitability of both principal neurons and interneurons during associative learning. Neuroscientist 21:372–384
Payne JD, Chambers AM, Kensinger EA (2012) Sleep promotes lasting changes in selective memory for emotional scenes. Front Integr Neurosci 6:108
Payne JD, Stickgold R, Swanberg K, Kensinger EA (2008) Sleep preferentially enhances memory for emotional components of scenes. Psychol Sci 19:781–788
Payne JD, Kensinger EA, Wamsley EJ, Spreng RN, Alger SE, Gibler K, Schacter DL, Stickgold R (2015) Napping and the selective consolidation of negative aspects of scenes. Emotion 15: 176–186
Perogamvros L, Schwartz S (2012) The roles of the reward system in sleep and dreaming. Neurosci Biobehav Rev 36:1934–1951
Plihal W, Born J (1997) Effects of early and late nocturnal sleep on declarative and procedural memory. J Cogn Neurosci 9:534–547
Powell RA, Cheung JS, Nielsen TA, Cervenka TM (1995) Temporal delays in incorporation of events into dreams. Percept Mot Skills 81:95–104
Propper RE, Stickgold R, Keeley R, Christman SD (2007) Is television traumatic? Dreams, stress, and media exposure in the aftermath of September 11, 2001. Psychol Sci 18:334–340
Rasch B, Buchel C, Gais S, Born J (2007) Odor cues during slow-wave sleep prompt declarative memory consolidation. Science 315:1426–1429

Ruby P, Blochet C, Eichenlaub JB, Bertrand O, Morlet D, Bidet-Caulet A (2013) Alpha reactivity to first names differs in subjects with high and low dream recall frequency. Front Psychol 4:419
Ruby PM (2011) Experimental research on dreaming: state of the art and neuropsychoanalytic perspectives. Front Psychol 2:286
Rudoy JD, Voss JL, Westerberg CE, Paller KA (2009) Strengthening individual memories by reactivating them during sleep. Science 326:1079
Schredl M (2006) Factors affecting the continuity between waking and dreaming: emotional intensity and emotional tone of the waking-life event. Sleep Hypn 8:1–5
Schredl M (2008) Dream recall frequency in a representative German sample. Percept Mot Skills 106:699–702
Schredl M (2010) Characteristics and contents of dreams. Int Rev Neurobiol 92:135–154
Schredl M, Fulda S (2005) Reliability and stability of dream recall frequency. Dreaming 15: 240–244
Schwartz S, Maquet P (2002) Sleep imaging and the neuro-psychological assessment of dreams. Trends Cogn Sci 6:23–30
Skaggs WE, McNaughton BL (1996) Replay of neuronal firing sequences in rat hippocampus during sleep following spatial experience. Science 271:1870–1873
Stickgold R, Walker MP (2013) Sleep-dependent memory triage: evolving generalization through selective processing. Nat Neurosci 16:139–145
Stickgold R, Malia A, Maguire D, Roddenberry D, O'Connor M (2000) Replaying the game: hypnagogic images in normals and amnesics. Science 290:350–353
Thompson LT, Moyer JR Jr, Disterhoft JF (1996) Transient changes in excitability of rabbit CA3 neurons with a time course appropriate to support memory consolidation. J Neurophysiol 76:1836–1849
Valli K, Revonsuo A, Palkas O, Ismail KH, Ali KJ, Punamaki RL (2005) The threat simulation theory of the evolutionary function of dreaming: evidence from dreams of traumatized children. Conscious Cogn 14:188–218
van Rijn E, Eichenlaub JB, Lewis PA, Walker MP, Gaskell MG, Malinowski JE, Blagrove M (2015) The dream-lag effect: selective processing of personally significant events during rapid eye movement sleep, but not during slow wave sleep. Neurobiol Learn Mem 122:98–109
Wagner U, Gais S, Born J (2001) Emotional memory formation is enhanced across sleep intervals with high amounts of rapid eye movement sleep. Learn Mem 8:112–119
Wagner U, Hallschmid M, Rasch B, Born J (2006) Brief sleep after learning keeps emotional memories alive for years. Biol Psychiatry 60:788–790
Walker MP (2009) The role of sleep in cognition and emotion. Ann N Y Acad Sci 1156:168–197
Walker MP, van der Helm E (2009) Overnight therapy? the role of sleep in emotional brain processing. Psychol Bull 135:731–748
Walker MP, Brakefield T, Morgan A, Hobson JA, Stickgold R (2002) Practice with sleep makes perfect: sleep-dependent motor skill learning. Neuron 35:205–211
Wamsley EJ (2014) Dreaming and offline memory consolidation. Curr Neurol Neurosci Rep 14:433
Wamsley EJ, Stickgold R (2011) Memory, sleep and dreaming: experiencing consolidation. Sleep Med Clin 6:97–108
Wamsley EJ, Perry K, Djonlagic I, Reaven LB, Stickgold R (2010a) Cognitive replay of visuomotor learning at sleep onset: temporal dynamics and relationship to task performance. Sleep 33:59–68
Wamsley EJ, Tucker M, Payne JD, Benavides JA, Stickgold R (2010b) Dreaming of a learning task is associated with enhanced sleep-dependent memory consolidation. Curr Biol 20: 850–855

Is Dreaming Related to Sleep-Dependent Memory Consolidation?

Michael Schredl

Abstract As testing the functions of dreams directly is not possible, empirical dream research has focused on three areas providing indirect support for a relationship between dreaming and sleep-dependent memory consolidation: (1) Correlation between the activity of the sleeping brain and dreaming, (2) Effects of waking-life on dream content (continuity hypothesis), and (3) Effects of dreams on subsequent daytime behavior and performance. Findings indicate that dream content might be related to reactivation processes, with dream content reflecting learning in waking life, and a dream training effect on waking performance provided some support for the claim that dreaming is influenced by brain processes related to sleep dependent memory consolidation. The research in this area is, however, just at its beginning.

Keywords Dreaming · Function of dreaming · Sleep-dependent memory consolidation · Continuity hypothesis · Lucid dreaming

Introduction

Dreaming has been defined as subjective experiences during sleep that are only accessible if the dreamer has a recollection of those experiences after awakening (Schredl 2014). Research indicates that dreaming is very likely present during all sleep stages; sleep onset, normal sleep, slow wave sleep, and REM sleep (Nielsen 2000; Schredl et al. 2013). With regard to the possible functions of dreaming, it is crucial to differentiate between physiological processes during sleep and the

---

M. Schredl (✉)
Sleep Laboratory, Medical Faculty Mannheim, Central Institute of Mental Health, Heidelberg University, PO Box 12 21 20, 68072 Mannheim, Germany
e-mail: Michael.Schredl@zi-mannheim.de

N. Axmacher and B. Rasch (eds.), *Cognitive Neuroscience of Memory Consolidation*, Studies in Neuroscience, Psychology and Behavioral Economics, DOI 10.1007/978-3-319-45066-7_11

psychological level of dreaming. Although dreaming is related to activation of the sleeping brain it does not reflect the total brain activity during REM sleep or other sleep stages and, thus, the functions of dreaming must not parallel the functions of sleep in general and REM sleep in particular.

One well-established function of sleep is memory consolidation (Rasch and Born 2013). Animal research (Blanco et al. 2015; Ramanathan et al. 2015; Wilson and McNaughton 1994) and research in humans (Deuker et al. 2013; Rasch et al. 2007) showed that sleep-dependent memory consolidation involves some form of replay of the learning task during sleep (see e.g. the chapters by Schreiner, Lehmann and Rasch, and by Deuker, Zhang and Axmacher). As this happens on a cellular or system level, consciousness in the form of dreaming might not be involved. This can be compared to the waking state: the brain is doing a lot of things on a physiological level and only part of this activity is related to subjective experiencing. So the question as to whether dreaming plays a part in the memory consolidation function is a very interesting one (see also the chapter by Eichenlaub, Cash and Blagrove). In the following, essential problems in researching this question and data that might provide indirect support will be addressed and reviewed.

Over the years many possible functions of dreaming have been postulated: guardian of sleep (Freud 1991), reverse learning hypothesis (Crick and Mitchison 1983), mastery hypothesis (Wright and Koulack 1987), mood regulation hypothesis (Kramer 1993), threat simulation theory (Revonsuo 2000), and the protoconsciousness theory (Hobson 2009). All these theories postulate that dreaming is of importance to subsequent waking life in one form or another. However, whether or not these theories are plausible is one part of the story; the other part is the question as to whether they can be tested empirically.

In order to illustrate the methodological issue of testing possible functions of dreaming, a study by Cartwright et al. (1984) will be briefly reviewed. The authors found that divorcing women who dream about their ex-husbands are more psychologically adapted after one year than women who only dreamed about other topics. One interpretation could be that working through the divorce issue within in the dream (dreaming of the ex) serves an adaptive function. But one could also argue that the women who reported ex-husband dreams began to think about these dreams and, as a result of these thinking processes, were able to cope better with the stressful divorce. I.e., it is impossible to differentiate between the effect of the dreamed dream and the effect of the recalling and reporting of the dream since, for this, the waking mind is necessarily involved. We cannot know whether unremembered dreams serve any function.

Despite this major problem in testing theories of dream function directly, empirical dream research might provide indirect evidence that dreams play a role in memory consolidation processes. Three aspects will be addressed: (1) Correlations between the activity of the sleeping brain and dreaming, (2) Effects of waking-life on dream content (continuity hypothesis), and (3) Effects of dreams on subsequent daytime behavior and performance.

Brain Activity and Dream Content

As one part of the consolidation process involves some form of replay of the learning task in the brain (Blanco et al. 2015; Deuker et al. 2013; Ramanathan et al. 2015; Wilson and McNaughton 1994), a basic prerequisite for the role of dreaming in memory consolidation is that dream content is in some form related to brain activity. Several EEG studies (for review see: Erlacher and Schredl 2008) showed a correlation between dreamed activity and the respective brain areas, e.g., talking during the dream was related to activity in the Broca area whereas listening with activity in the Wernicke area (Hong et al. 1996). A combined fMRI-EEG study (Dresler et al. 2011) in one lucid dreamer showed that hand clenching in the dream was related to brain activation in the motor cortex. In addition, Horikawa et al. (2013) were able to correlate brain activity patterns to specific pictures experienced in sleep-onset dreams, i.e., if the persons were looking at a scene, an object, or a person in the dream, the brain activation patterns were different. Studies looking at dream correlates of hippocampal and/or amygdala activity in REM sleep or other sleep stages, however, are still lacking.

An indirect approach to test whether dreams might reflect reactivation processes in the brain was adopted by Schredl et al. (2014). Since Rasch et al. (2007) demonstrated that odor cues presented during slow wave sleep can enhance memory consolidation—presumably via reactivation (the same odor cues were presented during learning the declarative task)—the paradigm was as follows: Two picture series (city themes and rural themes, 15 pictures from each series presented twice) were associated—in a cross-over design—with two different odors. The dreams stimulated with the corresponding odor were compared with the dreams experienced without stimulation. For rural scenes, the effect was significant, i.e., after the odor stimulation during the presentation of rural pictures the dreams of this REM period contained rural topics more often (Schredl et al. 2014). The related effect for city scenes, however, was not found. In a small pilot study, Smith and Hanke (2004) presented auditory stimuli (clicks) while learning a procedural task (mirror tracing). The subgroup (N = 5) who heard a replay of the clicks during active REM sleep reported longer dreams and more often topics related to the task like 'Driving' or 'Competitive sports', indicating that the clicks reactivated the task-related brain pattern and the subjective experiences related to it (driving as a popular metaphor for keeping the pencil between the lines of the figures they traced in the mirror).

Even though the research in this area is still at its beginning, the findings indicate that the basic condition for dreams playing a role in memory consolidation is fulfilled; there is a correlation between brain activity and dream content which might also reflect reactivation of brain patterns related to the learning task.

Effect of Waking-Life on Dream Content (The Continuity Hypothesis)

On the one hand, several researchers (Hobson 2009; Nielsen and Stenstrom 2005; Revonsuo 2000) have stressed the fact that dreaming is a perfect simulation of waking-life reality which seems especially important for procedural memory, e.g., learning motor skills. On the other hand, dreams rarely replay episodic memories with all the details experienced in the corresponding waking-life situations (Fosse et al. 2003; Malinowski and Horton 2014b). Wamsley (2014) pointed out this doesn't necessarily imply that dreaming plays no role in memory consolidation as other processes (in addition to exact replay) like integrating information or extract generalizations are also important in consolidating new information into long-term memory stores.

Empirical research (Schredl 2003) has shown that waking-life experiences, concerns, and preoccupations affect dreams, especially emotionally salient experiences (Malinowski and Horton 2014a; Schredl 2006). For example, it has been shown that sport students dream much more often about sports than psychology students (Erlacher and Schredl 2004; Schredl and Erlacher 2008). Playing computer games also shows up in sleep onset mentation and REM dreams (Gackenbach et al. 2011; Kusse et al. 2012; Stickgold et al. 2000). I.e., the findings support the notion that learning new tasks in waking life is reflected in dreams. However, so-called cognitive activities like reading, writing, arithmetic, and working with a computer appear relatively rarely in dreams and are underrepresented compared to other activities like driving a car or talking with friends, even in students who devote quite some time in their waking lives to these activities (Hartmann 2000; Schredl and Hofmann 2003).

Social interactions play a far more important role in dreams (Hall and Van de Castle 1966; Schredl and Hofmann 2003; Schweickert 2007) and this predominance of social interactions lead some theorists to postulate that dreams might play a role in acquiring and rehearsing social skills because those skills had been of importance from an evolutionary perspective (Revonsuo et al. 2015). Furthermore, dreams also include topics that have never been experienced in waking life, e.g., flying without any adequate means (Schredl 2008, 2011; Schredl and Piel 2007). This lead to the hypothesis that memory consolidation processes affect dream content but not every dream is closely linked to this function of sleep (Payne and Nadel 2004; Wamsley 2014).

To summarize, dream content studies indicate that waking-life experiences related to learning are present in dreams and, thus, support the idea that dreaming might reflect memory consolidation processes of the sleeping brain.

Effect of Dreams on Subsequent Daytime Behavior and Performance

This section will review the studies that linked task-related dream content to retest performance. Moreover, lucid dreaming studies investigating the effect deliberate training of skills while dreaming has on subsequent performance will be presented.

Over the years anecdotes regarding dreams inspiring artists and providing solutions for complex scientific problems (e.g., August Kekulé and the ring structure of benzene) have been reported (Barrett 2001) indicating that the nocturnal brain tries to complete the tasks of the waking brain. Schredl and Erlacher (2007) found in "normal" persons that about 8% of the dreams are creative, i.e., include something helpful for the dreamer. One math student participating in this study reported the following: "Within a dream an error in a computational formula of my master thesis came to my mind which I was able to resolve in the dream by creating a new formula." These accounts support the notion that dreams might be related to some aspects of memory consolidation during sleep.

A small pilot study (N = 6) presented an emotional short story prior to the second night in the sleep lab (Fiss et al. 1977). As expected the dreams of this night included more references to the story compared to the dreams of the preceding night and there was a high correlation between incorporations of the story and the recall performance in the morning, i.e., persons who dreamed about the story recalled it better (Fiss et al. 1977). The research group of Joseph de Koninck conducted several studies in students who participated in a six-week French course, a very intensive program including leisure activities that included spoken French (De Koninck et al. 1990; De Koninck et al. 1988; De Koninck et al. 2012; Sabourin et al. 2006). Interestingly, the participants who first dreamed about French were those showing the best performance in the French test at the end of the program.

Cipolli et al. (2005) investigated factors that explain why some parts of the dream reported in the night immediately after the awakening from REM sleep are forgotten when the participants were asked to report the dream a second time in the morning. Interestingly, they found that dream topics that were present in two or more dreams were more often included in the recollection of the night dreams compared to topics that were only present once in their dream. Cipolli et al. (2005) interpreted this finding as showing that dreams reflect consolidation, i.e., those topics that are processed several times are more stable and could be easier recalled in the morning.

A procedural task was investigated by De Koninck et al. (1996). The participants wore goggles that inverted the visual field for several days, i.e., everything and everyone was seen upside-down. They were carefully monitored and had to complete different tasks like writing. During two nights dreams were collected in the sleep lab via REM awakenings. Four out of eight participants had at least one dream reflecting the learning task, i.e., seeing something upside-down. The participant with four of those dreams outperformed all other participants in the retest at the end of the study. This was interpreted as showing that dreams reflect the process of reorganization of perceptual structures in the brain (De Koninck et al. 1996). Another procedural task (mirror tracing) was applied by Schredl and Erlacher (2010). During the evening the participants were trained to trace different figures they only saw in a mirror. In the following night they were awakened from REM sleep to obtain dream reports. The retest of the mirror tracing was carried out in the morning. Only one out of 71 dream reports included a direct reference to the task. Whereas task-related content was not related to morning performance, persons with more bizarre and negatively toned dreams showed the lowest increments from evening to morning performance, indicating that disturbed dreaming might impede sleep-dependent memory consolidation and/or these dreams reflect participants' worries about their performing well in the experiment (Schredl and Erlacher 2010). The computer game "Doom" (first-person shooter) was played 1 h before sleep and one hour in the morning (Pantoja et al. 2009); 17 of the 22 participants reported game-related dream content and the morning performances showed an inverse-U-shaped relationship to the incorporation of game elements. I.e., persons who dreamed rarely or very often about the task performed worse compared the person who had an intermediate number of game-related incorporations.

Lastly, Wamsley et al. (2010) used a maze task with subsequent naps. The four participants in the sleep condition who reported task-related dreams showed higher increases from pre-testing to post-testing that those (N = 46) without task-related dreams. However, it should be noted that two reports just included hearing the music presented during the training session at sleep onset and none of the dreams—due to the nap design (average total sleep time was about 45 min)—were obtained from REM sleep. Furthermore, the four participants with task-related dreams showed considerably worse performances at the pretest, i.e., this increased their chance of large improvements. One might speculate that trait aspects (bad performance in these kinds of tasks in dreamers very often incorporating daytime experiences in their dreams) may have played a role or that those dreams reflected the participants' worries about performing well as they had not done so in the pretest.

In sport science, the positive effect of mental practice, defined as cognitive rehearsal of a task in absence of overt physical movement, on performance has been repeatedly demonstrated (Driskell et al. 1994). As skilled lucid dreamers are able to perform tasks they planned to do prior to sleep (Stumbrys et al. 2014), research has focused on the effect of lucid dream practice on motor skills. Surveys in German athletes (N = 840) found that about 5% of the participants used lucid dreaming for training their skills in a way similar to the mental practice used in waking (Erlacher et al. 2011–2012). Interestingly, about 75% of these athletes reported positive effects. The following report of a female spring board diver illustrates this: "I practice complex twists and somersaults in my lucid dreams. When I do my lucid dream practice, the movements feel real, but if I want to I can slow down the whole sequence to focus on important details of the dive". (p. 12; Erlacher and Schredl

2008). The participants of Erlacher and Schredl (2010) were trained in a simple task prior to sleep, i.e., tossing a coin twenty times into a cup positioned at a distance of 2 m. The coin tossing was repeated in the morning. The seven lucid dreamers who were able to train during their dreams in the course of the night significantly outperformed those without dream training (Erlacher and Schredl 2010). This finding was replicated for a sequential finger-tapping task, sequences of e.g., "4–1–3–2–4" should be typed repeatedly as fast as possible; the 17 participants who trained the task during their lucid dreams showed about 20% increases in their performance, comparable to the effects of the mental practice control group (Stumbrys et al. 2016).

In summary, the studies relating task-related dream content to performance improvement and the lucid dream studies showed first promising results supporting the notion that dreaming can play a role in sleep-dependent memory consolidation.

Conclusions and Future Directions

Despite the problem of testing the functions of dreams directly, empirical dream research has provided some support for the claim that dreaming is influenced by brain processes related to sleep dependent memory consolidation (Payne and Nadel 2004; Smith 2010; Wamsley 2014). However, the number of studies in this field is still very small and a lot of questions are still to be answered. Smith (2010), for example, pointed out that children's dreams are not very elaborated even though they had to learn a lot in their waking life. Therefore, a promising topic for future research is the relationship between dream content and learning in children.

Most of the dream studies reviewed in this chapter focused on REM dreams but, since declarative memory consolidation is strongly linked to slow wave sleep (Rasch and Born 2013), it would be very interesting to correlate the contents of NREM dreams with memory performance in the morning. Another interesting approach would be to awaken participants from sleep if a reactivation pattern is detected by the MRI machine: if the dream content is in some form related to the task this would support the notion that dreams play a role in memory consolidation.

A similar paradigm could be used with sleep spindles that also play an important role in sleep-dependent memory consolidation (Nielsen et al. 2015). The research so far has indicated that some tasks might be more suitable than others; for example, dreams about mirror tracing were rare (Erlacher and Schredl 2010) whereas computer games were more effective (Pantoja et al. 2009). Furthermore, there might be a variety of factors that affect the relationship between dream content and subsequent memory performance, parallel to the memory consolidation research so far, e.g., performance levels at pretest, emotional involvement, task types and so on.

To conclude, there is evidence that dreams might reflect memory consolidation processes during sleep but many questions are still unanswered.

References

Barrett D (2001) The committee of sleep: how artists, scientists, and athletes use dreams for creative problem-solving—and how you can too. Crown, New York

Blanco W, Pereira CM, Cota VR, Souza AC, Renno-Costa C, Santos S, Dias G, Guerreiro AM, Tort AB, Neto AD, Ribeiro S (2015) Synaptic homeostasis and restructuring across the sleep-wake cycle. PLoS Comput Biol 11(5):e1004241

Cartwright RD, Lloyd S, Knight S, Trenholm I (1984) Broken dreams: a study of the effects of divorce and depression on dream content. Psychiatry 47:251–259

Cipolli C, Fagioli I, Mazzetti M, Tuozzi G (2005) Consolidation effect of the processing of declarative knowledge during human sleep: evidence from long-term retention of interrelated contents of mental sleep experiences. Brain Res Bull 65:97–104

Crick F, Mitchison G (1983) The function of dream sleep. Nature 304:111–114

De Koninck J, Christ G, Rinfret N, Proulx G (1988) Dreams during language learning: when and how is the new language integrated. Psychiatr J Univ Ottawa 13:72–74

De Koninck J, Christ G, Hebert G, Rinfret N (1990) Language learning efficiency, dreams and REM sleep. Psychiatr J Univ 15(2):91–92

De Koninck J, Prevost F, Lortie-Lussier M (1996) Vertical inversion of the visual field and REM sleep mentation. J Sleep Res 5:16–20

De Koninck J, Wong C, Hebert G (2012) Types of dream incorporations of language learning and learning efficiency. J Sleep Res Suppl 21(1):190

Deuker L, Olligs J, Fell J, Kranz TA, Mormann F, Montag C, Reuter M, Elger CE, Axmacher N (2013) Memory consolidation by replay of stimulus-specific neural activity. J Neurosci 33 (49):19373–19383

Dresler M, Koch SP, Wehrle R, Spoormaker VI, Holsboer F, Steiger A, Sämann PG, Obrig H, Czisch M (2011) Dreamed movement elicits activation in the sensorimotor cortex. Curr Biol 21:1833–1837

Driskell JE, Copper C, Moran A (1994) Does mental practice enhance performance? J Appl Psychol 79(4):481–492

Erlacher D, Schredl M (2004) Dreams reflecting waking sport activities: a comparison of sport and psychology students. Int J Sport Psychol 35:301–308

Erlacher D, Schredl M (2008) Do REM (lucid) dreamed and executed actions share the same neural substrate? Int J Dream Res 1:7–14

Erlacher D, Schredl M (2010) Practicing a motor task in a lucid dream enhances subsequent performance: a pilot study. Sport Psychol 24:157–167

Erlacher D, Stumbrys T, Schredl M (2011–2012). Frequency of lucid dreams and lucid dream practice in German athletes. Imagination Cogn Pers 31:237–2012

Fiss H, Kremer E, Litchman J (1977) The mnemonic fuction of dreaming. Sleep Res 6:122

Fosse MJ, Fosse R, Hobson JA, Stickgold RJ (2003) Dreaming and episodic memory: a functional dissociation? J Cogn Neurosci 15:1–9

Freud S (1991) The interpretation of dreams (Org.: Die Traumdeutung, 1900). Penguin Books, London

Gackenbach J, Rosie M, Bown J, Sample T (2011) Dream incorporation of video-game play as a function of interactivity and fidelity. Dreaming 21(1):32–50

Hall CS, Van de Castle RL (1966) The content analysis of dreams. Appleton-Century-Crofts, New York

Hartmann E (2000) We do not dream of the 3 R's: implications for the nature of dream mentation. Dreaming 10:103–110

Hobson JA (2009) REM sleep and dreaming: towards a theory of protoconsciousness. Nat Rev Neurosci 10(11):803–813

Hong CC, Jin Y, Potkin S, Buchsbaum M, Wu J, Callagham G, Nudleman K, Gillin JC (1996) Language in dreaming and regional EEG alpha power. Sleep 19:232–235

Horikawa T, Tamaki M, Miyawaki Y, Kamitani Y (2013) Neural decoding of visual imagery during sleep. Science 340(6132):639–642

Kramer M (1993) The selective mood regulatory function of dreaming: An update and revision. In: Moffitt A, Kramer M, Hoffmann R (eds) The functions of dreaming. State University of New York Press, Albany, pp 139–195

Kusse C, Shaffii-Le Bourdiec A, Schrouff J, Matarazzo L, Maquet P (2012) Experience-dependent induction of hypnagogic images during daytime naps: a combined behavioural and EEG study. J Sleep Res 21:10–20. doi:10.1111/j.1365-2869.2011.00939.x

Malinowski JE, Horton CL (2014a) Evidence for the preferential incorporation of emotional waking-life experiences into dreams. Dreaming 24(1):18–31

Malinowski JE, Horton CL (2014b) Memory sources of dreams: the incorporation of autobiographical rather than episodic experiences. J Sleep Res 23:441–447

Nielsen TA (2000) A review of mentation in REM and NREM sleep: "covert" REM sleep as a possible reconciliation of two opposing models. Behav Brain Sci 23:851–866

Nielsen TA, Stenstrom P (2005) What are the memory sources of dreaming? Nature 437: 1286–1289

Nielsen TA, O'Reilly C, Carr M, Dumel G, Godin I, Solomonova E, Lara-Carrasco J, Blanchette-Carrière C, Paquette T (2015) Overnight improvements in two REM sleep-sensitive tasks are associated with both REM and NREM sleep changes, sleep spindle features, and awakenings for dream recall. Neurobiol Learn Mem 122:88–97

Pantoja AL, Faber J, Rocha LH, Ferro D, Silvestre-Souza RC, Dias G, Araujo JF, Nicolelis MA, Ribeiro S (2009) Assessment of the adaptive value of dreams. Sleep Suppl 32:A421

Payne JD, Nadel L (2004) Sleep, dreams, and memory consolidation: the role of the stress hormone cortisol. Learn Mem 11:671–678

Ramanathan DS, Gulati T, Ganguly K (2015) Sleep-dependent reactivation of ensembles in motor cortex promotes skill consolidation. PLoS Biol 13(9):e1002263

Rasch B, Born J (2013) About sleep's role in memory. Physiol Rev 93(2):681–766

Rasch B, Buchel C, Gais S, Born J (2007) Odor cues during slow-wave sleep prompt declarative memory consolidation. Science 315(5817):1426–1429

Revonsuo A (2000) The reinterpretation of dreams: an evolutionary hypothesis of the function of dreaming. Behav Brain Sci 23:877–901

Revonsuo A, Tuominen J, Valli K (2015) The avatars in the machine. In: Metzinger TK, Windt JM (eds) Open MIND. MIND Group, Frankfurt am Main, pp 1–28

Sabourin C, Forest G, Hebert G, De Koninck J (2006) Intensive language learning and episodic memory in dreams. Sleep Suppl 29:A48

Schredl M (2003) Continuity between waking and dreaming: a proposal for a mathematical model. Sleep Hypn 5:38–52

Schredl M (2006) Factors affecting the continuity between waking and dreaming: emotional intensity and emotional tone of the waking-life event. Sleep Hypn 8:1–5

Schredl M (2008) Personality correlates of flying dreams. Imagination, Cogn Pers 27:129–137

Schredl M (2011) Frequency and nature of flying dreams in a long dream series. Int J Dream Res 4:31–34

Schredl M (2014) Sleep and dreaming. In: Bassetti CL, Dogas Z, Peigneux P (eds) ESRS european sleep medicine textbook. European Sleep Resarch Society, Regensburg, pp 63–71

Schredl M, Erlacher D (2007) Self-reported effects of dreams on waking-life creativity: an empirical study. J Psychol 141:35–46

Schredl M, Erlacher D (2008) Relationship between waking sport activities, reading and dream content in sport and psychology students. J Psychol 142:267–275

Schredl M, Erlacher D (2010) Is sleep-dependent memory consolidation of a visuo-motor task related to dream content? Int J Dream Res 3:74–79

Schredl M, Hofmann F (2003) Continuity between waking activities and dream activities. Conscious Cogn 12:298–308

Schredl M, Piel E (2007) Prevalence of flying dreams. Percept Mot Skills 105:657–660

Schredl M, Brennecke J, Reinhard I (2013) Does training increase NREM dream recall? A pilot study. Int J Dream Res 6:54–58

Schredl M, Hoffmann L, Sommer JU, Stuck BA (2014) Olfactory stimulation during sleep can reactivate odor-associated images. Chemosens Percept 7:140–146

Schweickert R (2007) Social networks of characters in dreams. In: Barrett D, McNamara P (eds) The new science of dreaming, vol 3., Cultural and theoretical perspectivesPraeger, Westport, pp 277–297

Smith C (2010) Sleep states, memory processing, and dreams. Sleep Med Clin 5(2):217–228

Smith C, Hanke J (2004) Memory processing reflected in dreams from rapid eye movement sleep. Sleep Suppl 27:A60

Stickgold R, Malia A, Maguire D, Roddenberry D, O'Connor M (2000) Replaying the game: hypnagogic images in normals and amnesics. Science 290:350–353

Stumbrys T, Erlacher D, Johnson M, Schredl M (2014) The phenomenology of lucid dreaming: an online survey. Am J Psychol 127(2):191–204

Stumbrys T, Erlacher D, Schredl M (2016) Effectiveness of motor practice in lucid dreams: a comparison with physical and mental practice. J Sports Sci 34(1):27–34

Wamsley EJ (2014) Dreaming and offline memory consolidation. Curr Neurol Neurosci Rep 14 (3):1–7

Wamsley EJ, Tucker M, Payne JD, Benavides JA, Stickgold R (2010) Dreaming of a learning task is associated with enhanced sleep-dependent memory consolidation. Curr Biol 20:850–855

Wilson MA, McNaughton BL (1994) Reactivation of hippocampal enseble memories during sleep. Science 265:676–679

Wright J, Koulack D (1987) Dreams and contemporary stress: a disruption-avoidance-adaptation model. Sleep 10:172–179

Part III
Mechanisms of Memory Consolidation on a Systems Physiology Level

Neuronal Oscillations and Reactivation Subserving Memory Consolidation

Til Ole Bergmann and Bernhard P. Staresina

Abstract Newly acquired memories are initially hippocampus-dependent and need to undergo a process of active system consolidation, during which they are redistributed to neocortical sites for long-term storage. This process is thought to take place during phases of quiet wakefulness and non-rapid-eye movement (NREM) sleep and is presumably based on the repeated reactivation of memory engrams (patterns of hippocampo-neocortical connections) which gradually drives the establishment of respective direct cortico-cortical connections. During NREM sleep (and similarly during quiet wakefulness), control via brainstem neuromodulatory systems (in particular the cholinergic one) enables a specific kind of oscillatory activity in the thalamo-neocortico-hippocampal system that facilitates memory reactivation. NREM oscillatory activity is characterized by the neocortical slow oscillation (SO; <1 Hz), the thalamic sleep spindle ($\sim$12–15 Hz) and the hippocampal ripple (>80 Hz). The intricate interaction of SOs, spindles and ripples constitutes a set of hierarchically nested oscillations, which provides the fine-tuned temporal and spatial structure that is required to orchestrate the reactivation of memory traces and the information flow between hippocampus and neocortex. In this chapter we (i) provide a conceptual introduction to system memory consolidation, (ii) describe the neuronal mechanisms thought to underlie the generation of and interaction between SOs, spindles and ripples, (iii) discuss how these oscillations presumably mediate memory reactivation and hippocampo-neocortical cross-talk, and (iv) outline new promising approaches to directly study the ongoing reactivation of memory representations in humans.

T.O. Bergmann (✉)
Institute for Medical Psychology and Behavioral Neurobiology,
Eberhard Karls University of Tübingen, Tübingen, Germany
e-mail: til.bergmann@uni-tuebingen.de

T.O. Bergmann
Department of Neurology & Stroke, and Hertie Institute for Clinical Brain Research,
University of Tübingen, Tübingen, Germany

B.P. Staresina (✉)
School of Psychology, University of Birmingham, Birmingham, UK
e-mail: b.staresina@bham.ac.uk

N. Axmacher and B. Rasch (eds.), *Cognitive Neuroscience of Memory Consolidation*, Studies in Neuroscience, Psychology and Behavioral Economics, DOI 10.1007/978-3-319-45066-7_12

Keywords Neuronal oscillation · Reactivation · Memory representation · Active system memory consolidation · Slow oscillation · Sleep spindle · Ripple · Hippocampus · Neocortex · Hippocampo-neocortical dialogue · Acetylcholine

Active System Memory Consolidation

According to *two-stage models of declarative memory formation*, newly acquired mnemonic representations initially rely strongly on the hippocampus, a key region of the medial temporal lobe (MTL) memory system and top-level zone of associative convergence in the brain. Over time, however, these representations are thought to rely less on the hippocampus and more on neocortical regions (Born et al. 2006; Buzsaki 1996; Frankland and Bontempi 2005; McClelland et al. 1995) (see also chapter by Genzel and Wixted). During initial *encoding*, plastic changes of synaptic weights within the hippocampus allow to rapidly form new associations between the distributed neocortical activation patterns (*representations*) that together constitute (represent) the mnemonic event (Eichenbaum 2004; Frankland and Bontempi 2005; Josselyn et al. 2015). During later memory *recollection*, these neocortical representations can then be conjointly triggered by a retrieval cue to reinstantiate an activity pattern sufficiently similar to the one representing the original mnemonic event (Eichenbaum 2004; Greenberg and Rubin 2003). In between encoding and retrieval, sometimes years apart, the associative memory trace (the *engram*) has to be maintained, which is believed to require an active process of *memory consolidation* (Müller and Pilzecker 1900). The initially labile memory trace is prone to decay and interference and needs to be stabilized for the long-term, a process taking place at both the *synaptic* and the *systems level* (Diekelmann and Born 2010; Dudai 2004; Frankland and Bontempi 2005). During *synaptic consolidation* cellular processes of long-term potentiation and depression stabilize the early encoding-related changes in synaptic strength within minutes to hours. In contrast, *system consolidation* takes days to years and requires a spatial reorganization and redistribution of memory engrams and a transformation of the constituting elements (i.e. associations) from neocortico-hippocampo-neocortical into neocortico-neocortical connections (Buzsaki 1996). Notably, during *system consolidation* the encoded information may also undergo more complex transformations such as the abstraction of relevant information (e.g. *semantization*) (Meeter and Murre 2004) and the integration into existing mnemonic networks (i.e. *schemas*) (Morris 2006; van Kesteren et al. 2012; Walker and Stickgold 2010; see chapters by Fernandez, by Genzel and Battaglia, and by Sekeres, Moscovitch and Winocur).

Memory Consolidation by Reactivation

The neuronal mechanisms underlying system level consolidation are thought to involve the repeated conjoint reactivation of those neocortical representations that constitute a memory engram via their common hippocampal associations (Frankland and Bontempi 2005). This synchronous activation may then lead to the establishment of direct neocortico-neocortical associations according to the laws of Hebbian plasticity (Buzsaki 1998; Frankland and Bontempi 2005). Importantly, the proposed mechanism of neuronal reactivation for system memory consolidation requires two conditions to be met. Firstly, the reactivation of target representations should not incidentally coincide with the activation of other neocortical representations to avoid interference with information processing and the formation of spurious associations. Secondly, the reactivation procedure needs to follow a precise temporal structure to ensure that the distributed neocortical representations are simultaneously activated within a few tens of milliseconds to enable spike timing dependent plasticity (STDP) (Dan and Poo 2004). How can these requirements be met? Firstly, to prevent interference, reactivation is presumably restricted to *offline periods* where active information processing is minimized, that is preferably during deep sleep or at least inactive resting wakefulness (Buzsaki 1996; McClelland et al. 1995). Secondly, in the absence of sensory input during these offline periods, a hierarchical nesting of spontaneous neuronal oscillations at multiple frequencies can provide the fine-grained temporal structure needed for STDP to occur (Dan and Poo 2004; Masquelier et al. 2009). While little is known about the potential oscillatory mechanisms underlying memory consolidation during resting wakefulness, there has been remarkable progress in the understanding of the intricate interplay of sleep-specific neuronal oscillations subserving system memory consolidation (Staresina et al. 2015).

Sleep Stages for Memory Consolidation

By now there is large consensus that sleep benefits the consolidation of various kinds of memories (Rasch and Born 2013; see chapter by Schönauer and Gais). However, the specific—and most likely complementary—functions of non-rapid-eye-movement (NREM) sleep, especially slow wave sleep (SWS), and rapid eye movement sleep (REM) are not yet fully understood (Diekelmann and Born 2010; Rasch and Born 2013). The *dual-process hypothesis* of memory consolidation assumes that NREM sleep facilitates the consolidation of declarative, hippocampus-dependent mnemonic contents, whereas non-declarative (e.g. procedural or emotional), hippocampus-independent memories particularly benefit from REM sleep (Maquet 2001). In contrast, the *sequential hypothesis* argues that the successive alternation of NREM and REM epochs is important for the consolidation of either mnemonic content, with both stages mediating complementary functions: During SWS, memories to be retained are distinguished from those to be downgraded, whereas during subsequent REM sleep memories are stored and integrated with preexisting memories (Giuditta 2014; Giuditta et al. 1995). The *active system consolidation hypothesis* reconciles these two perspectives (Born et al. 2006; Diekelmann and Born 2010; Rasch and Born 2013). It assumes that SWS and REM sleep play complementary roles for memory consolidation and that declarative and non-declarative memories differentially benefit from these sleep stages as they rely to varying degrees on their respective function. While during SWS hippocampo-neocortical memory traces are actively reactivated and reorganized (*system consolidation*), ensuing REM sleep is crucial for the stabilization of the reactivated memory traces (*synaptic consolidation*). It is possible that the initial reactivation of hippocampo-neocortical memory traces during SWS initiates early plastic processes but merely 'tags' the relevant neocortico-neocortical synapses for later consolidation (Frey and Morris 1998), while subsequent REM sleep is required for long-term plasticity to be established (Rasch and Born 2007, 2013).

Neuromodulatory Control of Oscillatory Brain States

So what are the neurobiological mechanisms mediating these divergent roles of NREM and REM sleep in memory consolidation? To begin with, the shifts between and maintenance of sleep stages are controlled via the activity of monoaminergic (noradrenergic, serotonergic, dopaminergic) and cholinergic neuromodulatory systems in the brain stem (and basal forebrain), which are thought to prominently contribute to processes of memory consolidation (Diekelmann and Born 2010; Rasch and Born 2013). These neuromodulatory systems do not only project abundantly into neocortex and hippocampus but also into the thalamus where they regulate the firing characteristics of *reticular thalamic* and *thalamocortical neurons*, and thereby exert a fundamental impact on the kind of neuronal oscillations generated throughout the thalamo-cortical system (Datta and Maclean 2007; Destexhe et al. 1994; Hill and Tononi 2005; McCormick 1992). Perhaps the most important neuromodulator in this context is *acetylcholine* (ACh). Cholinergic neurons of the *brain stem cholinergic system* (including pedunculopontine tegmental nucleus and laterodorsal pontine tegmentum) project mainly to the thalamus but also to the *basal forebrain cholinergic system* (nucleus basalis of Meynert, medial septal nucleus, and diagonal band of Broca) which in turn projects to neocortex and hippocampus (Datta and Maclean 2007) (Fig. 1).

During NREM sleep, cholinergic tone is at a minimum and noradrenergic activity is at intermediate levels, whereas during REM sleep, cholinergic tone is comparable to (neocortex) or even higher than during wakefulness (hippocampus) and noradrenergic and serotonergic levels are at a minimum (Datta and Maclean 2007; Diekelmann and Born 2010; Marrosu et al. 1995). Notably, cholinergic tone also varies within wakefulness, being higher during behavioral states of active wakefulness (exploration) than quiet wakefulness (rest) (Marrosu et al. 1995). Moreover, acetylcholine has been implicated in attentional top-down control via the regulation of local oscillatory brain states with low ACh causing neuronal synchronization and high ACh causing neuronal desynchronization in the neocortex (Harris and Thiele 2011).

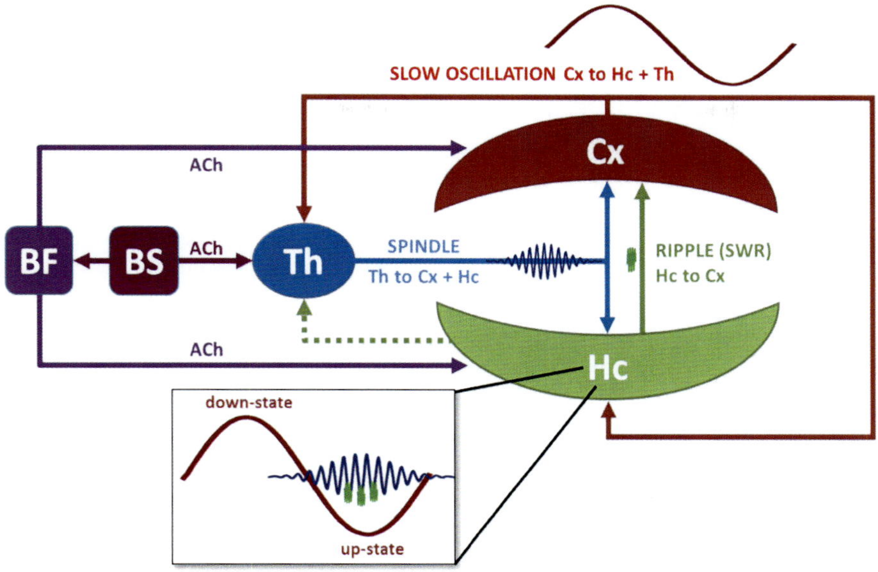

Fig. 1 Model of neuronal circuitry subserving the interaction of slow oscillations, spindles and ripples. Abbreviations: *Cx* neocortex, *Hc* hippocampus, *Th* thalamus, *BF* basal forebrain cholinergic system (including *medial septal nucleus*, *nucleus basalis of Meynert*, and *diagonal band of Broca*), *BS* brain stem cholinergic system (including *pedunculopontine tegmental nucleus* and *laterodorsal pontine tegmentum*), *ACh* acetylcholine. Note that also noradrenergic, serotonergic and dopaminergic brain stem nuclei are likely involved in the control of sleep stages but are left out here for the sake of clarity. Arrows depict the proposed direction of information flow mediated by the respective oscillations. *Bottom left insert* depicts phase-amplitude coupling (PAC) of slow oscillations, spindles and ripples in the human hippocampus. Note that the polarity of the EEG signal is inverted in depth recordings with respect to scalp recordings, such that the SO up-state corresponds to the signal trough in hippocampal recordings

Accordingly, in the thalamocortical system, the low cholinergic tone during NREM sleep and quiet wakefulness facilitates the emergence of synchronized neuronal activity. For NREM sleep, this pertains to the characteristic NREM defining slow oscillations, delta waves and spindles, which are crucially involved in system memory consolidation (Born et al. 2006; Diekelmann and Born 2010; Rasch and Born 2013). On the other hand, the high cholinergic tone during REM sleep and active wakefulness completely obliterates those oscillations and instead facilitates the faster beta and gamma oscillations characterizing these more alert brain states (Steriade 2004).

In the hippocampus, low cholinergic tone during NREM sleep and quiet wakefulness facilitates the occurrence of hippocampal *sharp wave ripple* (SWR) events (Hasselmo and McGaughy 2004; see chapter by Maier and Kempter), which are thought to represent memory reactivation and replay (Buzsaki 1996; Sadowski et al. 2011; Sutherland and McNaughton 2000). It has even been argued that cholinergic tone may regulate the direction (forward vs. backward) of replay during SWRs in different behavioral states such as active wakefulness, quiet wakefulness, and NREM sleep (Atherton et al. 2015). Conversely, high cholinergic drive from the basal forebrain during REM sleep and active wakefulness reduces the occurrence of SWRs and instead fosters the emergence of theta and coupled gamma oscillations in the hippocampus (Colgin 2016; Hutchison and Rathore 2015; Lisman and Jensen 2013), which during wakefulness is strongly associated with memory encoding (Colgin 2015, 2016; Düzel et al. 2010; Hasselmo 2005), also by synchronization with the medial prefrontal cortex (Anderson et al. 2010; Fuentemilla et al. 2014; Jensen 2005). During REM sleep, when cholinergic tone is at a maximum, theta (and coupled gamma) oscillations are ubiquitous in the hippocampus and have been suggested to contribute to encoding-like processes and synaptic consolidation (Diekelmann and Born 2010; Hutchison and Rathore 2015) (see also chapter by Schreiner, Lehmann and Rasch on the relevance of theta oscillations during NREM sleep). Also the REM sleep characteristic pontine-geniculo-occipital (PGO) waves depend on high cholinergic tone, are strongly linked to the presence of theta waves, specifically the theta synchronization between hippocampus and amygdala, and have been implicated in synaptic plasticity (Datta and Maclean 2007; Hutchison and Rathore 2015) and consolidation of emotional memory (Hutchison and Rathore 2015).

In summary, the cholinergic system appears to toggle the entire thalamo-cortico-hippocampal system between two modes, i.e. a high ACh *encoding mode* during active wakefulness and REM sleep that facilitates information flow in neocortio-hippocampal direction and thus memory encoding and synaptic plasticity, and a low ACh *consolidation mode* during quiet wakefulness and NREM sleep that promotes hippocampo-neocortical information flow and thus system memory consolidation (Diekelmann and Born 2010; Gais and Born 2004a, b; Hasselmo 1999; Hasselmo and McGaughy 2004; Rasch et al. 2006). Accordingly, this chapter will focus exclusively on the consolidation mode, mainly during NREM sleep but also during quiet wakefulness. In the next sections will thus discuss the neuronal mechanisms behind the most relevant neuronal oscillations of NREM sleep, their intricate interaction, and how they are thought to mediate system memory consolidation.

Functional Neuroanatomy for Neuronal Oscillations During NREM Sleep

As became clear in the previous section, it is not the sleep stages per se that mediate memory consolidation, as they are not the cause but the result of the above described processes. Neuromodulatory neurons located in the brain stem and basal forebrain control the characteristic oscillatory patterns which are then used to define and score sleep stages (Iber et al. 2007; Rechtschaffen and Kales 1968). Due to the historically evolved and somewhat arbitrary scoring rules applied to fixed 30 s intervals, the oscillatory (and presumably also the neuromodulatory) composition within certain sleep stages is usually rather heterogeneous. Therefore, when studying the neuronal mechanisms of memory consolidation during sleep, our focus should not be on the sleep stages but on the oscillations themselves, as it is those oscillations that eventually control information processing and plastic processes during sleep (Diekelmann and Born 2010; Rasch and Born 2013; Steriade and Timofeev 2003).

Neuronal information transfer requires a precise temporal structure, which during sleep (in the absence of external stimuli) has to be provided by spontaneous brain activity itself. NREM sleep is hallmarked by three cardinal types of neuronal oscillations which may provide the temporal scaffold for such information transfer (Buzsaki 1996; Diekelmann and Born 2010): (i) neocortical *slow oscillations* (<1 Hz) and *delta waves* (1–4 Hz), (ii) thalamo-cortical *sleep spindles* ($\sim$12–15 Hz) and (iii) hippocampal *ripples* ($\sim$80–100 Hz in humans, and at up to $\sim$200 Hz in rodents), often occurring as *sharp wave ripple* complexes (SWR). The next subsections will thus describe the origin and mechanisms of these oscillations as well as their interplay subserving system memory consolidation.

Slow Oscillations

The *slow oscillation* ($\sim$0.75 Hz, SO) is the hallmark rhythm of NREM sleep and defining for its deepest stage, termed slow wave sleep (SWS) (Iber et al. 2007; Rechtschaffen and Kales 1968). SOs can be measured with surface EEG recordings at basically all cortical sites, though with maximum amplitude over frontal channels. They reflect global fluctuations in cellular excitability resulting from alternating phases of joint hyperpolarization (down-states), associated with widespread neuronal silence, and depolarization (up-states), associated with massive synchronous neuronal firing of large neuron populations (Achermann and Borbely 1997; Steriade et al. 1993b). SOs emerge spontaneously in the neocortex, as observed in vivo even after cortical differentiation (Timofeev et al. 2000) or thalamectomy (Steriade et al. 1993a) as well as in vitro in cortical slices (Sanchez-Vives and McCormick 2000); but they may also be triggered via thalamic input (Crunelli and Hughes 2009). SOs can be recorded in all major types of neocortical neurons, including excitatory pyramidal and inhibitory interneurons (Contreras and Steriade 1995). The SO up-state is mainly composed by non-NMDA-mediated excitatory postsynaptic potentials (EPSPs), fast prepotentials, voltage-dependent persistent inward directed NA^+ currents, and also fast inhibitory postsynaptic potentials (IPSPs) of GABAergic inhibitory interneurons (Steriade 2006; Steriade and Timofeev 2003). The SO down-state is thus probably not produced by GABAergic inhibitory interneurons, but is due to disfacilitation, that is the removal of excitatory synaptic inputs possibly through extracellular Ca^{2+} depletion, and NA^+-induced outward directed K^+ currents (Steriade 2006; Steriade and Timofeev 2003). Despite the existence of local SOs (Nir et al. 2011), SOs typically reflect travelling waves that originate locally anywhere in the neocortex, though mainly in prefrontal regions (Massimini et al. 2004; Murphy et al. 2009), and eventually encompass the entire neocortex (Massimini et al. 2004), the hippocampus (Nir et al. 2011; Sharma et al. 2010; Staresina et al. 2015; Wolansky et al. 2006), as well as the thalamus where they trigger the release of sleep spindles (Mölle et al. 2002; Sirota et al. 2003; Steriade 2006).

Sometimes, SOs and *delta waves* (1–4 Hz) (Steriade 2006) are pooled together as *slow waves* (0.5–4.5 Hz) to denote *slow wave activity* (SWA) (Tononi and Cirelli 2003). However, there seem to be two kinds of delta waves: those generated by thalamocortical neurons and those generated in the neocortex (Steriade 2003). Here, with 'slow waves' we always refer to the neocortical variant and usually to the ~0.75 Hz slow oscillation, which appears to be most relevant for the hippocampo-neocortical dialogue during system memory consolidation (Achermann and Borbely 1997; Rasch and Born 2013).

Also, *K-complexes* are sometimes treated separately from SOs. The K-complex is characterized by a sharp high-amplitude hyperpolarization phase followed by moderate depolarization which is often accompanied by pronounced spindle activity. While K-complexes are a defining feature of light NREM sleep, their distinction from slow waves during deep NREM sleep is questionable as there is accumulating evidence for both events relying on the very same neuronal mechanism (Amzica and Steriade 1997; Cash et al. 2009; Colrain 2005). Here, we therefore do not differentiate between K-complexes and isolated slow waves combined with a sleep spindle.

Sleep Spindles

Sleep spindles (originally ~12–15 Hz) are transient (0.5–3 s) oscillatory events of waxing and waning amplitude, and constitute another hallmark characteristic and defining feature of NREM sleep (Iber et al. 2007; Rechtschaffen and Kales 1968) (see also chapter by McDevitt can colleagues). They can be observed with surface EEG recordings over the entire scalp, but are sometimes differentiated into two different classes according to their frequency and topography, i.e. the classical *fast spindles* (12–15 Hz or 13–16 Hz), peaking at centroparietal sites and the *slow spindles* (9–12 Hz or 10–13 Hz) with a more frontal topography (De Gennaro and Ferrara 2003; Mölle et al. 2011). The exact frequency boundaries between fast and slow spindles (see above) as well as their potential functional differentiation are still a matter of debate and ongoing research (Andrillon et al. 2011; De Gennaro and Ferrara 2003; Mölle et al. 2011; Peter-Derex et al. 2012), which is however beyond the scope of this chapter. Here, we will focus on the classical fast spindle (~12–15 Hz) which has more consistently been associated with system memory consolidation and (para-) hippocampal activity (Andrade et al. 2011; Bergmann et al. 2008, 2012a; Clemens et al. 2011; Schabus et al. 2007; Staresina et al. 2015).

Sleep spindles are generated by intra-thalamic negative feedback loops between excitatory glutamatergic thalamo-cortical (TC) neurons in the thalamic core and inhibitory GABAergic neurons located in the reticular thalamic nucleus (TRN) (Luthi 2014; Steriade 2003). When the depolarizing drive of brainstem neuromodulatory systems is withdrawn from the thalamus during NREM sleep, TRN cells slightly hyperpolarize and switch from regular firing into a burst firing mode (mediated by low-threshold voltage-gated Ca^{2+} channels, called T-channels), rhythmically inhibiting TC activity by GABA-A-receptor-mediated burst IPSPs; TC neurons in turn respond with excitatory rebound burst discharges re-exciting TRN neurons by burst EPSPS, recruiting more and more TRN and TC neurons into the rhythmic synchronized discharge pattern of waxing amplitude (Luthi 2014; Steriade 2006). TRN cells cause widespread recruitment of TC cells in both specific and non-specific thalamic nuclei which in turn produce strong synchronous excitatory thalamocortical volleys entraining both excitatory and inhibitory neurons in large parts of the neocortex and thereby create the spindle-like activity patterns that are observed in surface EEG recordings (Luthi 2014; Steriade 2006). While (paravalbumin-positive) *core cells* of specific thalamic relay nuclei project to neocortical layers IV and V, (calbindin-positive) *matrix cells* of non-specific thalamic nuclei project to layers I and VI; but both excite inhibitory TRN cells via axon collaterals and receive feedback from layer VI neocortical neurons (Jones 2001; Llinas et al. 2002; Pereira de Vasconcelos and Cassel 2014). These excitatory cortico-thalamic (CT) neurons projecting back from layer VI onto both TRN and TC cells, further support the evolution of the spindle and may even contribute (along with intrinsic thalamic processes) to its eventual termination via increasingly asynchronous firing during the waning phase (Astori et al. 2013; Bonjean et al. 2011; Gardner et al. 2013; Luthi 2014). Cortico-thalamic input may also initiate sleep spindles (Astori et al. 2013; Bal et al. 2000), as in the case of spontaneous SOs, where the massive rebound in neocortical firing during the beginning up-state following down-state hyperpolarization triggers a sleep spindle sequence in the thalamus (Mölle et al. 2002; Steriade 2003, 2006). The global synchronization of neocortical neurons during global SOs is thus partially responsible for the frequently observed global synchronization of spindles across neocortical sites (Contreras et al. 1996, 1997), but see Bonjean et al. (2012). However, locally restricted spindle activity exist as well (Andrillon et al. 2011), and it has been proposed that matrix cells, diffusively projecting into the neocortex, are responsible for more synchronized global spindles, whereas the topographically specific projections from core cells rather cause locally restricted neocortical spindle input (Bonjean et al. 2012).

Importantly, sleep spindles have also been observed in the hippocampus (Andrillon et al. 2011; Sarasso et al. 2014; Staresina et al. 2015), suggesting that thalamic neurons also drive hippocampal activity via direct thalamo-hippocampal projections. Notably, as shown in rats, the non-specific *ventral midline thalamic nuclei*, i.e. *reuniens nucleus* and *rhomboid nucleus*, do indeed provide direct input into the hippocampus as well as the prefrontal cortex (Pereira de Vasconcelos and Cassel 2014; Varela et al. 2014). The rhomboid nucleus mainly innervates the CA1 region of the dorsal hippocampus (analogous to human posterior hippocampus), whereas the reuniens nucleus innervates CA1 and subiculum of both dorsal and even more so ventral hippocampus (analog to human anterior hippocampus), from where it also receives direct feedback connections (Pereira de Vasconcelos and Cassel 2014; Varela et al. 2014). Like for the neocortex, rhomboid nucleus neurons project to both excitatory pyramidal cells and inhibitory GABAergic hippocampal interneurons (Cassel et al. 2013), thus providing a potential mechanism for entraining spindle activity in the output regions (CA1, subiculum) of the hippocampus in synchrony with spindle input to the neocortex. However, further research has yet to reveal whether neurons in these particular thalamic nuclei are indeed locked to the spindle rhythm, and an alternative route for spindles to reach the hippocampus could involve the typical neocortical-entorhinal-hippocampal input pathway (Buzsaki 2015).

Ripples

Ripples are high-frequency bursts, observed at ~200 Hz in the CA1 subregion of the rodent hippocampus (Buzsaki 2015; Buzsaki et al. 1992; O'Keefe 1976), but at lower frequencies of 80–140 Hz in non-human primates (Logothetis et al. 2012; Skaggs et al. 2007) and around 80–100 Hz in intracranial parahippocampal (Clemens et al. 2007, 2011) and hippocampal recordings in humans (Axmacher et al. 2008; Bragin et al. 1999; Le Van Quyen et al. 2008, 2010; Staba et al. 2002; Urrestarazu et al. 2007). Ripple activity is often superimposed on irregularly occurring high amplitude *sharp waves*, together forming so called *sharp wave ripple* (SWR) complexes, and although sharp waves and ripples are tightly coupled, they are separate events of distinct neuronal origin (Buzsaki 2015; Colgin 2016). Sharp waves are brief excitatory events transmitted from the CA3 to the CA1 subregion of the hippocampus (Buzsaki 1986), where they trigger ripple oscillations that are locally generated within CA1 by ripple-frequency spiking basket cell interneurons rhythmically inhibiting pyramidal cells (Buzsaki 2015; Colgin 2016; Sullivan et al. 2011).

SWRs in rodent CA1 have been observed across the entire longitudinal axis of the hippocampus; however, they do not synchronize the whole hippocampus but rather remain local or travel across small segments in either direction along the hippocampal axis (Buzsaki 2015; Patel et al. 2013). Due to their strong local synchronization of neuronal firing, hippocampal SWRs represent a powerful signal, which is preserved along the output network of the hippocampal system, i.e. the CA1-subiculum-entorhinal cortex pathway (Buzsaki 2015; Chrobak and Buzsaki 1996). Ripples may thus exert a strong impact on virtually any neocortical site depending on their location along the longitudinal axis of the hippocampus, as different connectivity profiles (and functions) have been revealed for the more anterior (ventral/temporal in rodents) and posterior (dorsal/septal in rodents) portions of the hippocampus (Poppenk et al. 2013; Ranganath and Ritchey 2012; Strange et al. 2014).

Importantly, hippocampal SWRs may be triggered by phasic input to the hippocampus, such as by SOs (from the neocortex via the entorhinal cortex) and spindles (either via the same route or alternatively directly from the non-specific thalamus), as reflected by their tight temporal relationship with SO up-states and spindle troughs facilitating SWR occurrence (Battaglia et al. 2004; Isomura et al. 2006; Mölle et al. 2006, 2009; Sirota et al. 2003; Sullivan et al. 2011). Possibly, SWRs may in turn contribute to spindle generation via the above described hippocampo-thalamic projections to the ventral midline thalamic nuclei. However, the exact temporal relationship between spindles and ripples is still unclear, with some rodent data suggesting spindles to precede ripples (Sirota et al. 2003) and other data ripples to precede spindles (Siapas and Wilson 1998). In accordance with the latter, intracranial recordings in humans via *foramen ovale* electrodes also found parahippocampal ripple activity to precede cortical spindles at a timescale of seconds (Clemens et al. 2007; Clemens et al. 2011), but at a timescale of milliseconds found parahippocampal ripples grouped in the troughs of both parahippocampal and centroparietal (but not frontal) spindles (Clemens et al. 2011). Likewise, our own data from intracranial hippocampal recordings in humans suggests that ripples occur during spindles, with the spindle onset preceding the ripple event but most ripple events preceding the spindle center (Staresina et al. 2015). More importantly, however, we found that human hippocampal ripples were grouped into the troughs of hippocampal spindles as previously reported for rodents (Sirota et al. 2003). Thus, the dynamics of how parietal and frontal cortical as well as (para-)hippocampal spindles interact with (para-)hippocampal ripples on a slower time scale, potentially initiating and driving one another, are still not completely understood. However, there is clear evidence now that hippocampal ripples are grouped by the phase of hippocampal spindles during their occurrence (Staresina et al. 2015), enabling the fine-tuned temporal structure of neuronal activity that is necessary for the reactivation and redistribution of memory representations during system consolidation which will be discussed in the next section.

Oscillatory Interactions Mediating Hippocampo-neocortical System Memory Consolidation

As outlined in the previous section, the cardinal oscillations of NREM sleep, that is SOs, spindles and ripples, are closely linked and influence each other in multiple ways. In the following we will discuss how exactly, building on the *active system consolidation hypothesis* (Born et al. 2006; Diekelmann and Born 2010; Rasch and Born 2013), we propose these oscillations interact to actively drive the reactivation and redistribution of hippocampo-neocortical memories during NREM sleep (Fig. 1).

Under reduced cholinergic tone during NREM sleep (controlled by the brain stem and basal forebrain), the human *neocortico-thalamo-hippocampal system* expresses a distinct and intricate set of hierarchically coupled oscillations, consisting of neocortical slow oscillations (SO; ~0.5–1 Hz), thalamic spindles (~12–15 Hz) and hippocampal ripples (~80–100 Hz) (Staresina et al. 2015). The superordinate organizing feature is the SO with its global up- and down-states, propagating through the entire neocortex and, via neocortico-thalamic and neocortico-entorhinal-hippocampal projections, also reaching thalamus and hippocampus, respectively. Thereby, on a comparatively slow time scale, excitability levels are synchronized across these brain regions, with down-states suppressing and subsequent up-states facilitating neuronal processing in the entire *neocortico-thalamo-hippocampal system*. By means of the SO up-states, transient windows for an efficient hippocampo-neocortical dialogue are established, during which the reactivation of hippocampo-neocortical connections, triggered by sharp wave ripples (SWRs), is facilitated for both the hippocampal and the neocortical parts of the mnemonic representation. Direct synaptic connections between those neocortical neurons forming the mnemonic representation are then strengthened due to their conjoint reactivation (Buzsaki 1998; Diekelmann and Born 2010; Fell and Axmacher 2011; Frankland and Bontempi 2005; Pereira de Vasconcelos and Cassel 2014). However, the scaling of temporal synchronization between hippocampus and neocortex achieved by the SO alone may be too coarse to allow high-fidelity information transfer, as relevant memory reactivation-related neocortical input signals are easily lost in the general background noise of SO related neocortical activation. Here, sleep spindles may play an essential role. Triggered by cortico-thalamic input during the SO, spindles are triggered in the thalamus and feed into both neocortex and hippocampus, as evident from the prominent grouping of spindle activity in the SO up-state observed in electrophysiological recordings from neocortex (Andrillon et al. 2011; Mölle et al. 2002, 2011; Steriade 2006), thalamus (Steriade 2006), parahippocampal cortex (Clemens et al. 2007, 2011), and the hippocampus (Andrillon et al. 2011; Sarasso et al. 2014; Staresina et al. 2015). In addition to the above cited direct evidence from intracranial recordings in humans, the occurrence of spindles in the hippocampus has also been inferred from concurrent EEG-fMRI studies revealing spindle-related activation in the hippocampus (Andrade et al. 2011; Bergmann et al. 2012a; Schabus et al. 2007). Strikingly, the size of the hippocampal BOLD response was even predicted on a spindle-by-spindle level by the amplitude of the cortical spindle (recorded with surface EEG), thus suggesting a tight relationship between cortical and hippocampal spindle input (Bergmann et al. 2012a).

The near-simultaneous arrival of spindle volleys in both hippocampus and neocortex (Staresina et al. 2015) is crucial, as it provides a more fine-grained level of inter-area synchronization compared to the SO. In the hippocampus, spindles are in turn grouping ripples in their more excitable troughs (Staresina et al. 2015), whereas in the neocortex the incoming excitatory spindle volleys may boost the impact of incoming ripples simultaneously arriving via the hippocampo-entorhinal-neocortical pathway. Specifically, the increased Ca^{2+} influx into neocortical neurons evoked by spindles may be involved in local processes of neocortical synaptic plasticity (Astori et al. 2013; Rosanova and Ulrich 2005; Sejnowski and Destexhe 2000; Timofeev et al. 2002), either by initiating them immediately during NREM sleep or by merely tagging them for synaptic consolidation during subsequent REM sleep (Diekelmann and Born 2010; Rasch and Born 2013; Ribeiro et al. 2007). It is thus possible that locally restricted spindles (Andrillon et al. 2011), possibly generated via a specific thalamic core rather than non-specific thalamic matrix cells (Bonjean et al. 2012), also provide a more topographically specific facilitation of concurrently arriving hippocampal ripple input to the neocortex. This idea is indirectly corroborated by findings of increased spindle power (Clemens et al. 2005, 2006; Cox et al. 2014) or spindle-related BOLD responses (Bergmann et al. 2012a) in neocortical brain regions previously involved in learning, but is complicated by the notion that hippocampal spindle input and ripple release is mainly mediated by non-specific thalamic nuclei producing global spindles (Pereira de Vasconcelos and Cassel 2014; Varela et al. 2014).

Notably, solitary sleep spindles, i.e. those occurring without concurrent SOs during NREM sleep, constitute a large proportion of overall NREM spindles and still exert the same phase-amplitude modulation on hippocampal ripples as do spindles grouped by SOs (Staresina et al. 2015). This raises the possibility that it is the spindles that play the most crucial role in mediating the hippocampo-neocortical cross-talk during system memory consolidation. However, there is also some evidence that spindles grouped by SOs may be more relevant for system memory consolidation than solitary spindles (Cox et al. 2012). If so, this may be owed to the increased neuronal excitation (Bergmann et al. 2012b) and the phasic increase in noradrenergic drive associated with SO up-states, which may transiently facilitate synaptic plasticity (Eschenko et al. 2011; Eschenko and Sara 2008). Interestingly, however, there is also some indirect EEG-fMRI evidence for spindle-related phasic activation of the noradrenergic *locus coeruleus* (Bergmann et al. 2012a).

On a final note, it is interesting that spindle power was found to be maximal at posterior hippocampal sites, gradually decreasing towards its middle portion and largely absent in the anterior hippocampus (Staresina et al. 2015). It is thus possible that spindle input is strongest for the posterior hippocampus and spindles then travel along the posterior-anterior axis, as shown for theta oscillations in rodents (Lubenov and Siapas 2009) and humans (Zhang and Jacobs 2015). Indeed, at least in rodents, the rhomboid nucleus of the non-specific midline thalamus has strong projections to the dorsal (posterior) but not the ventral (anterior) hippocampus; however, the reuniens nucleus projects to both and receives hippocampal feedback (Pereira de Vasconcelos and Cassel 2014; Varela et al. 2014). Also, the posterior hippocampus is generally more tightly coupled to the thalamus than the anterior hippocampus, based on fMRI resting state connectivity analyses (Zarei et al. 2013), but see Blessing et al. (2016). An intriguing question is whether a posterior-anterior spindle power gradient has any functional implications for the more recently proposed anatomical or functional segregation of the hippocampus along its longitudinal axis (Poppenk et al. 2013; Ranganath and Ritchey 2012; Strange et al. 2014).

In summary, the hierarchical nesting of SOs, spindles and ripples may indeed provide the neuronal machinery to mediate the temporally and spatially fine-tuned reactivation of hippocampo-neocortical connections which is required to modify connections within the neocortex for long-term storage. In fact, there is ample empirical evidence associating the behavioral effects of sleep-dependent memory consolidation with SOs (e.g., Huber et al. 2004; Mölle et al. 2004,), spindles (e.g., Clemens et al. 2005, 2006; Fogel and Smith 2006; Gais et al. 2002; Schabus et al. 2008, see chapter by McDevitt, Krishnan, Bazhenov & Mednick) and ripples (Axmacher et al. 2008; e.g., Peyrache et al. 2009; Wilson and McNaughton 1994, see chapter by Maier & Kempter). However, the detailed discussion of these is beyond the scope of this chapter (for comprehensive reviews see Born and Wilhelm 2012; Diekelmann et al. 2009; Rasch and Born 2013). Importantly, there is also some recent causal evidence from interventional studies. In humans, stimulation of the prefrontal cortex during NREM sleep by means of anodal transcranial direct current stimulation (TDCS) or alternating current stimulation (TACS) with an anodal DC offset increased both SO and spindle power during subsequent stimulation-free intervals and improved overnight memory retention (Marshall et al. 2006; Marshall et al. 2004; see chapter by Marshall). Further, SOs and associated sleep spindles as well as memory retention were augmented by closed-loop SO-upstate-triggered auditory stimulation (Ngo et al. 2013, 2015). Lastly, the selective detection and targeted disruption of hippocampal ripples by means of electrical stimulation impaired overnight memory consolidation in rats (Ego-Stengel and Wilson 2009; Girardeau et al. 2009). Future studies employing targeted brain stimulation approaches may further elucidate the causal relevance of specific oscillatory events. While in humans these studies will largely be limited to non-invasive transcranial brain stimulation in combination with electrophysiology (Bergmann et al. 2016), novel optogenetic stimulation techniques in rodents can now be used to study the reactivation of specific memory engrams (Josselyn et al. 2015; Santoro and Frankland 2014).

Reactivation of Memory Representations During NREM Sleep and Quiet Wakefulness

While the above-mentioned oscillations may be the mechanistic vehicles of memory transfer and consolidation, it is ultimately the mnemonic representations themselves that need to be strengthened. The first question is thus whether encoded representations indeed re-emerge during *offline periods*, i.e. either NREM sleep or quiet wakefulness, and whether such offline reinstatement would benefit later memory performance (see also chapter by Zhang, Deuker and Axmacher). If so, the key question is whether such offline reinstatement during NREM sleep is directly linked to the emergence of SOs, spindles and/or ripples, and whether other oscillations are involved during quiet wakefulness.

The first evidence towards the existence of offline reactivation in humans was furnished by a positron emission tomography (PET) study showing that hippocampal recruitment during a virtual navigation task re-emerged during a subsequent SWS period, with the extent of activation predicting behavioral performance after the rest period (Peigneux et al. 2004). A follow-up fMRI study suggested that offline reactivation may not be restricted to sleep, but could also be observed during performance of an unrelated low-level task immediately after learning, again with the extent of the persistent encoding activation predicting later memory performance (Peigneux et al. 2006). Although these findings suggest that learning-related activation extends well into post-learning periods, it is unclear whether individual learning experiences/representations are reactivated during offline periods and whether such reactivation would be of functional relevance for later memory performance. The recent advent of multivariate analytical tools in neuroimaging has provided a potentially sensitive method for capturing the reinstatement of individual learning experiences during offline periods (Norman et al. 2006). In brief, multivariate representational patterns (e.g., voxel intensities across an anatomically defined region in a fMRI study) can be derived for a particular encoding event, akin to a 'neural fingerprint' of a given learning experience. During the post-learning rest period, the extent to which this pattern is reinstated as a function of subsequent memory performance can then be assessed. Indeed, using this analytical approach, we were recently able to show that the reinstatement of individual learning experiences during a 2-minute post-learning rest period predicts subsequent memory for those experiences (Staresina et al. 2013). Importantly, the reward associated with newly encoded material appears to be predictive for its subsequent reinstatement: Gruber et al. (2016) asked their participants to make simple semantic judgments on objects, with half of the trials yielding a higher or lower monetary reward for correct responses, respectively. Critically, voxel-wise hippocampal fMRI patterns during a subsequent rest period resembled the high-reward objects more than the low-reward objects. Finally, Tambini and colleagues showed that hippocampo-neocortical connections rather than hippocampal networks alone constitute memory traces, as not only post-learning rest fMRI patterns within the hippocampus resembled the patterns during learning (Tambini and Davachi 2013), but also the connectivity between the hippocampus and regions involved in processing the learning task increased during post-learning rest (Tambini et al. 2010).

Is reinstatement of learning experiences triggered by the canonical sleep signatures discussed above? In rodents, there is ample evidence that hippocampal sharp wave ripples during NREM sleep are linked to the reactivation and even sequential *replay* of previous task-related neuronal patterns, mostly of place cells in the hippocampus after spatial navigation (Buzsaki 1996; Girardeau and Zugaro 2011; Sutherland and McNaughton 2000; Wilson and McNaughton 1994). In fact, a recent study even showed that also those motor cortical ensembles related to motor skill learning were replayed during subsequent slow wave and spindle events (Ramanathan et al. 2015). In humans, however, no single study has been able to link SOs, spindles or ripples to the reinstatement of individual learning experiences to date. That said, a number of findings provide converging evidence in favor of such a role. First, the coherence among scalp electrodes was found to increase at SO up-states during post-learning sleep (Mölle et al. 2004), perhaps mimicking the increase in functional connectivity between the hippocampus and neocortical regions mentioned above (Tambini et al. 2010). Moreover, an increase in local SOs during post-learning sleep has been observed over cortical sites involved in the initial task performance (Huber et al. 2004). Similarly, fast spindles were shown to be more prevalent over regions initially involved in a learning task (Clemens et al. 2005, 2006; Cox et al. 2014). Finally, a simultaneous EEG-fMRI study was able to show that post-learning sleep spindles were accompanied by an increase in BOLD activation in those regions involved during learning, that is both the hippocampus and face- and scene-specific inferior temporal cortices (fusiform face area and parahippocampal place area, respectively) (Bergmann et al. 2012a). Further demonstrating that spindle-triggered hippocampal activity was stronger in better learners, that study arguably provides the strongest link between spindles and memory-related reactivation of learning materials to date.

In sum, both SOs and spindles appear to be linked to memory-related reactivation during sleep. But this begs the question what the underlying mechanisms of offline reinstatement are during wake periods, where SOs and spindles are usually absent. Based on rodent data, the most likely candidate are hippocampal ripples. That is, ripple-triggered replay of hippocampal place cells occurs not only during SWS, but also during post-learning quiet wakefulness (Diba and Buzsaki 2007; Foster and Wilson 2006). Although less common, hippocampal ripple-triggered reactivation has also been reported in regions beyond the hippocampus (Pennartz et al. 2004), opening the possibility that hippocampal ripples could also coordinate reactivation in neocortical regions during quiet wakefulness periods in humans.

Acknowledgements T.O.B. was supported by the German Research Foundation (Deutsche Forschungsgemeinschaft, DFG), via TR-SFB 654 ("Plasticity and Sleep"), and by the Hertie Foundation (Gemeinnützige Hertie-Stiftung), via the Hertie Institute for Clinical Brain Research. B.P.S. was supported by a Sir Henry Dale Fellowship to B.P.S. jointly funded by the Wellcome Trust and the Royal Society (107672/Z/15/Z).

References

Achermann P, Borbely AA (1997) Low-frequency (<1 Hz) oscillations in the human sleep electroencephalogram. Neuroscience 81:213–222

Amzica F, Steriade M (1997) The K-complex: its slow (<1-Hz) rhythmicity and relation to delta waves. Neurology 49:952–959

Anderson KL, Rajagovindan R, Ghacibeh GA, Meador KJ, Ding M (2010) Theta oscillations mediate interaction between prefrontal cortex and medial temporal lobe in human memory. Cereb Cortex 20:1604–1612

Andrade KC, Spoormaker VI, Dresler M, Wehrle R, Holsboer F, Samann PG, Czisch M (2011) Sleep spindles and hippocampal functional connectivity in human NREM sleep. J Neurosci 31:10331–10339

Andrillon T, Nir Y, Staba RJ, Ferrarelli F, Cirelli C, Tononi G, Fried I (2011) Sleep spindles in humans: insights from intracranial EEG and unit recordings. J Neurosci 31:17821–17834

Astori S, Wimmer RD, Luthi A (2013) Manipulating sleep spindles—expanding views on sleep, memory, and disease. Trends Neurosci 36:738–748

Atherton LA, Dupret D, Mellor JR (2015) Memory trace replay: the shaping of memory consolidation by neuromodulation. Trends Neurosci 38:560–570

Axmacher N, Elger CE, Fell J (2008) Ripples in the medial temporal lobe are relevant for human memory consolidation. Brain 131:1806–1817

Bal T, Debay D, Destexhe A (2000) Cortical feedback controls the frequency and synchrony of oscillations in the visual thalamus. J Neurosci 20:7478–7488

Battaglia FP, Sutherland GR, McNaughton BL (2004) Hippocampal sharp wave bursts coincide with neocortical "up-state" transitions. Learn Mem 11:697–704

Bergmann TO, Molle M, Marshall L, Kaya-Yildiz L, Born J, Siebner HR (2008) A local signature of LTP- and LTD-like plasticity in human NREM sleep. Eur J Neurosci 27:2241–2249

Bergmann TO, Molle M, Diedrichs J, Born J, Siebner HR (2012a) Sleep spindle-related reactivation of category-specific cortical regions after learning face-scene associations. Neuroimage 59:2733–2742

Bergmann TO, Molle M, Schmidt MA, Lindner C, Marshall L, Born J, Siebner HR (2012b) EEG-guided transcranial magnetic stimulation reveals rapid shifts in motor cortical excitability during the human sleep slow oscillation. J Neurosci 32:243–253

Bergmann TO, Karabanov A, Hartwigsen G, Thielscher A, Siebner HR (2016) Combining non-invasive transcranial brain stimulation with neuroimaging and electrophysiology: current approaches and future perspectives. Neuroimage

Blessing EM, Beissner F, Schumann A, Brunner F, Bar KJ (2016) A data-driven approach to mapping cortical and subcortical intrinsic functional connectivity along the longitudinal hippocampal axis. Hum Brain Mapp 37:462–476

Bonjean M, Baker T, Lemieux M, Timofeev I, Sejnowski T, Bazhenov M (2011) Corticothalamic feedback controls sleep spindle duration in vivo. J Neurosci 31:9124–9134

Bonjean M, Baker T, Bazhenov M, Cash S, Halgren E, Sejnowski T (2012) Interactions between core and matrix thalamocortical projections in human sleep spindle synchronization. J Neurosci 32:5250–5263

Born J, Wilhelm I (2012) System consolidation of memory during sleep. Psychol Res 76:192–203

Born J, Rasch B, Gais S (2006) Sleep to remember. Neuroscientist 12:410–424

Bragin A, Engel J Jr, Wilson CL, Fried I, Buzsaki G (1999) High-frequency oscillations in human brain. Hippocampus 9:137–142

Buzsaki G (1986) Hippocampal sharp waves: their origin and significance. Brain Res 398:242–252

Buzsaki G (1996) The hippocampo-neocortical dialogue. Cereb Cortex 6:81–92

Buzsaki G (1998) Memory consolidation during sleep: a neurophysiological perspective. J Sleep Res 7(Suppl 1):17–23

Buzsaki G (2015) Hippocampal sharp wave-ripple: a cognitive biomarker for episodic memory and planning. Hippocampus 25:1073–1188

Buzsaki G, Horvath Z, Urioste R, Hetke J, Wise K (1992) High-frequency network oscillation in the hippocampus. Science 256:1025–1027

Cash SS, Halgren E, Dehghani N, Rossetti AO, Thesen T, Wang C, Devinsky O, Kuzniecky R, Doyle W, Madsen JR, Bromfield E, Eross L, Halasz P, Karmos G, Csercsa R, Wittner L, Ulbert I (2009) The human K-complex represents an isolated cortical down-state. Science 324:1084–1087

Cassel JC, Pereira de Vasconcelos A, Loureiro M, Cholvin T, Dalrymple-Alford JC, Vertes RP (2013) The reuniens and rhomboid nuclei: neuroanatomy, electrophysiological characteristics and behavioral implications. Prog Neurobiol 111:34–52

Chrobak JJ, Buzsaki G (1996) High-frequency oscillations in the output networks of the hippocampal-entorhinal axis of the freely behaving rat. J Neurosci 16:3056–3066

Clemens Z, Fabo D, Halasz P (2005) Overnight verbal memory retention correlates with the number of sleep spindles. Neuroscience 132:529–535

Clemens Z, Fabo D, Halasz P (2006) Twenty-four hours retention of visuospatial memory correlates with the number of parietal sleep spindles. Neurosci Lett 403:52–56

Clemens Z, Mölle M, Eross L, Barsi P, Halasz P, Born J (2007) Temporal coupling of parahippocampal ripples, sleep spindles and slow oscillations in humans. Brain 130:2868–2878

Clemens Z, Molle M, Eross L, Jakus R, Rasonyi G, Halasz P, Born J (2011) Fine-tuned coupling between human parahippocampal ripples and sleep spindles. Eur J Neurosci 33:511–520
Colgin LL (2015) Theta-gamma coupling in the entorhinal-hippocampal system. Curr Opin Neurobiol 31:45–50
Colgin LL (2016) Rhythms of the hippocampal network. Nat Rev Neurosci 17:239–249
Colrain IM (2005) The K-complex: a 7-decade history. Sleep 28:255–273
Contreras D, Steriade M (1995) Cellular basis of EEG slow rhythms: a study of dynamic corticothalamic relationships. J Neurosci 15:604–622
Contreras D, Destexhe A, Sejnowski TJ, Steriade M (1996) Control of spatiotemporal coherence of a thalamic oscillation by corticothalamic feedback. Science 274:771–774
Contreras D, Destexhe A, Sejnowski TJ, Steriade M (1997) Spatiotemporal patterns of spindle oscillations in cortex and thalamus. J Neurosci 17:1179–1196
Cox R, Hofman WF, Talamini LM (2012) Involvement of spindles in memory consolidation is slow wave sleep-specific. Learn Mem 19:264–267
Cox R, Hofman WF, de Boer M, Talamini LM (2014) Local sleep spindle modulations in relation to specific memory cues. Neuroimage 99:103–110
Crunelli V, Hughes SW (2009) The slow (<1 Hz) rhythm of non-REM sleep: a dialogue between three cardinal oscillators. Nat Neurosci
Dan Y, Poo MM (2004) Spike timing-dependent plasticity of neural circuits. Neuron 44:23–30
Datta S, Maclean RR (2007) Neurobiological mechanisms for the regulation of mammalian sleep-wake behavior: reinterpretation of historical evidence and inclusion of contemporary cellular and molecular evidence. Neurosci Biobehav Rev 31:775–824
De Gennaro L, Ferrara M (2003) Sleep spindles: an overview. Sleep Med Rev 7:423–440
Destexhe A, Contreras D, Sejnowski TJ, Steriade M (1994) Modeling the control of reticular thalamic oscillations by neuromodulators. NeuroReport 5:2217–2220
Diba K, Buzsaki G (2007) Forward and reverse hippocampal place-cell sequences during ripples. Nat Neurosci 10:1241–1242
Diekelmann S, Born J (2010) The memory function of sleep. Nat Rev Neurosci 11:114–126
Diekelmann S, Wilhelm I, Born J (2009) The whats and whens of sleep-dependent memory consolidation. Sleep Med Rev 13(5):309–321
Dudai Y (2004) The neurobiology of consolidations, or, how stable is the engram? Annu Rev Psychol 55:51–86
Düzel E, Penny WD, Burgess N (2010) Brain oscillations and memory. Curr Opin Neurobiol 20:143–149
Ego-Stengel V, Wilson MA (2009) Disruption of ripple-associated hippocampal activity during rest impairs spatial learning in the rat. Hippocampus 20:1–10
Eichenbaum H (2004) Hippocampus: cognitive processes and neural representations that underlie declarative memory. Neuron 44:109–120
Eschenko O, Sara SJ (2008) Learning-dependent, transient increase of activity in noradrenergic neurons of locus coeruleus during slow wave sleep in the rat: brain stem-cortex interplay for memory consolidation? Cereb Cortex 18:2596–2603
Eschenko O, Magri C, Panzeri S, Sara SJ (2011) Noradrenergic neurons of the locus coeruleus are phase locked to cortical up-down states during sleep. Cereb Cortex
Fell J, Axmacher N (2011) The role of phase synchronization in memory processes. Nat Rev Neurosci 12:105–118
Fogel SM, Smith CT (2006) Learning-dependent changes in sleep spindles and Stage 2 sleep. J Sleep Res 15:250–255
Foster DJ, Wilson MA (2006) Reverse replay of behavioural sequences in hippocampal place cells during the awake state. Nature 440:680–683
Frankland PW, Bontempi B (2005) The organization of recent and remote memories. Nat Rev Neurosci 6:119–130
Frey U, Morris RG (1998) Synaptic tagging: implications for late maintenance of hippocampal long-term potentiation. Trends Neurosci 21:181–188

Fuentemilla L, Barnes GR, Duzel E, Levine B (2014) Theta oscillations orchestrate medial temporal lobe and neocortex in remembering autobiographical memories. Neuroimage 85(Pt 2):730–737

Gais S, Born J (2004a) Declarative memory consolidation: mechanisms acting during human sleep. Learn Mem 11:679–685

Gais S, Born J (2004b) Low acetylcholine during slow-wave sleep is critical for declarative memory consolidation. Proc Natl Acad Sci U S A 101:2140–2144

Gais S, Mölle M, Helms K, Born J (2002) Learning-dependent increases in sleep spindle density. J Neurosci 22:6830–6834

Gardner RJ, Hughes SW, Jones MW (2013) Differential spike timing and phase dynamics of reticular thalamic and prefrontal cortical neuronal populations during sleep spindles. J Neurosci 33:18469–18480

Girardeau G, Zugaro M (2011) Hippocampal ripples and memory consolidation. Curr Opin Neurobiol

Girardeau G, Benchenane K, Wiener SI, Buzsaki G, Zugaro MB (2009) Selective suppression of hippocampal ripples impairs spatial memory. Nat Neurosci 12:1222–1223

Giuditta A (2014) Sleep memory processing: the sequential hypothesis. Front Syst Neurosci 8:219

Giuditta A, Ambrosini MV, Montagnese P, Mandile P, Cotugno M, Grassi Zucconi G, Vescia S (1995) The sequential hypothesis of the function of sleep. Behav Brain Res 69:157–166

Greenberg DL, Rubin DC (2003) The neuropsychology of autobiographical memory. Cortex 39:687–728

Gruber MJ, Ritchey M, Wang SF, Doss MK, Ranganath C (2016) Post-learning hippocampal dynamics promote preferential retention of rewarding events. Neuron 89:1110–1120

Harris KD, Thiele A (2011) Cortical state and attention. Nat Rev Neurosci 12:509–523

Hasselmo ME (1999) Neuromodulation: acetylcholine and memory consolidation. Trends Cogn Sci 3:351–359

Hasselmo ME (2005) What is the function of hippocampal theta rhythm?—Linking behavioral data to phasic properties of field potential and unit recording data. Hippocampus 15:936–949

Hasselmo ME, McGaughy J (2004) High acetylcholine levels set circuit dynamics for attention and encoding and low acetylcholine levels set dynamics for consolidation. Prog Brain Res 145:207–231

Hill S, Tononi G (2005) Modeling sleep and wakefulness in the thalamocortical system. J Neurophysiol 93:1671–1698

Huber R, Ghilardi MF, Massimini M, Tononi G (2004) Local sleep and learning. Nature 430:78–81

Hutchison IC, Rathore S (2015) The role of REM sleep theta activity in emotional memory. Front Psychol 6:1439

Iber C, Ancoli-Israel S, Chesson A, Quan SF (2007) The AASM manual for the scoring of sleep and associated events: rules, terminology and technical specifications. American Academy of Sleep Medicine, Westchester, IL

Isomura Y, Sirota A, Ozen S, Montgomery S, Mizuseki K, Henze DA, Buzsaki G (2006) Integration and segregation of activity in entorhinal-hippocampal subregions by neocortical slow oscillations. Neuron 52:871–882

Jensen O (2005) Reading the hippocampal code by theta phase-locking. Trends Cogn Sci 9:551–553

Jones EG (2001) The thalamic matrix and thalamocortical synchrony. Trends Neurosci 24:595–601

Josselyn SA, Kohler S, Frankland PW (2015) Finding the engram. Nat Rev Neurosci 16:521–534

Le Van Quyen M, Bragin A, Staba R, Crepon B, Wilson CL, Engel J Jr (2008) Cell type-specific firing during ripple oscillations in the hippocampal formation of humans. J Neurosci 28:6104–6110

Le Van Quyen M, Staba R, Bragin A, Dickson C, Valderrama M, Fried I, Engel J (2010) Large-scale microelectrode recordings of high-frequency gamma oscillations in human cortex during sleep. J Neurosci 30:7770–7782

Lisman JE, Jensen O (2013) The theta-gamma neural code. Neuron 77:1002–1016

Llinas RR, Leznik E, Urbano FJ (2002) Temporal binding via cortical coincidence detection of specific and nonspecific thalamocortical inputs: a voltage-dependent dye-imaging study in mouse brain slices. Proc Natl Acad Sci U S A 99:449–454

Logothetis NK, Eschenko O, Murayama Y, Augath M, Steudel T, Evrard HC, Besserve M, Oeltermann A (2012) Hippocampal-cortical interaction during periods of subcortical silence. Nature 491:547–553

Lubenov EV, Siapas AG (2009) Hippocampal theta oscillations are travelling waves. Nature 459:534–539

Luthi A (2014) Sleep spindles: where they come from, what they do. Neuroscientist 20:243–256

Maquet P (2001) The role of sleep in learning and memory. Science 294:1048–1052

Marrosu F, Portas C, Mascia MS, Casu MA, Fa M, Giagheddu M, Imperato A, Gessa GL (1995) Microdialysis measurement of cortical and hippocampal acetylcholine release during sleep-wake cycle in freely moving cats. Brain Res 671:329–332

Marshall L, Mölle M, Hallschmid M, Born J (2004) Transcranial direct current stimulation during sleep improves declarative memory. J Neurosci 24:9985–9992

Marshall L, Helgadottir H, Mölle M, Born J (2006) Boosting slow oscillations during sleep potentiates memory. Nature 444:610–613

Masquelier T, Hugues E, Deco G, Thorpe SJ (2009) Oscillations, phase-of-firing coding, and spike timing-dependent plasticity: an efficient learning scheme. J Neurosci 29:13484–13493

Massimini M, Huber R, Ferrarelli F, Hill S, Tononi G (2004) The sleep slow oscillation as a traveling wave. J Neurosci 24:6862–6870

McClelland JL, McNaughton BL, O'Reilly RC (1995) Why there are complementary learning systems in the hippocampus and neocortex: insights from the successes and failures of connectionist models of learning and memory. Psychol Rev 102:419–457

McCormick DA (1992) Neurotransmitter actions in the thalamus and cerebral cortex and their role in neuromodulation of thalamocortical activity. Prog Neurobiol 39:337–388

Meeter M, Murre JM (2004) Consolidation of long-term memory: evidence and alternatives. Psychol Bull 130:843–857

Mölle M, Marshall L, Gais S, Born J (2002) Grouping of spindle activity during slow oscillations in human non-rapid eye movement sleep. J Neurosci 22:10941–10947

Mölle M, Marshall L, Gais S, Born J (2004) Learning increases human electroencephalographic coherence during subsequent slow sleep oscillations. Proc Natl Acad Sci U S A 101:13963–13968

Mölle M, Yeshenko O, Marshall L, Sara SJ, Born J (2006) Hippocampal sharp wave-ripples linked to slow oscillations in rat slow-wave sleep. J Neurophysiol 96:62–70

Mölle M, Eschenko O, Gais S, Sara SJ, Born J (2009) The influence of learning on sleep slow oscillations and associated spindles and ripples in humans and rats. Eur J Neurosci 29:1071–1081

Mölle M, Bergmann TO, Marshall L, Born J (2011) Fast and slow spindles during the sleep slow oscillation: disparate coalescence and engagement in memory processing. Sleep 34:1411–1421

Morris RG (2006) Elements of a neurobiological theory of hippocampal function: the role of synaptic plasticity, synaptic tagging and schemas. Eur J Neurosci 23:2829–2846

Müller GE, Pilzecker A (1900) Experimentelle Beiträge zur Lehre vom Gedächtnis. Z Psychol Ergänzungsband 1:1–300

Murphy M, Riedner BA, Huber R, Massimini M, Ferrarelli F, Tononi G (2009) Source modeling sleep slow waves. Proc Natl Acad Sci U S A 106:1608–1613

Ngo HV, Martinetz T, Born J, Molle M (2013) Auditory closed-loop stimulation of the sleep slow oscillation enhances memory. Neuron 78:545–553

Ngo HV, Miedema A, Faude I, Martinetz T, Molle M, Born J (2015) Driving sleep slow oscillations by auditory closed-loop stimulation-a self-limiting process. J Neurosci 35:6630–6638

Nir Y, Staba RJ, Andrillon T, Vyazovskiy VV, Cirelli C, Fried I, Tononi G (2011) Regional slow waves and spindles in human sleep. Neuron 70:153–169

Norman KA, Polyn SM, Detre GJ, Haxby JV (2006) Beyond mind-reading: multi-voxel pattern analysis of fMRI data. Trends Cogn Sci 10:424–430
O'Keefe J (1976) Place units in the hippocampus of the freely moving rat. Exp Neurol 51:78–109
Patel J, Schomburg EW, Berenyi A, Fujisawa S, Buzsaki G (2013) Local generation and propagation of ripples along the septotemporal axis of the hippocampus. J Neurosci 33:17029–17041
Peigneux P, Laureys S, Fuchs S, Collette F, Perrin F, Reggers J, Phillips C, Degueldre C, Del Fiore G, Aerts J, Luxen A, Maquet P (2004) Are spatial memories strengthened in the human hippocampus during slow wave sleep? Neuron 44:535–545
Peigneux P, Orban P, Balteau E, Degueldre C, Luxen A, Laureys S, Maquet P (2006) Offline persistence of memory-related cerebral activity during active wakefulness. PLoS Biol 4:e100
Pennartz CM, Lee E, Verheul J, Lipa P, Barnes CA, McNaughton BL (2004) The ventral striatum in off-line processing: ensemble reactivation during sleep and modulation by hippocampal ripples. J Neurosci 24:6446–6456
Pereira de Vasconcelos A, Cassel JC (2014) The nonspecific thalamus: A place in a wedding bed for making memories last? Neurosci Biobehav Rev 54:175–196
Peter-Derex L, Comte JC, Mauguiere F, Salin PA (2012) Density and frequency caudo-rostral gradients of sleep spindles recorded in the human cortex. Sleep 35:69–79
Peyrache A, Khamassi M, Benchenane K, Wiener SI, Battaglia FP (2009) Replay of rule-learning related neural patterns in the prefrontal cortex during sleep. Nat Neurosci 12:919–926
Poppenk J, Evensmoen HR, Moscovitch M, Nadel L (2013) Long-axis specialization of the human hippocampus. Trends Cogn Sci 17:230–240
Ramanathan DS, Gulati T, Ganguly K (2015) Sleep-dependent reactivation of ensembles in motor cortex promotes skill consolidation. PLoS Biol 13:e1002263
Ranganath C, Ritchey M (2012) Two cortical systems for memory-guided behaviour. Nat Rev Neurosci 13:713–726
Rasch B, Born J (2007) Maintaining memories by reactivation. Curr Opin Neurobiol 17:698–703
Rasch B, Born J (2013) About sleep's role in memory. Physiol Rev 93:681–766
Rasch BH, Born J, Gais S (2006) Combined blockade of cholinergic receptors shifts the brain from stimulus encoding to memory consolidation. J Cogn Neurosci 18:793–802
Rechtschaffen A, Kales A (1968) A manual of standardized terminology, techniques and scoring system for sleep stages of human subjects. United States Government Printing Office, Washington, DC
Ribeiro S, Shi X, Engelhard M, Zhou Y, Zhang H, Gervasoni D, Lin SC, Wada K, Lemos NA, Nicolelis MA (2007) Novel experience induces persistent sleep-dependent plasticity in the cortex but not in the hippocampus. Front Neurosci 1:43–55
Rosanova M, Ulrich D (2005) Pattern-specific associative long-term potentiation induced by a sleep spindle-related spike train. J Neurosci 25:9398–9405
Sadowski JH, Jones MW, Mellor JR (2011) Ripples make waves: binding structured activity and plasticity in hippocampal networks. Neural Plast 2011:960389
Sanchez-Vives MV, McCormick DA (2000) Cellular and network mechanisms of rhythmic recurrent activity in neocortex. Nat Neurosci 3:1027–1034
Santoro A, Frankland PW (2014) Chasing the trace. Neuron 84:243–246
Sarasso S, Proserpio P, Pigorini A, Moroni F, Ferrara M, De Gennaro L, De Carli F, Russo GL, Massimini M, Nobili L (2014) Hippocampal sleep spindles preceding neocortical sleep onset in humans. Neuroimage 86:425–432
Schabus M, Dang-Vu TT, Albouy G, Balteau E, Boly M, Carrier J, Darsaud A, Degueldre C, Desseilles M, Gais S, Phillips C, Rauchs G, Schnakers C, Sterpenich V, Vandewalle G, Luxen A, Maquet P (2007) Hemodynamic cerebral correlates of sleep spindles during human non-rapid eye movement sleep. Proc Natl Acad Sci U S A 104:13164–13169
Schabus M, Hoedlmoser K, Pecherstorfer T, Anderer P, Gruber G, Parapatics S, Sauter C, Kloesch G, Klimesch W, Saletu B, Zeitlhofer J (2008) Interindividual sleep spindle differences and their relation to learning-related enhancements. Brain Res 1191:127–135
Sejnowski TJ, Destexhe A (2000) Why do we sleep? Brain Res 886:208–223

Sharma AV, Wolansky T, Dickson CT (2010) A comparison of sleep-like slow oscillations in the hippocampus under ketamine and urethane anaesthesia. J Neurophysiol 104:932–939
Siapas AG, Wilson MA (1998) Coordinated interactions between hippocampal ripples and cortical spindles during slow-wave sleep. Neuron 21:1123–1128
Sirota A, Csicsvari J, Buhl D, Buzsaki G (2003) Communication between neocortex and hippocampus during sleep in rodents. Proc Natl Acad Sci U S A 100:2065–2069
Skaggs WE, McNaughton BL, Permenter M, Archibeque M, Vogt J, Amaral DG, Barnes CA (2007) EEG sharp waves and sparse ensemble unit activity in the macaque hippocampus. J Neurophysiol 98:898–910
Staba RJ, Wilson CL, Bragin A, Fried I, Engel J Jr (2002) Quantitative analysis of high-frequency oscillations (80-500 Hz) recorded in human epileptic hippocampus and entorhinal cortex. J Neurophysiol 88:1743–1752
Staresina BP, Alink A, Kriegeskorte N, Henson RN (2013) Awake reactivation predicts memory in humans. Proc Natl Acad Sci U S A 110:21159–21164
Staresina BP, Bergmann TO, Bonnefond M, van der Meij R, Jensen O, Deuker L, Elger CE, Axmacher N, Fell J (2015) Hierarchical nesting of slow oscillations, spindles and ripples in the human hippocampus during sleep. Nat Neurosci 18:1679–1686
Steriade M (2003) The corticothalamic system in sleep. Front Biosci 8:d878–d899
Steriade M (2004) Acetylcholine systems and rhythmic activities during the waking–sleep cycle. Prog Brain Res 145:179–196
Steriade M (2006) Grouping of brain rhythms in corticothalamic systems. Neuroscience 137:1087–1106
Steriade M, Timofeev I (2003) Neuronal plasticity in thalamocortical networks during sleep and waking oscillations. Neuron 37:563–576
Steriade M, Nunez A, Amzica F (1993a) Intracellular analysis of relations between the slow (<1 Hz) neocortical oscillation and other sleep rhythms of the electroencephalogram. J Neurosci 13:3266–3283
Steriade M, Nunez A, Amzica F (1993b) A novel slow (<1 Hz) oscillation of neocortical neurons in vivo: depolarizing and hyperpolarizing components. J Neurosci 13:3252–3265
Strange BA, Witter MP, Lein ES, Moser EI (2014) Functional organization of the hippocampal longitudinal axis. Nat Rev Neurosci 15:655–669
Sullivan D, Csicsvari J, Mizuseki K, Montgomery S, Diba K, Buzsaki G (2011) Relationships between Hippocampal Sharp Waves, Ripples, and Fast Gamma Oscillation: Influence of Dentate and Entorhinal Cortical Activity. J Neurosci 31:8605–8616
Sutherland GR, McNaughton B (2000) Memory trace reactivation in hippocampal and neocortical neuronal ensembles. Curr Opin Neurobiol 10:180–186
Tambini A, Davachi L (2013) Persistence of hippocampal multivoxel patterns into postencoding rest is related to memory. Proc Natl Acad Sci U S A 110:19591–19596
Tambini A, Ketz N, Davachi L (2010) Enhanced brain correlations during rest are related to memory for recent experiences. Neuron 65:280–290
Timofeev I, Grenier F, Bazhenov M, Sejnowski TJ, Steriade M (2000) Origin of slow cortical oscillations in deafferented cortical slabs. Cereb Cortex 10:1185–1199
Timofeev I, Grenier F, Bazhenov M, Houweling AR, Sejnowski TJ, Steriade M (2002) Short- and medium-term plasticity associated with augmenting responses in cortical slabs and spindles in intact cortex of cats in vivo. J Physiol 542:583–598
Tononi G, Cirelli C (2003) Sleep and synaptic homeostasis: a hypothesis. Brain Res Bull 62:143–150
Urrestarazu E, Chander R, Dubeau F, Gotman J (2007) Interictal high-frequency oscillations (100-500 Hz) in the intracerebral EEG of epileptic patients. Brain 130:2354–2366
van Kesteren MT, Ruiter DJ, Fernandez G, Henson RN (2012) How schema and novelty augment memory formation. Trends Neurosci 35:211–219
Varela C, Kumar S, Yang JY, Wilson MA (2014) Anatomical substrates for direct interactions between hippocampus, medial prefrontal cortex, and the thalamic nucleus reuniens. Brain Struct Funct 219:911–929

Walker MP, Stickgold R (2010) Overnight alchemy: sleep-dependent memory evolution. Nat Rev Neurosci 11:218; author reply 218

Wilson MA, McNaughton BL (1994) Reactivation of hippocampal ensemble memories during sleep. Science 265:676–679

Wolansky T, Clement EA, Peters SR, Palczak MA, Dickson CT (2006) Hippocampal slow oscillation: a novel EEG state and its coordination with ongoing neocortical activity. J Neurosci 26:6213–6229

Zarei M, Beckmann CF, Binnewijzend MA, Schoonheim MM, Oghabian MA, Sanz-Arigita EJ, Scheltens P, Matthews PM, Barkhof F (2013) Functional segmentation of the hippocampus in the healthy human brain and in Alzheimer's disease. Neuroimage 66:28–35

Zhang H, Jacobs J (2015) Traveling theta waves in the human hippocampus. J Neurosci 35:12477–12487

The Role of Sleep Spindles in Sleep-Dependent Memory Consolidation

Elizabeth A. McDevitt, Giri P. Krishnan, Maxim Bazhenov and Sara C. Mednick

Abstract A well-established literature supports a critical role of sleep for learning and memory (Abel et al. in Curr Biol 23(17);R774–R788, 2013; Diekelmann and Born in Nat Rev Neurosci 11(2):114–126, 2010; Rasch and Born in Physiol Rev 93 (2):681–766, 2013; Tononi and Cirelli in Sleep Med Rev 10(1):49–62, 2014). Studies have demonstrated memory improvements following a period of sleep compared to an equivalent time awake, and specific sleep features have been shown to correlate with improvements in discrete memory domains. For example, overnight procedural motor learning correlates with the amount of stage 2 sleep (Walker et al. in Neuron 35(1):205–211, 2002; Learn Mem 10(4):275–284, 2003), non-hippocampal dependent perceptual learning correlates with the product of the amount of slow wave sleep (SWS) and rapid eye movement (REM) sleep (Mednick et al. in Nat Neurosci 6(7):697–698, 2003; Stickgold et al. in J Cogn Neurosci 12 (2):246–254, 2000), and implicit priming also appears to depend on REM sleep (Cai et al. in Proc Natl Acad Sci U S A 106(25);10130–10134, 2009). One feature of sleep that is widely implicated in memory processing is the sleep spindle, short (0.5–3 s) bursts of oscillatory activity in the frequency range of approximately 12–15 Hz (Spindles have also been defined as slow as 8–12 Hz, with some

This work was supported by Bazhenov: Office of Naval Research/Multidisciplinary University Research Initiative Grant N000141310672 and National Institutes of Health Grants R01 EB009282, R01 MH099645; Mednick and Bazehnov: R01AG046646; Mednick: NSF BCS1439210; McDevitt: NSF fellowship.

E.A. McDevitt · S.C. Mednick (✉)
Department of Psychology, University of California, Riverside, USA
e-mail: smednick@ucr.edu

E.A. McDevitt
e-mail: mcdevitt@ucr.edu

G.P. Krishnan · M. Bazhenov
Department of Medicine, University of California, San Diego, USA
e-mail: gkrishnan@ucsd.edu

M. Bazhenov
e-mail: mbazhenov@ucsd.edu

N. Axmacher and B. Rasch (eds.), *Cognitive Neuroscience of Memory Consolidation*, Studies in Neuroscience, Psychology and Behavioral Economics, DOI 10.1007/978-3-319-45066-7_13

indication that these slow spindles may be physiologically distinct from alpha frequency, which oscillates in the same frequency range (8–12 Hz) but has a different spatial distribution (Manshanden et al. in Clin Neurophysiol 113 (12):1937–1947, 2002). However, more data is required to determine the distinctiveness of these two signals. For the purpose of this chapter, we will primarily discuss spindles defined as ~12–15 Hz.). This chapter aims to (1) summarize correlational and causal evidence supporting the role of sleep spindles in memory processing; and (2) describe spindle dynamics and how they may be related to proposed mechanisms of sleep-dependent consolidation.

Keywords Sleep spindles · Slow oscillations · Declarative memory · Procedural memory · Systems consolidation · Phase amplitude coupling

An Introduction to Sleep Effects on Memory Consolidation

Sleep can be separated into four stages characterized by stereotypic electrical activity. The four stages progress in structured cycles from light Stages 1 and 2 through deep SWS (formerly Stages 3 and 4) and into REM sleep. Together, Stages 1, 2 and SWS are often referred to as non-REM (NREM) sleep. Stage 1 is briefly observed at sleep onset, and can be identified by the presence of slow rolling eye movements and a disappearance of alpha (8–12 Hz) activity over occipital regions. Stage 2 sleep is more synchronized than Stage 1 and is characterized by sigma activity (12–15 Hz, i.e., spindles) and occasional, high-amplitude K-complex signals. SWS is named for the high amplitude, slow wave activity [slow oscillations (0.5–1 Hz) and delta (1–4 Hz)] that predominates. REM sleep is characterized by fast, low-amplitude EEG similar to waking, as well as increased heart rate, increased cortical blood flow, muscle paralysis and its eponymous rapid eye movements.

Sleep spindles are an electrophysiological hallmark of NREM sleep (Fig. 1). In addition to duration and frequency criteria, spindles are defined as phasic events distinct from general background 12–15 Hz sigma activity with a distinct waxing and waning morphology (Astori et al. 2013; Gennaro and Ferrara 2003; Weiner and Dang-Vu 2016). Present in both Stage 2 and SWS, spindles occur along side several other features of NREM sleep such as vertex sharp waves, K-complexes, slow oscillations, or superimposed on delta oscillations during SWS. Additionally, two kinds of spindles can be differentiated by distinct spatio-temporal dynamics. "Slow" spindles (<12 Hz, with a spectral peak ~10.2 Hz) predominate over frontal sites and are more pronounced during SWS than Stage 2 sleep, whereas "fast" spindles (>12 Hz, with a peak spectral frequency ~13.4 Hz) are more densely distributed over parietal and central sites (Mölle et al. 2011). However the functional difference between these spindle types is unclear (Andrillon et al. 2011; Timofeev and Bazhenov 2005).

The need to understand the mechanisms and properties of sleep spindles is driven by their role in memory and learning. An open question in memory research

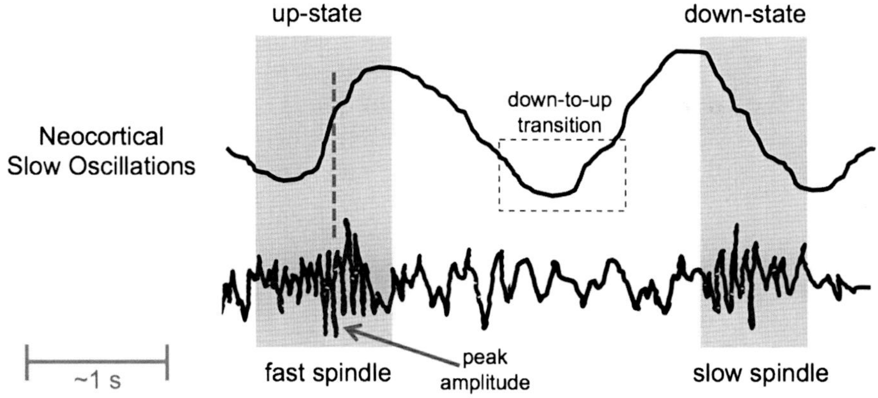

Fig. 1 The coordination of sleep spindles and slow oscillations during sleep. Neocortical slow oscillations (<1 Hz) are characterized by up and down states that reflect periods of neuronal spiking and neuronal silence, respectively. Sleep spindles are phasic events with a distinct waxing and waning morphology. Spindles are sometimes classified as "slow" (<12 Hz) or "fast" (>12 Hz) depending on their peak frequency. During sleep, spindles often occur during the down-to-up transition of the slow oscillation, with the peak amplitude of the spindle aligned with the slow oscillation up-state (represented by the dashed vertical line). This pattern of coupling may be specific to fast spindles, with slow spindles more likely to follow fast spindles and occur during the slow oscillation down-state

asks how the human brain learns new information without overwriting previously stored memories, the so-called "stability-plasticity" problem. For declarative memory (i.e., conscious or explicit recall of episodic and semantic memories), theoretical models propose two separate memory stores that interact in a process of systems consolidation—a fast-learning, temporary store and a slow-learning, long-term store. The hippocampus and neocortex are the hypothesized neural structures associated with the temporary and long-term stores, respectively. New information is originally encoded concurrently in both stores. During subsequent periods of consolidation, successive reactivation, or "replay", of this network is presumed to allow new memories to become strengthened and integrated with pre-existing memories in the long-term store, as well as becoming less reliant on the fast-learning store. Off-line periods when no encoding is happening, such as sleep, are thought to be ideal times for replay to occur since no new, incoming information will interfere with consolidation (Mednick et al. 2011).

Neural replay has been observed in studies of rodent spatial memory,[1] where place cells that are activated in sequence together during spatial learning tend to fire

[1]Although best demonstrated in the hippocampus, it is possible that neural replay also occurs at the level of cortex, independent of the hippocampus (e.g., both temporary and long-term stores are within cortex). However, for the purpose of this review we will detail neuronal and behavioral evidence of replay involving the hippocampus, although it is likely that replay is a general mechanism of systems consolidation across memory systems and not specific to hippocampal-dependent memories.

in a similar sequence, and at a faster, time-compressed rate, during subsequent SWS (Lee and Wilson 2002; Wilson and McNaughton 1994). Replay-like activity has also been observed during periods of immobility, or quiet wakefulness (Foster and Wilson 2006), and REM sleep (Louie and Wilson 2001), but the replay dynamics are different than those observed during SWS. Specifically, studies have shown: (1) hippocampal replay during SWS in rats is coordinated with firing patterns in the visual cortex (Ji and Wilson 2007); (2) the hippocampus and cortex appear to communicate during sleep by means of hippocampal sharp waves and ripples (Buzsáki 1989), during which place cells are reactivated (Diba and Buzsáki 2007); and (3) these events are temporally correlated with spindles in the medial prefrontal cortex during SWS (Siapas and Wilson 1998).

The temporal coupling of thalamic sleep spindles and hippocampal ripples along with neocortical slow oscillations (<1 Hz) is proposed to be a key mechanism underlying the hippocampal-neocortical dialogue characteristic of systems consolidation. Generation of spindles in the thalamus and sharp wave-ripples in hippocampus is suppressed during the hyperpolarizing down-state of the slow oscillation, followed by a rebound in spindle and sharp wave-ripple activity during the succeeding depolarizing up-state. Thus, slow oscillations are thought to provide a top-down temporal frame for these oscillatory events, though it is not clear whether this thalamo-hippocampal-cortical circuit is driven by the thalamus, cortex, or a combination of both brain areas (Crunelli and Hughes 2010; Lemieux et al. 2014). Regarding the coupling of spindles and ripples, it has been demonstrated that individual ripple events are nested in the trough of succeeding spindles (Staresina et al. 2015; Timofeev and Bazhenov 2005) (see also chapters by Bergmann and Staresina and by Maier and Kempter). These "spindle-ripple" events might represent a bottom-up mechanism where reactivated hippocampal memory information (coded in ripples) is passed to spindles, which then reach neocortical networks via the slow oscillation. Thus, spindles appear to be one critical component of a complex interaction between several electrophysiological events that together provide a mechanistic explanation for memory reactivation during sleep. Given that spindles are easily detected and measured using scalp electroencephalogram (EEG), they provide a convenient and non-invasive method to examine one of the neural correlates of consolidation in humans.

Sleep Spindles and Human Memory: Correlational Evidence

Although a plethora of studies in the past two decades have indicated that sleep spindles play a functional role in memory processing during sleep, it has also become apparent that the relationship between spindles and memory is quite complex and may be moderated by any number of factors including memory domain, task difficulty, initial skill level of the individual, sleep stage (Stage 2 vs. SWS), spindle frequency (fast vs. slow), scalp derivation (frontal vs.

centro-parietal), and other spindle characteristics (number, density, amplitude, duration, power). In the following section we discuss the neuronal correlates of sleep spindles and examine correlational evidence for the role of spindles in memory consolidation.

Neuronal Mechanisms of Sleep Spindles

The presence of spindle oscillations after decortication provides strong evidence for the thalamic origin of this activity (Contreras et al. 1996; Morison and Bassett 1945; Timofeev and Steriade 1996). Studies suggest that the minimal substrate contributing to the generation of spindle oscillations is generated in the thalamus as a result of the interaction between thalamic reticular (RE) and relay (TC) cells (Steriade et al. 1985, 1990; Steriade and Deschenes 1984; Steriade and Llinas 1988; von Krosigk et al. 1993). According to this hypothesis, RE inhibitory neurons fire a spike burst that elicits an inhibitory post-synaptic potential (IPSP) in TC neurons. At the end of this IPSP, the TC neurons generate rebound spike-bursts that in turn excite RE neurons, which then generate spike-bursts, starting the next cycle of spindle oscillations. However, this minimal model may not describe all the mechanisms involved in spindle generation because (a) spindles can be generated in isolated RE nucleus (Steriade et al. 1987); and (b) during the early 3–4 IPSPs composing the spindle, many TC neurons do not display rebound spike-bursts (Bazhenov et al. 2000), suggesting that the reciprocal TC-RE connections are not contributing to the early phase of a spindle sequence. The simplest computational model sufficient to generate the spindle oscillations includes two reciprocally coupled RE neurons and two TC cells providing excitation to and receiving inhibition from RE neurons (Destexhe et al. 1996). Persistence of spindle-like activity in the isolated RE nuclei suggests a mechanism for spindle initiation, with activity of RE cells initiating a new sequence of spindle oscillations (Bazhenov et al. 2000).

Termination of spindles depends both on intrinsic and network mechanisms. The first includes Ca^{2+} accumulation leading to cyclic adenosine monophosphate (cAMP) upregulation of hyperpolarization-activated non-specific cation current (Ih) and, following TC neurons depolarization, making rebound spike-bursts impossible (Bal and McCormick 1996; Budde et al. 1997; Luthi et al. 1998). The second includes the desynchronizing effect of cortico-thalamic projections (Andersen and Andersson 1968; Bonjean et al. 2011; Timofeev 2001), based on dissimilarity of intrinsic responses in different cortical and TC neurons. An extensive review of the intrinsic and synaptic mechanisms of spindle oscillations can be found in Timofeev and Bazhenov (2005). These thalamic mechanisms of spindle generation correspond to fast spindles, while the mechanisms of slow spindle generation remain to be investigated. Slow spindles could originate from the neocortex. At least, upon stimulation, isolated neocortical slabs are able to generate oscillations with frequencies around 10 Hz (Timofeev et al. 2002), i.e. the frequency range of slow spindles.

Learning Modulates Spindle Activity During Subsequent Sleep

Early studies reporting an association between sleep spindles and memory in humans found that spindle activity during non-REM sleep is sensitive to previous learning experience (Gais et al. 2002; Meier-Koll et al. 1999). Gais et al. (2002) showed that compared with a non-learning task, memorizing unrelated word pairs (learning task) increased spindle density during Stage 2 and spindle density also correlated with recall performance, both before and after sleep, but the overnight effect was not assessed. A separate study found that encoding difficulty moderated the relationship between learning and spindles, such that difficult encoding (more abstract word pairs) resulted in a significant increase in the slow spindle frequency range (11.25–13.75 Hz) during Stage 2 sleep, whereas easy encoding (more concrete word pairs) did not alter sleep spindle activity compared to a control condition (Schmidt et al. 2006).

Similar learning-dependent increases in spindle activity have also been observed for motor memory. Fogel and Smith (2006) tested four procedural tasks before and after sleep and found that spindle densities were increased following learning and that the change in performance on the four learning tasks accounted for 98% of the variability in the change in spindle density. A second study expanded these results and showed that following learning a Pursuit Rotor task (but not a mirror tracing task), Stage 2 and SWS spindle densities as well as Stage 2 spindle duration were increased compared to a baseline night (Fogel et al. 2007). In both of these studies, the tasks were re-tested one week after learning and thus the immediate overnight gains in performance could not be examined. Additionally, studies have shown that the degree to which spindles correlate with motor learning may be moderated by task complexity (Fogel et al. 2007; Smith et al. 2004) and skill level of the experimental subject (Schmidt et al. 2006).

Correlations Between Sleep Spindles and Memory Improvement

Declarative Memory

A large number of studies have established that declarative memory shows less forgetting after a sleep period compared with wake, and that performance benefits are correlated with spindles (Clemens et al. 2005; Cox et al. 2012; Genzel et al. 2009; Schabus et al. 2004) (see also chapter by Schönauer and Gais). More recent studies have also found that spindles correlate with overnight integration of new memories with existing knowledge (Tamminen et al. 2010, 2013). Some variation has been reported in frequency of spindles that correlate with improvement. For example, Holz et al. (2012) reported that overnight retention of a word-list task was correlated with sigma activity, a result exclusively driven by a correlation with slow sigma activity (12–14 Hz) whereas there was no significant correlation observed with fast sigma activity (14–16 Hz). However, many others have found the opposite, with memory performance correlating with fast spindle activity but not slow spindle activity (e.g., Saletin et al. 2011; van der Helm et al. 2011) or even no association in spite of large sample sizes (Ackermann et al. 2015, Sleep). More research is needed to understand the extent to which these divisions distinguish different types of spindles either functionally or topographically.

Additionally, although sleep is a global phenomenon in many respects, sleep spindles may exert their beneficial influence for declarative memory consolidation in a regionally specialized manner. Saletin et al. (2011) tested a directed forgetting task in a nap paradigm and found that subjects who napped recalled significantly more words that were cued to-be-remembered than subjects who spent an equivalent amount of time awake. There was no such sleep-related enhancement for words cued to-be-forgotten. This selective enhancement of to-be-remembered words was correlated with fast sleep spindle (13.5–15 Hz) density during NREM sleep. Specifically, spindles over the posterior parietal regions were positively correlated with the proportion of to-be-remembered words recalled, yet negatively correlated with words cued for forgetting at frontal locations. These results argue against models suggesting that sleep uniformly decreases forgetting or enhances learning (Tononi and Cirelli 2006), and suggest that spindles can benefit memory consolidation in a specific and selective manner (see also chapter by Rauss and Born).

Procedural Skills

Several studies have demonstrated correlations between motor learning and sleep spindles. Nishida and Walker (Nishida and Walker 2007) capitalized on the known lateralized, offline plastic changes observed across a night of sleep using a motor-sequence task. They found that when subjects trained with their left hand and then took a nap, motor-skill memory improvements were correlated with the difference in spindle activity between the "learning" (contralateral, right hemisphere) and "nonlearning" hemisphere. This lateralization difference score was suggested to represent the homeostatic difference in spindle activity following learning, thereby providing evidence of regionally specific spindle associations with offline motor-skill learning in central locations. In another study, subjects were either trained on a motor-sequence task or a control task followed by a night of sleep (Morin et al. 2008). Subjects in the motor task condition showed increased number and duration of spindles during NREM sleep, and higher EEG power in the sigma (13 Hz) and beta (18–20 Hz) frequency bands, than the control task condition. However, none of these physiological changes during post-training sleep were correlated with overnight gains. A follow-up study found that specifically fast (13–15 Hz) but not slow (11–13 Hz) spindle densities were increased following motor learning compared to control (Barakat et al. 2011). Additionally, the difference in fast spindle density between the control and experimental nights was correlated with overnight performance gain on the motor but not control task, suggesting that fast spindles, which are most prominent over central and parietal derivations, are implicated in motor sequence learning and consolidation.

Experimentally Manipulating Spindles

It is possible that more spindles cause better memory consolidation, or that deeper encoding of information prior to sleep increases subsequent spindle activity (Fogel and Smith 2006; Gais et al. 2002), or even that individuals with better memory ability naturally have more spindles (Fogel and Smith 2011). For example, Tamaki et al. (2009) used a mirror-tracing task and found (1) mean amplitude and duration of fast spindles was greater on the learning night than nonlearning night; and (2) fast spindle density, amplitude, and duration were all positively correlated with overnight improvement on the learning night. However, upon closer inspection, fast spindle activity on the nonlearning night was also correlated with motor performance, indicating that individuals who had greater spindle activity always had better performance, and this effect was not specific to the learning night. Therefore, differences in fast spindle activity may be related to native motor learning ability. In the declarative domain, Lustenberger et al. (2015) highlighted how spindle activity is oftentimes positively associated with baseline encoding performance (indicating better learning abilities in individuals who have more spindle activity), and subjects with good encoding performance may have the least amount of overnight improvement, resulting in negative spindle associations with overnight performance. To better understand the role of spindles in memory consolidation, it is necessary to move beyond correlations between sleep features and memory improvement and discover critical mechanisms of memory consolidation by perturbing sleep and testing the effects on memory performance.

Important advances have been made by utilizing experimental methods that manipulate sleep features in order to test their effects on memory, such as transcranial stimulation to enhance (Marshall et al. 2006) or inhibit (Marshall et al. 2011) slow wave activity (see also chapter by Campos Beltran and Marshall), and pharmacology. Using pharmacological intervention, Mednick et al. (2013) increased sleep spindles with zolpidem (Ambien), and decreased sleep spindles with sodium oxybate (Xyrem) during a nap (Fig. 2). Declarative and motor memory performance was tested before and after the pharmacologically altered nap using a word paired-associates task and motor sequence task, respectively. They showed that pharmacologically enhancing sleep spindles with zolpidem in healthy adults produced exceptional declarative memory performance beyond that seen with sleep alone (placebo) or sleep with the comparison drug (SO), which showed decreased memory performance. Spindles were correlated with declarative memory improvement in all three conditions, suggesting that spindles are a neural correlate of verbal memory consolidation, and when enhanced produce a systematic increase in verbal memory retention.

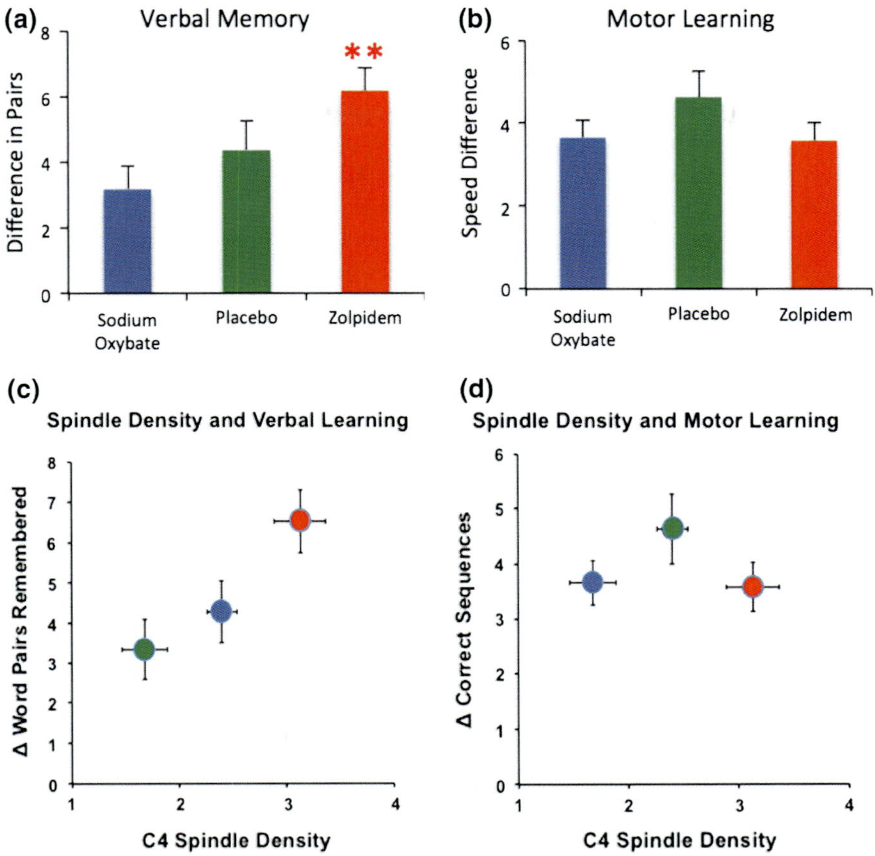

Fig. 2 Pharmacologically enhanced spindles boost verbal learning but not motor learning. Verbal memory was increased with zolpidem (**a**), and there were no differences in motor learning across drug conditions (**b**). **c**, **d** Memory performance improvement plotted against spindle density for naps with different drug conditions—zolpidem (*red circle*), placebo (*green circle*), and sodium oxybate (*blue circle*). Zolpidem enhanced sleep spindle density, and spindle density was related to verbal memory performance (**c**), but not motor learning (**d**). Statistical comparisons were between drugs conditions and placebo. ** indicates p-value < .005. Data are from Mednick et al. (2013)

Contrary to predictions, motor learning was not altered by the drug intervention and motor performance was not correlated with spindles in the zolpidem condition. A similar null result in motor learning was found by enhancing slow wave activity and spindles via transcranial application of slow oscillations (0.75 Hz) (Marshall et al. 2006). These results indicate that spindles may not be directly related to motor learning and suggest that there might be a third variable related to sleep-dependent motor improvement heretofore unexamined. Alternatively, spindles may be more related to recovery from motor fatigue than learning per se, as suggested by results showing no sleep-dependent effects on motor learning when controlling for fatigue and time of day (Rickard et al. 2008). Taken together, these results point toward a tighter coupling between sleep spindles and declarative verbal memory than motor learning. This difference may be attributable to the stronger dependence on hippocampal processing in the verbal task due to the associative nature of the task (Henke et al. 1999). In the case of hippocampal-dependent processing, spindles may be causally related to memory improvement by their direct role in hippocampal replay, whereas in non-hippocampal-dependent tasks, spindles may be a marker of a yet undefined process that correlates with consolidation.

In contrast with the Mednick et al. (2013) findings, Rasch et al. (2009) showed that administration of selective re-uptake inhibitors of serotonin (SSRI) and norepinephrine (SNRI) both decreased the amount of time spent in REM sleep, while the SNRI significantly increased amount of time spent in Stage 2 sleep, as well as the number and density of fast spindles (>13 Hz). Performance on a motor-sequence task was improved in both drug conditions compared to placebo, and overnight gains in performance were significantly correlated with the change in number and density of spindles between drug and placebo nights. Subjects were also tested on a declarative paired-associates task. Given the spindle enhancement with SNRI, one would predict improvement of declarative memory with SNRI. However, surprisingly they found no difference in number of words recalled in either SSRI or SNRI condition compared to placebo.

Using the same pharmacological intervention as Mednick et al. (2013), Kaestner et al. (2013) examined the role of sleep spindles in emotional memory consolidation (see also chapter by Cunningham and Payne). In this study, subjects encoded a full range of emotional pictures (negative, positive and neutral valence, as well as high and low arousal) before a nap with zolpidem or placebo. The authors reported that memory can be experimentally biased toward negative and highly arousing stimuli after a sleep period with pharmacologically elevated sleep spindles. Specifically, naps with zolpidem demonstrated increased sleep spindles and greater memory performance for negative and high-arousal stimuli compared with placebo or sodium oxybate. Both hypnotics elevated SWS, but only zolpidem increased spindle density, whereas sodium oxybate decreased spindle density. Thus, the increase in memory performance relative to placebo was more likely due to the increased spindle density rather than SWS. Importantly, total sleep time and all other measures of sleep were consistent across each drug manipulation suggesting that changes in memory performance were likely due to the drug manipulation of sleep spindle density.

Interaction Between Spindle and Slow Oscillations

Although spindles are often treated as discrete events, they do not occur in isolation from other oscillatory events. Current advancements in the field have begun to tease apart the complex relationship between spindles and other brain activity patterns, namely neocortical slow oscillations. Slow oscillations (<1 Hz) are characterized by up and down states that reflect active and silent periods of spiking in individual neurons (Steriade et al. 1993). During Stage 2 and SWS, spindles are often temporally coupled with slow oscillations. The spindle amplitude is observed to be highest during the up-state and lowest during the down-state of the slow oscillations (Contreras and Steriade 1995; Molle et al. 2002). Specifically, the peak of spindle amplitude often occurs during the beginning of the slow oscillation (i.e., the down-to-up-state transition) (see Fig. 1). These findings have been observed using scalp EEG (Molle et al. 2002) and intracellular recordings (Andrillon et al. 2011), and further supported by computational models (Bazhenov et al. 2002).

Co-occurring slow oscillations and spindles may be a key mechanism of memory consolidation during sleep (Mölle et al. 2011). In humans, spindle power during the up-state of the slow oscillations was increased following learning in a hippocampal-dependent task (Mölle et al. 2009). Ngo et al. (2013) used closed-loop stimulation to induce slow oscillations and observed an increase in spindles during the early phase of the slow oscillations, which was correlated with an increase in the memory retention task. Using the same pharmacological intervention as Mednick et al. (2013), who demonstrated increased spindle density using zolpidem, Niknazar et al. (2015) examined the relationship between the phase of the slow oscillation at which spindles occurred and memory performance (Fig. 3). In addition to boosting the rate of spindle events, zolpidem also increased the temporal consistency of when spindles occurred relative to the phase of slow oscillations, as measured by the modulation index between slow oscillation phase and spindle power (Fig. 3a, b). Phase-amplitude coupling between different neural oscillations has been observed in wake (Canolty et al. 2006; Lakatos et al. 2005), and this study provides evidence that such coupling is also observed with spontaneously occurring oscillations during sleep. Further, performance on the verbal memory task was significantly correlated with the phase of the slow oscillation at which the spindle peak was observed in the zolpidem and placebo conditions, but not in sodium oxybate (Fig. 3c–e). This suggests that memory improvement was increased when spindles occurred during the rising phase of the slow oscillation. Interestingly, when slow oscillation power, sigma power and phase/amplitude timing were allowed to compete for variance in performance change using a regression framework, phase/amplitude timing accounted for the most variance, followed by sigma, and slow oscillatory power was not a significant predictor. Overall, these studies strongly suggest that for hippocampal-dependent memory tasks, there is a preferred phase for spindle timing during the slow oscillation up-state to increase memory consolidation.

The exact neural mechanism underlying the coupling between spindles and slow oscillations (and hippocampal sharp-wave ripples), and how it enhances memory consolidation, is not known. Sleep spindles result in massive increases in intracellular Ca^{2+} (Sejnowski and Destexhe 2000), which is required to induce long-term potentiation, and a coincidence of thalamic spindles with other sleep EEG events, such as hippocampal sharp waves (Battaglia et al. 2004), at the down-to-up transition phase of cortical slow oscillations may be necessary to form permanent memories. One hypothesis suggests that hippocampal ripples could initiate the thalamic inputs (Vertes et al. 2007), and hippocampus to cortical

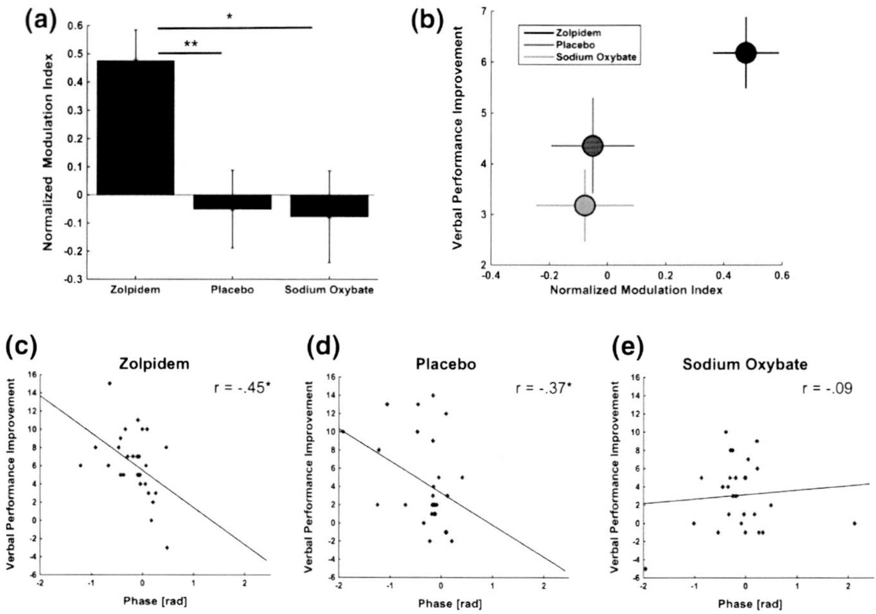

Fig. 3 Phase amplitude coupling between slow oscillations and spindles influences memory improvement during sleep. **a** Normalized modulation index measured for SO phase and spindle power was significantly increased with zolpidem compared to placebo and sodium oxybate. **b** Memory performance improvement plotted against normalized modulation index for naps with different drug conditions—zolpidem (*black circle*), placebo (*dark grey circle*), and sodium oxybate (*light grey circle*). **c–e** The SO phase at spindle peak (i.e., phase/amplitude timing) was correlated with memory improvement in both zolpidem and placebo conditions, suggesting a general mechanism of memory formation. The negative correlation indicates better memory performance was associated with negative SO phase, i.e., spindle peak occurring during the SO up-state. * indicates p-value < .05; ** indicates p-value < .005. Data are from Niknazar et al. (2015)

projections (Jay and Witter 1991) could selectively modulate the spindle activity through spindle-ripple coupling during the initiation of up-state. Such events could lead to the replay and therefore enhancement of selective memory during the up-states. Indeed, recent intracranial recordings in humans have demonstrated a tight coupling between thalamic spindles, hippocampal sharp wave ripples and cortical slow oscillations (Staresina et al. 2015).

An up-state of the slow oscillation is also a time period of high synchrony within the cortical network, marked by the high amplitude activity observed in the EEG. This synchronization may facilitate synaptic weight changes due to spike-timing-dependent plasticity (STDP). Any input, such as hippocampal ripples, that occurs during the later phase of the down-state of the slow oscillation, would influence the timing of the cortical cell firing during the following up-state. Indeed, a recent computational study suggests that the spatiotemporal pattern of the slow oscillation determines synaptic changes during slow wave sleep (Wei et al. 2016). Furthermore, it found that spindles preceding the slow oscillation (as occurs during the natural cycle of sleep state transitions) might influence the spatio-temporal pattern of slow oscillations and facilitate replay of selected memories.

Acknowledgements Research was supported by National Institutes of Health grants NSF Cognitive Neuroscience BCS1439210 (S.M.), R01 AG046646 (S.M. and M.B.), and R01 EB009282 (M.B.); Office of Naval Research/Multidisciplinary University Research Initiative Grant N000141310672; and National Science Foundation Graduate Research Fellowship Program (E.M.).

Glossary

Declarative memory: Conscious memory of facts and events. This type of memory is dependent on the hippocampus and other areas of the medial temporal lobe.

Non-declarative memory: Unconscious memories such as habits or skills. This type of memory is typically not dependent on the hippocampus.

Rapid eye movement (REM) sleep: A sleep stage characterized by rapid eye movements, low muscle tone, and rapid, low-voltage electroencephalogram (EEG) waveforms.

Slow wave sleep (SWS): Also referred to as deep sleep; previously Stage 3 and Stage 4. Slow, high amplitude delta activity (1–4 Hz) predominates the EEG during SWS.

Slow oscillations: Waveforms <1 Hz frequency with high voltage up and down states, which reflect periods of neuronal spiking and neuronal silence, respectively.

Spindles: Bursts of oscillatory activity visible on an EEG that typically occur during NREM sleep. Spindles typically consist of 12–15 Hz waveforms that occur for at least 0.5 s. Spindle density refers to the number of spindles per minute of sleep.

Sharp-wave ripple complexes: Composed of fast (~50–100 ms) bursts of spike activity (sharp waves) that are associated with high-frequency "ripples" (~150–200 Hz). These events are generated in the hippocampus.

Systems consolidation: The process that refers to the time-limited role of the hippocampus in declarative memory storage. Information is originally encoded in both hippocampal and cortical regions. Successive reactivation of this hippocampal-cortical network is presumed to allow new memories to be gradually integrated with pre-existing memories and become independent of the hippocampus.

Spike-timing-dependent plasticity: This refers to the process of change in synaptic weights, as a function of spike timing in pre- and post-synaptic neurons—there is an increase in synaptic weight when the pre-synaptic neuron fires prior to the post-synaptic neuron and a decrease in synaptic weight when the post-synaptic neuron fires prior to the pre-synaptic neuron.

Phase amplitude coupling: This refers to modulation of the amplitude of one oscillation by the phase of another oscillation, and provides information about the temporal relationship between oscillations. For example, such coupling is observed between the peak amplitude of spindle frequency and slow oscillation phase.

Modulation index: This index is one way to estimate the consistency of phase-amplitude coupling across various trials or events.

References

Abel T, Havekes R, Saletin JM, Walker MP (2013) Sleep, plasticity and memory from molecules to whole-brain networks. Curr Biol 23(17):R774–R788

Ackermann S, Hartmann F, Papassotiropoulos A, de Quervain DJ, Rasch B (2015) No associations between interindividual differences in sleep parameters and episodic memory consolidation. Sleep 38(6):951–959

Andersen P, Andersson S (1968) Physiological basis of the alpha rhythm, vol 1. Plenum Publishing Corporation

Andrillon T, Nir Y, Staba RJ, Ferrarelli F, Cirelli C, Tononi G, Fried I (2011) Sleep spindles in humans: insights from intracranial EEG and unit recordings. J Neurosci (The Official Journal of the Society for Neuroscience) 31(49):17821–17834

Astori S, Wimmer RD, Lüthi A (2013) Manipulating sleep spindles—expanding views on sleep, memory, and disease. Trends Neurosci 36(12):738–748

Bal T, McCormick DA (1996) What stops synchronized thalamocortical oscillations? Neuron 17 (2):297–308

Barakat M, Doyon J, Debas K, Vandewalle G, Morin A, Poirier G et al (2011) Fast and slow spindle involvement in the consolidation of a new motor sequence. Behav Brain Res 217 (1):117–121

Battaglia FP, Sutherland GR, McNaughton BL (2004) Hippocampal sharp wave bursts coincide with neocortical "up-state" transitions. Learn Mem (Cold Spring Harbor, NY), 11(6):697–704

Bazhenov M, Timofeev I, Steriade M, Sejnowski T (2000) Spiking-bursting activity in the thalamic reticular nucleus initiates sequences of spindle oscillations in thalamic networks. J Neurophysiol 84(2):1076–1087

Bazhenov M, Timofeev I, Steriade M, Sejnowski TJ (2002) Model of thalamocortical slow-wave sleep oscillations and transitions to activated states. J Neurosci 22(19):8691–8704

Bonjean M, Baker T, Lemieux M, Timofeev I, Sejnowski T, Bazhenov M (2011) Corticothalamic feedback controls sleep spindle duration in vivo. J Neurosci (The Official Journal of the Society for Neuroscience) 31(25):9124–9134

Budde T, Biella G, Munsch T, Pape HC (1997) Lack of regulation by intracellular Ca^{2+} of the hyperpolarization-activated cation current in rat thalamic neurones. J Physiol 503(1):79–85

Buzsáki G (1989) Two-stage model of memory trace formation: a role for "noisy" brain states. Neuroscience 31(3):551–570

Cai DJ, Mednick SA, Harrison EM, Kanady JC, Mednick SC (2009) REM, not incubation, improves creativity by priming associative networks. Proc Natl Acad Sci U S A 106 (25):10130–10134

Canolty RT, Edwards E, Dalal SS, Soltani M, Nagarajan SS, Kirsch HE et al (2006) High gamma power is phase-locked to theta oscillations in human neocortex. Science (New York, NY), 313 (5793):1626–1628

Clemens Z, Fabó D, Halász P (2005) Overnight verbal memory retention correlates with the number of sleep spindles. Neuroscience 132(2):529–535

Contreras D, Steriade M (1995) Cellular basis of EEG slow rhythms: a study of dynamic corticothalamic relationships. J Neurosci 15(1):604–622

Contreras D, Destexhe A, Sejnowski TJ, Steriade M (1996) Control of spatiotemporal coherence of a thalamic oscillation by corticothalamic feedback. Science 274(5288):771–774

Cox R, Hofman WF, Talamini LM (2012) Involvement of spindles in memory consolidation is slow wave sleep-specific. Learn Mem (Cold Spring Harbor, NY), 19(7):264–267

Crunelli V, Hughes SW (2010) The slow (< 1 Hz) rhythm of non-REM sleep: a dialogue between three cardinal oscillators. Nat Neurosci 13(1):9–17

Destexhe A, Contreras D, Steriade M, Sejnowski T, Huguenard J (1996) In vivo, in vitro, and computational analysis of dendritic calcium currents in thalamic reticular neurons. J Neurosci 16(1):169–185

Diba K, Buzsáki G (2007) Forward and reverse hippocampal place-cell sequences during ripples. Nat Neurosci 10(10):1241–1242

Diekelmann S, Born J (2010) The memory function of sleep. Nat Rev Neurosci 11(2):114–126

Fogel SM, Smith CT (2006) Learning-dependent changes in sleep spindles and Stage 2 sleep. J Sleep Res 15:250–255

Fogel SM, Smith CT (2011) The function of the sleep spindle: a physiological index of intelligence and a mechanism for sleep-dependent memory consolidation. Neurosci Biobehav Rev 35(5):1154–1165

Fogel SM, Smith CT, Cote KA (2007) Dissociable learning-dependent changes in REM and non-REM sleep in declarative and procedural memory systems. Behav Brain Res 180(1):48–61

Foster DJ, Wilson MA (2006) Reverse replay of behavioural sequences in hippocampal place cells during the awake state. Nature 440(7084):680–683

Gais S, Molle M, Helms K, Born J (2002) Learning-dependent increases in sleep spindle density. J Neurosci 22(15):6830–6834

Gennaro L De, Ferrara M (2003) Sleep spindles: an overview. Sleep Med Rev 7(5):423–440

Genzel L, Dresler M, Wehrle R, Grözinger M, Steiger A (2009) Slow wave sleep and REM sleep awakenings do not affect sleep dependent memory consolidation. Sleep 32(3):302–310

Henke K, Weber B, Kneifel S, Wieser HG, Buck A (1999) Human hippocampus associates information in memory. Proc Natl Acad Sci 96(10):5884–5889

Holz J, Piosczyk H, Feige B, Spiegelhalder K, Baglioni C, Riemann D, Nissen C (2012) EEG sigma and slow-wave activity during NREM sleep correlate with overnight declarative and procedural memory consolidation. J Sleep Res 21(6):612–619

Jay TM, Witter MP (1991) Distribution of hippocampal CA1 and subicular efferents in the prefrontal cortex of the rat studied by means of anterograde transport of Phaseolus vulgaris-leucoagglutinin. J Comp Neurol 313(4):574–586

Ji D, Wilson Ma (2007) Coordinated memory replay in the visual cortex and hippocampus during sleep. Nat Neurosci 10(1):100–107

Kaestner EJ, Wixted JT, Mednick SC (2013) Pharmacologically increasing sleep spindles enhances recognition for negative and high-arousal memories. J Cogn Neurosci 25(10):1597–1610

Lakatos P, Shah AS, Knuth KH, Ulbert I, Karmos G, Schroeder CE (2005) An oscillatory hierarchy controlling neuronal excitability and stimulus processing in the auditory cortex. J Neurophysiol 94(3):1904–1911

Lee AK, Wilson MA (2002) Memory of sequential experience in the hippocampus during slow wave sleep. Neuron 36(6):1183–1194

Lemieux M, Chen J-Y, Lonjers P, Bazhenov M, Timofeev I (2014) The impact of cortical deafferentation on the neocortical slow oscillation. J Neurosci (The Official Journal of the Society for Neuroscience) 34(16):5689–5703

Louie K, Wilson MA (2001) Temporally structured replay of awake hippocampal ensemble activity during rapid eye movement sleep. Neuron 29(1):145–156

Lustenberger C, Wehrle F, Tüshaus L, Achermann P, Huber R (2015) The multidimensional aspects of sleep spindles and their relationship to word-pair memory consolidation. Sleep 38 (7):1093–1103

Luthi A, Bal T, McCormick DA (1998) Periodicity of thalamic spindle waves is abolished by ZD7288, a Blocker of I_h. J Neurophysiol 79(6):3284–3289

Manshanden I, De Munck JC, Simon NR, Lopes da Silva FH (2002) Source localization of MEG sleep spindles and the relation to sources of alpha band rhythms. Clin Neurophysiol (Official Journal of the International Federation of Clinical Neurophysiology) 113(12):1937–1947

Marshall L, Helgadóttir H, Mölle M, Born J (2006) Boosting slow oscillations during sleep potentiates memory. Nature 444(7119):610–613

Marshall L, Kirov R, Brade J, Mölle M, Born J (2011) Transcranial electrical currents to probe EEG brain rhythms and memory consolidation during sleep in humans. PLoS ONE 6(2):e16905

Mednick SC, Nakayama K, Stickgold R (2003) Sleep-dependent learning: a nap is as good as a night. Nat Neurosci 6(7):697–698

Mednick SC, Cai DJ, Shuman T, Anagnostaras S, Wixted JT (2011) An opportunistic theory of cellular and systems consolidation. Trends Neurosci 34(10):504–514

Mednick SC, McDevitt EA, Walsh JK, Wamsley E, Paulus M, Kanady JC, Drummond SPa (2013) The critical role of sleep spindles in hippocampal-dependent memory: a pharmacology study. J Neurosci (The Official Journal of the Society for Neuroscience) 33(10):4494–4504

Meier-Koll A, Bussmann B, Schmidt C, Neuschwander D (1999) Walking through a maze alters the architecture of sleep. Percept Mot Skills 88:1141–1159

Molle M, Marshall L, Gais S, Born J (2002) Grouping of spindle activity during slow oscillations in human non-rapid eye movement sleep. J Neurosci 22(24):10941–10947

Mölle M, Eschenko O, Gais S, Sara SJ, Born J (2009) The influence of learning on sleep slow oscillations and associated spindles and ripples in humans and rats. Eur J Neurosci 29(5):1071–1081

Mölle M, Bergmann TO, Marshall L, Born J (2011) Fast and slow spindles during the sleep slow oscillation: disparate coalescence and engagement in memory processing. Sleep 34 (10):1411D–1421D

Morin A, Doyon J, Dostie V, Barakat M, Tahar AH, Korman M et al (2008) Motor sequence learning increases sleep spindles and fast frequencies in post-training sleep. Sleep 31(8):1149–1156

Morison RS, Bassett DL (1945) Electrical activity of the thalamus and basal ganglia in decorticate cats. J Neurophysiol 8(5):309–314

Ngo H-VV, Martinetz T, Born J, Mölle M (2013) Auditory closed-loop stimulation of the sleep slow oscillation enhances memory. Neuron 78(3):545–553

Niknazar M, Krishnan GP, Bazhenov M, Mednick SC (2015) Coupling of thalamocortical sleep oscillations are important for memory consolidation in humans. PLoS ONE 10(12):e0144720

Nishida M, Walker MP (2007) Daytime naps, motor memory consolidation and regionally specific sleep spindles. PLoS ONE 2(4):e341

Rasch B, Born J (2013) About sleep's role in memory. Physiol Rev 93(2):681–766

Rasch B, Pommer J, Diekelmann S, Born J (2009) Pharmacological REM sleep suppression paradoxically improves rather than impairs skill memory. Nat Neurosci 12(4):396–397

Rickard TC, Cai DJ, Rieth CA, Jones J, Ard MC (2008) Sleep does not enhance motor sequence learning. J Exp Psychol Learn Mem Cogn 34(4):834–842

Saletin JM, Goldstein AN, Walker MP (2011) The role of sleep in directed forgetting and remembering of human memories. Cerebral Cortex (New York, NY : 1991), 21(11):2534–25341

Schabus M, Gruber G, Parapatics S, Sauter C, Klösch G, Anderer P et al (2004) Sleep spindles and their significance for declarative memory consolidation. Sleep 27(8):1479–1485

Schmidt C, Peigneux P, Muto V, Schenkel M, Knoblauch V, Münch M et al (2006) Encoding difficulty promotes postlearning changes in sleep spindle activity during napping. J Neurosci (The Official Journal of the Society for Neuroscience) 26(35):8976–8982

Sejnowski TJ, Destexhe A (2000) Why do we sleep? Brain Res 886(1–2):208–223

Siapas AG, Wilson MA (1998) Coordinated interactions between hippocampal ripples and cortical spindles during slow-wave sleep. Cell Press 21:1123–1128

Smith CT, Aubrey JB, Peters KR (2004) Different roles for REM and stage 2 sleep in motor learning: A proposed model. Psychol Belg 44(1–2):81–104

Staresina BP, Bergmann TO, Bonnefond M, van der Meij R, Jensen O, Deuker L et al (2015) Hierarchical nesting of slow oscillations, spindles and ripples in the human hippocampus during sleep. Nat Neurosci 18(11):1679–1686

Steriade M, Deschenes M (1984) The thalamus as a neuronal oscillator. Brain Res Rev 8(1):1–63

Steriade M, Llinas RR (1988) The functional states of the thalamus and the associated neuronal interplay. Physiol Rev 68(3):649–742

Steriade M, Deschenes M, Domich L, Mulle C (1985) Abolition of spindle oscillations in thalamic neurons disconnected from nucleus reticularis thalami. J Neurophysiol 54(6):1473–1497

Steriade M, Domich L, Oakson G, Deschenes M (1987) The deafferented reticular thalamic nucleus generates spindle rhythmicity. J Neurophysiol 57(1):260–273

Steriade M, Jones EG, Llinás RR (1990) Thalamic oscillations and signaling. Wiley, Oxford

Steriade M, Nunez A, Amzica F (1993) A novel slow (<1 Hz) oscillation of neocortical neurons in vivo: depolarizing and hyperpolarizing components. J Neurosci 13(8):3252–3265

Stickgold R, Whidbee D, Schirmer B, Patel V, Hobson Ja (2000) Visual discrimination task improvement: A multi-step process occurring during sleep. J Cogn Neurosci 12(2):246–254

Tamaki M, Matsuoka T, Nittono H, Hori T (2009) Activation of fast sleep spindles at the premotor cortex and parietal areas contributes to motor learning: a study using sLORETA. Clin Neurophysiol (Official Journal of the International Federation of Clinical Neurophysiology) 120(5):878–886

Tamminen J, Payne JD, Stickgold R, Wamsley EJ, Gaskell MG (2010) Sleep spindle activity is associated with the integration of new memories and existing knowledge. J Neurosci (The Official Journal of the Society for Neuroscience) 30(43):14356–14360

Tamminen J, Lambon Ralph Ma, Lewis Pa (2013) The role of sleep spindles and slow-wave activity in integrating new information in semantic memory. J Neurosci (The Official Journal of the Society for Neuroscience) 33(39):15376–15381

Timofeev I (2001) Contribution of intrinsic and synaptic factors in the desynchronization of thalamic oscillatory activity. Thalamus Relat Syst 1(1):53–69

Timofeev I, Bazhenov M (2005) Mechanisms and biological role of thalamocortical oscillations (Columbus F, ed). Trends Chronobiol Res

Timofeev I, Steriade M (1996) Low-frequency rhythms in the thalamus of intact-cortex and decorticated cats. J Neurophysiol 76(6):4152–4168

Timofeev I, Grenier F, Bazhenov M, Houweling AR, Sejnowski TJ, Steriade M (2002) Short- and medium-term plasticity associated with augmenting responses in cortical slabs and spindles in intact cortex of cats in vivo. J Physiol 542(2):583–598

Tononi G, Cirelli C (2006) Sleep function and synaptic homeostasis. Sleep Med Rev 10(1):49–62

Tononi G, Cirelli C (2014) Sleep and the price of plasticity: from synaptic and cellular homeostasis to memory consolidation and integration. Neuron 81(1):12–34

van der Helm E, Gujar N, Nishida M, Walker MP (2011) Sleep-dependent facilitation of episodic memory details. PLoS ONE 6(11):e27421

Vertes RP, Hoover WB, Szigeti-Buck K, Leranth C (2007) Nucleus reuniens of the midline thalamus: link between the medial prefrontal cortex and the hippocampus. Brain Res Bull 71 (6):601–609

von Krosigk M, Bal T, McCormick D (1993) Cellular mechanisms of a synchronized oscillation in the thalamus. Science 261(5119):361–364

Walker MP, Brakefield T, Morgan A, Hobson JAA, Stickgold R (2002) Practice with sleep makes perfect: Sleep-dependent motor skill learning. Neuron 35(1):205–211

Walker MP, Brakefield T, Seidman J, Morgan A, Hobson JA, Stickgold R (2003) Sleep and the time course of motor skill learning. Learn Mem (Cold Spring Harbor, NY), 10(4):275–284

Wei Y, Krishnan GP, Bazhenov M (2016) Synaptic mechanisms of memory consolidation during sleep slow oscillations. J Neurosci 36(15):4231–4247

Weiner OM, Dang-Vu TT (2016) Spindle oscillations in sleep disorders: a systematic review. Neural Plasticity 2016:1–30

Wilson M, McNaughton B (1994) Reactivation of hippocampal ensemble memories during sleep. Science 265(July):5–8

Hippocampal Sharp Wave/Ripple Complexes—Physiology and Mechanisms

Nikolaus Maier and Richard Kempter

Abstract Hippocampal sharp wave/ripple complexes (SWRs) are rhythmic electrographic activities that appear strictly coupled with resting behavioral waking states of the brain and slow-wave sleep. Several lines of evidence now support their involvement in the formation and long-term consolidation of hippocampus-dependent memories. During SWRs, the hippocampal networks experience a sharp and transient ($\sim$50–100 ms) increase in neuronal activity that is temporally coherent across both hippocampi; further, sharp wave-associated ripple oscillations express a remarkably high oscillation frequency of $\sim$120–250 cycles per second; finally, SWRs are irregular in occurrence. Despite around three decades of research into the mechanistic underpinnings of this phenomenon, a coherent theory of various aspects—e.g. its initiation and termination, and the precise synchronization of thousands of neurons at millisecond precision—is still unavailable. Here, we will outline the current understanding of the implications and the mechanisms that govern SWRs, from both a physiological and a network-theoretical perspective. We will put special emphasis on the contributing neuronal populations and will discuss unresolved aspects.

Keywords Hippocampus · Local field potential · Large irregular activity · Ripple oscillation · Sharp wave/ripple complex · Slow wave sleep · Declarative memory · Consolidation

N. Maier (✉)
Charité – Universitätsmedizin Berlin, Berlin, Germany
e-mail: nikolaus.maier@charite.de

R. Kempter (✉)
Humboldt-Universität zu Berlin, Berlin, Germany
e-mail: r.kempter@biologie.hu-berlin.de

N. Axmacher and B. Rasch (eds.), *Cognitive Neuroscience of Memory Consolidation*, Studies in Neuroscience, Psychology and Behavioral Economics, DOI 10.1007/978-3-319-45066-7_14

Behavioral Phenomenology of Sharp Wave/Ripple Complexes

Sharp wave/ripple complexes (SWRs) are one of the hallmark electrographic signatures observed in the local field potential (LFP) of the hippocampus. They occur as transient and large (up to 2 mV) amplitude deflections—sharp waves—of $\sim$50–100 ms duration, with negative polarity in the dendritic fields of pyramidal cells. Sharp waves appear in conjunction with oscillations—ripples—at $\sim$120–250 Hz. As for various brain rhythms, SWRs seem to be a phenomenon biologically conserved across different members of the animal class of *Mammalia* (for review, see Buzsáki and Moser 2013): they occur in rodents such as rat (Buzsáki 1986; Suzuki and Smith 1987; Buzsáki et al. 1992), mouse (Buzsáki et al. 2003), and rabbit (Nokia et al. 2010). SWRs have also been observed in the bat (Ulanovsky and Moss 2007), in carnivores such as cat (Kanamori 1985), and in primates as for example monkey (Skaggs et al. 2007; Logothetis et al. 2013) and human (Bragin et al. 1999; Axmacher et al. 2008). A recent study reported SWRs also in a reptile, the central bearded dragon (Shein-Idelson et al. 2016). The presence of SWRs across these different vertebrates may support their biological significance, and we thus first turn to experimental evidence on the correlation between SWRs and associated behaviors.

With electrophysiological recordings, Vanderwolf (1969) identified a brain state linked to 'automatic', or 'consummatory' (Routtenberg 1968), behaviors, such as fur grooming, face washing, licking and chewing. These behaviors were associated with an electrographic pattern referred to as 'large irregular activity' (LIA) in the hippocampus. In contrast, during more voluntary types of movement, e.g. walking, the hippocampal EEG displayed a more regular pattern, theta rhythm at $\sim$3–10 Hz (Vanderwolf 1969; Sainsbury 1970). Electrographic signatures similar to LIA, although not termed as such, had been shown before in rabbit (Stumpf 1965) and dog (Yoshii et al. 1966). Augmenting this initial description of LIA, O'Keefe and Nadel (1978) observed ripples as a spindle-like oscillatory signature in the LFP of rats during behaviors associated with LIA (Fig. 1). The availability of more powerful computational resources in the early 1980s enabled more fine-grained analyses of electrophysiological data in both the time and the frequency domains (Leung et al. 1982). With these resources at hand, Buzsáki and colleagues were the first to define the hippocampal sharp wave (SPW) as the dominating electrographic hallmark during LIA (Buzsáki et al. 1983), and they also identified the temporal association of SPWs and ripple oscillations (Buzsáki et al. 1992; Ylinen et al. 1995, see also Suzuki and Smith 1987; Fig. 1).

Theta- and LIA-dominated states alternate not only during wakefulness, but also during sleep: Ripple-enriched periods during LIA are prominent during slow-wave sleep (SWS) while the theta rhythm is paramount during phases of rapid eye movement (REM) paradoxical sleep (O'Keefe and Nadel 1978). However, this strict and state-dependent separation of SWRs and theta rhythm has been scrutinized with the finding of 'exploratory ripples' that occur intermingled with theta oscillations during walking in rats (O'Neill et al. 2006).

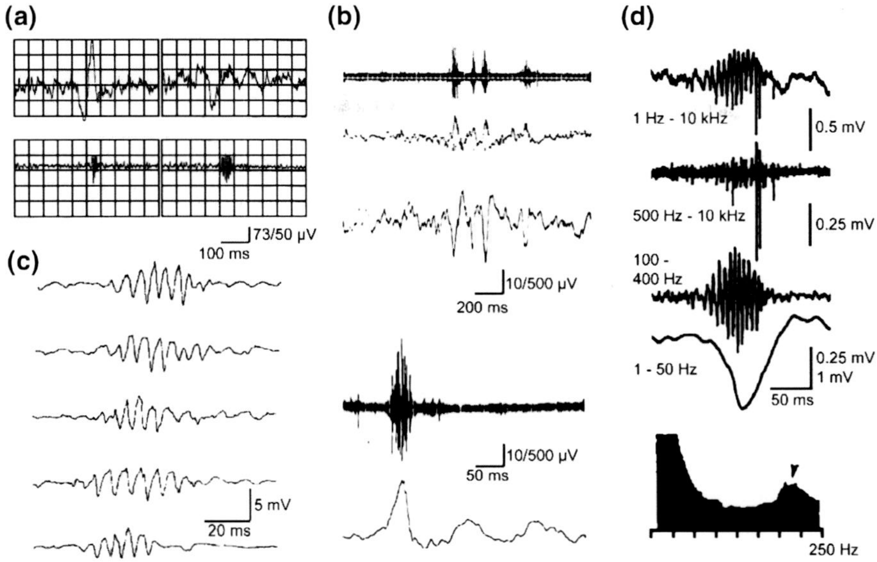

Fig. 1 Early displays of SWRs recorded from rats. **a** From O'Keefe and Nadel (1978), with permission, their Fig. 14. *Upper* and *lower* sweeps were recorded simultaneously with separate electrodes. Ripple-associated LFP deflections (*upper*) recorded from stratum radiatum, and associated ripples (*lower*) recorded with another electrode from the pyramidal cell layer (80–800 Hz bandpass filtered). **b** Two example recordings from Buzsáki et al. (1983), with permission, their Fig. 13. *Top* upper sweep, ripple-associated multi-unit activity from CA1 pyramidal cell layer; *below* simultaneously recorded SPW signatures in stratum oriens and radiatum. *Bottom* another example from the same recording (multiunit activity, top, and LFP SPW, below). **c** From Suzuki and Smith (1988), with permission, their Fig. 2. Several example ripples, after bandpass-filtering at 0.1–1 kHz. **d** From Buzsáki et al. (1992), their Figs. 1 and 4. *Top to bottom* Wide-band recording of a SWR; 500 Hz high-pass filtered trace to isolate multiunit activity; band-pass filtered version to highlight the ripple oscillation; low-pass (50 Hz) filtered trace that shows the associated sharp wave, recorded in stratum radiatum at a position 200 µm below the pyramidal cell layer. *Bottom* Power spectrum of the LFP signal recorded during awake immobility; y-axis: arbitrary units

Functional Implications

In agreement with the idea that the hippocampus forms a 'cognitive map' representing a subject's environment (Tolman 1948; O'Keefe and Nadel 1978), cells in the hippocampus respond with increased firing at particular locations of the environment. Since their discovery (O'Keefe and Dostrovsky 1971) these *place cells* were supposed to constitute neuronal 'assemblies' (Hebb 1949) that form episodic-like memories in the spatial context. SWRs were hypothesized to play an important role in the consolidation of these memories (Buzsáki 1989; Hasselmo 1999). Experimental evidence supporting this hypothesis came from work demonstrating that place cells are re-activated during SWS (Wilson and McNaughton 1994); more specifically, this reactivation was linked to SWRs

(Lee and Wilson 2002). Several experiments were designed to causally tighten this link: For humans, Axmacher et al. (2008) showed a correlation between the load of successfully learnt items and ripple density in the rhinal cortex during post-learning sleep. This is in agreement with an animal study showing an increase in occurrence of hippocampal SWRs in post-learning sleep (Eschenko et al. 2008). Also, transcranial direct current stimulation above the frontolateral cortices, or acoustic stimulation, can enhance slow oscillations in post-learning SWS and facilitates declarative memory formation in humans (Marshall et al. 2006; Ngo et al. 2013) (see also chapter by Campos-Beltrán and Marshall), indirectly supporting the role of SWRs during consolidation. The most direct evidence so far was drawn in studies on rats: the delivery of short electric pulses to the hippocampal system can terminate SWRs; and when such electric pulses were selectively applied during SWS following behavioral training, it caused an impairment of spatial memory (Girardeau et al. 2009; Ego-Stengel and Wilson 2010). In contrast to SWR-associated replay of assemblies during SWS, the replay during awake SWRs might underlie the planning of future action, as shown by behavioral tests in spatial alternation tasks where ripples were electrically suppressed (Jadhav et al. 2012); planning is necessarily based on recent experience and thus involves and relies on functional working memory (Buzsáki 2015).

Together, these studies support the suggested connection between SWRs and hippocampus-dependent memory consolidation. But is there also a link between SWRs and synaptic plasticity, a suggested key condition for learning at the cellular and neuronal network levels (Hebb 1949; Takeuchi et al. 2014) (see also chapter by Kreutzmann and colleagues)? Indeed, a recent study in behaving rats provided evidence for a connection between SWRs and synaptic plasticity involved in spatial learning (Girardeau et al. 2014). The authors first replicated the earlier findings of an increased SWR incidence in post-learning SWS (Eschenko et al. 2008). When they suppressed SWRs in this period, they observed a significant elevation in the cumulative number of SWRs, compared to control without suppression. This might indicate a homeostatic regulation of the system to express a sufficient load of SWRs to support consolidation. Intriguingly, the upregulation of SWRs required NMDA receptors (NMDARs), and could be blocked by NMDAR antagonists applied systemically before the spatial training. Together, these findings emphasize that spatial learning can induce synaptic plasticity that translates into elevated SWR activity to assist memory consolidation. This supports earlier results demonstrating that NMDAR dependent potentiation of synaptic transmission in CA3 is paralleled with an increased SWR occurrence in vitro (Behrens et al. 2005).

Anatomical Substrates for SWR Generation in CA1

Before we disentangle the specific activity patterns of those neuron types in the CA1 area that actively contribute to the generation of SWRs, we should define the most important constituting elements in this network.

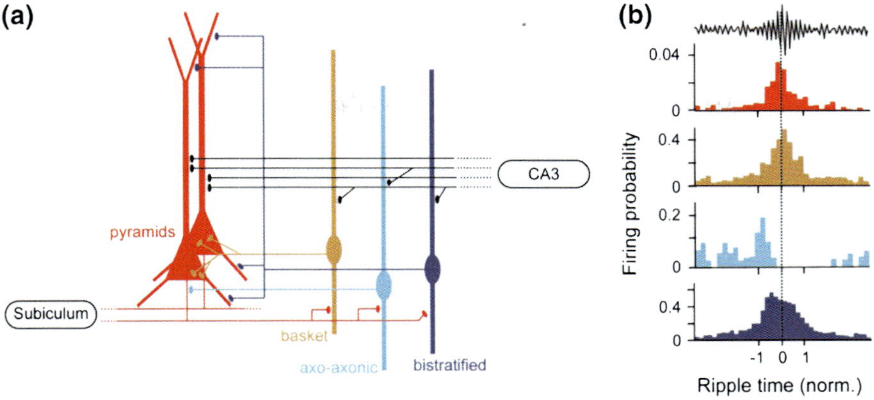

Fig. 2 Neuronal network elements of hippocampal area CA1 and their activity during SWRs. Modified with permission from Somogyi et al. (2014), their Fig. 1. **a** Pyramidal neurons and three main interneuron types involved in ripple generation within CA1, and their connections. **b** Spike time histograms illustrating the timing with respect to the ripple maximum, and discharge probability of involved neuron types

Until recently, it was assumed that CA1 principal neurons make up a fairly homogenous population; however, these cells are different with respect to anatomy, function, and electrophysiology (Lee et al. 2014; Graves et al. 2012; Arszovszki et al. 2014; Thome et al. 2014; Malik et al. 2015). Pyramidal cells in CA1 are recurrently coupled by chemical synapses at a rate of ~1% (Deuchars and Thomson 1996), and they can also exhibit electrical coupling (Schmitz et al. 2001; Mercer et al. 2006). Valero et al. (2015) suggested that deep and superficial CA pyramids are differentially inhibited during SWRs.

For interneurons in CA1, more than 20 neuron types are known (Somogyi et al. 2014): Immunohistochemical, morphological, and electrophysiological differences have been identified. In the context of SWR mechanisms, two major groups of interneuron should be mainly considered because of their recruitment into the active network. These are (i) cells that target the perisomatic domains of pyramidal neurons, and include parvalbumin-expressing, fast-discharging basket and axo-axonic interneurons (Klausberger et al. 2003) (Fig. 2a); (ii) inhibitory interneurons that project onto the dendritic areas of principal neurons, mainly bistratified and oriens-lacunosum moleculare (O-LM) interneurons, both co-expressing somatostatin and parvalbumin (Klausberger et al. 2003, 2004; Varga et al. 2012; Pangalos et al. 2013).

Electrophysiological Phenomenology of SWRs

The local network in area CA1 receives feed-forward input from CA3 (Fig. 2a). This input can give rise to the sharp wave (SPW) in the stratum radiatum of CA1, i.e., a prominent negative voltage deflection in the local field potential (Fig. 1).

Simultaneously with the SPW, the ripple oscillation emerges with strongest appearance in the CA1 pyramidal cell layer (Buzsáki 1986; Buzsáki et al. 1992). Interestingly, the amplitude of the SPW component in stratum pyramidale can vary considerably across individual events even at a given recording site within or close to the pyramidal cell layer, where SPWs may display positive or negative polarity. However, ripples can also occur isolated, i.e., without any obvious SPW signature; such trains of oscillatory cycles can emerge out of the baseline in the LFP (Ramirez-Villegas et al. 2015).

According to the layout mentioned above, the suppression of excitatory synaptic input arriving from CA3 should diminish the generation of SWRs. This concept was challenged in a CA3-TeTX transgenic mouse line where glutamate release from Schaffer collaterals can be inducibly suppressed, and, indeed, hippocampus-dependent memory formation was impaired when CA3 output was blocked (Nakashiba et al. 2008). However, SWRs were still present, but with lower ripple oscillation frequency ($\sim$120 Hz compared to $\sim$150 Hz in control) (Nakashiba et al. 2009). These results suggest that input different from that arriving via the Schaffer collaterals can drive SWRs in CA1. This might include input from subcortical areas (Logothetis et al. 2013) or input from the entorhinal cortex that arrives in CA1 via the temporo-ammonic pathway (Isomura et al. 2006).

What are the activation patterns of the different neuronal populations in CA1 during SWRs? In the rat hippocampus, an estimated fraction of $\sim$10% of the total neuronal population is activated during a given SWR episode (Csicsvari et al. 2000; Buzsáki 2006), rendering this network pattern the most coherent type of hippocampal neuronal network oscillations. Comparison of the activity of excitatory and inhibitory neurons during SWRs revealed marked differences (Fig. 2b): up to 30% of pyramidal neurons in CA1 participate during these episodes and show the strongest relative increase in discharge frequency with respect to non-SWR periods ($\sim$9-fold). Concurrently, up to about 70% of GABAergic interneurons located in, or close to the principal layer of CA1 are activated and express an average 4-fold increase in discharge rate during SWR over non-SWR epochs (Csicsvari et al. 1999, 2000). However, in absolute terms, the mean firing rate during SWRs is considerably lower for principal neurons compared with interneurons ($\sim$10 spikes/s vs. up to 100 spikes/s) (Csicsvari et al. 1999, 2000).

While these numbers are based on wire tetrode- and silicon probe simultaneous measurements of a large number of neurons, recordings from identified, single neurons provide a more differentiated view. Recent advancement in the methodology allow for intracellular recordings from individual neurons in non-anesthetized, drug-free, behaving (head-fixed or freely moving) mice. Such recordings from CA1 pyramidal cells revealed a SWR-associated train of small-amplitude ripple-associated voltage fluctuations riding on a wave of depolarization of several millivolts, followed by a prominent hyperpolarization that could last several hundreds of milliseconds (Maier et al. 2011; English et al. 2014; Hulse et al. 2016). Pyramidal neurons can fire at rates of up to $\sim$40 spikes/s during a ripple (English et al. 2014; Hulse et al. 2016). For CA1 principal neurons it was further shown that superficial (radiatum-directed) calbindin-expressing and deep (oriens-directed) calbindin-negative cells are differentially recruited into the active network during SWRs; superficial pyramids preferentially receive an excitatory synaptic drive, while deep principal neurons experience a net inhibitory input (Valero et al. 2015). These observations support a local network in area CA1 that favors inhibitory signaling of parvalbumin-expressing inhibitory basket cells onto deep compared with superficial pyramids (Lee et al. 2014). In agreement with this, single-neuron juxtacellular recordings in rats and mice demonstrate strong discharge of parvalbumin-expressing bistratified and basket cells during SWRs, with firing rates of up to 130 spikes/s in basket cells (Lapray et al. 2012; Varga et al. 2012, 2014; Katona et al. 2014), in contrast to CCK-expressing basket cells that show only weak coupling with SWRs (Klausberger et al. 2005). Another class of interneuron whose activity is modulated by SWRs is the axo-axonic interneuron, specialized to target the axon initial segments of pyramidal neurons (Somogyi 1977). These cells increase their discharge rate before, but stop firing during SWRs (Klausberger et al. 2003; Varga et al. 2014). Finally, oriens lacunosum-moleculare (O-LM) interneurons, whose axons terminate in the most apical dendritic field of CA1, discharge in about 50% of SWRs and fire both delayed and at low rates (<30 spikes/s) during these events (Varga et al. 2012; Katona et al. 2014).

In summary, this rich phenomenology of SWRs in vivo provides considerable insight into, but also constraints about underlying processes. However, revealing the principles also requires approaches that allow the detailed study of mechanism in defined experimental conditions, for example in the hippocampal slice preparation.

An in Vitro Model System to Study SWRs

The in vitro slice preparation of rodent hippocampus has turned out to be a useful model system to study synaptic, cellular, and network mechanisms underlying SWRs (for review, see Butler and Paulsen 2015). In the following we will compare properties of SWRs found in vivo and in vitro.

The presence of spontaneous SWR activity in hippocampal slices is remarkable per se, and its basis not fully understood. However, several findings in vivo are in line with SWRs as a primarily intrinsic phenomenon of the hippocampus, although SWRs are modulated by cortical and subcortical inputs (Battaglia et al. 2004; Logothetis et al. 2013; Ishikawa et al. 2014; Vandecasteele et al. 2014). Indeed, in de-afferentiation experiments, where input pathways to the hippocampus were severed, SPWs were the dominating electrographic signature, and they were largely uncoupled from behavioral modulation (Buzsáki et al. 1983). Furthermore, in isolates of the entire hippocampus in vitro, SWRs can be readily observed (Wu et al. 2002). It is therefore feasible that SWRs represent the 'default' activity pattern in the hippocampus, and that input conveying extrahippocampal modulation can shift the system to non-SWR activity states [but see Goutagny et al. (2009) and Pietersen et al. (2009)].

Fundamental features of SWRs are similar in vitro and in vivo: The average rate of occurrence is alike [range in vivo: 0.01–3/s (Buzsáki 1986); mean ± SEM in vitro: 2.8 ± 0.2/s and 2.7 ± 1.1/s (Papatheodoropoulos and Kostopoulos 2002; Maier et al. 2003)]. (ii) The range of ripple oscillation frequency is comparable [in vivo: 150–250 Hz (Buzsáki et al. 1992); in vitro: 210 ± 16 Hz (Maier et al. 2003)]. (iii) SWRs are initiated in CA3 and propagate to CA1, subiculum, and the deep layers of the entorhinal cortex in both conditions (Buzsáki 1986; Chrobak and Buzsáki 1994, 1996; Wu et al. 2002; Maier et al. 2003, 2009). (iv) When sampled along the somato-dendritic axis within CA1, SWRs express a typical amplitude profile, which is identical in recordings from living animals and from slices (Buzsáki et al. 1983; Maier et al. 2009; Fig. 3). (v) The discharge patterns of interneuron activation during SWRs is comparable in both conditions (Bähner et al. 2011; Lapray et al. 2012; Varga et al. 2012; Pangalos et al. 2013; Katona et al. 2014). (vi) Pharmacological manipulations modify SWRs similarly in vivo and in vitro (Table 1). Finally, (vii) the phenomenon of intraripple frequency accommodation, where the peak of the ripple frequency in the first half of the event is followed by a monotonic decrease during the second half, is observed in vivo (Ponomarenko et al. 2004; Nguyen et al. 2009; Sullivan et al. 2011) as well as in vitro (Donoso et al. 2017).

However, in vivo and in vitro SWRs are also distinct in several aspects: (i) In vivo, when recorded in the vicinity of the pyramidal cell layer, the polarity of sharp waves can fluctuate between positive or negative deflections (Ramirez-Villegas et al. 2015). In contrast, the in vitro SWRs appear comparably stereotypical, with some variation in amplitude and duration, however rarely with different polarity (but see Hofer et al. 2015). (ii) The occurrence rate of SWRs shows a largely clock-like periodicity in slices, while it is strictly depending on the behavioral states and more irregular in vivo (Buzsáki 1986). (iii) The recruitment of pyramidal neurons into the active network during SWRs differs in both conditions: while in behaving mice CA1 pyramidal cells discharge regularly and with rates of up to 40 spikes/s during SWRs (see Section "Electrophysiological phenomenology" above), the discharge rate is considerably lower in vitro (e.g. 0.038 spikes/s in Bähner et al. 2011). (iv) Further, SWR-associated antidromic (ectopic) spikes as observed in the in vitro preparation (Papatheodoropoulos 2008; Bähner et al. 2011; Hofer et al. 2015) have not been identified so far in vivo (English et al. 2014; Hulse et al. 2016). (v) Even though ripples in vivo and in vitro roughly fall into the same frequency band, more careful comparison within the same experimental model (mouse) reveals a higher frequency of ripples recorded in slices (range: 160–240 Hz) compared to in vivo (range: 127–147 Hz; both from Maier et al. 2011). (vi) The slice model of SWRs is relying on the ventral to mid hippocampus (Papatheodoropoulos and Kostopoulos 2002), while in vivo SWRs have been mostly studied in dorsal, but also in ventral hippocampus (Patel et al. 2013; Ciocchi et al. 2015). Finally, (vii) the amplitudes of SPWs can be different (e.g. Fig. 3b vs. Fig. 3e).

The mentioned differences can be likely explained by the lack of extra-hippocampal input (i–iii) and the smaller size of the active network (iii, v, vi)

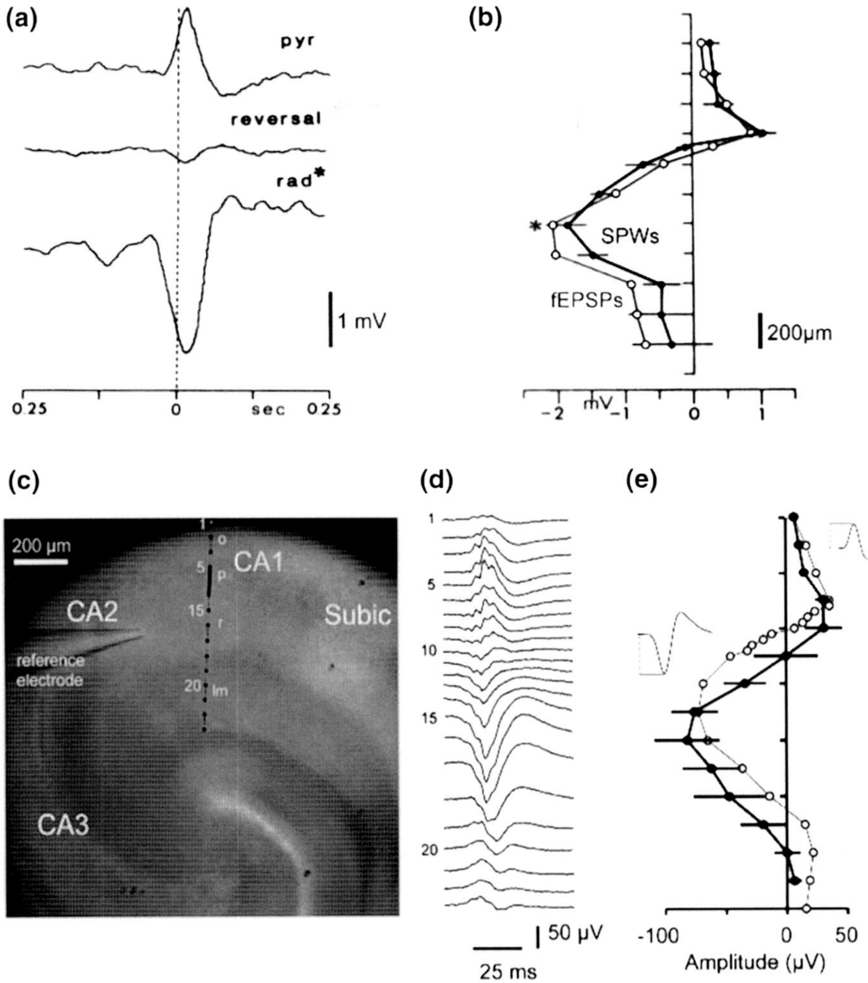

Fig. 3 Comparison of spatial characteristics of SWRs in vivo (**a**, **b**) and in vitro (**c**–**e**). **a** Averaged (128 events) and low-pass filtered (1–30 Hz) SWRs in vivo from three different depths (*pyr* pyramidal cell layer; *rad* stratum radiatum). SWRs were aligned to multi-unit activity recorded by a reference electrode in the pyramidal cell layer. **b** Laminar amplitude profiles of SWRs (*filled circles*) and of Schaffer collateral evoked field potentials (*open circles*, *grey line*). For SWRs, each point is an average of 30 events. *Horizontal bars* indicate standard error of the mean. **c** Serial multisite LFP recordings reveal the laminar profile of SWRs in a slice of ventral hippocampus. Infrared-differential interference contrast video image displaying the recording area, the reference electrode positioned in CA1 radiatum close to CA2, while SWRs were recorded at different laminar positions with a second electrode in mid-CA1 (positions indicated as *black dots*, 10–100 μm steps). Abbreviations: *o* oriens; *p* pyramidale; *r* radiatum; *lm* lacunosum-moleculare; *Subic* subiculum. **d** LFP averages from the recording sites indicated in (**c**), each representing 20 events, triggered by the signal sampled with the reference electrode. At each recording site, the first significant maximum from baseline was determined as the SWR peak (see insets in **e**). **e** Amplitude profile of the experiment shown in **c**, **d** (*open circles*, *thin line*), and averaged amplitude depth profile representing profiles from five independent recordings (*five slices, closed circles, thick line*). *Horizontal bars* indicate standard error of the mean. Panels **a** and **b** from Buzsáki et al. 1983, their Fig. 14, modified with permission. Panels **c**–**e** from Maier et al. 2009, their Fig. 5

in the slice model. Artifacts due to slicing and recovery, and possible synaptic re-organization (Kirov et al. 2004) may contribute to (iv–vi). Nevertheless, basic features of SWR in vitro and in vivo are often remarkably similar. Thus, data obtained from in vitro preparations may be useful to understand also SWRs in vivo: Acute brain slices give us access to many details that are not available in vivo, for example properties of synaptic transmission and connectivity, which are essential for understanding network mechanisms, and provide the basis for biophysical, bottom-up computational models of SWRs.

Mechanisms Underlying the Dynamics of SWRs

Despite the huge amount of available data on SWRs and their likely physiological relevance, the basic mechanisms underlying this phenomenon remain largely enigmatic, both in vitro and in vivo. To discuss some ideas, we first focus on sharp waves, which are population bursts that originate in the CA3 network. Later, we separately summarize hypotheses on how ripple oscillations may emerge, which are most prominent in the CA1 area.

Mechanisms underlying sharp waves: SPWs arise in the hippocampal CA3 area and spread into the CA1 area (Csicsvari et al. 2000). Mechanisms that govern the dynamics of these events in CA3 remain unresolved. As mentioned before, possibly weak external input (e.g., from entorhinal cortex, dentate gyrus, and/or subcortical nuclei) may control the recall of specific activity patterns in the CA3 network. Constraints for mechanisms that control SPWs in CA3 (e.g. Hájos et al. (2013) for slices) are that pyramids have low activity outside SPWs, and only a small portion of them spike once during SPWs (see also Fig. 2). Most interneurons increase activity, and parvalbumin-expressing basket cells are the most active cells. What are the principles that govern the onset/initiation, the termination, and the incidence of SPWs?

Principles of sharp wave initiation: Bazelot et al. (2016) showed in vitro that a single action potential (AP) discharge of a CA3 pyramidal cell can induce a SPW. Interestingly, this pyramidal cell discharge selectively triggered several putative interneuron spikes at short (2–3 ms) delays, which supports the idea that the initiation of SPWs in vitro involves interneurons. Similarly, Ellender et al. (2010), Schlingloff et al. (2014), and Kohus et al. (2016) demonstrated that activation of fast spiking basket cells can induce SPWs. However, it remains puzzling how small changes in the activity of single interneurons could contribute to the initiation of SPWs, which requires a tremendous increase of network activity in only a few milliseconds.

Termination of sharp waves: SPWs are terminated after only 50–100 ms by processes hitherto unclear. Candidate mechanisms for termination of SPWs in CA3 include short-term plasticity at the active synapses. Parvalbumin-expressing interneurons, for example, show a frequency-dependent depression of synaptic responses (Kohus et al. 2016). Furthermore, SPWs might be stopped by increased

Table 1 Pharmacological modulation of SWRs; comparison of in vivo and in vitro results

Target	Drug/Action	Reference	Model	Administration	SWR incidence	Ripple	
						Power	Frequency
NMDA receptor	MK-801/antagonist	Girardeau et al. (2014)	in vivo, rat	i.p.	==	↑[a]	n.r.
	CPP/antagonist	Kouvaros et al. (2015)	in vitro, rat, spont.	Bath	↓	↑	↑[b]
$GABA_A$ receptor	Bicuculline/antagonist	Suzuki and Smith (1988)	in vivo, rat	i.p.	==	↑[c]	↓[d]
	Picrotoxin/antagonist	Stark et al. (2014)	in vivo, mouse, opt.-ind.	Intrahipp.	↓	↓	↓[e]
	Gabazine/antagonist	Nimmrich et al. (2005)	in vitro, mouse, spont.	Bath	↓	n.r.	n.r.
$GABA_A$ receptor (unspecific at α subunits)	Diazepam/pos. modulator	Buzsáki (1986)	in vivo, rat	i.p.	↓	==[f]	n.r.
	Diazepam/pos. modulator	Suzuki and Smith (1988)	in vivo, rat	i.p.	↓	n.r.	n.r.
	Diazepam/pos. modulator	Ponomarenko et al. (2004)	in vivo, rat	i.p.	↓	↓	↓
	Diazepam/pos. modulator	Koniaris et al. (2011)	in vitro, rat, spont.	Bath	↓	↓	↓
	Diazepam/pos. modulator	Viereckel et al. (2013)	in vitro, mouse, spont.	Bath	↓	n.r.	==
$GABA_A$ receptor (α1 subunit)	Zolpidem/pos. modulator	Ponomarenko et al. (2004)	in vivo, rat	i.p.	↓	↓[g]	↓[g]
	Zolpidem/pos. modulator	Koniaris et al. (2011)	in vitro, rat, spont.	Bath	==	↑/↓[h]	==/↓[i]

(continued)

Table 1 (continued)

Target	Drug/Action	Reference	Model	Administration	SWR incidence	Ripple	
						Power	Frequency
Adenosine receptor	Adenosine/agonist	Maier et al. (2012)	in vivo, mouse	Intrahipp.	↓	↓	n.r.
	Adenosine/agonist	Maier et al. (2012)	in vitro, mouse, spont.	Bath	↓	↓	n.r.
CB1 receptor	CP55,940/agonist	Robbe et al. (2006)	in vivo, rat	i.p./intrahipp.	↓	↓	n.r.
	CP55,940/agonist	Maier et al. (2012)	in vivo, mouse	i.p.	↓	↓	n.r.
	CP55,940/agonist	Maier et al. (2012)	in vitro, mouse, spont.	Bath	↓	↓	n.r.
	WIN55,212-2/agonist	Maier et al. (2012)	in vitro, mouse, spont.	Bath	↓	↓	n.r.
5-HT receptors	Serotonin/agonist	ul Haq et al. (2016)	in vitro, rat, stim.-ind.	Bath	↓	↓	==
5-HT_{1A} receptor	WAY100635/antagonist	Ponomarenko et al. (2003)	in vivo, rat	i.c.v.	↓	==	↓

↑ increase; ↓ decrease; == no effect; *n.r.* not reported; *i.p.* intraperitoneally; *i.c.v.* intracerebroventricularly; *intrahipp.* intrahippocampal injection; *opt.-ind.* optogenetically induced; *stim.-ind.* stimulus train induced; [a]5% ↑ of SWR amplitude; [b]4% ↑ in ripple frequency; [c]in subepileptic condition: SPW amplitude and ripple amplitude; [d]in some ripples longer interwave periods (4–8 ms → 8–20 ms); [e]inferred from time-frequency spectrogram (their Fig. 4B); [f]no change in SPW amplitude; [g]↓ in ripple amplitude (–0.02 ± 0.01 mV) and ↓ in ripple frequency (–2.8 ± 0.6 Hz); [h]concentration-dependent: 0.1 μM: ↑, 10 μM: ↓; [i]concentration-dependent: 0.1 μM: ==, 1 or 10 μM: ↓ by 5% of control inhibitory feedback via activation of fast interneuronal synaptic inhibition (as modeled in Taxidis et al. 2012, 2013), by extrasynaptic $GABA_A$ or $GABA_B$ receptors (Hollnagel et al. 2014), intrinsic neuronal conductances (Zhang et al. 2006; Fano et al. 2012), refractory processes (Traub and Wong 1982; Omura et al. 2015), or bursting mechanisms (Dur-e-Ahmad et al. 2012). However, there is no unified framework that fully explains the termination of SPWs.

Incidence of sharp waves: Low doses of $GABA_A$ receptor antagonists increase inter-SPW intervals both in vitro (Nimmrich et al. 2005; Ellender et al. 2010) and in vivo (Stark et al. 2014), supporting the critical role of inhibition in SPW incidence. However, to explain delays between successive SPWs in the range of about a second or even longer, a mechanism with a time constant of that order is required. A possible candidate that determines inter-SPW intervals is recovery from short-term depression of inhibitory synapses: Parvalbumin-positive basket cells discharge typically several times during a SPW, which depresses their synapses onto pyramidal cells (Szabó et al. 2010) and onto other parvalbumin-positive basket cells (Bartos et al. 2002). Kohus et al. (2016) showed that the recovery of transmission at parvalbumin-positive basket cell synapses could be an essential component of the refractory mechanism between SPWs. In line with this refractoriness, a spontaneous SPW did not restart within a 200–300 ms time window after an optogenetically induced SPW (Schlingloff et al. 2014). Together, however, it does remain counterintuitive that the recovery of *inhibition* increases the *excitability* of the CA3 network to ultimately allow the emergence of the next SPW.

To conclude, many details about the dynamics of sharp waves have been clarified recently, including the key role of parvalbumin-positive interneurons, but a coherent picture is missing. Early models emphasized the importance of mutual excitation between pyramidal neurons in CA3 (Traub and Wong 1982). Computational neuronal network models may help to understand how the spontaneous and fast emergence of large network events can be controlled by a balance of recurrent inhibition and excitation (Omura et al. 2015; Chenkov et al. 2017) that can cope with the opposing demands of stability of network activity and strong and fast amplification at the beginning of SPWs. Thus, a specific model for the dynamics of sharp waves in CA3 is not available yet, which requires also more details on the connectivity among pyramidal cells and various interneurons (Kohus et al. 2016). In contrast to this lack of understanding of sharp waves, relatively concise views are available on possible mechanisms underlying the high-frequency synchronization during the ripple oscillation in CA1, to which we turn in the next section.

Mechanisms underlying ripple oscillations. Sharp waves arise in the CA3 network, but ripple oscillations ($\sim$200 Hz) are weaker there (Sullivan et al. 2011). However, via the Schaffer collaterals the sharp wave in CA3 provides strong excitation to CA1, where it evokes SWRs. Because ripple oscillations are locally generated, and are strongest in the CA1 pyramidal cell layer, and, in particular, are not coherent between CA3 and CA1 (Sullivan et al. 2011), we restrict our attention in this section to the CA1 region of the hippocampus.

Current hypotheses on mechanisms underlying the generation of ripple oscillations agree that ripples are due to interactions between neurons at the network

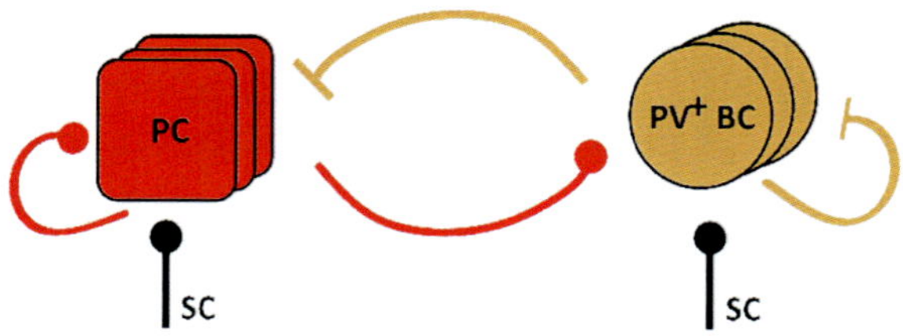

Fig. 4 Basic layout of the CA1 microcircuit to generate ripple oscillations. The local network receives activation via Schaffer collaterals (SCs). Pyramidal cells (PCs) are recurrently connected —via chemical and possibly electrical synapses—and project onto interneurons, for example parvalbumin-positive basket cells (PV^+ BCs). These again are recurrently connected by chemical and electrical synapses, and exert feed-forward and feed-back inhibitory control onto PCs

level (Fig. 4). However, the available computational models fundamentally differ on how such oscillatory activity emerges. Models can be roughly classified in excitation-first and inhibition-first models. Excitation-first models suggest that ripple oscillations are generated in a population of recurrently coupled principal cells. The activity in a population of principal cells then passes its oscillatory activity on to other neurons and networks, for example they can entrain inhibitory cells via local excitatory connections from principal cells to interneurons. The proposed generative mechanisms in excitation-first models depend, on the one hand, on the propagation of activity between pyramidal cells via axonal gap junctions (Draguhn et al. 1998; Traub et al. 1999; Schmitz et al. 2001). This model is supported by the observation that ripple-like oscillations can persist in the absence of chemical synaptic transmission (Draguhn et al. 1998; Nimmrich et al. 2005). The gap-junction model (Traub and Bibbig 2000; Traub and Whittington 2010) predicts antidromic spikes, as observed in vitro (Papatheodoropoulos 2008; Bähner et al. 2011), and can account for synaptic currents (potentials) observed during SWRs (Maier et al. 2011; Hulse et al. 2016). Excitation-first models are also supported, on the other hand, by models that predict ripple-synchronization via chemical synapses combined with supralinear dendritic integration (Memmesheimer 2010; Jahnke et al. 2015). Such an amplification, which could lead also to fast dendritic spikes, has been observed in slice experiments (Ariav et al. 2003; Gasparini et al. 2004; Gasparini and Magee 2006).

In contrast, inhibition-first models suggest that ripple oscillations are generated by a recurrent interneuron network (Buzsáki et al. 1992; Buzsáki and Chrobak 1995; Ylinen et al. 1995). When an interneuron network experiences sufficiently strong excitation, which does not need to be oscillatory, such a network can quickly synchronize and start oscillating at ripple frequency. The activity in a population of inhibitory cells can then pass its oscillatory activity on to other neurons and local networks, for example by entraining excitatory cells via local inhibitory connections. An interesting feature of ripple-oscillating interneuron networks is that they can be in a sparsely synchronized state (Brunel and Wang 2003; Maex and De Schutter 2003). In this state, the network frequency critically depends on the transmission latency and the rise time of the inhibitory postsynaptic current; the network frequency is, however, independent of the amplitude and the decay time of the GABAergic conductance; local interaction between pyramidal cells and interneurons may slow down the oscillation (Brunel and Wang 2003). Perisomatically targeting interneurons such as parvalbumin-expressing basket cells do show the proposed short-latency recurrent interaction among them (Bartos et al. 2002).

Inhibition-first models have recently received experimental (Stark et al. 2014; Gulyás and Freund 2015) and theoretical (Taxidis et al. 2012) support. Furthermore, a computational network model of basket cells in CA1 (Donoso et al. 2017) confirmed that the simulated ripple frequency does not depend on the amplitude and decay time constant of the GABAergic postsynaptic conductance, as observed experimentally when GABAergic modulators were applied (Papatheodoropoulos et al. 2007; Koniaris et al. 2011; Viereckel et al. 2013). Moreover, Donoso et al. could explain the mechanisms underlying the phenomenon of intra-ripple frequency accommodation (Ponomarenko et al. 2004; Nguyen et al. 2009; Sullivan et al. 2011) and the dependence of ripple frequency on the sharp wave amplitude (Sullivan et al. 2011; Hulse et al. 2016). However, inhibition-first models cannot explain, for example, the existence of ripples when all synaptic transmission is blocked (Draguhn et al. 1998; Nimmrich et al. 2005).

To summarize, recent work has significantly advanced our understanding of the potential mechanisms underlying ripple oscillations. However, experimental tests of the basic model classes have not yet been able to distinguish between excitation-first and inhibition-first models; possibly both excitatory and inhibitory networks contribute to the generation of ripple oscillations.

Summary and Outlook

In this chapter, we aimed at providing an up-to-date and concise overview of diverse aspects of the hippocampal SWR phenomenon. However, it lies in the complex nature of the subject that many aspects could be only superficially touched, or had to be omitted entirely. For some of these topics the reader may be referred to other recent work (Buzsáki 2015; Colgin 2016).

A largely unexplored field we have not addressed is the connection of SWRs and disease. And in turn, a better understanding of alterations in SWRs paralleled by impaired function in neuropsychiatric disease models will most likely help us to obtain deeper insight also into the significance of sharp waves and/or ripple oscillations. Recent work on models of schizophrenia, dementia, and epilepsy has already begun to fulfill this expectation (Cunningham et al. 2012; Jefferys et al. 2012; Suh et al. 2013; Aivar et al. 2014; Karlócai et al. 2014; Simon et al. 2014; Witton et al. 2014; Nicole et al. 2016; see also the chapter by Baker and Zeman).

To conclude this review, we note that early research on hippocampal network oscillations (Vanderwolf 1969; Buzsáki et al. 1983) has triggered a collective effort to unveil the behavioral correlates, the mechanisms, and the functional significance of SWRs. And indeed, much insight has been obtained, on behavioral coupling (occurrence during awake immobility and SWS), on cellular activity (differential contributions of the diverse interneuron types), and on function of SWRs (ripple-associated reactivation of neuronal ensembles; see Buzsáki 2015). A strong *correlative* link between SWR activity and memory consolidation has been drawn, but only recently we have begun to better understand the *causal* connections (Ego-Stengel and Wilson 2010; Girardeau and Zugaro 2011; Jadhav et al. 2012; Girardeau et al. 2014) (see also the chapter by Talamini). One of the most interesting open questions relates to the association of SWRs and synaptic plasticity. The modification of synaptic strength is potentially a constituting factor underlying learning and memory (Martin et al. 2000; Takeuchi et al. 2014). Buzsáki (1989) has provided a framework for SWRs contributing to synaptic plasticity, suggesting that the 'detonating' property as a requirement of a 'Hebbian' synapse (Andersen and Løyning 1962; McNaughton and Morris 1987) is functionally provided by the massive neuronal activity associated with SWRs. Indeed, in area CA1, the impinging concerted population bursts of CA3 cells during SWRs can be the trigger to induce long-term enhancement of synaptic transmission (King et al. 1999; Sadowski et al. 2016). The ripple oscillation may be a means to synchronize activity among many neurons at a millisecond precision, which may further boost plasticity, potentially also in hippocampal output brain areas (Chrobak and Buzsáki 1996; Siapas and Wilson 1998; Sirota et al. 2003; Jadhav et al. 2016). To test this possibility, a separation of sharp waves and ripples would be favorable, which may be possible once the underlying mechanisms are known. If interneuron networks generate ripple oscillations, the excitatory recurrent connectivity among pyramidal cells might be responsible for the replay of previously stored assemblies. In this way, interneurons and pyramidal cells could have complementary roles. However, it may well be that ripple oscillations are an epi-phenomenon that arises as a result of fast feedback to control the stability of hippocampal activity bursts.

Acknowledgements The authors thank Aarti Swaminathan, André Holzbecher, Barbara Imbrosci, Benedikt Salmen, Dietmar Schmitz, José Donoso, Nikolay Chenkov, Roberta Evangelista and Roger D. Traub for helpful comments on the manuscript.

References

Aivar P, Valero M, Bellistri E, Menendez de la Prida L (2014) Extracellular calcium controls the expression of two different forms of ripple-like hippocampal oscillations. J Neurosci 34: 2989–3004

Andersen P, Løyning Y (1962) Interaction of various afferents on CA1 neurons and dentate granule cells. Colloq Int CNRS 107:23–45

Ariav G, Polsky A, Schiller J (2003) Submillisecond precision of the input-output transformation function mediated by fast sodium dendritic spikes in basal dendrites of CA1 pyramidal neurons. J Neurosci 23:7750–7758

Arszovszki A, Borhegyi Z, Klausberger T (2014) Three axonal projection routes of individual pyramidal cells in the ventral CA1 hippocampus. Front Neuroanat 8:53

Axmacher N, Elger CE, Fell J (2008) Ripples in the medial temporal lobe are relevant for human memory consolidation. Brain 131:1806–1817

Bähner F, Weiss EK, Birke G, Maier N, Schmitz D, Rudolph U, Frotscher M, Traub RD, Both M, Draguhn A (2011) Cellular correlate of assembly formation in oscillating hippocampal networks in vitro. Proc Natl Acad Sci U S A 108:E607–E616

Bartos M, Vida I, Frotscher M, Meyer A, Monyer H, Geiger JP, Jonas P (2002) Fast synaptic inhibition promotes synchronized gamma oscillations in hippocampal interneuron networks. Proc Natl Acad Sci U S A 99:13222–13227

Battaglia FP, Sutherland GR, McNaughton BL (2004) Hippocampal sharp wave bursts coincide with neocortical "up-state" transitions. Learn Mem 11:697–704

Bazelot M, Telenczuk MT, Miles R (2016) Single CA3 pyramidal cells trigger sharp waves in vitro by exciting interneurones. J Physiol 594:2565–2577

Behrens CJ, van den Boom LP, de Hoz L, Friedman A, Heinemann U (2005) Induction of sharp wave-ripple complexes in vitro and reorganization of hippocampal networks. Nat Neurosci 8:1560–1567

Bragin A, Engel JJ, Wilson CL, Fried I, Buzsáki G (1999) High-frequency oscillations in human brain. Hippocampus 9:137–142

Brunel N, Wang X-J (2003) What determines the frequency of fast network oscillations with irregular neural discharges? I. Synaptic dynamics and excitation-inhibition balance. J Neurophysiol 90:415–430

Butler JL, Paulsen O (2015) Hippocampal network oscillations – recent insights from in vitro experiments. Curr Opin Neurobiol 31:40–44

Buzsáki G (1986) Hippocampal sharp waves: their origin and significance. Brain Res 398:242–252

Buzsáki G (1989) Two-stage model of memory trace formation: a role for "noisy" brain states. Neuroscience 31:551–570

Buzsáki G (2006) Rhythms of the brain. Oxford University Press, Inc. Oxford, New York

Buzsáki G (2015) Hippocampal sharp wave-ripple: a cognitive biomarker for episodic memory and planning. Hippocampus 25:1073–1188

Buzsáki G, Chrobak JJ (1995) Temporal structure in spatially organized neuronal ensembles: a role for interneuronal networks. Curr Opin Neurobiol 5:504–510

Buzsáki G, Moser EI (2013) Memory, navigation and theta rhythm in the hippocampal-entorhinal system. Nat Neurosci 16:130–138

Buzsáki G, Leung L-WS, Vanderwolf CH (1983) Cellular bases of hippocampal EEG in the behaving rat. Brain Res 287:139–171

Buzsáki G, Horváth Z, Urioste R, Hetke J, Wise K (1992) High-frequency network oscillation in the hippocampus. Science 256:1025–1027

Buzsáki G, Buhl DL, Harris KD, Csicsvari J, Czéh B, Morozov A (2003) Hippocampal network patterns of activity in the mouse. Neuroscience 116:201–211

Chenkov N, Sprekeler H, Kempter R (2017) Memory replay in balanced recurrent networks. PLOS Comp Biol (under review)

Chrobak JJ, Buzsáki G (1994) Selective activation of deep layer (V-VI) retrohippocampal cortical neurons during hippocampal sharp waves in the behaving rat. J Neurosci 14:6160–6170

Chrobak JJ, Buzsáki G (1996) High-frequency oscillations in the output networks of the hippocampal-entorhinal axis of the freely behaving rat. J Neurosci 16:3056–3066

Ciocchi S, Passecker J, Malagon-Vina H, Mikus N, Klausberger T (2015) Selective information routing by ventral hippocampal CA1 projection neurons. Science 348:560–563

Colgin LL (2016) Rhythms of the hippocampal network. Nat Rev Neurosci 17:239–249

Csicsvari J, Hirase H, Czurkó A, Mamiya A, Buzsáki G (1999) Oscillatory coupling of hippocampal pyramidal cells and interneurons in the behaving Rat. J Neurosci 19:274–287

Csicsvari J, Hirase H, Mamiya A, Buzsáki G (2000) Ensemble patterns of hippocampal CA3-CA1 neurons during sharp wave-associated population events. Neuron 28:585–594

Cunningham MO, Roopun A, Schofield IS, Whittaker RG, Duncan R, Russell A, Jenkins A, Nicholson C, Whittington MA, Traub RD (2012) Glissandi: transient fast electrocorticographic oscillations of steadily increasing frequency, explained by temporally increasing gap junction conductance. Epilepsia 53:1205–1214

Deuchars J, Thomson AM (1996) CA1 pyramid-pyramid connections in rat hippocampus in vitro: dual intracellular recordings with biocytin filling. Neuroscience 74:1009–1018

Donoso JR, Maier N, Schmitz D, Kempter R (2017) Hippocampal ripple oscillations and inhibition-first network models (in preparation)

Draguhn A, Traub RD, Schmitz D, Jefferys JGR (1998) Electrical coupling underlies high-frequency oscillations in the hippocampus in vitro. Nature 394:189–192

Dur-e-Ahmad M, Nicola W, Campbell SA, Skinner FK (2012) Network bursting using experimentally constrained single compartment CA3 hippocampal neuron models with adaptation. J Comput Neurosci 33:21–40

Ego-Stengel V, Wilson MA (2010) Disruption of ripple-associated hippocampal activity during rest impairs spatial learning in the rat. Hippocampus 20:1–10

Ellender TJ, Nissen W, Colgin LL, Mann EO, Paulsen O (2010) Priming of hippocampal population bursts by individual perisomatic-targeting interneurons. J Neurosci 30:5979–5991

English DF, Peyrache A, Stark E, Roux L, Vallentin D, Long MA, Buzsáki G (2014) Excitation and inhibition compete to control spiking during hippocampal ripples: Intracellular study in behaving mice. J Neurosci 34:16509–16517

Eschenko O, Ramadan W, Mölle M, Born J, Sara SJ (2008) Sustained increase in hippocampal sharp-wave ripple activity during slow-wave sleep after learning. Learn Mem 15:222–228

Fano S, Çalişkan G, Heinemann U (2012) Differential effects of blockade of ERG channels on gamma oscillations and excitability in rat hippocampal slices. Eur J Neurosci 36:3628–3635

Gasparini S, Magee JC (2006) State-dependent dendritic computation in hippocampal CA1 pyramidal neurons. J Neurosci 26:2088–2100

Gasparini S, Migliore M, Magee JC (2004) On the initiation and propagation of dendritic spikes in CA1 pyramidal neurons. J Neurosci 24:11046–11056

Girardeau G, Zugaro M (2011) Hippocampal ripples and memory consolidation. Curr Opin Neurobiol 21:452–459

Girardeau G, Benchenane K, Wiener SI, Buzsáki G, Zugaro MB (2009) Selective suppression of hippocampal ripples impairs spatial memory. Nat Neurosci 12:1222–1223

Girardeau G, Cei A, Zugaro M (2014) Learning-induced plasticity regulates hippocampal sharp wave-ripple drive. J Neurosci 34:5176–5183

Goutagny R, Jackson J, Williams S (2009) Self-generated theta oscillations in the hippocampus. Nat Neurosci 12:1491–1493

Graves AR, Moore SJ, Bloss EB, Mensh BD, Kath WL, Spruston N (2012) Hippocampal pyramidal neurons comprise two distinct cell types that are countermodulated by metabotropic receptors. Neuron 76:776–789

Gulyás AI, Freund TT (2015) Generation of physiological and pathological high frequency oscillations: The role of perisomatic inhibition in sharp-wave ripple and interictal spike generation. Curr Opin Neurobiol 31:26–32

Hájos N, Karlócai MR, Németh B, Ulbert I, Monyer H, Szabó G, Erdélyi F, Freund TF, Gulyás AI (2013) Input-output features of anatomically identified CA3 neurons during hippocampal sharp wave/ripple oscillation in vitro. J Neurosci 33:11677–11691

Hasselmo ME (1999) Neuromodulation: acetylcholine and memory consolidation. Trends Cogn Sci 3:351–359

Hebb DO (1949) The organization of behavior. Wiley, New York

Hofer KT, Kandrács Á, Ulbert I, Pál I, Szabó C, Héja L, Wittner L (2015) The hippocampal CA3 region can generate two distinct types of sharp wave-ripple complexes, in vitro. Hippocampus 25:169–186

Hollnagel JO, Maslarova A, ul Haq R, Heinemann U (2014) GABAB receptor dependent modulation of sharp wave-ripple complexes in the rat hippocampus in vitro. Neurosci Lett 574:15–20

Hulse BK, Moreaux LC, Lubenov EV, Siapas AG (2016) Membrane potential dynamics of CA1 pyramidal neurons during hippocampal ripples in awake mice. Neuron 89:800–813

Ishikawa D, Matsumoto N, Sakaguchi T, Matsuki N, Ikegaya Y (2014) Operant conditioning of synaptic and spiking activity patterns in single hippocampal neurons. J Neurosci 34:5044–5053

Isomura Y, Sirota A, Özen S, Montgomery S, Mizuseki K, Henze DA, Buzsáki G (2006) Integration and segregation of activity in entorhinal-hippocampal subregions by neocortical slow oscillations. Neuron 52:871–882

Jadhav SP, Kemere C, German PW, Frank LM (2012) Awake hippocampal sharp-wave ripples support spatial memory. Science 336:1454–1458

Jadhav SP, Rothschild G, Roumis DK, Frank LM (2016) Coordinated excitation and inhibition of prefrontal ensembles during awake hippocampal sharp-wave ripple events. Neuron 90:113–127

Jahnke S, Timme M, Memmesheimer R (2015) A unified dynamic model for learning, replay, and sharp-wave/ripples. J Neurosci 35:16236–16258

Jefferys JGR, Menendez de la Prida L, Wendling F, Bragin A, Avoli M, Timofeev I, Lopes da Silva FH (2012) Mechanisms of physiological and epileptic HFO generation. Prog Neurobiol 98:250–264

Kanamori N (1985) A spindle-like wave in the cat hippocampus: a novel vigilance level-dependent electrical activity. Brain Res 334:180–182

Karlócai MR, Kohus Z, Káli S, Ulbert I, Szabó G, Máté Z, Freund TF, Gulyás AI (2014) Physiological sharp wave-ripples and interictal events in vitro: what's the difference? Brain 137:463–485

Katona L, Lapray D, Viney TJ, Oulhaj A, Borhegyi Z, Micklem BR, Klausberger T, Somogyi P (2014) Sleep and movement differentiates actions of two types of somatostatin-expressing GABAergic interneuron in rat hippocampus. Neuron 82:872–886

King C, Henze DA, Leinekugel X, Buzsáki G (1999) Hebbian modification of a hippocampal population pattern in the rat. J Physiol 521:159–167

Kirov SA, Petrak LJ, Fiala JC, Harris KM (2004) Dendritic spines disappear with chilling but proliferate excessively upon rewarming of mature hippocampus. Neuroscience 127:69–80

Klausberger T, Magill PJ, Márton LF, Roberts JDB, Cobden PM, Buzsáki G, Somogyi P (2003) Brain-state- and cell-type-specific firing of hippocampal interneurons in vivo. Nature 421:844–848

Klausberger T, Márton LF, Baude A, Roberts JDB, Magill PJ, Somogyi P (2004) Spike timing of dendrite-targeting bistratified cells during hippocampal network oscillations in vivo. Nat Neurosci 7:41–47

Klausberger T, Marton LF, O'Neill J, Huck JHJ, Dalezios Y, Fuentealba P, Suen WY, Papp E, Kaneko T, Watanabe M, Csicsvari J, Somogyi P (2005) Complementary roles of cholecystokinin- and parvalbumin-expressing GABAergic neurons in hippocampal network oscillations. J Neurosci 25:9782–9793

Kohus Z, Káli S, Rovira L, Schlingloff D, Papp O, Freund TF, Hájos N, Gulyás AI (2016) Properties and dynamics of inhibitory synaptic communication within the CA3 microcircuits of pyramidal cells and interneurons expressing parvalbumin or cholecystokinin. J Physiol 594:3745–3774

Koniaris E, Drimala P, Sotiriou E, Papatheodoropoulos C (2011) Different effects of zolpidem and diazepam on hippocampal sharp wave-ripple activity in vitro. Neuroscience 175:224–234

Kouvaros S, Kotzadimitriou D, Papatheodoropoulos C (2015) Hippocampal sharp waves and ripples: effects of aging and modulation by NMDA receptors and L-type Ca^{2+} channels. Neuroscience 298:26–41

Lapray D, Lasztoczi B, Lagler M, Viney TJ, Katona L, Valenti O, Hartwich K, Borhegyi Z, Somogyi P, Klausberger T (2012) Behavior-dependent specialization of identified hippocampal interneurons. Nat Neurosci 15:1265–1271

Lee AK, Wilson MA (2002) Memory of sequential experience in the hippocampus during slow wave sleep. Neuron 36:1183–1194

Lee SH, Marchionni I, Bezaire M, Varga C, Danielson N, Lovett-Barron M, Losonczy A, Soltesz I (2014) Parvalbumin-positive basket cells differentiate among hippocampal pyramidal cells. Neuron 82:1129–1144

Leung LWS, Lopes da Silva FH, Wadman WJ (1982) Spectral characteristics of the hippocampal EEG in the freely moving rat. Electroencephalogr Clin Neurophysiol 54:203–219

Logothetis NK, Eschenko O, Murayama Y, Augath M, Steudel T, Evrard HC, Besserve M, Oeltermann A (2013) Hippocampal-cortical interaction during periods of subcortical silence. Nature 491:547–553

Maex R, De Schutter E (2003) Resonant synchronization in heterogeneous networks of inhibitory neurons. J Neurosci 23:10503–10514

Maier N, Nimmrich V, Draguhn A (2003) Cellular and network mechanisms underlying spontaneous sharp wave-ripple complexes in mouse hippocampal slices. J Physiol 550:873–887

Maier N, Morris G, Johenning FW, Schmitz D (2009) An approach for reliably investigating hippocampal sharp wave-ripples in vitro. PLoS ONE 4:e6925

Maier N, Tejero-Cantero Á, Dorrn AL, Winterer J, Beed PS, Morris G, Kempter R, Poulet JFA, Leibold C, Schmitz D (2011) Coherent phasic excitation during hippocampal ripples. Neuron 72:137–152

Maier N, Morris G, Schuchmann S, Korotkova T, Ponomarenko A, Böhm C, Wozny C, Schmitz D (2012) Cannabinoids disrupt hippocampal sharp wave-ripples via inhibition of glutamate release. Hippocampus 22:1350–1362

Malik R, Dougherty KA, Parikh K, Byrne C, Johnston D (2015) Mapping the electrophysiological and morphological properties of CA1 pyramidal neurons along the longitudinal hippocampal axis. Hippocampus 26:341–361

Marshall L, Helgadóttir H, Mölle M, Born J (2006) Boosting slow oscillations during sleep potentiates memory. Nature 444:610–613

Martin SJ, Grimwood PD, Morris RGM (2000) Synaptic plasticity and memory: an evaluation of the hypothesis. Annu Rev Neurosci 23:649–711

McNaughton BL, Morris RGM (1987) Hippocampal synaptic enhancement and information storage within a distributed memory system. Trends Neurosci 10:408–415

Memmesheimer RM (2010) Quantitative prediction of intermittent high-frequency oscillations in neural networks with supralinear dendritic interactions. Proc Natl Acad Sci U S A 107: 11092–11097

Mercer A, Bannister AP, Thomson AM (2006) Electrical coupling between pyramidal cells in adult cortical regions. Brain Cell Biol 35:13–27

Nakashiba T, Young JZ, McHugh TJ, Buhl DL, Tonegawa S (2008) Transgenic inhibition of synaptic transmission reveals role of CA3 output in hippocampal learning. Science 319:1260–1264

Nakashiba T, Buhl DL, McHugh TJ, Tonegawa S (2009) Hippocampal CA3 output is crucial for ripple-associated reactivation and consolidation of memory. Neuron 62:781–787

Ngo HV, Martinetz T, Born J, Mölle M (2013) Auditory closed-loop stimulation of the sleep slow oscillation enhances memory. Neuron 78:545–553

Nguyen DP, Kloosterman F, Barbieri R, Brown EN, Wilson MA (2009) Characterizing the dynamic frequency structure of fast oscillations in the rodent hippocampus. Front Integr Neurosci 3:11

Nicole O, Hadzibegovic S, Gajda J, Bontempi B, Bem T, Meyrand P (2016) Soluble amyloid beta oligomers block the learning-induced increase in hippocampal sharp wave-ripple rate and impair spatial memory formation. Sci Rep 6:22728

Nimmrich V, Maier N, Schmitz D, Draguhn A (2005) Induced sharp wave-ripple complexes in the absence of synaptic inhibition in mouse hippocampal slices. J Physiol 563:663–670

Nokia MS, Penttonen M, Wikgren J (2010) Hippocampal ripple-contingent training accelerates trace eyeblink conditioning and retards extinction in rabbits. J Neurosci 30:11486–11492

O'Keefe J, Dostrovsky J (1971) The hippocampus as a spatial map. Preliminary evidence from unit activity in the freely-moving rat. Brain Res 34:171–175

O'Keefe J, Nadel L (1978) The hippocampus as a cognitive map. Oxford University Press, Oxford

O'Neill J, Senior T, Csicsvari J (2006) Place-selective firing of CA1 pyramidal cells during sharp wave/ripple network patterns in exploratory behavior. Neuron 49:143–155

Omura Y, Carvalho XMM, Inokuchi K, Fukai XT (2015) A lognormal recurrent network model for burst generation during hippocampal sharp waves. J Neurosci 35:14585–14601

Pangalos M, Donoso JR, Winterer J, Zivkovic AR, Kempter R, Maier N, Schmitz D (2013) Recruitment of oriens-lacunosum-moleculare interneurons during hippocampal ripples. Proc Natl Acad Sci U S A 110:4398–4403

Papatheodoropoulos C (2008) A possible role of ectopic action potentials in the in vitro hippocampal sharp wave-ripple complexes. Neuroscience 157:495–501

Papatheodoropoulos C, Kostopoulos G (2002) Spontaneous, low frequency (~2–3 Hz) field activity generated in rat ventral hippocampal slices perfused with normal medium. Brain Res Bull 57:187–193

Papatheodoropoulos C, Sotiriou E, Kotzadimitriou D, Drimala P (2007) At clinically relevant concentrations the anaesthetic/amnesic thiopental but not the anticonvulsant phenobarbital interferes with hippocampal sharp wave-ripple complexes. BMC Neurosci 8:60

Patel J, Schomburg EW, Berényi A, Fujisawa S, Buzsáki G (2013) Local generation and propagation of ripples along the septotemporal axis of the hippocampus. J Neurosci 33: 17029–17041

Pietersen ANJ, Patel N, Jefferys JGR, Vreugdenhil M (2009) Comparison between spontaneous and kainate-induced gamma oscillations in the mouse hippocampus in vitro. Eur J Neurosci 29:2145–2156

Ponomarenko AA, Knoche A, Korotkova TM, Haas HL (2003) Aminergic control of high-frequency (~200 Hz) network oscillations in the hippocampus of the behaving rat. Neurosci Lett 348:101–104

Ponomarenko AA, Korotkova TM, Sergeeva OA, Haas HL (2004) Multiple GABAA receptor subtypes regulate hippocampal ripple oscillations. Eur J Neurosci 20:2141–2148

Ramirez-Villegas JF, Logothetis NK, Besserve M (2015) Diversity of sharp-wave-ripple LFP signatures reveals differentiated brain-wide dynamical events. Proc Natl Acad Sci U S A 112:E6379-6387

Robbe D, Montgomery SM, Thome A, Rueda-Orozco PE, McNaughton BL, Buzsáki G (2006) Cannabinoids reveal importance of spike timing coordination in hippocampal function. Nat Neurosci 9:1526–1533

Routtenberg A (1968) Hippocampal correlates of consummatory and observed behavior. Physiol Behav 3:533–535

Sadowski JHLP, Jones MW, Mellor JR (2016) Sharp-wave ripples orchestrate the induction of synaptic plasticity during reactivation of place cell firing patterns in the hippocampus. Cell Rep 14:1916–1929

Sainsbury RS (1970) Hippocampal activity during natural behavior in the guinea pig. Physiol Behav 5:317–324

Schlingloff D, Káli S, Freund TF, Hájos N, Gulyás AI (2014) Mechanisms of sharp wave initiation and ripple generation. J Neurosci 34:11385–11398

Schmitz D, Schuchmann S, Fisahn A, Draguhn A, Buhl EH, Petrasch-Parwez E, Dermietzel R, Heinemann U, Traub RD (2001) Axo-axonal coupling: a novel mechanism for ultrafast neuronal communication. Neuron 31:831–840

Shein-Idelson M, Ondracek JM, Liaw H, Reiter S, Laurent G (2016) Slow waves, sharp waves, ripples, and REM in sleeping dragons. Science 352:590–595

Siapas AG, Wilson MA (1998) Coordinated interactions between hippocampal ripples and cortical spindles during slow-wave sleep. Neuron 21:1123–1128

Simon A, Traub RD, Vladimirov N, Jenkins A, Nicholson C, Whittaker RG, Schofield I, Clowry GJ, Cunningham MO, Whittington MA (2014) Gap junction networks can generate both ripple-like and fast ripple-like oscillations. Eur J Neurosci 39:46–60

Skaggs WE, McNaughton BL, Permenter M, Archibeque M, Vogt J, Amaral DG, Barnes CA (2007) EEG sharp waves and sparse ensemble unit activity in the macaque hippocampus. J Neurophysiol 98:898–910

Sirota A, Csicsvari J, Buhl D, Buzsáki G (2003) Communication between neocortex and hippocampus during sleep in rodents. Proc Natl Acad Sci U S A 100:2065–2069

Somogyi P (1977) A specific "axo-axonal" interneuron in the visual cortex of the rat. Brain Res 136:345–350

Somogyi P, Katona L, Klausberger T, Lasztóczi B, Viney TJ (2014) Temporal redistribution of inhibition over neuronal subcellular domains underlies state-dependent rhythmic change of excitability in the hippocampus. Philos Trans R Soc Lond B Biol Sci 369:20120518

Stark E, Roux L, Eichler R, Senzai Y, Royer S, Buzsáki G (2014) Pyramidal cell-interneuron interactions underlie hippocampal ripple oscillations. Neuron 83:467–480

Stumpf C (1965) The fast component in the electrical activity of rabbit's hippocampus. Electroencephalogr Clin Neurophysiol 18:477–486

Suh J, Foster DJ, Davoudi H, Wilson MA, Tonegawa S (2013) Impaired hippocampal ripple-associated replay in a mouse model of schizophrenia. Neuron 80:484–493

Sullivan D, Csicsvari J, Mizuseki K, Montgomery S, Diba K, Buzsáki G (2011) Relationships between hippocampal sharp waves, ripples, and fast gamma oscillation: influence of dentate and entorhinal cortical activity. J Neurosci 31:8605–8616

Suzuki SS, Smith GK (1987) Spontaneous EEG spikes in the normal hippocampus. I. Behavioral correlates, laminar profiles and bilateral synchrony. Electroencephalogr Clin Neurophysiol 67:348–359

Suzuki SS, Smith GK (1988) Spontaneous EEG spikes in the normal hippocampus. V. Effects of ether, urethane, pentobarbital, atropine, diazepam and bicuculline. Electroencephalogr Clin Neurophysiol 70:84–95

Szabó GG, Holderith N, Gulyás AI, Freund TF, Hájos N (2010) Distinct synaptic properties of perisomatic inhibitory cell types and their different modulation by cholinergic receptor activation in the CA3 region of the mouse hippocampus. Eur J Neurosci 31:2234–2246

Takeuchi T, Duszkiewicz AJ, Morris RGM (2014) The synaptic plasticity and memory hypothesis: encoding, storage and persistence. Philos Trans R Soc Lond B Biol Sci 369:20130288

Taxidis J, Coombes S, Mason R, Owen MR (2012) Modeling sharp wave-ripple complexes through a CA3-CA1 network model with chemical synapses. Hippocampus 22:995–1017

Taxidis J, Mizuseki K, Mason R, Owen MR (2013) Influence of slow oscillation on hippocampal activity and ripples through cortico-hippocampal synaptic interactions, analyzed by a cortical-CA3-CA1 network model. Front Comput Neurosci 7:3

Thome C, Kelly T, Yanez A, Schultz C, Engelhardt M, Cambridge SB, Both M, Draguhn A, Beck H, Egorov AV (2014) Axon-carrying dendrites convey privileged synaptic input in hippocampal neurons. Neuron 83:1418–1430

Tolman EC (1948) Cognitive maps in rats and men. Psychol Rev 55:189–208

Traub RD, Bibbig A (2000) A model of high-frequency ripples in the hippocampus based on synaptic coupling plus axon-axon gap junctions between pyramidal neurons. J Neurosci 20:2086–2093

Traub RD, Whittington MA (2010) Cortical oscillations in health and disease. Oxford University Press, Oxford

Traub RD, Wong RKS (1982) Cellular mechanism of neuronal synchronization in epilepsy. Science 216:745–747

Traub RD, Schmitz D, Jefferys JGR, Draguhn A (1999) High-frequency population oscillations are predicted to occur in hippocampal pyramidal neuronal networks interconnected by axoaxonal gap junctions. Neuroscience 92:407–426

ul Haq R, Anderson ML, Hollnagel JO, Worschech F, Sherkheli MA, Behrens CJ, Heinemann U (2016) Serotonin dependent masking of hippocampal sharp wave ripples. Neuropharmacology 101:188–203

Ulanovsky N, Moss CF (2007) Hippocampal cellular and network activity in freely moving echolocating bats. Nat Neurosci 10:224–233

Valero M, Cid E, Averkin RG, Aguilar J, Sanchez-Aguilera A, Viney TJ, Gomez-Dominguez D, Bellistri E, de la Prida LM (2015) Determinants of different deep and superficial CA1 pyramidal cell dynamics during sharp-wave ripples. Nat Neurosci 18:1281–1290

Vandecasteele M, Varga V, Berényi A, Papp E, Barthó P, Venance L, Freund TF, Buzsáki G (2014) Optogenetic activation of septal cholinergic neurons suppresses sharp wave ripples and enhances theta oscillations in the hippocampus. Proc Natl Acad Sci U S A 111:13535–13540

Vanderwolf CH (1969) Hippocampal electrical activity and voluntary movement in the rat. Electroencephalogr Clin Neurophysiol 26:407–418

Varga C, Golshani P, Soltesz I (2012) Frequency-invariant temporal ordering of interneuronal discharges during hippocampal oscillations in awake mice. Proc Natl Acad Sci U S A 109:E2726–E2734

Varga C, Oijala M, Lish J, Szabo GG, Bezaire M, Marchionni I, Golshani P, Soltesz I (2014) Functional fission of parvalbumin interneuron classes during fast network events. eLife 3:e04006

Viereckel T, Kostic M, Bähner F, Draguhn A, Both M (2013) Effects of the GABA-uptake blocker NNC-711 on spontaneous sharp wave-ripple complexes in mouse hippocampal slices. Hippocampus 23:323–329

Wilson MA, McNaughton BL (1994) Reactivation of hippocampal ensemble memories during sleep. Science 265:676–679

Witton J, Staniaszek LE, Bartsch U, Randall AD, Jones MW, Brown JT (2014) Disrupted hippocampal sharp-wave ripple-associated spike dynamics in a transgenic mouse model of dementia. J Physiol 594:4615–4630

Wu C, Shen H, Luk WP, Zhang L (2002) A fundamental oscillatory state of isolated rodent hippocampus. J Physiol 540:509–527

Ylinen A, Bragin A, Nádasdy Z, Jandó G, Szabó I, Sik A, Buzsáki G (1995) Sharp wave-associated high-frequency oscillation (200 Hz) in the intact hippocampus: network and intracellular mechanisms. J Neurosci 15:30–46

Yoshii N, Shimokochi M, Miyamoto K, Ito M (1966) Studies on the neural basis of behavior by continuous frequency analysis of EEG. Prog Brain Res 21:217–250

Zhang L, Sheppy E, Wu C (2006) Regulation of hippocampal sharp waves by Ca^{2+}-dependent slow after hyperpolarization. Crit Rev Neurobiol 18:159–167

Replay in Humans—First Evidence and Open Questions

Hui Zhang, Lorena Deuker and Nikolai Axmacher

> *......how are common subevents copied out of the memory during the codon formation for new classificatory units? ...... When a subevent cluster of sufficient size and importance has been formed, this centre will (perhaps during sleep) call the information out from the memory during a period when codon formation is possible.*
>
> —David Marr (1971)

Abstract Memory consolidation serves to stabilize initially fragile memory traces. Rodent studies suggest that consolidation relies on replay of previously acquired stimulus-specific activity patterns. This replay is coupled to hippocampal sharp wave-ripple (SWR) events and sleep spindles. More recently, the application of multivariate analysis methods has allowed identifying stimulus-specific "engram patterns" in humans as well. These analyses have been applied to various modalities including functional magnetic resonance imaging (fMRI) and intracranial EEG (iEEG). A few initial studies suggest that engram patterns are indeed replayed after learning in humans, during awake resting state, tasks, and sleep. Here, we review these studies and point to open questions. It has been repeatedly shown that the extent of engram pattern replay predicts later memory performance, and that replay occurs during both awake resting state and sleep. On the other hand, cuing of specific memories improves memory consolidation selectively during sleep. Brain stimulation may disrupt consolidation on a behavioral level, but its effect on replay of engram patterns has not been shown yet. Finally, replay has been indirectly linked to sleep spindles, while its relationship to SWRs remains to be investigated. To summarize, the investigation of engram pattern replay in the human brain is an emerging field with still many open questions.

Keywords Replay · Engram · Intracranial EEG · Sharp-wave ripples

H. Zhang · L. Deuker · N. Axmacher (✉)
Department of Neuropsychology, Institute of Cognitive Neuroscience, Faculty of Psychology, Ruhr University Bochum, Universitaetsstrasse 150, 44801 Bochum, Germany
e-mail: nikolai.axmacher@rub.de

N. Axmacher and B. Rasch (eds.), *Cognitive Neuroscience of Memory Consolidation*, Studies in Neuroscience, Psychology and Behavioral Economics, DOI 10.1007/978-3-319-45066-7_15

Introduction

While watching a car pass by, you notice that the car is a two-door mini cooper colored in orange. Meanwhile, a labile transient trace about having seen the car is generated in your brain. After the car has disappeared from your field of view, this transient trace is either consolidated, and eventually becomes a stable representation of the experience, or fades out of your brain (Fig. 1). We experience a large number of events in daily life, yet only some of them are remembered later on. Why do we remember some events but not others? Forgetting can be due to shallow processing during encoding or inappropriate cuing during retrieval. Another possible reason is failed consolidation during the time interval between encoding and retrieval. Consolidation is a theoretical construct which has attracted more and more research interest in recent years (see chapter by Genzel & Wixted).

Here, we assume that the replay of previously established neuronal activity patterns is critical for memory consolidation. Many researchers have studied replay in animals after it was reported for the first time (Pavlides and Winson 1989; Skaggs et al. 2007; Dave and Margoliash 2000; Nokia et al. 2010; Buhry et al. 2011). In these experiments, rodents are typically trained to navigate in a corridor or an open field while place cells firing to a set of temporally-sequenced place fields along the animal's path are recorded. During subsequent non-locomotion periods of waking state and sleep, these place cells fire in the same order in a temporally compressed manner while maintaining the temporal sequence they exhibited during navigation (for review, Carr et al. 2011; Axmacher et al. 2009). Even before the first experimental study reported replay in rodents, David Marr had already proposed the idea that neurons that are active during encoding are reactivated afterwards during a consolidation period (Marr 1971). However, few empirical studies so far have directly investigated replay in humans, and crucial questions remain. In this chapter, we will first review the existing evidence for replay in humans and then specify which issues still need to be addressed.

What Is Replay and What Is Being Replayed?

According to the engram theory of memory proposed by Richard Semon around 100 years ago (Semon 1921, 1923), memories are formed as biophysical and biochemical changes of the brain, e.g., strengthened synaptic connectivity or synchronized firing pattern of neurons. Semon described four characteristics of the engram (Schacter 2001), which have recently been thoroughly reviewed by Josselyn et al. (2015). They are *persistence*, *ecphory*, *content*, and *dormancy*. We would like to add one more feature, which is *uniqueness*. If we can distinguish an SUV and a minivan at a behavioral level, the corresponding engram patterns differ at the neuronal level as well. From the reverse perspective, if engrams of two encounters of a car are the same, we would recognize the car as the same. For

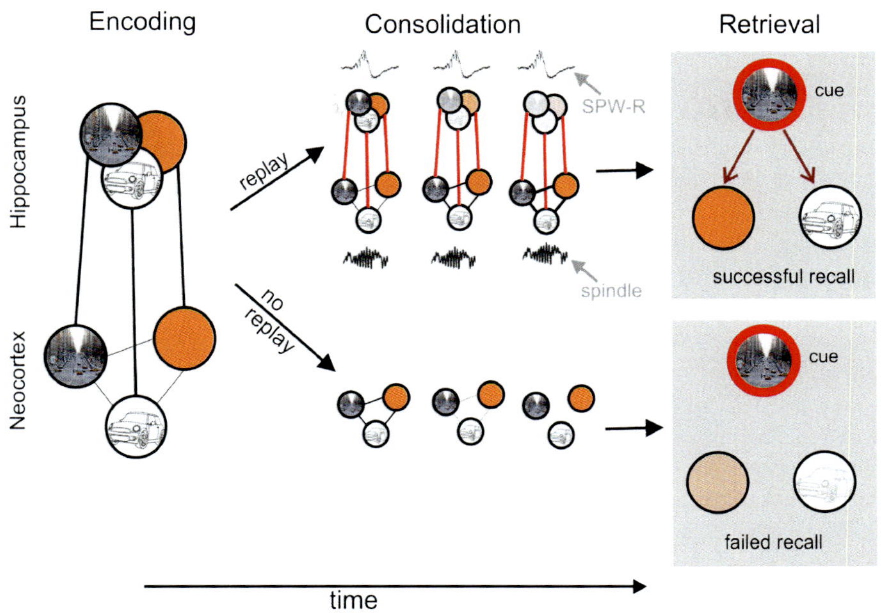

Fig. 1 When you experience an episode, like seeing an *orange* mini cooper on the street, a specific neural engram pattern is established. When this perception is transformed into an episodic memory trace, the hippocampus binds together the different elements of that episode (here, for example, street, mini cooper and the color *orange*) by establishing links to already existing neocortical representations. Thus, the hippocampus functions like an index card in a library catalogue, listing which elements belong to an episode and where they can be found. After memory formation, the hippocampus may replay this engram pattern. The replay process may be triggered by sharp-wave ripple (SWR) complexes, thereby also activating the associated neocortical links, which in turn strengthens connections between the neocortical representations. This process has been described as memory consolidation. In the end, the original hippocampal memory trace may be weakened or even be lost (according to standard consolidation theory (Alvarez and Squire 1994); multiple trace theory (Nadel and Moscovitch 1997) suggests that hippocampal traces remain; see chapters by Genzel and Wixted, by Sekeres, Moscovitch and Winocur and by Cheng), but the memory is represented by strong connections between the neocortical representations of the elements. So, when the memory is cued by activating one element (e.g. the cue "street", when someone asks you: "What did you see on the street that day?"), the strong connections to the other elements enable reactivation of engrams of other elements, thereby recollecting the entire episode. Without replay, however, hippocampal as well as neocortical representations become weaker and fade over time. Cueing one element will then not lead to successful retrieval of the associated elements, and the memory for that episode is lost

instance, Xue et al. (2010) reported an fMRI study in which participants viewed pictures of faces, with each face repeated multiple times. They found that brain patterns were more similar when viewing the same face compared with different ones, indicating stimulus-specific or unique representations. Furthermore, the level of similarity across repetitions was positively related to later memory performance for that face. An intracranial EEG (iEEG) study from our lab (Zhang et al. 2015) reported similar results while epilepsy patients were watching videos of navigating through virtual houses during encoding. During the subsequent testing period, patients were asked to find their way out of the same house. We found that the topographical patterns of brain activity in the gamma frequency band (around 80 Hz) were more similar when viewing the same, rather than different, video sequences. Interestingly, this was only true for when patients were able to remember the route in the virtual house. Together, these and other findings show that stimulus-specific representations of events (unique engrams) can be identified in humans on the level of distributed neural activity, i.e., via "engram patterns" (Watrous et al. 2015).

Given that memories are formed as engrams, we define replay as reactivation of memory engrams without the experience of the actual stimuli. Figure 1 illustrates this replay theory in a simplified way. As shown in the figure, a person first sees a '*mini cooper*', colored in '*orange*', on the '*street*'. During memory formation, in both hippocampus and neocortex, engrams consisting of representations of *different elements (mini cooper, orange color, street)* are formed. After memory formation, the engram pattern of the stimuli might be replayed when triggered by sharp-wave ripple (SWR) complexes (Kudrimoti et al. 1999; Nadasdy et al. 1999; Lee and Wilson 2002; Karlsson and Frank 2009) that are temporally linked to sleep spindles (Clemens et al. 2007; see chapters by Bergmann and Staresina on sleep-related oscillations and by Maier and Kempter on sharp waves and ripples). Thus, during replay engrams of elements are stabilized and connections between different elements are enhanced. In the meantime, the neocortex becomes more and more involved in engram representation of each item, and the engram in the hippocampus gradually fades out. As a result, when cued by '*street*', people can remember the '*orange*' '*mini cooper*'. Without replay of engram patterns, the engram—and the connection between different elements—fades out.

In general, replay can be divided into two formats—*intentional* and *spontaneous* replay. Intentional replay occurs when people voluntarily intend to repeat previously learned items, e.g., when a telephone number from a phone book is rehearsed in order to dial it. Intentional replay occurs during waking state and involves conscious effort. On the other hand, spontaneous replay occurs when previously acquired memory engrams reappear without any voluntary efforts. This may happen during both waking state and sleep. In this chapter, we will focus on the literature on spontaneous replay in humans.

Methods to Study Replay in Humans

As mentioned above, both the engram theory of memory and the idea of replay were proposed quite some time ago. However, empirical studies on the two phenomena were scarce until recently due to several reasons, including constraints of tools for data acquisition and limitations of analysis methods. Technological advancements have therefore had a significant role in enhancing the development of neuroscience in general, and the study of replay mechanisms for memory consolidation in particular.

An ideal tool to study replay, and one which may link human cognitive neuroscience with animal electrophysiology, consists of invasive recordings in epilepsy patients (Engel et al. 2005). These recordings provide either intracranial EEG (iEEG) data from clinical depth electrodes ("macroelectrodes" with a diameter of around 1–1.5 mm), electrocorticography (ECoG) data from subdural strip or grid electrodes, or local field potential and multi- or single unit data from "microelectrodes" (with a diameter of around 40 μm). Due to obvious ethical issues, these invasive methods can only be applied to a limited number of patients in a small number of hospitals around the world. In addition to being rare, patients implanted with macro- or microelectrodes by definition have a severe and often longstanding neurological disorder (e.g., a hippocampal sclerosis) and often take antiepileptic medications, both of which likely affect the results of these studies. Therefore, non-invasive neuroimaging tools, such as fMRI, MEG, and scalp EEG, offer an alternative to study replay. The difficulty with these methods is to identify the neuronal signature of an engram, so that its later reoccurrence can be detected.

It is known that our brain processes information via networks, or assemblies, comprising large numbers of neurons (Abeles 1982; Cohen et al. 1993; Palm 1990; Waydo et al. 2006). However, conventional mass-univariate approaches to data analysis neglect information that is represented by patterns across multiple voxels. In recent years, new multivariate approaches have been developed, such as pattern classification (Rissman and Wagner 2012; Haynes 2015) and representational similarity analysis (RSA; Kriegeskorte et al. 2008; Haxby et al. 2001). Both methods allow one to estimate item-specific activity, which is a prerequisite for studying replay of unique engrams.

First Evidence of Replay in Humans

Only a small number of studies have investigated replay in humans. At present, there are three fundamental lines of research to evaluate replay during memory consolidation. The first is searching for replay of item-specific engram patterns during memory consolidation; the second is using cuing experiments to study whether enhancing replay increases memory consolidation; and the third is applying electrical or magnetic stimulation to test how interrupting replay harms memory consolidation. In Sections "Replay During Sleep" and "Replay During the Waking Period", we describe studies showing replay of engrams during a resting period, related to the first line of research mentioned above. In Sections "Cue Triggered Replay Enhances Memory and Interrupting Replay Disrupts Memory", we address research lines 2 and 3 and review studies attempting to increase or decrease replay via cues or stimulation, respectively.

Replay During Sleep

Peigneux, Maquet and colleagues are among the pioneers to study replay during sleep (Maquet et al. 2000; Peigneux et al. 2004, 2006; see also chapter by Schönauer and Gais on the role of sleep for memory consolidation). In one of their studies (Peigneux et al. 2004), participants performed either a spatial navigation task or a serial reaction time task (SRTT) before sleep in the PET scanner. The spatial navigation task involved the hippocampus, while the SRTT did not. They found that, during the following sleep period, activity of the hippocampus was higher when participants had performed the spatial navigation task before sleep than when they had performed the SRTT. Furthermore, activation of the hippocampus during slow wave sleep correlated positively with performance on a spatial retrieval task the next day. A similar fMRI study, conducted by Bergmann and colleagues (Bergmann et al. 2012), had participants perform either a landmark-face association task or a visuomotor control task, before sleep (see also chapter by Bergmann and Staresina). Following the association task, the authors observed increased activation of category-specific face and landmark areas (the fusiform face area and the parahippocampal cortex, respectively), and of the hippocampus, compared with a control task. This replay was also temporally coupled with sleep spindles.

Although the studies mentioned above showed a replay-like effect, they did not directly assess whether the pattern replayed during the sleep period matched the engram pattern during memory formation. An fMRI study from our lab (Deuker et al. 2013) provides direct evidence for replay of item-specific engram patterns during the resting state and sleep. In this study, participants learned two sets of picture-location associations before and after an afternoon nap in the MRI scanner. A pattern classifier was trained to dissociate item-specific engram patterns, after which the re-occurrence of these engram patterns was tested on data acquired during a resting period including both waking state and sleep. We found that item-specific engram patterns learned before sleep were replayed more often than would be expected by chance. Most importantly, individual engrams that were replayed more often were afterwards remembered more accurately, indicating that replay is indeed behaviorally relevant.

Replay During the Waking Period

There is also evidence of replay during awake resting states after learning. Tambini and Davachi (2013) reported an fMRI study in which they measured multivoxel hippocampal patterns during the separate encoding of either object-face or scene-face associations. They found that task-specific patterns persisted into a post-encoding rest period. Again, the extent of pattern replay was positively related to later memory performance. In another fMRI study by Staresina et al. (2013), participants were shown a list of unique object-scene pairs during encoding. The authors found that the encoding engram pattern was replayed at a higher level if participants could successfully remember the object-scene pair in a cued recall task. One interesting aspect of this study was that during the interval between encoding and retrieval, participants performed a distractor task by judging odd/even numbers. Thus, the replay of encoding patterns is spontaneous rather than intentionally rehearsed. Together, these studies show that (1) unique engram patterns can be identified in human fMRI data, (2) replay of these engram patterns occurs spontaneously after encoding during both awake resting state and sleep, and (3) replay facilitates later memory, as would be expected for a neural correlate of memory consolidation.

Cue Triggered Replay Enhances Memory

Given that replay of previous engram patterns correlates with behavioral measures of memory consolidation, one may assume a causal role of replay for memory consolidation. In line with this idea, researchers have used cued recall tasks that selectively improve memory by presenting cues during sleep (Rasch et al. 2007; Rudoy et al. 2009; Diekelmann et al. 2011; van Dongen et al. 2012; Schreiner and Rasch 2015; see also chapters by Talamini and by Schreiner, Lehmann and Rasch). In a study by Rudoy and colleagues, participants were asked to learn fifty object-location pairs. Each pair was coupled with a specific sound. During the following non-rapid eye movement (non-REM) sleep period, half of the sound cues were presented to the participants through headphones. After waking up, participants viewed all previously learned objects and positioned each of them at their original location. The researchers found that objects for which sound cues had been presented during sleep were positioned more accurately than objects without sound cues during sleep. Van Dongen and colleagues, using a similar paradigm in an fMRI study, found that the right parahippocampal cortex was more active during periods of non-REM sleep when sound cues were presented than periods where sound cues were not presented (van Dongen et al. 2012) (see also chapter by Fernandez). During a subsequent retrieval session, they observed an inter-individual correlation between parahippocampal and medial prefrontal cortex connectivity and object-location memory. These studies show that cueing increases memory consolidation, and suggest that this is via an effect on replay. Indeed, presentation during sleep of odors that were associated with visual items presented in one hemifield specifically increased sleep spindle amplitudes over the contralateral hemisphere (Cox et al. 2014; see chapter by Talamini). However, none of these studies directly examined engram pattern replay. Future studies need to test the assumption that cueing during sleep indeed triggers replay of unique engram patterns.

Interrupting Replay Disrupts Memory

In rodents, replay is triggered by SWR events, which are oscillatory patterns in the mammalian hippocampus during immobility and slow wave sleep (O'Keefe 1976). Interrupting these SWR events impairs memory performance in rodents (Girardeau and Zugaro 2011; Girardeau et al. 2009; Ego-Stengel and Wilson 2010). In human studies, there is no direct evidence for this so far. However, studies have shown that interference with post-encoding processes, in task-related regions, impairs later performance (Muellbacher et al. 2002; Robertson et al. 2005). In a study conducted by Muellbacher and colleagues, participants were trained on a motor task, specifically involving the primary motor cortex (M1). Repetitive transcranial magnetic stimulation (rTMS) was applied to M1 and control regions between training sessions. The study found that the M1-rTMS condition resulted in reduced performance compared with control-rTMS. Another study by Robertson and colleagues showed that TMS on the primary motor cortex, specifically during waking and following a motor learning task, impairs learning performance (Robertson et al. 2005). However, none of the stimulation studies reported above directly examined whether replay of engram patterns was interrupted. This should be tested in future studies. Furthermore, in addition to these interference studies, artificially improving memory consolidation via electric or magnetic stimulation would be important as well, especially for clinical purposes (Lee et al. 2013). Ideally, new methods that can flexibly turn target neuron activities on and off without inducing any harm to human subjects are needed in this line of research.

Open Questions

As discussed, many aspects of replay still require scientific investigation. For the remainder of this chapter, we would like to discuss some open questions regarding replay, which are testable and may offer more insight in future studies.

The Difference Between Spontaneous and Intentional Replay

No study to date has addressed the difference between spontaneous and intentional replay. From the reviewed literature, however, it is likely that some form of replay supports different memory-related functions. During short-term memory, an intentional form of replay may support the rehearsal of previously presented information (LaRocque et al. 2013; Lepsien and Nobre 2007; Polania et al. 2012). With regards to cued or free long-term memory recall, the reinstatement of the engram of a stimulus may be necessary (Staudigl et al. 2015; Staresina et al. 2012; Polyn et al. 2005), and this could occur both intentionally and spontaneously (e.g., via ecphory; Waldhauser et al. 2016). Finally—and in the context of this chapter, most importantly—replay may support long-term memory consolidation. Notably, while all these processes involve "replay" of some previously established engram pattern, they may differ with respect to the format of this representation. Intentional replay involves vivid rehearsal of previously seen stimuli, which may rely on engram patterns within both early and associative sensory areas (Farah 1989; Mellet et al. 1998). By contrast, spontaneous replay during sleep is probably related to the integration of novel information into existing networks (e.g., Takashima et al. 2009). Thus, it may rely not only on replay of detailed perceptual representations within early sensory areas (Deuker et al. 2013), but also on replay of more conceptual and abstract representations within higher sensory areas.

Cued Replay and Spontaneous Replay During Wakefulness and Sleep

As mentioned above, the relationship between cued recall (see Section "Cue triggered replay enhances memory") and spontaneous replay (reviewed in Sections "Replay during sleep" and "Replay during the waking period") is still unclear. In particular, studies using cued recall have shown that the presentation of a cue during sleep, but not during waking, improves later memory performance (Diekelmann et al. 2011; Schreiner and Rasch 2015). By contrast, experiments investigating the spontaneous replay of engram patterns have shown beneficial effects of replay during waking periods as well (Deuker et al. 2013; Tambini and Davachi 2013; Staresina et al. 2013). There are several possible explanations for this discrepancy. First, during waking the replayed engram pattern is susceptible to external input, which may modify or even destabilize the original engram (Rodriguez-Ortiz and Bermudez-Rattoni 2007) in cueing experiments, as reactivation during waking has been shown to trigger reconsolidation processes under certain circumstances (see chapter by Kessler, Blackwell & Kehyayan). By contrast, spontaneous waking state replay may occur, and may support consolidation, when the relevant brain areas are not involved in an ongoing task, which protects brain activities from being interrupted (Tambini and Davachi 2013; Staresina et al. 2013). Second, cue triggered replay of the engram is initiated from the primary sensory cortex by processing the cue, which propagates from bottom-up to higher brain regions. By contrast, spontaneous replay may be initiated from higher level brain regions (e.g., triggered by hippocampal SWRs; Axmacher et al. 2008). Third, the mechanisms underlying memory consolidation may differ between the waking state and sleep state—for example, memory consolidation during waking may depend on strengthening of individual memories, while consolidation during sleep may also occur if larger networks are activated that represent categorically related information (Oudiette et al. 2013). It may be that cueing during waking actually triggers such larger networks and thus does not improve memory consolidation, whereas spontaneous replay is more specific and therefore plays a beneficial role (Deuker et al. 2013; Staresina et al. 2013). This speculative idea needs to be further tested, though.

Network Replay in Humans Versus Sparse Replay in Rodents

In the rodent literature, replay was first described as the temporally compressed sequential firing of a set of place cells, which are mainly located in the hippocampus. In human studies, by contrast, researchers usually investigate "engram patterns" across large brain regions. The obvious reason for this difference is that recording methods differ between rodents and humans. In rodent studies, a limited number of electrodes are implanted in each animal. Thus, the reported results are relatively sparse compared to the total number of neurons in the brain. By contrast, human studies typically use methods such as fMRI or intracranial EEG, which cover large brain areas. Furthermore, replay in rodents occurs as temporally sparse events linked to sleep spindles and ripples, whereas it is considered a more sustained process in most human studies. No study to date has linked replay of engram patterns to hippocampal ripples in humans. In addition, it would be tremendously important to perform simultaneous recordings of single neurons, intracranial EEG and fMRI—both in humans and animals—to bridge the gap between the two research approaches (see Logothetis et al. 2012; Kaplan et al. 2016). This would allow one to test the relationship between sparsely replayed neurons in the hippocampus and replay of engram patterns in the neocortex.

Conclusion

In conclusion, there is now first evidence that engram patterns are spontaneously replayed during both sleep and resting state in humans, and that this replay supports memory consolidation. Other studies have demonstrated that cueing during sleep facilitates, and interrupting activity impairs, memory consolidation; but the relationships to engram replay still need to be tested directly. Other open questions concern the relationships between intentional and spontaneous replay, between sparse replay of single cell sequences and of engram patterns, between replay during waking and sleep, and the role of SWRs for replay.

References

Abeles M (1982) Local cortical circuites: an electrophysiological study. Springer, Berlin

Alvarez P, Squire LR (1994) Memory consolidation and the medial temporallobe: a simple network model. Proc Natl Acad Sci USA 91(15):7041–7045

Axmacher N, Elger CE, Fell J (2008) Ripples in the medial temporal lobe are relevant for human memory consolidation. Brain 131(Pt 7):1806–1817

Axmacher N et al (2009) Memory processes during sleep: beyond the standard consolidation theory. Cell Mol Life Sci 66(14):2285–2297

Bergmann TO et al (2012) Sleep spindle-related reactivation of category-specific cortical regions after learning face-scene associations. Neuroimage 59(3):2733–2742

Buhry L, Azizi AH, Cheng S (2011) Reactivation, replay, and preplay: how it might all fit together. Neural Plast 2011:203462

Carr MF, Jadhav SP, Frank LM (2011) Hippocampal replay in the Awake state: a potential substrate for memory consolidation and retrieval. Nat Neurosci 14(2):147–153

Clemens Z et al (2007) Temporal coupling of parahippocampal ripples, sleep spindles and slow oscillations in humans. Brain 130(Pt 11):2868–2878

Cohen NJ, Eichenbaum H (1993) Memory, amnesia, and the hippocampal system. MIT Press, Cambridge, MA

Cox R et al (2014) Local sleep spindle modulations in relation to specific memory cues. Neuroimage 99:103–110

Dave AS, Margoliash D (2000) Song replay during sleep and computational rules for sensorimotor vocal learning. Science 290(5492):812–816

Deuker L et al (2013) Memory consolidation by replay of stimulus-specific neural activity. J Neurosci 33(49):19373–19383

Diekelmann S et al (2011) Labile or stable: opposing consequences for memory when reactivated during waking and sleep. Nat Neurosci 14(3):381–386

Ego-Stengel V, Wilson MA (2010) Disruption of ripple-associated hippocampal activity during rest impairs spatial learning in the rat. Hippocampus 20(1):1–10

Engel AK et al (2005) Invasive recordings from the human brain: Clinical insights and beyond. Nat Rev Neurosci 6(1):35–47

Farah MJ (1989) The neural basis of mental imagery. Trends Neurosci 12(10):395–399

Girardeau G, Zugaro M (2011) Hippocampal ripples and memory consolidation. Curr Opin Neurobiol 21(3):452–459

Girardeau G et al (2009) Selective suppression of hippocampal ripples impairs spatial memory. Nat Neurosci 12(10):1222–1223

Haxby JV et al (2001) Distributed and overlapping representations of faces and objects in ventral temporal cortex. Science 293(5539):2425–2430

Haynes JD (2015) A primer on pattern-based approaches to FMRI: principles, pitfalls, and perspectives. Neuron 87(2):257–270

Josselyn SA, Kohler S, Frankland PW (2015) Finding the engram. Nat Rev Neurosci 16(9): 521–534

Kaplan R et al (2016) Hippocampal sharp-wave ripples influence selective activation of the default mode network. Curr Biol 26(5):686–691

Karlsson MP, Frank LM (2009) Awake replay of remote experiences in the hippocampus. Nat Neurosci 12(7):913–918

Kriegeskorte N, Mur M, Bandettini P (2008) Representational similarity analysis—connecting the branches of systems neuroscience. Front Syst Neurosci 2:4

Kudrimoti HS, Barnes CA, McNaughton BL (1999) Reactivation of hippocampal cell assemblies: effects of behavioral state, experience, and EEG dynamics. J Neurosci 19(10):4090–4101

LaRocque JJ et al (2013) Decoding attended information in short-term memory: an EEG study. J Cogn Neurosci 25(1):127–142

Lee AK, Wilson MA (2002) Memory of sequential experience in the hippocampus during slow wave sleep. Neuron 36(6):1183–1194

Lee H, Fell J, Axmacher N (2013) Electrical engram: how deep brain stimulation affects memory. Trends Cogn Sci 17(11):574–584

Lepsien J, Nobre AC (2007) Attentional modulation of object representations in working memory. Cereb Cortex 17(9):2072–2083

Logothetis NK et al (2012) Hippocampal-cortical interaction during periods of subcortical silence. Nature 491(7425):547–553

Maquet P et al (2000) Experience-dependent changes in cerebral activation during human REM sleep. Nat Neurosci 3(8):831–836

Marr D (1971) Simple memory: a theory for archicortex. Philos Trans R Soc Lond B Biol Sci 262(841):23–81

Mellet E et al (1998) Reopening the mental imagery debate: lessons from functional anatomy. Neuroimage 8(2):129–139

Muellbacher W et al (2002) Early consolidation in human primary motor cortex. Nature 415(6872):640–644

Nadasdy Z et al (1999) Replay and time compression of recurring spike sequences in the hippocampus. J Neurosci 19(21):9497–9507

Nadel L, Moscovitch M (1997) Memory consolidation, retrograde amnesia and the hippocampal complex. Curr Opin Neurobiol 7(2):217–227

Nokia MS, Penttonen M, Wikgren J (2010) Hippocampal ripple-contingent training accelerates trace eyeblink conditioning and retards extinction in rabbits. J Neurosci 30(34):11486–11492

O'Keefe J (1976) Place units in the hippocampus of the freely moving rat. Exp Neurol 51(1): 78–109

Oudiette D et al (2013) The role of memory reactivation during wakefulness and sleep in determining which memories endure. J Neurosci 33(15):6672–6678

Palm G (1990) Cell assemblies as a guideline for brain research. Concepts Neurosci 1:133–147

Pavlides C, Winson J (1989) Influences of hippocampal place cell firing in the awake state on the activity of these cells during subsequent sleep episodes. J Neurosci 9(8):2907–2918

Peigneux P et al (2004) Are spatial memories strengthened in the human hippocampus during slow wave sleep? Neuron 44(3):535–545

Peigneux P et al (2006) Offline persistence of memory-related cerebral activity during active wakefulness. PLoS Biol 4(4):e100

Polania R, Paulus W, Nitsche MA (2012) Noninvasively decoding the contents of visual working memory in the human prefrontal cortex within high-gamma oscillatory patterns. J Cogn Neurosci 24(2):304–314

Polyn SM et al (2005) Category-specific cortical activity precedes retrieval during memory search. Science 310(5756):1963–1966

Rasch B et al (2007) Odor cues during slow-wave sleep prompt declarative memory consolidation. Science 315(5817):1426–1429

Rissman J, Wagner AD (2012) Distributed representations in memory: insights from functional brain imaging. Annu Rev Psychol 63:101–128

Robertson EM, Press DZ, Pascual-Leone A (2005) Off-line learning and the primary motor cortex. J Neurosci 25(27):6372–6378

Rodriguez-Ortiz CJ, Bermudez-Rattoni F (2007) Memory reconsolidation or updating consolidation? In: Bermudez-Rattoni F (Ed.) Neural plasticity and memory: from genes to brain imaging, Boca Raton (FL)

Rudoy JD et al (2009) Strengthening individual memories by reactivating them during sleep. Science 326(5956):1079–1079

Schacter DL (2001) Forgotten ideas, neglected pioneers: Richard Semon and the story of memory. illustrated ed. 2001. Psychology Press, UK

Schreiner T, Rasch B (2015) Boosting vocabulary learning by verbal cueing during sleep. Cereb Cortex 25(11):4169–4179

Semon R (1921) The Mneme. George Allen & Unwin, London

Semon R (1923) Mnemic psychology. George Allen & Unwin, London

Skaggs WE et al (2007) EEG sharp waves and sparse ensemble unit activity in the macaque hippocampus. J Neurophysiol 98(2):898–910

Staresina BP et al (2012) Episodic reinstatement in the medial temporal lobe. J Neurosci 32(50):18150–18156

Staresina BP et al (2013) Awake reactivation predicts memory in humans. Proc Natl Acad Sci U S A 110(52):21159–21164

Staudigl T et al (2015) Temporal-pattern similarity analysis reveals the beneficial and detrimental effects of context reinstatement on human memory. J Neurosci 35(13):5373–5384

Takashima A et al (2009) Shift from hippocampal to neocortical centered retrieval network with consolidation. J Neurosci 29(32):10087–10093

Tambini A, Davachi L (2013) Persistence of hippocampal multivoxel patterns into postencoding rest is related to memory. Proc Natl Acad Sci U S A 110(48):19591–19596

van Dongen EV et al (2012) Memory stabilization with targeted reactivation during human slow-wave sleep. Proc Natl Acad Sci USA 109(26):10575–10580

Waldhauser GT, Braun V, Hanslmayr S (2016) Episodic memory retrieval functionally relies on very rapid reactivation of sensory information. J Neurosci 36(1):251–260

Watrous AJ et al (2015) More than spikes: common oscillatory mechanisms for content specific neural representations during perception and memory. Curr Opin Neurobiol 31:33–39

Waydo S et al (2006) Sparse representation in the human medial temporal lobe. J Neurosci 26 (40):10232–10234

Xue G et al (2010) Greater neural pattern similarity across repetitions is associated with better memory. Science 330(6000):97–101

Zhang H et al (2015) Gamma power reductions accompany stimulus-specific representations of dynamic events. Curr Biol 25(5):635–640

Cortico-Hippocampal Circuits for Memory Consolidation: The Role of the Prefrontal Cortex

Lisa Genzel and Francesco P. Battaglia

Abstract Memory is made up of multiple interacting systems, intervening at different times during the lifetime of the memory, and organizing information in different ways. One system is centered on the hippocampus and it is key for episodic memory and initial memory acquisition; while another system is based in the neocortex and is involved in storage of remote memories and semantic memory. These two stores communicate during sleep and other quiet periods, when neural patterns related to previously acquired memories are replayed. This reactivation is thought to engage plasticity processes in many brain areas, therefore enabling memory consolidation. We review here some of the experimental evidence on memory replay and dynamical interactions between cortex and hippocampus during sleep, with a focus on the prefrontal cortex, one of the key cortical areas for memory.

Keyword Hippocampus · Prefrontal cortex · Schema · Memory · Consolidation · System · Sleep

Cortico-Hippocampal Interactions for a Dual Memory System

The idea that memory in the brain does not consist in a single store of information that gets encoded and later retrieved in the same form as it was originally encoded, is a very old one. The existence of anterograde and retrograde amnesia gradient in patients with medial temporal lobe or neocortical lesions gave rise to the 'two memory systems', corresponding respectively to the hippocampus and neocortex

L. Genzel (✉)
Centre for Cognitive and Neural Systems, University of Edinburgh,
1 George Square, Edinburgh EH89JZ, UK
e-mail: lgenzel@ed.ac.uk

F.P. Battaglia
Donders Institute for Brain Cognition and Behavior Radboud Universiteit,
Heyendaalseweg 135, 6525AJ Nijmegen, The Netherlands
e-mail: F.Battaglia@science.ru.nl

N. Axmacher and B. Rasch (eds.), *Cognitive Neuroscience of Memory Consolidation*, Studies in Neuroscience, Psychology and Behavioral Economics, DOI 10.1007/978-3-319-45066-7_16

(Bayley et al. 2005). David Marr was the first to give this theory a computational vest (Marr 1970, 1971). A key tenet, that was made explicit by McClelland et al. (1995), was that this split between two systems was a way to address contrasting requirements for fast, flexible storage of new information and maintenance of previously acquired memories. 'Offline' processes after memory encoding would be necessary for the transfer of information from the hippocampal to the neocortical system.

An update to this theory (reviewed in Battaglia et al. 2012) included the further assumption that, with memory consolidation, memory is reprocessed and transitions from an 'episodic' form (a series of records from an autobiographical timeline) to a more 'semantic' encoding, a store of facts and regularities about the world that can be flexibly updated by new information and is useful to guide behavior.

These offline processes are supported by the rich neural activity during sleep, when previous memories are reactivated throughout the brain. In this chapter, we will give a short overview of this activity and the interactions between hippocampus and neocortex in this context, with an emphasis on the role of the prefrontal cortex (PFC), a critical area for memory, with strong links to the hippocampus.

The Prefrontal Cortex: A Cortical Hub for Memory

Besides its role in executive function, the PFC has a key role in memory. PFC is important for the retrieval of remote memories, and under those conditions it appears to play a 'hub' role coordinating the activity of other cortical areas, similar to that played by the hippocampus for recent memories (see e.g. for reviews Nieuwenhuis and Takashima 2011; Preston and Eichenbaum 2013). The PFC is also peculiar in the neocortex because of a direct pathway from the hippocampus (reviewed in Thierry et al. 2000), which bypasses the relay areas in the entorhinal cortex and the parahippocampal cortices, mediating most of the interaction between the hippocampus and the neocortex. Return pathways from PFC to the hippocampus exist via the parahippocampal and entorhinal cortex, as well as through the midline thalamic nuclei (Vertes 2006; Varela et al. 2013), such as the nucleus reuniens, which has been found to play an important role in memory (Ito et al. 2015).

From the physiological point of view, the PFC shows a very rich pattern of interactions with the hippocampus, which is likely to be very important for memory consolidation (see also chapter by Fernandez). We will review it in what follows, but first we need to give a short account of the hippocampal activity patterns of sleep, to fully understand how these interact with cortical function.

The Hippocampal Sharp Wave: A Critical Activity Pattern for Memory

Just as cortex, the hippocampus displays a radically different temporal organization of activity during active behavior on one side when memory is effectively encoded and used, and sleep and other inactive periods, when memory is consolidated on the other side. A clear demarcation has been defined between ‘two stages’ of hippocampal activity (Buzsaki 1989), that are more amenable to, respectively, the encoding of new memory, and the reprocessing of already acquired information. The active state is dominated by the theta rhythm (6–10 Hz), which plays an important role in the sequential organization of neuronal activity, most clearly via the so called ‘theta phase precession’ of hippocampal place cells (O’Keefe and Recce 1993; Skaggs et al. 1996). Theta oscillations are also propagated to many cortical areas, in particular PFC (Siapas et al. 2005; Sirota et al. 2008), and are thought to be an important carrier of information exchange between the hippocampus, cortex, and other structures (Battaglia et al. 2011; see also chapter by Schreiner, Lehmann and Rasch).

The sharp wave (Buzsaki 2015) is the most important process of the ‘offline’ state (see also chapter by Maier and Kempter). A sharp wave is a burst of activity generated in the CA3 subfield of the hippocampus, which is rich in recurrent excitatory connectivity. These synapses endow that neural network with a great deal of positive feedback, making them prone to explosive bouts of activation, initiated by transient fluctuations in activity or synchrony involving small groups of pyramidal cells and interneurons. Sharp waves terminate a few tens of milliseconds later, with mechanisms still to be clarified, but probably involving the role of inhibitory interneurons and hyperpolarizing membrane currents.

Excitations spread from CA3 to the rest of the hippocampus. In CA1, the depolarization wave caused by a sharp wave engenders both a deflection in the LFP, and high-frequency oscillations (150–200 Hz) termed *ripples* (O’Keefe 1976; Buzsáki et al. 1992). Ripples are often the most recognizable sign of sharp waves in electrophysiological recordings, and are therefore commonly used to detect them in experiments. Because of that, the name ‘sharp wave/ripple complex’ (SWR) is frequently used to refer to these events.

SWR are thought to be important for memory processes for two main reasons: first, that their origins in the recurrent network of CA3, and the role of the excitatory feedback suggest that SWR may be a manifestation of attractor dynamics, and may correspond to transient ‘retrieval’ of patterns engraved in the synaptic matrix, as would be predicted by attractor neural network theory (Amit 1989; Mongillo et al. 2008, see also the chapter by Cheng). Second, the potent, synchronous input may be ideal for effectively spreading the hippocampal input to the neocortex. While afferents from the hippocampus, reach (directly or via parahippocampal cortices and other structures) many cortical structures, they still represent a minor proportion of inputs to cortical cells. Moreover, hippocampal inputs also excite directly cortical interneurons, at least in PFC (Tierney et al. 2004), thus causing feed-forward inhibition alongside excitation. It is therefore likely that SWR, as a concentrated, synchronous pattern of activity, are a much-needed help for the hippocampus to effectively influence the cortical state.

Indeed, during SWR, many cortical neurons show transient increases in activity (Sirota et al. 2003; Battaglia et al. 2004; Peyrache et al. 2011; Jadhav et al. 2016).

Such increases are widespread throughout the cortex (Battaglia et al. 2004), and are very prominent in PFC (Peyrache et al. 2011), reflecting the great degree of connectivity between that structure and hippocampus. Interestingly, in PFC, together with excitatory responses, also a subset of neurons showing decreased activity during SWR is observed (Jadhav et al. 2016), probably reflecting the recruitment of PFC interneurons by hippocampal inputs. Recent data obtained with technically challenging combinations of electrophysiology and functional magnetic resonance imaging (fMRI) (Logothetis et al. 2012), show that indeed SWRs are concurrent with overall activation in cortical activity in macaque monkeys, reaching most areas, except, notably, primary visual cortex, while at the same time many subcortical structures are down-regulated. The activation is most evident in the areas participating in the default mode network (Kaplan et al. 2016), pointing both at high level of connection between the hippocampus and this cortical network, and a likely involvement of the default mode network in memory reprocessing.

Together, these results suggest that the SWR may be a valid 'carrier' for input from the hippocampus to reach the cortex, but what is the information 'payload' that is being transported? For sure, hippocampal activity during SWR presents a high degree of variability from one event to the other (Buzsaki 2015). In each SWR event, 15–30% of hippocampal principal cells are activated, with different ensembles being recruited at each SWR. Correspondingly, the LFP correlates of sharp waves vary distinctly to one event to next. The LFP signature of each SWR event is informative about the set of neurons that activated (Reichinnek et al. 2010; Taxidis et al. 2015). Different LFP waveforms during SWR events correlated with differential patterns of cortical activation as measured with fMRI (Ramirez-Villegas et al. 2015), suggesting that not all hippocampal ensemble activations are equally successful at eliciting a brain-wide response.

Hippocampal Replay and Sharp Waves

SWRs show a rich variety of activity patterns, ideally suited for transporting information. It has been known for a long time that SWR represent a preferential state for memory replay in the hippocampus. Kudrimoti et al. (1999) showed that the activity correlation structure during SWR brings more resemblance with that measured during previous active behavior, compared with inter-SWR periods. The result has been replicated in many behavioral contexts and with many analytical techniques. It has soon become apparent that memory replay during SWR took the form of neural activity sequences (Lee and Wilson 2002): a set of neuron activates sequentially, in the same order in which it activates during activate behavior. A case in point is that of hippocampal place cells: as an animal runs on a track, it traverses the place fields of many hippocampal neurons, one after the other, giving rise to a consistent order of activation as the trajectory is repeated. These activity sequences may be used as a template, and compared with activity during sleep in order to find consistent reactivations of the same sequences. Those tend to happen during SWRs. Sharp waves tend to take place in temporally tight series of at least 2–3 (doublets, triplets), especially during sleep. Consecutive SWRs may replay different chunks of a long activation sequence, for example related to the traversal of a long track, and the activity it elicits in hippocampal place cells. (Davidson et al. 2009).

What are the mechanisms that support the expression of memory replay (see also chapter by Zhang, Deuker and Axmacher)? As delineated above, replay may depend on attractor dynamics in CA3, a hypothesis supported by the fact that replay coincides with increased oscillatory coherence in the hippocampus in the slow gamma (20–50 Hz), linking CA3 and CA1 (Carr et al. 2012). For replay to be relevant for memory consolidation, one would assume that its expression depends on Hebbian plasticity. This hypothesis is yet to be demonstrated directly, however, there are several lines of evidence that may support this idea: NMDA blockade disrupts the stability of hippocampal place fields, which is thought to relate to memory consolidation (Kentros et al. 1998), and blocks the experience-related place field expansion, that is taken as an index of intra-hippocampal plasticity (Ekstrom et al. 2001). While an experiment that clearly show that replay is abolished in the absence of NMDA receptors is still lacking, probably for technical reasons (but see Dragoi and Tonegawa 2013, see also chapter by Cheng), the obvious prediction is that replay is as affected as the aspects of place cells dynamics described above. Yet, a pressing question is how much the patterns that emerge from spontaneous activity during sleep and other circumstances reflect new plastic changes, and how much they are the consequence of pre-existing network wiring, expressed regardless of intervening experience. It has been shown that sequences in spontaneous activity before any experience with a particular track show similarity with the sequences activated as mice explore that track for the first time (Dragoi and Tonegawa 2011). While the statistical assessment of this preplay phenomenon has been the subject of some controversy (Silva et al. 2015), rigid expression of sequences even disconnected from spatial cues has been shown (Villette et al. 2015) in normal mice, and in a mouse model of tau neurodegeneration (Cheng and Ji 2013). This suggests that intrinsic connectivity may indeed play a role in determining the fine structure of hippocampal activity patterns, even in the absence of external information, or synaptic plasticity.

Another interesting question is the role played by neuromodulation in influencing SWRs and replay. For example, activation of dopaminergic Ventral Tegmental Area (VTA) neurons at encoding increases the chances of subsequent replay of the tagged experience (McNamara et al. 2014). During sleep, activation of Locus Coeruleus (LC) during ripples disrupts the coupling between the SWR and cortical oscillations and impacts memory consolidation (Novitskaya et al. 2016). The frequency of SWR (possibly affecting replay as well) is regulated by the activation of raphe nucleus serotonergic cells, with stimulation in that nucleus depressing SWR and memory consolidation and inhibition increasing SWR occurrences (Wang et al. 2015). Thus, these neural mechanisms of memory consolidation appear to be under the control of multiple neuromodulatory systems that may control, and filter the memories that get encoded and consolidated.

Hippocampal replay during SWRs seems like an ideal mechanism to initiate systems memory consolidation, by enabling continued Hebbian plasticity which will reinforce the reactivated memory trace (Sadowski et al. 2016). This hypothesis found support in the finding that selective disruption of hippocampal activity during SWR by electrical stimulation disrupts memory consolidation (Girardeau et al. 2009; Ego-Stengel and Wilson 2010). The same stimulations, applied at times non-coinciding with SWRs did not cause any memory deficit. An even more direct demonstration of the importance for memory consolidation of memory replay, and of the neural plasticity it generates was given by the manipulation by de Lavilleon et al. (2015), which resulted in the generation of an artificial memory. After characterizing the place correlates of the activity of one hippocampal place cells, the researchers detected the spontaneous activation of the same cells during sleep, and used it to trigger the stimulation of the medial forebrain bundle, a fiber complex conveying neuromodulatory reward-related inputs. The result was the induction of plasticity linking the neural representation of a place (the location of the trigger cell's place field), with neural activity normally associated with reward. As a result, mice showed, upon waking up, a place preference for the tagged location, as if that was related to a hedonic memory.

It should be added here that hippocampal SWRs occur also during wakefulness, during quiet periods, or during brief pauses in exploratory behavior. Replay has been observed in those situations as well, with some distinctive features: for example, the 'fidelity' of the replay to the template sequence is higher in awake replay (Carr et al. 2011), and the template sequence may be replayed in the 'forward' or 'reverse' direction (Foster and Wilson 2006; Diba and Buzsáki 2007). Awake replay may have functions different from memory consolidation: during brief stops at the fork of a T-maze, spontaneously replayed sequences 'sweep' the two choice arms (Johnson and Redish 2007). This phenomenon has been posited to provide a substrate for 'vicarious trial and error' (Redish 2016), and as input for brain simulations (probably involving other structures, such as the PFC together with the hippocampus), of the possible alternative decisions and their projected outcome.

Indeed, spontaneous activity does not always reflect *verbatim* template sequences from previously occurred behavior: spontaneous 'sweeps' may describe 'shortcuts' in a maze (Gupta et al. 2012), that were never experienced by the animal before. Furthermore, 'sweeps' in pauses during behavior reflected future trajectory planning (Pfeiffer and Foster 2013). Consistently, SWR disruption during wakefulness seem to affect working memory, rather than memory consolidation processes (Jadhav et al. 2012). Thus, the hippocampal circuit that encodes spatial episodic memories, is also involved in the maintenance of information to be used for decision making. Together, these results suggest that the same hippocampal machinery may support different functions, from long-term memory consolidation to planning of future actions (Schacter et al. 2012; Preston and Eichenbaum 2013; Moscovitch et al. 2016).

Cortical Correlates of Hippocampal SWR and Replay

During sleep, the cortex and the hippocampus engage in a complex dialogue, pointing at bi-directional interaction, with a prominent role of SWR in propagating hippocampal influences to large cortical territories.

We have already mentioned the transient increases in activity of cortical neurons observed throughout the cortex. Also the overall dynamics of cortical activity couples to SWR: during NREM sleep, the cortex underlies an alternation of states of elevated activity (UP) state, and relative silence (DOWN states; Genzel et al. 2014). In light NREM (stage 2), short DOWN states are interspersed in longer periods of ongoing activity. The transient silencing of the cortex is accompanied by a LFP deflection named the slow oscillation (lasting 100–300 ms), and it is often followed by a bout of spindle oscillations. These are faster (9–20 Hz) oscillations generated in the interaction between the cortex and the thalamus. In the rats, SWRs tend to occur preferentially at a given phase of the spindle oscillation (Siapas and Wilson 1998). However, there is a tendency for hippocampal ripples to precede spindle episodes, rather than to co-occur with them. This order of activation was observed over multiple time scales, from the 100s of milliseconds that are characteristic of spindles, to the tens of milliseconds (Peyrache et al. 2011), suggesting that the tighter link is between the SWR and the UP-DOWN-UP transition (and the connected slow oscillation). Furthermore, during spindles, prefrontal interneurons, receiving monosynaptic connections from the hippocampus (Tierney et al. 2004), are strongly activated during SWR (Peyrache et al. 2011), whereas excitatory neurons do not show any sign of recruitment. It has been hypothesized that spindles play an active role in memory consolidation (Diekelmann and Born 2010), a hypothesis backed by a number of results on correlation between spindle density and subsequent memory performance (see also the chapter by McDevitt, Krishnan, Bazhenov and Mednick). However, a clear, direct association between spindles and the neural circuits correlates of memory consolidation in cerebral cortex has not been shown yet. The link of SWRs with spindles, at least in prefrontal cortex, may be mediated by the role played by the slow oscillation in initiating the spindles (Genzel et al. 2014).

In humans, due to the larger size of the brain, sleep spindles have been shown to be quite local phenomena (Andrillon et al. 2011; Nir et al. 2011) and can be sub-characterized by their frequency signature and main anatomical occurrence. More specifically, there are two types of spindles: slower ($\sim$9–13 Hz) spindles that mainly occur in the pre-frontal cortex and faster ($\sim$13–16 Hz) spindles in the parietal cortex (Buzsaki 2015). In humans especially the fast, parietal spindles have often been shown to increase in number and amplitude after a learning episode and correlate with memory performance increases over the sleep period (Gais et al. 2002; Clemens et al. 2005, 2006; Genzel et al. 2009, 2012). Interestingly, the dichotomy between fast and slow spindles may also be present in the co-occurrence with ripples and memory replay. Prefrontal spindles tend to occur after ripples and after memory replay can be measured (Peyrache et al. 2009; Maingret et al. 2016), in contrast the faster, parietal spindles have often been shown to have individual ripples occurring in their troughs (Clemens et al. 2007; Mölle et al. 2009; Clemens et al. 2011; Staresina et al. 2015, see also the chapter by Bergmann and Staresina). Further parietal, motor cortex replay occurs during the sleep spindle instead of before it as seen in prefrontal cortex replay (Peyrache et al. 2009; Ramanathan et al. 2015). For now it remains unclear whether these two types of spindles and replay events just differ due to the time it takes for the information to travel from the prefrontal to the parietal cortex or if they represent similar but different mechanistic processes (Genzel and Robertson 2015).

Cortical Replay

Supporting the importance of sleep-related spontaneous activity for memory consolidation, memory replay has been found in neocortex as well, in shapes and forms very similar to what is observed in the hippocampus. Pairwise correlations between spike trains of neurons from multiple cortical areas of monkeys, recording during sleep, following a task have greater similarity to those recorded while the monkey was performing the task, compared to a control sleep period prior to the task (Hoffman and McNaughton 2002). These reactivated correlations were observed both within cortical areas, as well as in cell pairs spanning multiple areas.

Replay of cortical sequences was observed in the rat PFC after execution of an over-trained sequential task (Euston et al. 2007). As it is the case in the hippocampus (Lee and Wilson 2002), the replay of sequences takes place at a faster (5–10$\times$), compressed rate. This increased speed of replay of the cortical sequences may have important consequences for synaptic plasticity during sleep, leading to memory consolidation, as it shortens the time interval between spikes of neurons that are neighbors in the firing sequences. This may help engage spike-timing dependent plasticity (STDP), by bringing the times of pre- and post-synaptic spikes within the time window of 5–20 ms in which plasticity processes may be triggered (Sadowski et al. 2016). What causes the speedup in replay is not known yet, however it seems to be a state-dependent phenomenon. In the thalamus, it has been shown (Peyrache et al. 2015), that the speedup is restricted to NREM sleep, whereas during REM, firing sequences involving head-direction cells are still replayed in a coherent fashion, but at a slower rate than during wakefulness.

To support the idea that cortical memory replay contributes to systems consolidation, it is necessary to show that it is linked to the hippocampus replay, as the hippocampus is thought to be the source of the information to be consolidated in neocortex. A first piece of evidence in this sense was provided by the experiments of (Peyrache et al. 2009). Here, rats were performing an extra-dimensional set-shift task on a Y-maze. During the subsequent sleep, groups of neurons that were found to activate synchronously during behavior were reactivated, specifically during NREM sleep. The activation of these 'putative cell assemblies' was very irregular, mostly taking place in transient events, whose size was distributed according to a power-law, a sign of 'avalanche like' dynamics (Plenz and Thiagarajan 2007). In a neuronal avalanche, excitatory feedback between neurons causes explosive events involving a large fraction of neurons in the network, to be terminated by adaptation, synaptic depression or other 'resource exhaustion' mechanisms. This type of dynamics is optimal to permit the propagation of information spanning an entire cortical area, and possibly multiple areas, starting from an input of relatively limited potency such as the hippocampal afferents to neocortex. Importantly these replay transients were found to occur preferentially in concurrence with hippocampal sharp waves, providing a first demonstration of correlations between replay (or replay-related) activity in the hippocampus and neocortex. A demonstration of coordinated replay between hippocampal and neocortical neuronal ensembles is yet to be provided. A first hint in this sense comes from the data of Jadhav et al. (2016), showing that PFC neurons with behavioral correlates congruent with those of the hippocampal ensembles activated at the same time increase their firing rate during SWR. Conversely, those PFC neurons that have different behavioral correlates from those of hippocampal neurons activated during a SWR, tend to be suppressed during that SWR event. Thus, the 'content' of PFC replay may be driven by hippocampal activity. Peyrache, Khamassi et al. (2009) showed that such content is not a random sampling of the entirety of the previous experience: replay tended to concentrate on the activity patterns registered at the fork of the Y-maze. This is the point at which a decision on which arm to take was to be taken, according to a previously learned (and regularly changing) rule. The conclusion from these data is that during sleep, neural activity carrying the most relevant behavioral information is the one that is most likely to be replayed, and possibly to be consolidated. This may depend on stronger encoding during memory acquisition. At the maze's decision point, brain activity showed two features that may be important for plasticity induction and memory encoding: Oscillatory coherence in the theta frequency range (6–10 Hz) was specifically elevated at the maze fork, also as an increasing function of performance level. At the same time, increased synchrony was observed in PFC neuronal ensembles (Benchenane et al. 2010). As it is the case for temporal compression during replay, the tighter timing of spikes in the PFC (and PFC-hippocampus) network, may be conducive to increased synaptic plasticity (Benchenane et al. 2011). The saliency of a behavioral event, may be conveyed by neuromodulatory input: dopamine infusion in PFC mimicked the findings of increased coherence found at the maze fork, both in the local PFC network and between PFC and hippocampus, suggesting that an involvement of this neuromodulator may be at play also during active behavior (Benchenane et al. 2011).

Replay in PFC was also found to be related to the cortical oscillations: in particular, it peaked prior to cortical delta waves, and to spindle bouts. In other words, it shared the same temporal relationship to the cortical global patterns as the hippocampal sharp wave (Peyrache et al. 2009, 2011). Indeed, it has been suggested that the transition from DOWN to UP state represents a favorable state for the replay of informative neuronal sequences (Luczak et al. 2007), possibly as the neuronal silence in the DOWN state may transiently decrease noise in the system.

So far, we have considered the interchange between hippocampus and neocortex mostly in terms of hippocampus inputs to the neocortex. However, another interesting question is whether, and how, the neocortical state affects activity in the hippocampus during sleep, therefore realizing a complete two-way interchange. This would have very important theoretical implications, as it may lead to a scenario where the cortex can influence hippocampal replay and extract the information that is needed to accomplish the ongoing cortical consolidation processes.

Isomura et al. (2006) showed that hippocampal activity is modulated by the transitions between the UP and DOWN state. Furthermore, Hahn et al. (2006) demonstrated that the membrane potential of hippocampal neurons is also affected by the same transition (see also the chapter by Bergmann and Staresina). Together, these results suggest that consolidation-related spontaneous activity during sleep should better be seen as a loop, between cortex and hippocampus. The recent discovery of a novel cortical afferent from Anterior Cingulate cortex to the hippocampus (Rajasethupathy et al. 2015), a pathway that has been linked to memory retrieval, reinforces this idea and provides an interesting direction for future research.

Finally, a first line of evidence of the causal role of coordinated activity between the hippocampus and neocortex, PFC in particular, in memory consolidation has recently been provided by Maingret et al. (2016), who used the sharp-wave triggered technique in order to stimulate a delta wave/spindle cortex in PFC. In this way, the researchers were able to increase the degree of coordination between SWR and cortical slow oscillations, and an increase in memory retention was correspondingly observed.

Prefrontal Cortex and Schemas

The importance of the medial prefrontal cortex for memory is even more remarkable, when new learning occurs in the context of previous knowledge. Classically, in animal studies the memory tasks used are completely novel to the subject and in these types of tasks it takes weeks to months for systems consolidation to be completed (Squire et al. 2015). In contrast, when you initially allow rats to build up previous knowledge, i.e. a schema, adding new information can result in rapid systems consolidation. Schemas are thought to represent extensive cortical networks encoding overlapping information that has been extracted over multiple episodes. Instead of the hippocampus the medial prefrontal cortex functions as a binding hub for these networks, which allow for rapid updating of congruent information (Lewis and Durrant 2011; van Kesteren et al. 2012; Ghosh and Gilboa 2014; Squire et al. 2015).

Tse et al. taught rats a map of flavour-location associations and could show that after acquiring the initial map over a period of many weeks not only could rats learn new flavour-location associations in one trial, but this new information became hippocampal independent within 48 h (Tse et al. 2007). They went on to show that updating the map led to immediate upregulation of gene expression in the medial prefrontal cortex and pharmacological inhibition of this brain region during encoding made the animals forget the newly learned information (Tse et al. 2011; Wang et al. 2012). These seminal studies in rodents, led to a sequence of studies in humans, replicating the importance of the medial prefrontal cortex when learning content is congruent with previous knowledge (van Kesteren et al. 2010a, b; van Buuren et al. 2014; Wagner et al. 2015, see chapter by Fernandez). The hippocampus seems to be bypassed already during encoding in humans, when tasks can take advantage of very elaborate, world-knowledge schemas (van Kesteren et al. 2010a, b), in contrast to e.g. relatively newly learned spatial map schema that still requires the hippocampus during updating (Bethus et al. 2010; van Buuren et al. 2014).

Of note, the medial prefrontal cortex is also important when encoding novel information without any previous knowledge. Lesburguères et al. (2011) presented initial evidence for an AMPA- and NMDA receptor dependent "tagging process" in the cortex during encoding, which was crucial for the progressive hippocampal-driven rewiring of cortical networks supporting remote memory storage. In a seminal study Cowansage et al. (2014) then went on to show that when such neural ensembles in the retrosplenial cortex are activated by optogenetic techniques, memory retrieval can occur even with hippocampal inactivation at a time point when sensory cues are not sufficient for memory retrieval without hippocampal involvement (i.e., when the memory is still hippocampal dependent).

Hippocampus—Prefrontal Cortex and Psychiatric Disease

As presented in the previous paragraphs the hippocampal-prefrontal cortex axis is crucial for memory, especially in regard to systems consolidation. However, it is important to keep the wider picture of brain function itself in mind. Memories and previous knowledge are not only focused on the mnemonic function of remembering something for the sake itself, instead all our previous memories will also affect how we interpret new stimuli and the world around us. In this regard, it seems striking that the hippocampal-prefrontal cortex axis has been implicated as a common factor and perhaps vulnerability indicator across many psychiatric diseases. Cumulative evidence indicates that early-life stress accompanied by increased cortisol levels may have negative effects on the hippocampal-prefrontal cortex axis decreasing functional coupling. This circuit dysfunction may contribute to disease formation as well as disease symptomology (Godsil et al. 2013; Gardner et al. 2014).

Both depression and schizophrenia are characterized by deficient sleep-dependent memory consolidation (Manoach et al. 2004; Dresler et al. 2010a, b; Manoach et al. 2010; Dresler et al. 2011; Genzel et al. 2011; Wamsley et al. 2012) a process thought to involve the updating of cortical networks via hippocampally led replay (Genzel et al. 2014). More importantly, the decrease in overnight memory consolidation in these patients is predicted by the decrease in functional connectivity between the hippocampus and medial prefrontal cortex during encoding of the task (Genzel et al. 2015a, b). Furthermore, neocortical slow-oscillations coordinate the timing of hippocampal ripples and prefrontal spindles during NREM sleep. This coordination is selectively disrupted in a rat neurodevelopmental model of schizophrenia: fragmented NREM sleep and impaired slow-oscillation propagation in the model culminate in deficient ripple-spindle coordination and disrupted spike timing, indicating a decoupling of hippocampal and cortical circuits, which may underlie the deficiencies in memory consolidation (Phillips et al. 2012).

However, a dysfunctional hippocampal-prefrontal cortex circuit may not only decrease the amount of consolidation seen overnight, it could also potentially lead to maladaptive consolidation processes. For example, post-traumatic stress disorder (PTSD) is a possible result of maladaptive consolidation of traumatic events. Cardinal symptoms of PTSD are generalization of fear and increased general arousal, reminiscent of the qualitative changes seen in contextual fear memory in rodents when systems consolidation has occurred (Wiltgen et al. 2010; Winocur et al. 2013, see also chapter by Sekeres, Moscovitch and Winocur). Sleep fragmentation as well as norepinephrine and corticosterone in the medial prefrontal cortex and hippocampus predicted the development of PTSD in a mouse model (Polta et al. 2013; Kao et al. 2015). Thus perhaps sleep disturbances together with a dysfunctional hippocampal-prefrontal cortex circuit with maladaptive consolidation may contribute to the development of PTSD (Godsil et al. 2013; Herry and Johansen 2014; Genzel et al. 2015a, b).

Conclusion: Cortico-Hippocampal Loop and Memory Reprocessing

The last ten years of research have unveiled an extremely rich pattern of interaction between these cortical areas. Invasive experiments in animals have enabled us to study how global brain dynamics in different areas may interact and support memory consolidation. Ensemble neurophysiology has played a critical role in the progress in this field. While at this point we have an initial understanding of the global modes of activations in the different structures and how they interact, we still need to figure out the exact mechanisms of information exchange and reprocessing. A key problem for the next few years of research will be to understand how cortico-hippocampal exchange supports the update of semantic schemas: how is new information embedded in the "right place" (i.e. the appropriate synaptic connections, the appropriate cortical patterns)? Are there separate cortically initiated and hippocampally initiated replay cortex? Do they interact in any way? These are some of the questions that may motivate future research, and bring closer our physiological and cognitive understanding of memory.

References

Amit DJ (1989) Modeling brain function. Cambridge University Press, Cambridge

Andrillon T, Nir Y et al (2011) Sleep spindles in humans: insights from intracranial EEG and unit recordings. J Neurosci 31(49):17821–17834

Battaglia FP, Sutherland GR et al (2004) Hippocampal sharp wave bursts coincide with neocortical â€œup-stateâ€ transitions. Learn Mem 11(6):697–704

Battaglia FP, Benchenane K et al (2011) The hippocampus: hub of brain network communication for memory. Trends Cogn Sci 15(7):310–318
Battaglia FP, Borensztajn G et al (2012) Structured cognition and neural systems: from rats to language. Neurosci Biobehav Rev 36(7):1626–1639
Bayley PJ, Gold JJ et al (2005) The neuroanatomy of remote memory. Neuron 46(5):799–810
Benchenane K, Peyrache A et al (2010) Coherent theta oscillations and reorganization of spike timing in the hippocampal- prefrontal network upon learning. Neuron 66(6):921–936
Benchenane K, Tiesinga PH et al (2011) Oscillations in the prefrontal cortex: a gateway to memory and attention. Curr Opin Neurobiol 21:475–485
Bethus I, Tse D et al (2010) Dopamine and memory: modulation of the persistence of memory for novel hippocampal NMDA receptor-dependent paired associates. J Neurosci 30(5):1610–1618
Buzsaki G (1989) Two-stage model of memory trace formation: a role for "noisy" brain states. Neuroscience 31(3):551–570
Buzsaki G (2015) Hippocampal sharp wave-ripple: a cognitive biomarker for episodic memory and planning. Hippocampus 25(10):1073–1188
Buzsáki G, Horváth Z et al (1992) High-frequency network oscillation in the hippocampus. Science 256(5059):1025–1027
Carr MF, Jadhav SP et al (2011) Hippocampal replay in the awake state: a potential substrate for memory consolidation and retrieval. Nat Neurosci 14(2):147–153
Carr MF, Karlsson MP, Frank LM (2012) Transient slow gamma synchrony underlies hippocampal memory replay. Neuron 23; 75(4):700–713
Cheng J, Ji D (2013) Rigid firing sequences undermine spatial memory codes in a neurodegenerative mouse model. Elife 2:e00647
Clemens Z, Fabó DN et al (2005) Overnight verbal memory retention correlates with the number of sleep spindles. Neuroscience 132:529–535
Clemens Z, Fabó DN et al (2006) Twenty-four hours retention of visuospatial memory correlates with the number of parietal sleep spindles. Neurosci Lett 403:52–56
Clemens Z, Mölle M et al (2007) Temporal coupling of parahippocampal ripples, sleep spindles and slow oscillations in humans. Brain 130(11):2868–2878
Clemens Z, Mölle M et al (2011) Fine-tuned coupling between human parahippocampal ripples and sleep spindles. Eur J Neurosci 33(3):511–520
Cowansage KK, Shuman T et al (2014) Direct reactivation of a coherent neocortical memory of context. Neuron 84(2):432–441
Davidson TJ, Kloosterman F et al (2009) Hippocampal replay of extended experience. Neuron 63 (4):497–507
de Lavilleon G, Lacroix MM et al (2015) Explicit memory creation during sleep demonstrates a causal role of place cells in navigation. Nat Neurosci 18(4):493–495
Diba K, Buzsáki G (2007) Forward and reverse hippocampal place-cell sequences during ripples. Nat Neurosci 10(10):1241–1242
Diekelmann S, Born J (2010) The memory function of sleep. Nat Rev Neurosci 11(2):114–126
Dragoi G, Tonegawa S (2011) Preplay of future place cell sequences by hippocampal cellular assemblies. Nature 469(7330):397–401
Dragoi G, Tonegawa S (2013) Development of schemas revealed by prior experience and NMDA receptor knock-out. Elife 2:e01326
Dresler M, Genzel L et al (2010a) Off-line memory consolidation impairments in multiple sclerosis patients receiving high-dose corticosteroid treatment mirror consolidation impairments in depression. Psychoneuroendocrinology 35(8):1194–1202
Dresler M, Kluge M et al (2010b) Impaired off-line memory consolidation in depression. Eur Neuropsychopharmacol 20:553–561
Dresler M, Kluge M et al (2011) A double dissociation of memory impairments in major depression. J Psychiatr Res 45(12):1593–1599
Ego-Stengel V, Wilson MA (2010) Disruption of ripple-associated hippocampal activity during rest impairs spatial learning in the rat. Hippocampus 20(1):1–10

Ekstrom AD, Meltzer J et al (2001) NMDA receptor antagonism blocks experience-dependent expansion of hippocampal "place fields". Neuron 31(4):631–638
Euston DR, Tatsuno M et al (2007) Fast-forward playback of recent memory sequences in prefrontal cortex during sleep. Science 318(5853):1147–1150
Foster DJ, Wilson MA (2006) Reverse replay of behavioural sequences in hippocampal place cells during the awake state. Nature 440(7084):680–683
Gais S, Helms K et al (2002) Increased density of sleep spindles after extensive learning of a declarative memory task. J Sleep Res 11(Suppl 1):147
Gardner RJ, Kersante F et al (2014) Neural oscillations during non-rapid eye movement sleep as biomarkers of circuit dysfunction in schizophrenia. Eur J Neurosci 39(7):1091–1106
Genzel L, Robertson EM (2015) To replay, perchance to consolidate. PLoS Biol 13(10):e1002285
Genzel L, Dresler M et al (2009) Slow wave sleep and REM sleep awakenings do not affect sleep dependent memory consolidation. Sleep 32(3):302–310
Genzel L, Ali E et al (2011) Sleep-dependent memory consolidation of a new task is inhibited in psychiatric patients. J Psychiatr Res 45(4):555–560
Genzel L, Kiefer T et al (2012) Sex and modulatory menstrual cycle effects on sleep related memory consolidation. Psychoneuroendocrinology 37(7):987–989
Genzel L, Kroes MCW et al (2014) Light sleep vs. slow wave sleep in memory consolidation: a question of global versus local processes? Trends Neurosci 37(1):10–19
Genzel L, Dresler M et al (2015a) Medial prefrontal-hippocampal connectivity and motor memory consolidation in depression and schizophrenia. Biol Psychiatry 77(2):177–186
Genzel L, Spoormaker VI et al (2015b) The role of rapid eye movement sleep for amygdala-related memory processing. Neurobiol Learn Mem 122:110–121
Ghosh VE, Gilboa A (2014) What is a memory schema? A historical perspective on current neuroscience literature. Neuropsychologia 53:104–114
Girardeau G, Benchenane K et al (2009) Selective suppression of hippocampal ripples impairs spatial memory. Nat Neurosci 12(10):1222–1223
Godsil BP, Kiss JP et al (2013) The hippocampal-prefrontal pathway: the weak link in psychiatric disorders? Eur Neuropsychopharmacol 23(10):1165–1181
Gupta AS, van der Meer MAA et al. (2012) Segmentation of spatial experience by hippocampal theta sequences. Nat Neurosci 15(7):1032–1039
Hahn TTG, Sakmann B et al (2006) Phase-locking of hippocampal interneurons' membrane potential to neocortical up-down states. Nat Neurosci 9(11):1359–1361
Herry C, Johansen JP (2014) Encoding of fear learning and memory in distributed neuronal circuits. Nat Neurosci 17(12):1644–1654
Hoffman KL, McNaughton BL (2002) Coordinated reactivation of distributed memory traces in primate neocortex. Science 297(5589):2070–2073
Isomura Y, Sirota A et al (2006) Integration and segregation of activity in Entorhinal-Hippocampal subregions by neocortical slow oscillations. Neuron 52(5):871–882
Ito HT, Zhang S-J et al (2015) A prefrontal-thalamo-hippocampal circuit for goal-directed spatial navigation. Nature 522(7554):50–55
Jadhav SP, Kemere C et al. (2012) "Awake Hippocampal Sharp-Wave Ripples Support Spatial Memory." Science 336(6087):1454–1458
Jadhav SP, Rothschild G et al (2016) Coordinated excitation and inhibition of prefrontal ensembles during awake hippocampal sharp-wave ripple events. Neuron 90(1):113–127
Johnson A, Redish AD (2007) Neural ensembles in CA3 transiently encode paths forward of the animal at a decision point. J Neurosci 27(45):12176–12189
Kao CY, Stalla G et al (2015) Norepinephrine and corticosterone in the medial prefrontal cortex and hippocampus predict PTSD-like symptoms in mice. Eur J Neurosci 41(9):1139–1148
Kaplan R, Adhikari MH et al (2016) Hippocampal sharp-wave ripples influence selective activation of the default mode network. Curr Biol 26(5):686–691
Kentros C, Hargreaves E et al (1998) Abolition of long-term stability of new hippocampal place cell maps by NMDA receptor blockade. Science 280(5372):2121–2126

Kudrimoti HS, Barnes CA et al (1999) Reactivation of hippocampal cell assemblies: effects of behavioral state, experience, and EEG dynamics. J Neurosci 19(10):4090–4101

Lee AK, Wilson MA (2002) Memory of sequential experience in the hippocampus during slow wave sleep. Neuron 36(6):1183–1194

Lesburgueres E, Gobbo OL et al (2011) Early tagging of cortical networks is required for the formation of enduring associative memory. Science 331(6019):924–928

Lewis PA, Durrant SJ (2011) Overlapping memory replay during sleep builds cognitive schemata. Trends Cogn Sci 15(8):343–351

Logothetis NK, Eschenko O et al (2012) Hippocampal-cortical interaction during periods of subcortical silence. Nature 491(7425):547–553

Luczak A, Barthó P et al (2007) Sequential structure of neocortical spontaneous activity in vivo. Proc Natl Acad Sci U S A 104(1):347–352

Maingret N, Girardeau G et al (2016) "Hippocampo-cortical coupling mediates memory consolidation during sleep". Nat Neurosci (in press)

Manoach DS, Cain MS et al (2004) A failure of sleep-dependent procedural learning in chronic, medicated schizophrenia. Biol Psychiatry 56:951–956

Manoach DS, Thakkar KN et al (2010) Reduced overnight consolidation of procedural learning in chronic medicated schizophrenia is related to specific sleep stages. J Psychiatr Res 44 (2):112–120

Marr D (1970) A theory for cerebral neocortex. Proc Roy Soc Lond Ser B Biol Sci 176:161–234

Marr D (1971) A theory for archicortex. Philos Trans Roy Soc Lond Ser B Biol Sci 262:23–81

McClelland JL, McNaughton BL et al (1995) Why there are complementary learning systems in the hippocampus and neocortex: Insights form the successes and failures of connectionist models of learning and memory. Psychol Rev 102:419–457

McNamara CG, Tejero-Cantero A et al (2014) Dopaminergic neurons promote hippocampal reactivation and spatial memory persistence. Nat Neurosci 17(12):1658–1660

Mölle M, Eschenko O et al (2009) The influence of learning on sleep slow oscillations and associated spindles and ripples in humans and rats. Eur J Neurosci 29(5):1071–1081

Mongillo G, Barak O et al (2008) Synaptic theory of working memory. Science 319 (5869):1543–1546

Moscovitch M, Cabeza R et al (2016) Episodic memory and beyond: the hippocampus and neocortex in transformation. Annu Rev Psychol 67:105–134

Nieuwenhuis ILC, Takashima A (2011) The role of the ventromedial prefrontal cortex in memory consolidation. Behav Brain Res 218(2):325–334

Nir Y, Staba Richard J et al (2011) Regional slow waves and spindles in human sleep. Neuron 70 (1):153–169

Novitskaya Y, Sara SJ et al (2016) Ripple-triggered stimulation of the locus coeruleus during post-learning sleep disrupts ripple/spindle coupling and impairs memory consolidation. Learn Mem 23(5):238–248

O'Keefe J (1976) Place units in the hippocampus of the freely moving rat. Exp Neurol 51(1):78–109

O'Keefe J, Recce ML (1993) Phase relationship between hippocampal place units and the EEG theta rhythm. Hippocampus 3(3):317–330

Peyrache A, Khamassi M et al (2009) Replay of rule-learning related neural patterns in the prefrontal cortex during sleep. Nat Neurosci 12(7):919–926

Peyrache A, Battaglia FP et al (2011) Inhibition recruitment in prefrontal cortex during sleep spindles and gating of hippocampal inputs. Proc Natl Acad Sci 108(41):17207–17212

Peyrache A, Lacroix MM et al (2015) Internally organized mechanisms of the head direction sense. Nat Neurosci 18(4):569–575

Pfeiffer BE, Foster DJ (2013) Hippocampal place-cell sequences depict future paths to remembered goals. Nature 497(7447):74–79

Phillips K-G, Bartsch U et al (2012) Decoupling of sleep-dependent cortical and hippocampal interactions in a neurodevelopmental model of schizophrenia. Neuron 76(3):526–533

Plenz D, Thiagarajan TC (2007) The organizing principles of neuronal avalanches: cell assemblies in the cortex? Trends Neurosci 30(3):101–110

Polta SA, Fenzl T et al (2013) Prognostic and symptomatic aspects of rapid eye movement sleep in a mouse model of posttraumatic stress disorder. Front Behav Neurosci 7:60

Preston AR, Eichenbaum H (2013) Interplay of hippocampus and prefrontal cortex in memory. Curr Biol 23(17):R764–773

Rajasethupathy P, Sankaran S et al (2015) Projections from neocortex mediate top-down control of memory retrieval. Nature 526(7575):653–659

Ramanathan DS, Gulati T et al (2015) Sleep-dependent reactivation of ensembles in motor cortex promotes skill consolidation. PLoS Biol 13(9):e1002263

Ramirez-Villegas JF, Logothetis NK et al (2015) Diversity of sharp-wave-ripple LFP signatures reveals differentiated brain-wide dynamical events. Proc Natl Acad Sci U S A 112(46): E6379–6387

Redish AD (2016) Vicarious trial and error. Nat Rev Neurosci 17(3):147–159

Reichinnek S, Kuensting T et al (2010) Field potential signature of distinct multicellular activity patterns in the mouse hippocampus. J Neurosci 30(46):15441–15449

Sadowski JH, Jones MW et al (2016) Sharp-wave ripples orchestrate the induction of synaptic plasticity during reactivation of place cell firing patterns in the hippocampus. Cell Rep 14 (8):1916–1929

Schacter DL, Addis DR et al (2012) The future of memory: remembering, imagining, and the brain. Neuron 76(4):677–694

Siapas AG, Wilson MA (1998) Coordinated interactions between hippocampal ripples and cortical spindles during slow-wave sleep. Neuron 21(5):1123–1128

Siapas AG, Lubenov EV et al (2005) Prefrontal phase locking to hippocampal theta oscillations. Neuron 46(1):141–151

Silva D, Feng T et al (2015) Trajectory events across hippocampal place cells require previous experience. Nat Neurosci 18(12):1772–1779

Sirota A, Csicsvari J et al (2003) Communication between neocortex and hippocampus during sleep in rodents. Proc Natl Acad Sci 100(4):2065–2069

Sirota A, Montgomery S et al (2008) Entrainment of neocortical neurons and gamma oscillations by the hippocampal theta rhythm. Neuron 60(4):683–697

Skaggs WE, McNaughton BL et al (1996) Theta phase precession in hippocampal neuronal populations and the compression of temporal sequences. Hippocampus 6(2):149–172

Squire LR, Genzel L et al (2015) Memory consolidation. Cold Spring Harb Perspect Biol 7(8): a021766

Staresina B, Bergmann TO et al (2015) "Hierachical nesting of slow oscillations, spindles and ripples in the human hippocampus during sleep." Nat Neurosci

Taxidis J, Anastassiou CA et al (2015) Local field potentials encode place cell ensemble activation during hippocampal sharp wave ripples. Neuron 87(3):590–604

Thierry AM, Gioanni Y et al (2000) Hippocampo-prefrontal cortex pathway: anatomical and electrophysiological characteristics. Hippocampus 10(4):411–419

Tierney PL, Dégenètais E et al (2004) Influence of the hippocampus on interneurons of the rat prefrontal cortex. Eur J Neurosci 20(2):514–524

Tse D, Langston RF et al (2007) Schemas and memory consolidation. Science 316(5821):76–82

Tse D, Takeuchi T et al (2011) Schema-dependent gene activation and memory encoding in neocortex. Science 333(6044):891–895

van Buuren M, Kroes MC et al (2014) Initial investigation of the effects of an experimentally learned schema on spatial associative memory in humans. J Neurosci 34(50):16662–16670

van Kesteren MT, Fernandez G et al (2010a) Persistent schema-dependent hippocampal-neocortical connectivity during memory encoding and postencoding rest in humans. Proc Natl Acad Sci U S A 107(16):7550–7555

van Kesteren MT, Rijpkema M et al (2010b) Retrieval of associative information congruent with prior knowledge is related to increased medial prefrontal activity and connectivity. J Neurosci 30(47):15888–15894

van Kesteren MTR, Ruiter DJ et al (2012) How schema and novelty augment memory formation. Trends Neurosci 35(4):211–219

Varela C, Kumar S et al (2013) "Anatomical substrates for direct interactions between hippocampus, medial prefrontal cortex, and the thalamic nucleus reuniens." Brain Struct Funct 1–19

Vertes RP (2006) Interactions among the medial prefrontal cortex, hippocampus and midline thalamus in emotional and cognitive processing in the rat. Neuroscience 142(1):1–20

Villette V, Malvache A et al (2015) Internally recurring hippocampal sequences as a population template of spatiotemporal information. Neuron 88(2):357–366

Wagner IC, van Buuren M et al (2015) "Schematic memory components converge within angular gyrus during retrieval." Elife 4:e09668

Wamsley EJ, Tucker MA et al (2012) Reduced sleep spindles and spindle coherence in schizophrenia: mechanisms of impaired memory consolidation? Biol Psychiatry 71(2):154–161

Wang SH, Tse D et al (2012) Anterior cingulate cortex in schema assimilation and expression. Learn Mem 19(8):315–318

Wang DV, Yau HJ et al (2015) Mesopontine median raphe regulates hippocampal ripple oscillation and memory consolidation. Nat Neurosci 18(5):728–735

Wiltgen BJ, Zhou M et al (2010) The hippocampus plays a selective role in the retrieval of detailed contextual memories. Curr Biol 20(15):1336–1344

Winocur G, Moscovitch M et al (2013) Factors affecting graded and ungraded memory loss following hippocampal lesions. Neurobiol Learn Mem 106:351–364

Part IV
Modulation of Memory Consolidation

Stress and Memory Consolidation

Shira Meir Drexler and Oliver T. Wolf

Abstract This chapter presents stress modulation of learning and memory processes, focusing on the consolidation (and reconsolidation) of emotional memories in health and disease. A stressor is any kind of condition, which presents an environmental demand that exceeds the natural regulatory capacity of the individual. A stressor can be of a physical or psychological nature, tangible or mentally evoked. The subjective state of sensing these possibly adverse conditions is termed 'stress' and it leads to the activation of two systems: the sympathetic nervous system and the hypothalamus-pituitary-adrenal axis. Their end-products of (nor) adrenaline and glucocorticoids mediate different functions but also work in concert to promote an adaptive physiological and behavioral response to the challenge. Stress can either enhance or impair memory, and the timing of the stress relative to the task plays a major role in determining the direction of these effects. The adaptive stress response prioritizes consolidation of potentially dangerous events, therefore while consolidation is enhanced, retrieval is usually impaired. Additional factors, such as stimulus and context characteristics (e.g. emotionality and arousal), stress intensity and duration, also play a role. While in several circumstances can stress hormones lead to strong and persistent maladaptive or traumatic memories, their memory-enhancing and retrieval-impairing properties also make them potential adjuvants for treatment, e.g. in extinction-learning based therapies.

Keywords Emotional memory · Glucocorticoids · Memory reconsolidation · Memory retrieval · Noradrenaline

S. Meir Drexler · O.T. Wolf (✉)
Department of Cognitive Psychology, Institute of Cognitive Neuroscience, Ruhr-University Bochum, Universitätsstraße 150, 44801 Bochum, Germany
e-mail: Oliver.T.Wolf@ruhr-uni-bochum.de

S. Meir Drexler · O.T. Wolf
International Graduate School of Neuroscience, Ruhr-University Bochum, 44801 Bochum, Germany

N. Axmacher and B. Rasch (eds.), *Cognitive Neuroscience of Memory Consolidation*, Studies in Neuroscience, Psychology and Behavioral Economics, DOI 10.1007/978-3-319-45066-7_17

Do you remember when did you first hear about the terror attacks on Tuesday, September 11th, 2001? Your answer will probably be 'yes'. You might even remember exact details of the event, such as the time of day, who you were with and where, how you felt, what you thought and said. But do you remember what you had for lunch on Monday, September 10th, 2001, just one day before these events took place? Your answer will probably be 'no'. The reason is that not all memories are created equal. Even years later, stressful events are better remembered than neutral ones.

This chapter will present the stress response and its mediators. Stress modulation of learning and memory processes will be discussed, focusing on the consolidation (and reconsolidation) of emotional memories in health and disease (see also the chapters by Cunningham and Payne on consolidation of emotional memory, and by Kessler, Blackwell and Kehyayan on reconsolidation and posttraumatic stress disorder).

Stress Response

A stressor is any kind of condition, which presents an environmental demand that exceeds the natural regulatory capacity of the animal, in particular when predictability and controllability are at stake (Koolhaas et al. 2011). The stressor can be of a physical or psychological nature, tangible or mentally evoked (Joels and Baram 2009; Joels et al. 2006). It could be the presence of a predator or an aggressive conspecific, an environmental challenge (e.g. flood, earthquake, forest fire) or, for humans nowadays, an important exam or a short deadline at work. The subjective state of sensing these possibly adverse conditions is termed 'stress' and it leads to a complex response, involving a variety of modulators (among them neurotransmitters, peptides and steroid hormones). Different stressors require different responses, and so the nature of the stressor determines the neuronal populations that perceive a potential threat as well as the stress mediators involved in the adaptive response (Joels and Baram 2009). For instance, physical stressors (e.g. cold, blood loss) recruit the brain stem and hypothalamus (Ulrich-Lai and Herman 2009) while psychological stressors (e.g. public speech) recruit brain areas that are involved in emotions (prefrontal cortex (PFC) and amygdala), learning and memory (hippocampus) and decision-making (PFC) (de Kloet et al. 2005). These systems are not segregated and many stressors (e.g. car accident, rape) combine both physical and psychological aspects and responses (Joels and Baram 2009).

This chapter will present the two systems involved in the stress response: the sympathetic nervous system and the hypothalamus-pituitary-adrenal axis (see Fig. 1). Their end-products of (nor)adrenaline and glucocorticoids and their interactions will be the main focus. The two systems mediate different functions but also work in concert to promote an adaptive physiological and behavioral response to the challenge, while suppressing functions that are not of immediate necessity (e.g. growth and reproduction). As the systems are highly conserved among vertebrates,

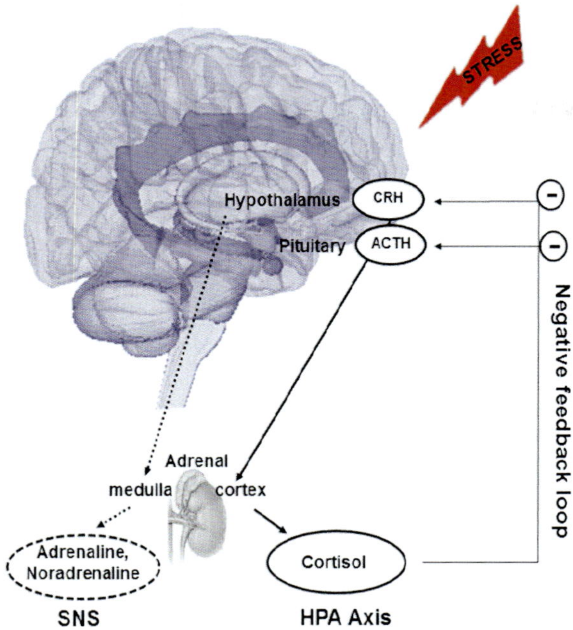

Fig. 1 *The stress response.* The sympathetic nervous system (SNS) and the hypothalamus-pituitary-adrenal (HPA) axis mediate different functions but also work in concert to promote an adaptive physiological and behavioral response to stressors. Their end-products are (nor)adrenaline and cortisol, respectively. Cortisol is also involved in a negative feedback loop, affecting the HPA axis. *ACTH* adrenocorticotropic hormone; *CRH* corticotrophin-releasing-hormone

the use of various animal models is rather common and so evidence from animal and humans studies will be presented interchangeably. For a detailed description of additional stress mediators, see Joels and Baram (2009).

The Sympathetic Nervous System

The sympathetic nervous system (SNS) is fast to respond when facing a threat. This system leads to the secretion of (nor)adrenaline (and other monoamines) from the adrenal medulla. After binding to G-protein coupled receptors in the membrane, they induce rapid but short lasting changes in the neuronal excitability. In some cases, secondary gene-mediated effects occur (Joels and Baram 2009), which are slow in onset but longer lasting. The mostly rapid SNS response promotes physiological (e.g. enhanced metabolism) and behavioral (e.g. increased arousal and vigilance) strategies that help the animal survive the initial phase of the stressful event.

The Hypothalamus-Pituitary-Adrenal Axis

While the SNS response changes neural activity quickly and transiently, the hypothalamus-pituitary-adrenal (HPA) axis mostly leads to a delayed but longer-lasting effect. Following the release of corticotrophin-releasing-hormone (CRH) from the paraventricular nucleus of the hypothalamus, the secretion of adrenocorticotropic hormone (ACTH) from the pituitary stimulates the adrenal cortex to release the steroid hormones glucocorticoids (GCs) to the general circulation. GCs (the main GC is cortisol in humans and corticosterone in rodents) are released in a pulsatile and circadian fashion, with peak concentrations shortly upon awakening and following stress exposure (Joels and Baram 2009; Kirschbaum and Hellhammer 1994). The degree of HPA activation after stress exposure depends on the severity, type and duration of the stressor but also varies between individuals. Genetic factors, personality traits, life history, age, hormonal and health status all affect the HPA response (Joels and Baram 2009; Joels et al. 2006; Kirschbaum and Hellhammer 1994). In addition, the reactivity of the HPA axis differs between males and females and is also altered during the female menstrual cycle (for a review on sex differences in HPA axis response, see Kudielka and Kirschbaum 2005; ter Horst et al. 2012). GCs regulate a wide variety of bodily functions that reinstate the homeostatic control after the temporary disturbance caused by stress. They play a major role in metabolism, and by mobilizing resources to provide energy they help to overcome the increased metabolic demand posed by the challenge. GCs regulate additional systems, such as the immune system, the cardiovascular system as well as affective and cognitive processes (Kudielka and Kirschbaum 2005).

GCs are lipophilic, and therefore can easily enter the brain (McEwen et al. 1968), where their actions facilitate behavior adaptation. In the brain, they bind to two receptor types: Mineralocorticoid (MR) and glucocorticoid (GR) receptors (de Kloet et al. 1998; Joels et al. 2006). Both receptors are co-localized in the hippocampus, amygdala and PFC (de Kloet et al. 2005; Joels and Baram 2009), brain areas that have a fundamental role in learning and memory (McGaugh 2000; Roozendaal 2002). MR are of high affinity, and are mainly present in limbic structures (Reul and de Kloet 1985). They become occupied and activated at lower concentrations, and mediate the initial GCs response to stress. For instance, they modulate appraisal of information and response selection (Lupien and McEwen 1997; Oitzl and de Kloet 1992). The GR, in contrast, are widely present in the brain, but due to their lower affinity become fully occupied only at times of high hormone concentration, e.g. at the circadian peak or following stress exposure (Reul and de Kloet 1985). GR contribute to the HPA negative feedback loop by terminating the stress response. In addition, they mediate the effects of stress on memory consolidation (de Kloet et al. 1998). Until recently, both MR and GR types were thought to lead to changes through gene expression with a delay of more than one hour. However, recent evidence has shown that both receptor types can also alter neuronal functions within minutes via non-genomic pathways (Joels et al. 2008; Joels and Karst 2012). Membrane-bound MR that reside in the plasma membrane, higher in affinity than the nuclear variant, were suggested to be involved in the fast cognitive effects of stress on memory and executive functions (Otte et al. 2015; Vogel et al. 2015), such as the stress-induced shift from 'cognitive' (i.e. goal-directed) to 'habit' (i.e. stimulus-response) memory system (Schwabe and Wolf 2013). Membrane-bound GR, which regulate the chromatin structure, can allow transient, but potentially stable, effects on transcriptional processes that maintain cellular memory (Roozendaal et al. 2010).

The Effects of Stress on Learning and Memory

How does stress affect memory? If you'd think of the example from the beginning of the chapter (or on any other stressful event you had experienced) you'd probably say that stress enhances memory. In contrast, you might think about a presentation you once held in front of your class, in which you were so stressed you could not remember the answer to an (otherwise simple) question. Indeed, the effects of stress on memory vary, and it can either enhance or impair memory, depending on the timing of the stress with regard to the memory task, on stress intensity and duration, as well as on task-related factors and individual characteristics (Lupien and McEwen 1997; McGaugh and Roozendaal 2002; Shors 2006; Wolf 2008). For additional reviews on GCs effects on memory consolidation and retrieval, see Roozendaal (2002); Wolf (2009).

Stress and Memory Consolidation

The protein-synthesis dependent process of memory consolidation at the cellular level is thought to be accomplished in the first minutes to hours after encoding (Dudai 2004; Kandel 2001). During this period, the memory trace can be affected by a variety of manipulations. Increasing GCs concentrations by stress induction or pharmacological administration after the learning experience enhances the memory for the particular event (de Kloet et al. 1999; Joels et al. 2006). This has been demonstrated in several species for various memory types: for instance, spatial learning (Oitzl et al. 2001) and passive avoidance (Bohus and de Kloet 1981) in rodents, and taste aversion in chicks (Sandi and Rose 1994). For example, rats that were trained in a Morris water maze (a spatial memory task) show elevated circulating GCs concentrations (Oitzl et al. 2001), which are more pronounced when the water temperature is lower (presumably more stressful for the animal compared to lukewarm water). This rise in GCs concentrations is positively correlated with a memory of the platform location in a subsequent test performed one day, or one week, later (Sandi et al. 1997). Preventing GR activity during water maze learning, either pharmacologically in rats (Oitzl and de Kloet 1992) or genetically in mice (Oitzl et al. 2001) reverses the GCs-mediated performance enhancement.

In humans, post-learning manipulations have demonstrated similar enhancing effects of stress and GCs on memory consolidation. In a typical design, such as demonstrated by Preuß and Wolf (2009), participants are presented with a new learning material (e.g. pictures or words of varying emotional valance). Immediately after learning (with or without immediate recall test), they are exposed to either the stress (e.g. psychosocial stress, cold pressor stress) or the control condition. On the next day, delayed memory recall is tested. While rising the GCs levels by stress exposure facilitates delayed recall in declarative memory tasks (Cahill et al. 2003; Preuß and Wolf 2009) inhibiting GCs activity using steroid synthesis inhibitor during learning of a task impairs the delayed (but not immediate) recall of the learned material (Lupien et al. 2002). The delayed, but not immediate, enhancing effect points to a post-encoding enhancement of memory consolidation by stress and GCs.

Yet not only timing matters. Other task- and stress-related factors play a role in the consequences of stress on memory consolidation (Joels et al. 2006). Even though some studies suggested that GCs enhance consolidation independent of arousal (Abercrombie et al. 2003; Maheu et al. 2004), there is ample evidence demonstrating that (under the same GCs conditions) emotional or arousing events tend to be better remembered than neutral ones. This will be discussed next.

Stimulus Emotionality and Arousal

GCs interact with other modulators (noradrenaline in particular) to enhance the consolidation of emotional and arousing experiences (McGaugh and Roozendaal 2002). In humans, Buchanan and Lovallo (2001) have shown that cortisol treatment prior to encoding of pictures of different emotionality results in enhanced memory for the emotional (whether negative or positive) pictures. In a similar manner, post-learning stress enhances the long-term memory for arousing slides, but not neutral slides (Cahill et al. 2003), and improves the recall of words, in particular emotional ones (Smeets et al. 2008). Noradrenergic arousal can be induced not only by the stimulus itself, but also by the context. For instance, the arousal level in rats is higher in a novel experimental context but decreases following habituation. Exposure of non-habituated (i.e. aroused) rats to a stressor enhanced the long-term memory in a non-aversive task of recognition memory. The effect was opposite (impaired consolidation) in habituated (non-aroused) rats (Maroun and Akirav 2008).

Roozendaal et al. (2006) demonstrated that noradrenergic activation in the basolateral amygdala (BLA) is necessary for GCs-induced effects on emotional memory formation. Unlike GCs, adrenaline does not readily cross the blood-brain barrier, and a peripheral-central pathway mediates its effects on the amygdala. Systemic adrenaline activates β-adrenoreceptors on vagal afferents that terminate in the nucleus of the solitary tract (NTS). These noradrenergic cell groups project directly to the amygdala and indirectly to the locus coeruleus, leading to noradrenaline secretion. In the BLA, the β-adrenoreceptors directly stimulate cAMP and cAMP-dependent protein kinase A (PKA). GR potentiate the efficacy of this pathway, and may also influence it via coupling with α_1-adrenoreceptors. In addition to interacting with the noradrenergic cascade at postsynaptic levels, GCs alter the levels of available noradrenaline via GR in the noradrenergic cell groups in the NTS. Administrating β-adrenoreceptor antagonist into the BLA blocks GCs-mediated memory enhancement (Roozendaal et al. 1996) while post-training agonists enhance memory consolidation (Liang et al. 1995). Evidence from recent years has also suggested a role for the endocannabinoid system, a lipid-based retrograde signaling system, in mediating this interaction (Atsak et al. 2012). The interaction between GCs and the noradrenergic system, and its contribution to emotional memory enhancement, is thoroughly described by Roozendaal (2002) and is also illustrated in Fig. 2.

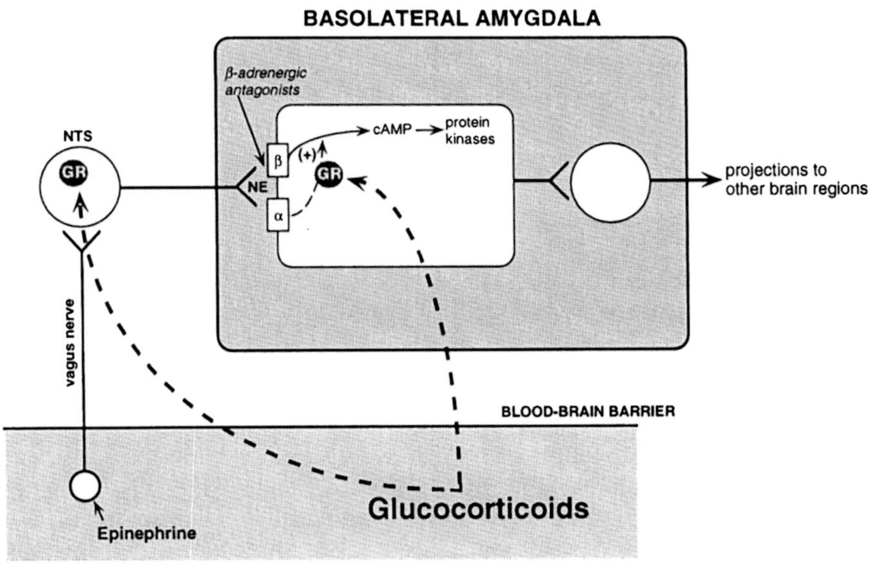

Fig. 2 *The interaction between GCs and the SNS in enhancing memory consolidation*. GCs and the noradrenergic system of the basolateral amygdala (BLA) interact at both presynaptic and postsynaptic sites. Unlike GCs, adrenaline does not readily cross the blood-brain barrier, and a peripheral-central pathway via the vagus nerve mediates its influences on the amygdala. Adrenaline activates β-adrenoreceptors on the NTS, which project directly to the BLA and indirectly to the locus coeruleus, leading to noradrenaline secretion. In the BLA, the β-adrenoreceptor directly stimulates cAMP and cAMP-dependent protein kinase A. GR potentiate this pathway efficacy, and also influence it via coupling with α_1-adrenoreceptors. In addition, GCs alter the levels of available noradrenaline via GR activation in NTS noradrenergic cell groups. *GCs* glucocorticoids; *GR* glucocorticoid receptors; *NTS* nucleus of the solitary tract. Reprinted from *Psychoneuroendocrinology*, 25, B. Roozendaal, "Glucocorticoids and the regulation of memory consolidation", 213–238. Copyright 2000 with permission from Elsevier Science

Stress Intensity

The intensity of stress is another factor determining its effects on memory consolidation. A nonlinear dose-response relationship of neurotransmitters and hormones is common, resulting from different receptor subtypes that operate with specific accessibility, affinity, desensitization and signaling cascades. A dose-response curve, inverted-U shaped, is well documented in the case of GCs (Joels 2006) and is supported by behavioral and electrophysiological studies in animals. For instance, while a moderate rise in GCs concentrations is positively correlated with spatial memory in the Morris water maze (Sandi et al. 1997), too high levels of stress (e.g. very low water temperature) do not lead to further improvement but impair performance (Kim and Diamond 2002). In the hippocampal CA1 pyramidal cells of rats, long-term potentiation (LTP) was found to be affected by GCs in a dose-depended fashion, responding to an inverted U-shaped curve (Diamond et al. 1992; Mesches et al. 1999). Using selective antagonists and agonists for MR and GR, several studies demonstrated that the underlying mechanism is the different affinity of the GCs receptors. Enchanting effects of GCs on memory consolidation were found to depend not only on saturated MR occupancy but also on low to moderate GR occupancy (de Kloet et al. 1998; Roozendaal 2000). However, the dose-dependent effects of GCs have been mainly demonstrated in animals (Joels 2006). Empirical evidence in humans is currently rather sparse. For a detailed review on the inverted U-shaped curve of GCs, see Joels (2006).

Stress Duration

The examples set above concern acute stress, in the context and around the time of the learning experience. The consequences might be significantly different in a brain that has been chronically exposed to stressors. Chronic hyper-(re)activity of the HPA axis can also occur in predisposed individuals and in association to many diseases as well as aging. This can result in dendritic atrophy, reduced neurogenesis and impaired synaptic plasticity in the hippocampus and in the medial PFC. In these cases, learning and memory performances are typically impaired (McEwen 2004; Sapolsky 1999). In the BLA, in contrast, chronic stress leads to robust dendritic growth, which is related to greater anxiety-like behavior (Roozendaal et al. 2009). In a similar way, hypertrophy in the dorsolateral striatum, seen in relation to chronic stress, possibly mediates the bias towards more habitual patterns in instrumental behaviors (Schwabe et al. 2012).

Timing: Consolidation Versus Retrieval

Many students might know this too well: Stress at the time of an exam might lead to a better memory of the stressing test experience itself (when recalled later), while impairing the retrieval of the study material during the exam. Indeed, timing is of critical importance in determining GCs effects on memory. In the short term, GCs and other stress-induced mediators facilitate the strengthening of synaptic contacts involved in the memory formation of the events that led to their release. At the same time, they initiate gene-mediated signals that suppresse any unrelated information from reaching the same brain areas. Indeed, long term memory retrieval is usually impaired by cortisol (de Quervain et al. 2009; Wolf 2009). In most cases this strategy is highly adaptive, prioritizing consolidation of potentially dangerous events over retrieval at times of stress (Diamond et al. 2005; Joels et al. 2006). However, its impairing effects on retrieval might negatively affect performance. In rats that already learned the location of an underwater platform in the Morris water maze, a footshock (i.e. stressor) or injection of corticosterone 30 min before a free swim test lead to performance impairment (de Quervain et al. 1998). In humans, similar impairing effects of stress and GCs were seen in declarative memory tasks (de Quervain et al. 2000; Kuhlmann et al. 2005). Neuroimaging studies have demonstrated that this GCs-induced impairment in declarative memory retrieval is associated with reduced activity of the medial temporal lobe, the hippocampus in particular (de Quervain et al. 2003; Oei et al. 2007).

Stress and Memory Reconsolidation

Stress and GCs have been demonstrated to enhance memory consolidation while impairing retrieval. Do they affect a memory that has been successfully retrieved? The traditional view on memory suggested that memory consolidation is a one-time event, completed shortly after acquisition (McGaugh 1966). This unidirectional view on memory was challenged by Misanin et al. (1968) who suggested that memory reactivation (i.e. retrieval) can cause the memory to re-enter a labile state until re-stabilization (reconsolidation) is completed. The reactivation-dependent lability period was found to last for up to 6 h post-retrieval (Kindt et al. 2009; Schiller et al. 2010), and was suggested to serve as an adaptive mechanism allowing memory update (Alberini 2011; Alberini and LeDoux 2013; Forcato et al. 2014). Various pharmacological agents have been found to affect memory reconsolidation, thereby revealing the mechanisms mediating memory formation and modulation after retrieval. For instance, Nader et al. (2000) demonstrated that reconsolidation is a protein-synthesis-dependent process, while Kindt et al. (2009) showed that reconsolidation of emotional memories is dependent on noradrenergic activity. Both studies pointed to a similarity between reconsolidation after retrieval and initial consolidation. The possible influence of GCs and stress on memory reconsolidation, however, have been investigated only recently.

Akirav and Maroun (2013) reviewed the different, often conflicting, effects of stress and GCs administration on memory reconsolidation. Several animal studies suggest an impairing effect of either stress induction or GCs administration on memory reconsolidation (Yang et al. 2013). However, both GR agonists (Abrari et al. 2008; Cai et al. 2006) and antagonists (Pitman et al. 2011) were found to impair reactivated memories. The human literature had mainly focused on the effects of stress on reactivated declarative memories. The studies demonstrated either an enhancement (Schwabe and Wolf 2010; Zhao et al. 2009) or impairment (Bos et al. 2014; Coccoz et al. 2011, 2013) of reactivated memories, with conflicting results with regard to the effect on strong emotional memories. Recently, however, Meir Drexler et al. (2015) demonstrated an enhancing effect of cortisol on the reconsolidation of reactivated fear memories in healthy men. The fear conditioning paradigm is a model for stress–and trauma-related disorders, and is often used to investigate the emotional and cognitive mechanisms of aversive memories (Pull 2007). The results of the study suggest a mechanism for emotional memory persistence, and could contribute to the understanding of the persistence of emotional memories in several psychiatric disorders.

Figure 3 provides a summary of the timing-dependent effects of stress on the various memory processes.

Relevance for Psychopathology and Treatment

Due to the enhanced consolidation of highly emotional and stressful events, strong memories are common following an aversive experience. This is a very adaptive mechanism, yet even emotional memories weaken over time. In several circumstances, however, extremely aversive events can lead to maladaptive and traumatic memories. This is seen in post-traumatic stress disorder (PTSD) and anxiety disorders (e.g. phobias). PTSD is characterized by re-experiencing the evet, avoidance of stimuli associated with it, and hyper-arousal (American Psychiatric Association 2013; Yehuda 2002). Re-experiencing symptoms include intrusive daytime recollections, nightmares and flashbacks in which the traumatic event is retrieved. The traumatic memories often keep their vividness and ability to evoke distress for decades or even a lifetime after the event. Anxiety disorders, such as phobias, are characterized by persistent fear that is excessive or unreasonable, cued by the presence or anticipation of a specific stimulus or context (American Psychiatric Association 2013). Exposure to phobic stimuli provokes retrieval of stimulus-associated fear memory that leads to the fear response (de Quervain and Margraf 2008; Fehm and Margraf 2002; Rapee and Heimberg 1997). The strength of the fear memory is a result of over-consolidation due to action of stress hormones at the time of the event (Pitman 1989). In these cases, the aversive event trace remains easily reactivable to an aversive cue or even spontaneously (de Quervain

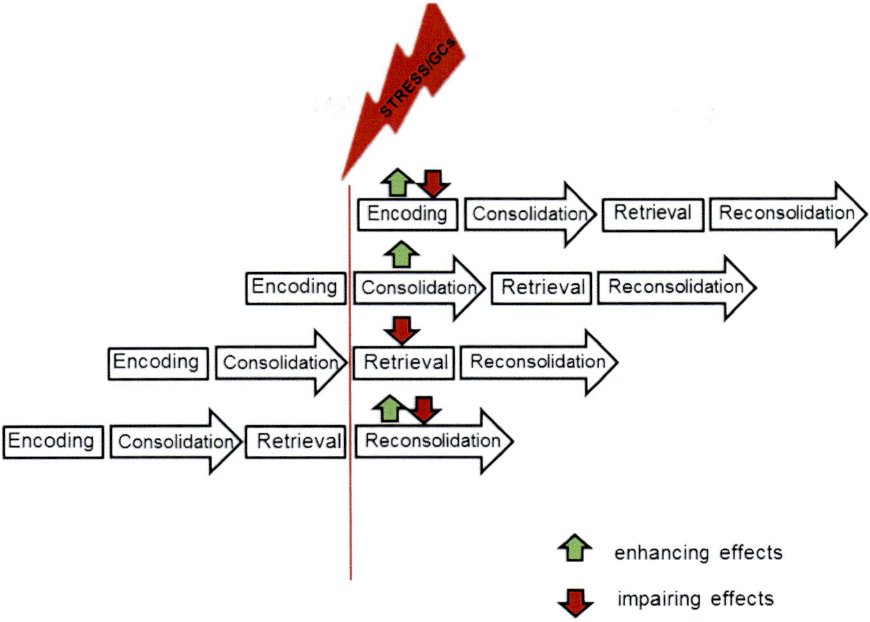

Fig. 3 *The effects of stress and glucocorticoids (GCs) on memory processes.* Stress and GCs effects on memory processes depend on the specific memory phase. Stress induction shortly before encoding typically enhances memory (even though the findings are somewhat heterogeneous). This effect is modulated by the exact timing and the emotionality/relevance of the material. Stress induction after encoding (at the beginning of consolidation) has memory enhancing properties (illustrated with the *green arrow pointing upwards*). Stress before memory retrieval, in contrast, leads to an impairment (illustrated with the *red arrow pointing downwards*). The possible influence of GCs and stress on memory reconsolidation have been investigated only recently with mixed results

and Margraf 2008). The persistence of the memories in the long-term is a possible result of repeated retrievals and enhanced reconsolidation of the fear memory trace at the presence of elevated GCs concentrations (Meir Drexler et al. 2015).

GCs, that can led to robust and maladaptive memories due to their enhancing effect on emotional memory consolidation, can also provide the remedy. Extinction learning occurs when a conditioned responding (e.g. fear) to a stimulus (e.g. spider) is decreased when the reinforcer is omitted (Quirk and Mueller 2008). Extinction is a new learning that creates a fear-inhibiting memory, and is the suggested mechanism underlying various cognitive-behavioral therapies (e.g. exposure therapy) that successfully reduce learned fears (Rachman 1989). As a new learning, it requires consolidation. Due to their memory-enhancing properties, GCs can be used to facilitate the new safety learning in extinction-based therapies (de Quervain and Margraf 2008). In addition, as a result of their retrieval-impairing properties, GCs could partly interrupt the vicious cycle of spontaneous retrieving and reconsolidation of traumatic memories, thereby promote the process of forgetting, a spontaneous process occurring when memory in not reactivated (de Quervain and Margraf 2008). For a review on GCs, their role in stress-related disorders and their potential for treatment, see de Quervain and Margraf (2008); de Quervain et al. (2009).

Conclusion

Stress leads to the activation of two systems: the sympathetic nervous system and the hypothalamus-pituitary-adrenal axis. Their end-products of (nor)adrenaline and glucocorticoids mediate different functions but also work in concert to promote an adaptive physiological and behavioral response to the challenge. Stress can either enhance or impair memory, and the timing of the stress relative to the task plays a major role in determining the direction of these effects. The adaptive stress response prioritizes consolidation of potentially dangerous events, therefore retrieval during or shortly after stress exposure is usually impaired. Additional factors, such as stimulus and context characteristics (e.g. emotionality and arousal), stress intensity and duration, also play a role. While in several circumstances stress hormones can lead to strong and persistent maladaptive or traumatic memories, their memory-enhancing and retrieval-impairing properties also make them potential adjuvants for treatment, e.g. in extinction-learning based therapies.

References

Abercrombie HC, Kalin NH, Thurow ME, Rosenkranz MA, Davidson RJ (2003) Cortisol variation in humans affects memory for emotionally laden and neutral information. Behav Neurosci 117:505–516

Abrari K, Rashidy-Pour A, Semnanian S, Fathollahi Y (2008) Administration of corticosterone after memory reactivation disrupts subsequent retrieval of a contextual conditioned fear memory: dependence upon training intensity. Neurobiol Learn Mem 89:178–184

Akirav I, Maroun M (2013) Stress modulation of reconsolidation. Psychopharmacology 226:747–761

Alberini CM (2011) The role of reconsolidation and the dynamic process of long-term memory formation and storage. Front Behav Neurosci 5

Alberini CM, LeDoux JE (2013) Memory reconsolidation. Curr Biol 23:R746–R750

American Psychiatric Association (2013) Diagnostic and statistical manual of mental disorders, 5th edn. American Psychiatric Publishing, Arlington, VA

Atsak P, Roozendaal B, Campolongo P (2012) Role of the endocannabinoid system in regulating glucocorticoid effects on memory for emotional experiences. Neuroscience 204:104–116

Bohus B, de Kloet ER (1981) Adrenal steroids and extinction behavior: antagonism by progesterone, deoxycorticosterone and dexamethasone of a specific effect of corticosterone. Life Sci 28:433–440

Bos MGN, Schuijer J, Lodestijn F, Beckers T, Kindt M (2014) Stress enhances reconsolidation of declarative memory. Psychoneuroendocrinology 46: 102–113

Buchanan TW, Lovallo WR (2001) Enhanced memory for emotional material following stress-level cortisol treatment in humans. Psychoneuroendocrinology 26:307–317

Cahill L, Gorski L, Le K (2003) Enhanced human memory consolidation with post-learning stress: interaction with the degree of arousal at encoding. Learn Mem 10:270–274
Cai WH, Blundell J, Han J, Greene RW, Powell CM (2006) Postreactivation glucocorticoids impair recall of established fear memory. J Neurosci 26:9560–9566
Coccoz V, Maldonado H, Delorenzi A (2011) The enhancement of reconsolidation with a naturalistic mild stressor improves the expression of a declarative memory in humans. Neuroscience 185:61–72
Coccoz V, Sandoval AV, Stehberg J, Delorenzi A (2013) The temporal dynamics of enhancing a human declarative memory during reconsolidation. Neuroscience 246:397–408
de Kloet ER, Vreugdenhil E, Oitzl MS, Joels M (1998) Brain corticosteroid receptor balance in health and disease. Endocr Rev 19:269–301
de Kloet ER, Oitzl MS, Joels M (1999) Stress and cognition: are corticosteroids good or bad guys? Trends Neurosci 22:422–426
de Kloet ER, Joels M, Holsboer F (2005) Stress and the brain: from adaptation to disease. Nat Rev Neurosci 6:463–475
de Quervain DJ, Margraf J (2008) Glucocorticoids for the treatment of post-traumatic stress disorder and phobias: a novel therapeutic approach. Eur J Pharmacol 583:365–371
de Quervain DJ, Roozendaal B, McGaugh JL (1998) Stress and glucocorticoids impair retrieval of long-term spatial memory. Nature 394:787–790
de Quervain DJ, Roozendaal B, Nitsch RM, McGaugh JL, Hock C (2000) Acute cortisone administration impairs retrieval of long-term declarative memory in humans. Nat Neurosci 3:313–314
de Quervain DJ, Henke K, Aerni A, Treyer V, Mcgaugh JL, Berthold T et al (2003) Glucocorticoid-induced impairment of declarative memory retrieval is associated with reduced blood flow in the medial temporal lobe. Eur J Neurosci 17:1296–1302
de Quervain DJ, Aerni A, Schelling G, Roozendaal B (2009) Glucocorticoids and the regulation of memory in health and disease. Front Neuroendocrinol 30:358–370
Diamond DM, Bennett MC, Fleshner M, Rose GM (1992) Inverted-U relationship between the level of peripheral corticosterone and the magnitude of hippocampal primed burst potentiation. Hippocampus 2:421–430
Diamond DM, Park CR, Campbell AM, Woodson JC (2005) Competitive interactions between endogenous LTD and LTP in the hippocampus underlie the storage of emotional memories and stress-induced amnesia. Hippocampus 15:1006–1025
Dudai Y (2004) The neurobiology of consolidations, or, how stable is the engram? Annu Rev Psychol 55:51–86
Fehm L, Margraf J (2002) Thought suppression: specificity in agoraphobia versus broad impairment in social phobia? Behav Res Ther 40:57–66
Forcato C, Fernandez RS, Pedreira ME (2014) Strengthening a consolidated memory: the key role of the reconsolidation process. J Physiol Paris 108:323–333
Joels M (2006) Corticosteroid effects in the brain: U-shape it. Trends Pharmacol Sci 27:244–250
Joels M, Baram TZ (2009) The neuro-symphony of stress. Nat Rev Neurosci 10:459–466
Joels M, Karst H (2012) Corticosteroid effects on calcium signaling in limbic neurons. Cell Calcium 51:277–283
Joels M, Pu Z, Wiegert O, Oitzl MS, Krugers HJ (2006) Learning under stress: how does it work? Trends Cogn Sci 10:152–158
Joels M, Karst H, DeRijk R, de Kloet ER (2008) The coming out of the brain mineralocorticoid receptor. Trends Neurosci 31:1–7
Kandel ER (2001) The molecular biology of memory storage: a dialog between genes and synapses. Biosci Rep 21:565–611
Kim JJ, Diamond DM (2002) The stressed hippocampus, synaptic plasticity and lost memories. Nat Rev Neurosci 3:453–462
Kindt M, Soeter M, Vervliet B (2009) Beyond extinction: erasing human fear responses and preventing the return of fear. Nat Neurosci 12:256–258

Kirschbaum C, Hellhammer DH (1994) Salivary cortisol in psychoneuroendocrine research: recent developments and applications. Psychoneuroendocrinology 19:313–333

Koolhaas JM, Bartolomucci A, Buwalda B, de Boer SF, Flugge G, Korte SM et al (2011) Stress revisited: a critical evaluation of the stress concept. Neurosci Biobehav Rev 35:1291–1301

Kudielka BM, Kirschbaum C (2005) Sex differences in HPA axis responses to stress: a review. Biol Psychol 69:113–132

Kuhlmann S, Piel M, Wolf OT (2005) Impaired memory retrieval after psychosocial stress in healthy young men. J Neurosci 25:2977–2982

Liang KC, Chen LL, Huang TE (1995) The role of amygdala norepinephrine in memory formation: involvement in the memory enhancing effect of peripheral epinephrine. Chin J Physiol 38:81–91

Lupien SJ, McEwen BS (1997) The acute effects of corticosteroids on cognition: integration of animal and human model studies. Brain Res Brain Res Rev 24:1–27

Lupien SJ, Wilkinson CW, Briere S, Menard C, Ng Ying Kin NM, Nair NP (2002) The modulatory effects of corticosteroids on cognition: studies in young human populations. Psychoneuroendocrinology 27:401–416

Maheu FS, Joober R, Beaulieu S, Lupien SJ (2004) Differential effects of adrenergic and corticosteroid hormonal systems on human short- and long-term declarative memory for emotionally arousing material. Behav Neurosci 118:420–428

Maroun M, Akirav I (2008) Arousal and stress effects on consolidation and reconsolidation of recognition memory. Neuropsychopharmacology 33:394–405

McEwen BS (2004) Protection and damage from acute and chronic stress: allostasis and allostatic overload and relevance to the pathophysiology of psychiatric disorders. Ann N Y Acad Sci 1032:1–7

McEwen BS, Weiss JM, Schwartz LS (1968) Selective retention of corticosterone by limbic structures in rat brain. Nature 220:911–912

McGaugh JL (1966) Time-dependent processes in memory storage. Science 153:1351–1358

McGaugh JL (2000) Memory–a century of consolidation. Science 287:248–251

McGaugh JL, Roozendaal B (2002) Role of adrenal stress hormones in forming lasting memories in the brain. Curr Opin Neurobiol 12:205–210

Meir Drexler S, Merz CJ, Hamacher-Dang TC, Tegenthoff M, Wolf OT (2015) Effects of cortisol on reconsolidation of reactivated fear memories. Neuropsychopharmacology 40:3036–3043

Mesches MH, Fleshner M, Heman KL, Rose GM, Diamond DM (1999) Exposing rats to a predator blocks primed burst potentiation in the hippocampus in vitro. J Neurosci 19: RC18

Misanin JR, Miller RR, Lewis DJ (1968) Retrograde amnesia produced by electroconvulsive shock after reactivation of a consolidated memory trace. Science 160:554–555

Nader K, Schafe GE, LeDoux JE (2000) Fear memories require protein synthesis in the amygdala for reconsolidation after retrieval. Nature 406:722–726

Oei NY, Elzinga BM, Wolf OT, de Ruiter MB, Damoiseaux JS, Kuijer JP et al (2007) Glucocorticoids decrease hippocampal and prefrontal activation during declarative memory retrieval in young men. Brain Imaging Behav 1:31–41

Oitzl MS, de Kloet ER (1992) Selective corticosteroid antagonists modulate specific aspects of spatial orientation learning. Behav Neurosci 106:62–71

Oitzl MS, Reichardt HM, Joels M, de Kloet ER (2001) Point mutation in the mouse glucocorticoid receptor preventing DNA binding impairs spatial memory. Proc Natl Acad Sci USA 98: 12790–12795

Otte C, Wingenfeld K, Kuehl LK, Kaczmarczyk M, Richter S, Quante A et al (2015) Mineralocorticoid receptor stimulation improves cognitive function and decreases cortisol secretion in depressed patients and healthy individuals. Neuropsychopharmacology 40:386–393

Pitman RK (1989) Post-traumatic stress disorder, hormones, and memory. Biol Psychiatry 26:221–223

Pitman RK, Milad MR, Igoe SA, Vangel MG, Orr SP, Tsareva A et al (2011) Systemic mifepristone blocks reconsolidation of cue-conditioned fear; propranolol prevents this effect. Behav Neurosci 125:632–638

Preuß D, Wolf OT (2009) Post-learning psychosocial stress enhances consolidation of neutral stimuli. Neurobiol Learn Mem 92:318–326
Pull CB (2007) Combined pharmacotherapy and cognitive-behavioural therapy for anxiety disorders. Curr Opin Psychiatry 20:30–35
Quirk GJ, Mueller D (2008) Neural mechanisms of extinction learning and retrieval. Neuropsychopharmacology 33:56–72
Rachman S (1989) The return of fear—review and prospect. Clin Psychol Rev 9:147–168
Rapee RM, Heimberg RG (1997) A cognitive-behavioral model of anxiety in social phobia. Behav Res Ther 35:741–756
Reul JM, de Kloet ER (1985) Two receptor systems for corticosterone in rat brain: microdistribution and differential occupation. Endocrinology 117:2505–2511
Roozendaal B (2000) Glucocorticoids and the regulation of memory consolidation. Psychoneuroendocrinology 25:213–238
Roozendaal B (2002) Stress and memory: opposing effects of glucocorticoids on memory consolidation and memory retrieval. Neurobiol Learn Mem 78:578–595
Roozendaal B, Portillo-Marquez G, Mcgaugh JL (1996) Basolateral amygdala lesions block glucocorticoid-induced modulation of memory for spatial learning. Behav Neurosci 110:1074–1083
Roozendaal B, Okuda S, Van der Zee EA, McGaugh JL (2006) Glucocorticoid enhancement of memory requires arousal-induced noradrenergic activation in the basolateral amygdala. Proc Natl Acad Sci USA, 103: 6741–6746
Roozendaal B, McEwen BS, Chattarji S (2009) Stress, memory and the amygdala. Nat Rev Neurosci 10:423–433
Roozendaal B, Hernandez A, Cabrera SM, Hagewoud R, Malvaez M, Stefanko DP et al (2010) Membrane-associated glucocorticoid activity is necessary for modulation of long-term memory via chromatin modification. J Neurosci 30:5037–5046
Sandi C, Rose SP (1994) Corticosterone enhances long-term retention in one-day-old chicks trained in a weak passive avoidance learning paradigm. Brain Res 647:106–112
Sandi C, Loscertales M, Guaza C (1997) Experience-dependent facilitating effect of corticosterone on spatial memory formation in the water maze. Eur J Neurosci 9:637–642
Sapolsky RM (1999) Glucocorticoids, stress, and their adverse neurological effects: relevance to aging. Exp Gerontol 34:721–732
Schiller D, Monfils MH, Raio CM, Johnson DC, LeDoux JE, Phelps EA (2010) Preventing the return of fear in humans using reconsolidation update mechanisms. Nature 463:U49–U51
Schwabe L, Wolf OT (2010) Stress impairs the reconsolidation of autobiographical memories. Neurobiol Learn Mem 94:153–157
Schwabe L, Wolf OT (2013) Stress and multiple memory systems: from 'thinking' to 'doing'. Trends Cogn Sci 17:60–68
Schwabe L, Joels M, Roozendaal B, Wolf OT, Oitzl MS (2012) Stress effects on memory: an update and integration. Neurosci Biobehav Rev 36:1740–1749
Shors TJ (2006) Stressful experience and learning across the lifespan. Annu Rev Psychol 57:55–85
Smeets T, Otgaar H, Candel I, Wolf OT (2008) True or false? Memory is differentially affected by stress-induced cortisol elevations and sympathetic activity at consolidation and retrieval. Psychoneuroendocrinology 33:1378–1386
ter Horst JP, de Kloet ER, Schachinger H, Oitzl MS (2012) Relevance of stress and female sex hormones for emotion and cognition. Cell Mol Neurobiol 32:725–735
Ulrich-Lai YM, Herman JP (2009) Neural regulation of endocrine and autonomic stress responses. Nat Rev Neurosci 10:397–409
Vogel S, Klumpers F, Krugers HJ, Fang Z, Oplaat KT, Oitzl MS et al (2015) Blocking the mineralocorticoid receptor in humans prevents the stress-induced enhancement of centromedial amygdala connectivity with the dorsal striatum. Neuropsychopharmacology 40:947–956
Wolf OT (2008) The influence of stress hormones on emotional memory: relevance for psychopathology. Acta Psychol 127:513–531
Wolf OT (2009) Stress and memory in humans: twelve years of progress? Brain Res 1293:142–154

Yang C, Liu JF, Chai BS, Fang Q, Chai N, Zhao LY et al (2013) Stress within a restricted time window selectively affects the persistence of long-term memory. Plos One, 8

Yehuda R (2002) Post-traumatic stress disorder. N Engl J Med 346:108–114

Zhao LY, Zhang XL, Shi J, Epstein D, Lu L (2009) Psychosocial stress after reactivation of drug-related memory impairs later recall in abstinent heroin addicts. Psychopharmacology 203:599–608

Electric Stimulation to Improve Memory Consolidation During Sleep

Diana Campos-Beltrán and Lisa Marshall

Abstract During the last decade the interest in the manipulation of learning and memory by non-invasive techniques in humans has increased dramatically. Many studies focus on sleep as a beneficial or even necessary state for the consolidation of many types of memories. For manipulation methods of transcranial electric stimulation, TMS, deep brain stimulation, cued reactivation, sensory stimulation, especially auditory stimulation have been employed. Techniques closely comparable to the non-invasive human methods have also been developed in rodents. In addition optogenetic tools have enabled the functional causality of very specific pathways to be investigated in a complimentary way. This chapter focuses on effects induced by weak electric stimulation on consolidation during sleep but includes also manipulations aside from weak electric stimulation in a complimentary fashion. As of recent the variability in the efficiency of weak electric stimulation has come into the spotlight. Specifically, the relevance not only of the technical parameters of stimulation, but also of the electrophysiologically defined 'brain state' at the time of stimulation, as well as cognitive features of the individual per se have been addressed.

Keywords Memory consolidation · Sleep · Weak brain electric stimulation · Oscillatory-tDCS · Slow oscillations

Introduction

The common principle of weak electric stimulation (WES) is that the applied current (or field) induces only neuromodulatory responses in the targeted area, i.e. imposing effects comparable to endogenous subthreshold extracellular activity (Priori 2003; Nitsche et al. 2008; Filmer et al. 2014). Briefly, WES mechanisms of action have been linked mostly towards changes in transmembrane polarization

D. Campos-Beltrán · L. Marshall (✉)
Institute of Experimental and Clinical Pharmacology and Toxicology, University of Lübeck, Ratzeburger Allee 160, 23562 Lübeck, Germany
e-mail: Diana.Campos@pharma.uni-luebeck.de

N. Axmacher and B. Rasch (eds.), *Cognitive Neuroscience of Memory Consolidation*, Studies in Neuroscience, Psychology and Behavioral Economics, DOI 10.1007/978-3-319-45066-7_18

affecting the probability of action potential generation and modifying spike-timing (Frohlich and McCormick 2010; Weiss and Faber 2010). WES has been shown to modify changes in the concentration of neurotransmitters and plasticity-related molecules, astrocyte function, and cerebral blood flow (Pelletier and Cicchetti 2015; Monai et al. 2016).

We use the term "weak electric stimulation" here as the broadest definition for stimulation considered to affect neuronal excitability without primarily being strong enough to generate action potentials of a neuron at resting membrane potential level. Notably, this definition stems primarily from studies on cellular responses and associated pharmacological interventions in the 1960s (reviewed in Priori 2003; Nitsche et al. 2008; Marshall and Born 2011) and has with the help of modeling and empirical observations been transferred to the level of transcranial usage. We choose in most instances to use this broader term which encompasses (constant) transcranial direct current stimulation (tDCS), transcranial alternating current stimulation (tACS) as well as oscillatory-tDCS, as the most common basic non-invasive stimulation modes (Marshall and Born 2011; Herrmann et al. 2013). It also encompasses low intensity electric stimulation procedures which have been applied directly to the brain, e.g. epidurally in humans (Del Felice et al. 2015), transcranial or cortical applications of current or use of electric fields in animals (Ozen et al. 2010; Dockery et al. 2011; Binder et al. 2014a; b; Greenberg et al. 2016). The original concept behind applying weak electric stimulation at a specific frequency as in tACS and oscillatory-tDCS is that the endogenous network already revealing a dominant oscillatory mode is most susceptible to the applied depolarizing and/or hyperpolarizing stimulation (Frohlich and McCormick 2010; Ozen et al. 2010; Frohlich 2015). Although tDCS can induce post-stimulation effects (or 'after-effects') if given sufficiently long (e.g. >10 min), for protocols of oscillatory WES indications towards and against post-stimulation entrainment can be found (Marshall et al. 2006; Marshall and Binder 2013; Helfrich et al. 2014; Struber et al. 2015; Vossen et al. 2015). Entrainment refers here to neuronal firing patterns reflecting the waveform of the applied field, i.e. consistent changes in activity (MUA, LFP etc.) occurring at a defined phase of the applied stimulation signal.

A Brief Outline on Modeling Transcranial-WES

Original studies used a 3 layered head model to estimate endogenous field strengths and calculate current strengths necessary to be applied in order to influence the brain. Later, MRI enabled finite element modeling (FEM) to estimate the resultant spread of current delivered by the electrode within the brain based on location, electrode size, shape and amplitude of the applied constant current (Wagner et al. 2007; Datta et al. 2009). The first FEMs used the conductivity of the three-four layers: skin, skull and cerebrospinal fluid and brain, both for humans and rodents (Miranda et al. 2006; Gasca et al. 2011). More recently 6 layer anisotropic FEM has been developed for two and multiple high density electrode configurations (Edwards et al. 2013; Wagner et al. 2014).

Questions most frequently addressed by FE and computational modeling include not only the different tissue conductivities between the stimulation electrode and target tissue, but also cell type specific susceptibility (Radman et al. 2009), the relationship between region of interest and electrode montage, cortical folding and effect of electrode shape on current distribution (Edwards et al. 2013; Woods et al. 2016).

Calculation and usage of FEMs are not at all trivial. Not only are they computationally demanding, but optimal benefit would require MRIs of each individual subject noting the exact topography of electrode placement. Employed stimulation protocols are therefore often empirical, using parameters shown to be efficient in earlier studies. Often stimulation positions and stimulation intensities are selected for which published FEM have been shown to target functionally the selected cortical region. Furthermore, magnitude of the resultant electric field per se cannot a priori be expected to correlate with a certain physiological, moreover cognitive effect. For instance an increment in the duration of anodal tDCS over the motor cortex was found to decrease cortical excitability (Monte-Silva et al. 2013). Also, an increase in magnitude of cathodal tDCs over the motor cortex from 1 to 2 mA was reported to shift its effect from decreased to increased cortical excitability (Batsikadze et al. 2013). Using a very fine-grained FEM, it was shown moreover that over the visual cortex an increase in applied current from 1 to 2 mA under otherwise constant conditions produced an above threshold electric field as opposed to a subthreshold field at the lower intensity (Neuling et al. 2012; Herrmann et al. 2013). Also the brain state during which stimulation is applied can significantly affect cortical responsiveness, as investigated particularly with tACS and oscillatory-tDCS (Kanai et al. 2008; Bergmann et al. 2009; Kirov et al. 2009; Marshall and Born 2011).

Effects on Memory Consolidation in Humans

In brief, memory consolidation represents a process in which information is transferred into an engram which can be retrieved at a later time point (see also chapters by Genzel and Wixted, by Sekeres, Moscovitch and Winocur, and by Rauss and Born). Even within one species cellular and systems consolidation underlie different rules dependent upon memory content. Although successful learning is a prerequisite for encoding, we shall deal here only with manipulations with the specific goal of modulating the consolidation process. Within the last decades the specific role of sleep on memory consolidation has been intensely investigated. Detailed information on cellular processes within the hippocampus have led to a large body of experimental research and theoretical developments on hippocampus-dependent memory consolidation during sleep, which is also reflected in usage of weak electric stimulation (Marshall and Born 2011; Marshall and Binder 2013; Rasch and Born 2013). As reviewed in other chapters of this book the neocortical sleep slow oscillation, thalamo-cortical spindle rhythms, hippocampal sharp wave ripples and their temporal fine-tuning are considered the main brain rhythms associated with hippocampus-dependent memory consolidation in sleep, with at least the slow oscillation also relevant for procedural memory consolidation (Pennartz et al. 2004; Albouy et al. 2013a; Lustenberger et al. 2016) (see chapters by Bergmann and Staresina and by Maier and Kempter). A concept underlying the efficiency of oscillatory WES on memory is that endogenous extracellular field fluctuations are not only a result of previous use-dependent activity, but that the endogenous field fluctuations themselves are of functional relevance (Jefferys 1995; Anastassiou et al. 2011). This concept has determined selected time points for oscillatory WES (Marshall et al. 2004, 2006).

For these reasons a similar protocol applying oscillatory WES at the mean frequency of the sleep slow oscillation during NREM sleep of the first sleep cycle over the (dorsolateral) frontal cortex, at a time and location where the endogenous sleep slow oscillation is most pronounced, has been used repeatedly to modify sleep and hippocampus-dependent memory consolidation. Details for seven of these experiments are presented in a recent meta-analysis by (Tables 1 and 2 in Barham et al. 2016). From the 13 experiments included in total in the meta-analysis, it is concluded that transcranial WES can enhance or disrupt declarative memory consolidation. Memory disruption being shown by 2 studies in which slow oscillatory activity during the first post-learning NREM sleep period was disrupted by applying either a current oscillating in the theta frequency range (5 Hz) or a direction of current flow discordant to the endogenous rhythmic activity (Marshall et al. 2011; Garside et al. 2015). All of the above studies used anodal polarity, aimed toward increasing excitability of the underlying network rhythm. In the one study in which endogenous EEG could be analyzed during a total of 20 min of constant stimulation EEG power below 3 Hz was enhanced as compared to stimulation free intermittent intervals of the equal total length (Marshall et al. 2004). Stimulation artifacts during oscillatory WES hamper corresponding analyses. As indicated above, thalamo-cortical and hippocampal events are grouped during sleep and their phase-dependent coupling has been found to interact with memory consolidation (Niknazar et al. 2015; Pereira de Vasconcelos and Cassel 2015). Initial findings in our laboratory using cross frequency phase amplitude coupling (Canolty et al. 2006) between the applied slow oscillatory stimulation and endogenous spindle activity did indicate that during anodal slow oscillatory stimulation fast spindles at centro-posterior locations can be coupled to the maximum depolarizing phase of stimulation during the third 5 min block of stimulation in NREM sleep. Interestingly slow spindles at Fz similarly coupled to the maximum anodal current. During cathodal stimulation fast spindles behaved opposite by coupling to the peak of minimal current after 2 blocks of 5 min transcranial slow oscillatory tDCS (Fig. 1). The stimulation blocks at other time points of the sleep cycle did not reveal any systematic coupling. Together these results indicate that oscillations of higher frequencies can couple to the ongoing weak oscillatory signal, but this coupling may not reflect the same pattern as is typical for the endogenous brain electric activity.

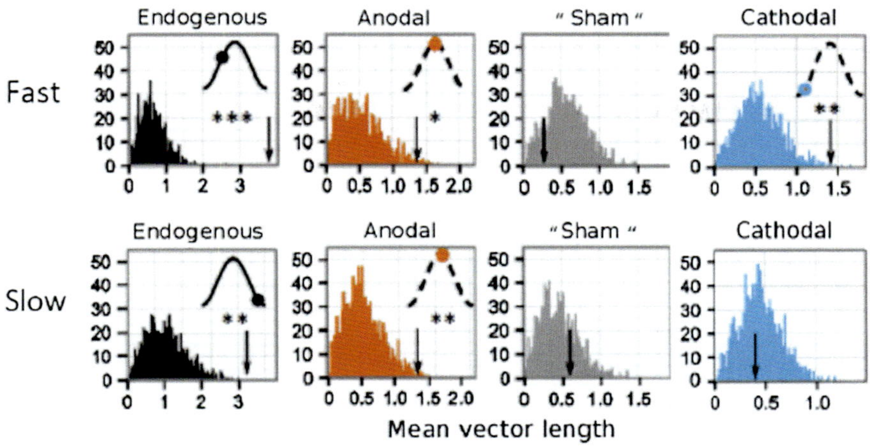

Fig. 1 Phase amplitude cross frequency coupling. Coupling of fast spindles at Pz (*top*) and slow spindles at Fz (*bottom*) to the endogenous slow oscillation. Anodal slow oscillatory-tDCS (SO-tDCS), a virtually generated sham SO-tDCS signal and cathodal SO-tDCS (from *left to right*). Histograms show the distribution of 1000 surrogate mean vectors of the shuffle statistic for each stimulation block. The length of the actual mean vectors is indicated by *arrows*. If the mean vectors are significantly different from zero, they are marked with *asterisks* and the phase to which the amplitude of the spindle frequency band is coupled is shown as a *dot* on *top* of a sketched oscillation (sketches depict positivity up). $*p < 0.05$, $** \; p < 0.01$, $*** \; p < 0.001$, shuffle statistic. (Aumann & Marshall, unpublished results)

Studies in humans specifically targeting with WES procedural memory during the consolidation phase, i.e., a time period subsequent to learning are few. Savic and Meier (2016) reported on 6 studies published up until now on tDCS and consolidation of implicit motor memories (Savic and Meier, 2016). A clear post-learning manipulation was conducted only by Ferrucci et al. 2013 for a short 35 min consolidation interval and by Nitsche et al. (2010) (Ferrucci et al. 2013). The efficiency of tDCS applied applied for 20 min over the cerebellum and a non-cephalic return electrode enhanced both general motor skill learning as well as most importantly sequence specific learning. Nitsche et al. (2010) found that anodal tDCS applied over the premotor cortex during REM sleep was associated with increased motor performance as compared to sham, but efficiency was limited to a time period immediately after tDCS, thus stabilization as assessed by performance measured hours later was not found. Lustenberger and colleagues were recently successful in selectively enhancing sleep spindles in NREM stage 2 using a feedback-controlled current oscillating in the spindle frequency range. The stimulation induced changes in spindle activity also correlated with enhancement in motor memory consolidation (Lustenberger et al., 2016), underscoring the specific relevance of spindle activity for procedural memory consolidation.

For an explicitly learned serial finger tapping task anodal tDCS over the contralateral M1 region was also effective in improving early consolidation during wakefulness (Tecchio et al. 2010). It would be interesting to test whether this performance boost by WES is stable, i.e. maintained when tested several hours later as compared to sham stimulation. A stabilization of the initial boost in motor memory consolidation was reported specifically subsequent to sleep as compared to wakefulness (Nettersheim et al. 2015). Sleep did not lead to a gain in performance.

Effects on Memory Consolidation and Neuroplasticity in Laboratory Animals

In contrast, only two studies in animals have applied weak electric stimulation to modify not only brain electric activity but also sleep-associated memory consolidation (Binder et al. 2014a, b). These studies applied a current oscillating at a frequency of the slow oscillation in rats (~ 1.5 Hz) between 0 and a maximal anodal current. This slow oscillation-tDCS (SO-tDCS) protocol was chosen to be most closely similar in mode, intensity and location to that applied in humans by Marshall et al. 2006. Electrodes were placed in the skull; anodes were applied bilaterally over the prefrontal cortex (AP +3.9 mm from bregma, L $\pm$ 2 mm from midline) and return electrodes over the rat interparietal bone. Maximal current at each bilateral electrode pair was 5.6 μA (0.29 mA/cm^2 current density under the electrode). Stimulation duration was however much shorter (30 s) than in humans to accommodate for the shorter NREM sleep epochs in rodents as compared to humans and was applied after the first occurrence of a 60 s stable NREM sleep period. Rats received daily across 12 days SO-tDCS during sleep after performance in a radial arm maze task. Binder et al. measured a reduced number of errors of re-entries into baited arms within the first four days. This was taken to indicate enhanced rule learning in the stimulation vs. sham group of animals, a function attributed to the prefrontal cortex (Euston et al. 2012). Furthermore, during the last 4 days of the experiment theta EEG power was relatively lower in the stimulation as compared to the sham group during acute stimulation suggesting possible long-term effects (Binder et al. 2014a).

Another study by Binder et al. revealed that SO-tDCS applied during NREM sleep subsequent to an initial sample trial was beneficial for performance on an object place recognition test. In this hippocampus-dependent task, subjects typically explore in the test trial an object longer if it is positioned at a new location as compared to the previous sample trial. However, increased preference for the newly positioned object can only occur if the subject recognizes this location as new. A 24 h retention interval between sample and test trials was used. After a 24-h retention interval the preference index for the displaced object in the test trial was only greater than chance, when animals received SO-tDCS during NREM sleep (Binder et al. 2014b). There was a tendency towards the enhancement of endogenous SO activity (0.8–2 Hz) in the first 10 s after stimulation. Anodal SO-tDCS was similar to the preceding study, however maximal current strength was higher and the prefrontal location differed slightly (AP: 2.5 mm). Both studies indicate at the behavioral and electrophysiological levels effects involving the hippocampal-mPFC network. The primary location at which WES is effective is presumably cortical as suggested by the direct modulation of a mPFC function in Binder et al. 2014a.

Two further studies investigated the effects of oscillatory stimulation over the frontal cortex on neuronal activity in rodents (Ozen et al. 2010; Greenberg et al. 2016). Ozen and colleagues applied weak alternating sinusoid electric fields in anesthetized and behaving rats which led to an intensity-dependent entrainment of neuronal firing during stimulation in widespread cortical areas like the hippocampus. Ozen used a 3-pole placement of stimulation electrodes: two electrodes bilaterally over the temporal bone and one midline on the calvarium above the olfactory bulb (Ozen et al. 2010). With this configuration, they tested different stimulation intensities during multiple epochs of a maximum of 1 min using the same polarity over the temporal locations and the opposite polarity over the most rostral electrode (Ozen et al. 2010). A voltage gradient of 1 mV/mm sufficed to phase-bias neuronal firing. Furthermore, at all intensities tested (0.4–2 V) 1.25 Hz WES entrained neurons (25–50%) during natural sleep, but not during exploration (Ozen et al. 2010). These findings are further evidence, that similarity between endogenous oscillations and stimulation frequencies is important for stimulation efficiency (Ali et al. 2013).

Another study conducted by Greenberg used a 3 pole stimulation configuration adapted from Ozen and colleagues. A central pole electrode was located anteriorly on the midline while the remaining two electrodes were positioned posteriorly on each hemisphere, through which 1.67 Hz sinusoidal electric fields of different voltages and durations were applied (Greenberg et al. 2016). When an oscillatory field of 7.28 $\pm$ 0.69 V peak to peak was applied to anaesthetized rats for 3–6 min, an increase in the slow activity in the hippocampus, the hippocampal sharp-wave ripples and cortical spindles was measured. These effects could lead to a better communication between these structures.

Optogenetic stimulation represent a further method to modulate selected pathways involved in predominant brain rhythms of sleep (Rolls et al. 2011; David et al. 2013; Kuki et al. 2013) making it an interesting tool to affect activity and sleep-associated consolidation performance. Kuki et al. demonstrated for the first time that optogenetics is also able to entrain in a frequency-dependent manner slow oscillations in anesthetized transgenic rats expressing channelrhodopsin-2 as measured by cortical local field potentials (LFP; Kuki et al. 2013). They placed the optical probes unilaterally into the caudal part of the motor cortex. Before the stimulation, LFP was distributed around 0.8 Hz and after the 1 Hz opto-stimulation pulse trains the LFP was concentrated at 1 Hz in both hemispheres (Kuki et al. 2013). David and colleagues demonstrated the importance of thalamic output in tuning the frequency of neocortical sleep slow oscillations (David et al. 2013). First, thalamic output was pharmacologically blocked by tetrodotoxin leading to a reduction in the NREM sleep slow waves and a suppression of spindles. Then, activating channelrhodopsin-2 expressing thalamocortical neurons in rats by applying brief light pulses at frequencies from 0.75 to 1.5 Hz EEG slow oscillations became entrained. A causal relationship between sleep spindles and the stability of NREM sleep is suggested from findings of Kim et al. This group used photostimulation of the thalamic reticular nucleus in mice to generate spindle oscillations similar to spontaneous sleep spindles which led to an increased duration of NREM sleep (Kim et al. 2012) (see also chapter by McDevitt and colleagues).

Finally, by optogenetically silencing medial septum GABAergic neurons in mice using an inhibitory opsin (Archaerhodopsin) theta activity during REM sleep was selectively attenuated which erased novel object place recognition and impaired fear conditioned contextual memory (Boyce et al. 2016) reviving a relevant contribution of REM sleep to memory consolidation.

Some General Concluding Remarks

Studies on oscillatory WES serve to understand systems, cellular and molecular processes involved in the effects of weak extracellular fields. These extracellular field effects are of at least two-fold interest: Firstly, they may simulate the impact of endogenous brain activity and reveal processes associated with rhythmic brain activity; secondly, parameters measured invasively in animals after WES are important for translational research and therapy, for instance to investigate long-term effects of chronic WES application.

Particularly in the context of memory consolidation and neuroplasticity it is important to underscore effects of WES not only on neuronal networks and their coupling but also on neurotransmitter activity such as dopamine, neurotropic signals, intracellular Ca^{2+} concentration of astrocytes and cerebral blood flow (Stagg et al. 2009; Fritsch et al. 2010; Stagg et al. 2013; Filmer et al. 2014; Pelletier and Cicchetti 2015; Leffa et al. 2016; Monai et al. 2016; Woods et al. 2016).

For translational research the question of stimulus intensity is relevant (Neuling et al. 2012). Liebetanz et al., showed brain lesions in rats when a current of 500 μA with a current density higher than 14.29 mA/cm2 was applied for more than 10 min through the skull (Liebetanz et al. 2009). In humans typical current densities at the electrode surface range from about 0.03 to 0.52 mA/cm^2 (Marshall et al. 2006; Nitsche et al. 2008) using current intensities from 0.250 to 2 mA. In behaving animals current densities from 0.3 to 2.04 mA/cm^2 have been reported with current intensities ranging from a total of 10 to 400 μA (de Souza Custodio et al. 2013; Binder et al. 2014a).

In addition to these diverse technical parameters differential susceptibility (or WES efficiency) dependent upon age, health, brain state and individual capabilities and network features not discussed here need to be considered (Berryhill et al. 2014; Prehn-Kristensen et al. 2014; Li et al. 2015; Munz et al. 2015; Westerberg et al. 2015).

Acknowledgements This work was funded by the Deutsche Forschungsgemeinschaft (TR/SFB 654 and SPP1665), and NIH/NSF/BMBF/CRCNS: US-German Collaboration in Computational Neuroscience (German Ministry of Education and Research BMBF, grant 01GQ1008).

References

Albouy G, Sterpenich V, Vandewalle G, Darsaud A, Gais S, Rauchs G, Desseilles M, Boly M, Dang-Vu T, Balteau E, Degueldre C, Phillips C, Luxen A, Maquet P (2013) Interaction between hippocampal and striatal systems predicts subsequent consolidation of motor sequence memory. PLoS ONE 8:e59490

Ali MM, Sellers KK, Frohlich F (2013) Transcranial alternating current stimulation modulates large-scale cortical network activity by network resonance. J Neurosci: Official J Soc Neurosci 33:11262–11275

Anastassiou CA, Perin R, Markram H, Koch C (2011) Ephaptic coupling of cortical neurons. Nat Neurosci 14:217–223

Barham MP, Enticott PG, Conduit R, Lum JA (2016) Transcranial electrical stimulation during sleep enhances declarative (but not procedural) memory consolidation: evidence from a meta-analysis. Neurosci Biobehav Rev 63:65–77

Batsikadze G, Moliadze V, Paulus W, Kuo MF, Nitsche MA (2013) Partially non-linear stimulation intensity-dependent effects of direct current stimulation on motor cortex excitability in humans. J Physiol 591:1987–2000

Bergmann TO, Groppa S, Seeger M, Molle M, Marshall L, Siebner HR (2009) Acute changes in motor cortical excitability during slow oscillatory and constant anodal transcranial direct current stimulation. J Neurophysiol 102:2303–2311

Berryhill ME, Peterson DJ, Jones KT, Stephens JA (2014) Hits and misses: leveraging tDCS to advance cognitive research. Front Psychol 5:800

Binder S, Rawohl J, Born J, Marshall L (2014a) Transcranial slow oscillation stimulation during NREM sleep enhances acquisition of the radial maze task and modulates cortical network activity in rats. Front Behav Neurosci 7:220

Binder S, Berg K, Gasca F, Lafon B, Parra LC, Born J, Marshall L (2014b) Transcranial slow oscillation stimulation during sleep enhances memory consolidation in rats. Brain stimulation 7:508–515

Boyce R, Glasgow SD, Williams S, Adamantidis A (2016) Causal evidence for the role of REM sleep theta rhythm in contextual memory consolidation. Science (New York, NY) 352: 812–816

Canolty RT, Edwards E, Dalal SS, Soltani M, Nagarajan SS, Kirsch HE, Berger MS, Barbaro NM, Knight RT (2006) High gamma power is phase-locked to theta oscillations in human neocortex. Science (New York, NY) 313:1626–1628

Datta A, Bansal V, Diaz J, Patel J, Reato D, Bikson M (2009) Gyri-precise head model of transcranial direct current stimulation: improved spatial focality using a ring electrode versus conventional rectangular pad. Brain stimulation 2:201–207

David F, Schmiedt JT, Taylor HL, Orban G, Di Giovanni G, Uebele VN, Renger JJ, Lambert RC, Leresche N, Crunelli V (2013) Essential thalamic contribution to slow waves of natural sleep. J Neurosci: Official J Soc Neurosci 33:19599–19610

de Souza Custodio JC, Martins CW, Lugon MD, Fregni F, Nakamura-Palacios EM (2013) Epidural direct current stimulation over the left medial prefrontal cortex facilitates spatial working memory performance in rats. Brain Stimulation 6:261–269

Del Felice A, Magalini A, Masiero S (2015) Slow-oscillatory transcranial direct current stimulation modulates memory in temporal lobe epilepsy by altering sleep spindle generators: a possible rehabilitation tool. Brain stimulation 8:567–573

Dockery CA, Liebetanz D, Birbaumer N, Malinowska M, Wesierska MJ (2011) Cumulative benefits of frontal transcranial direct current stimulation on visuospatial working memory training and skill learning in rats. Neurobiol Learn Mem 96:452–460

Edwards D, Cortes M, Datta A, Minhas P, Wassermann EM, Bikson M (2013) Physiological and modeling evidence for focal transcranial electrical brain stimulation in humans: a basis for high-definition tDCS. NeuroImage 74:266–275

Euston DR, Gruber AJ, McNaughton BL (2012) The role of medial prefrontal cortex in memory and decision making. Neuron 76:1057–1070

Ferrucci R, Brunoni AR, Parazzini M, Vergari M, Rossi E, Fumagalli M, Mameli F, Rosa M, Giannicola G, Zago S, Priori A (2013) Modulating human procedural learning by cerebellar transcranial direct current stimulation. Cerebellum (London, England) 12:485–492

Filmer HL, Dux PE, Mattingley JB (2014) Applications of transcranial direct current stimulation for understanding brain function. Trends Neurosci 37:742–753

Fritsch B, Reis J, Martinowich K, Schambra HM, Ji Y, Cohen LG, Lu B (2010) Direct current stimulation promotes BDNF-dependent synaptic plasticity: potential implications for motor learning. Neuron 66:198–204

Frohlich F (2015) Experiments and models of cortical oscillations as a target for noninvasive brain stimulation. Prog Brain Res 222:41–73

Frohlich F, McCormick DA (2010) Endogenous electric fields may guide neocortical network activity. Neuron 67:129–143

Garside P, Arizpe J, Lau CI, Goh C, Walsh V (2015) Cross-hemispheric alternating current stimulation during a nap disrupts slow wave activity and associated memory consolidation. Brain stimulation 8:520–527

Gasca F, Marshall L, Binder S, Schlaefer A, Hofmann UG, Schweikard A (2011) Finite element simulation of transcranial current stimulation in realistic rat head model. Neural Eng 3

Greenberg A, Whitten TA, Dickson CT (2016) Stimulating forebrain communications: Slow sinusoidal electric fields over frontal cortices dynamically modulate hippocampal activity and cortico-hippocampal interplay during slow-wave states. NeuroImage 133:189–206

Helfrich RF, Schneider TR, Rach S, Trautmann-Lengsfeld SA, Engel AK, Herrmann CS (2014) Entrainment of brain oscillations by transcranial alternating current stimulation. Curr Biol: CB 24:333–339

Herrmann CS, Rach S, Neuling T, Struber D (2013) Transcranial alternating current stimulation: a review of the underlying mechanisms and modulation of cognitive processes. Front Human Neurosci 7:279

Jefferys JG (1995) Nonsynaptic modulation of neuronal activity in the brain: electric currents and extracellular ions. Physiol Rev 75:689–723

Kanai R, Chaieb L, Antal A, Walsh V, Paulus W (2008) Frequency-dependent electrical stimulation of the visual cortex. Curr Biol: CB 18:1839–1843

Kim A, Latchoumane C, Lee S, Kim GB, Cheong E, Augustine GJ, Shin HS (2012) Optogenetically induced sleep spindle rhythms alter sleep architectures in mice. Proc Natl Acad Sci USA 109:20673–20678

Kirov R, Weiss C, Siebner HR, Born J, Marshall L (2009) Slow oscillation electrical brain stimulation during waking promotes EEG theta activity and memory encoding. Proc Natl Acad Sci USA 106:15460–15465

Kuki T, Ohshiro T, Ito S, Ji ZG, Fukazawa Y, Matsuzaka Y, Yawo H, Mushiake H (2013) Frequency-dependent entrainment of neocortical slow oscillation to repeated optogenetic stimulation in the anesthetized rat. Neurosci Res 75:35–45

Leffa DT, de Souza A, Scarabelot VL, Medeiros LF, de Oliveira C, Grevet EH, Caumo W, de Souza DO, Rohde LA, Torres IL (2016) Transcranial direct current stimulation improves short-term memory in an animal model of attention-deficit/hyperactivity disorder. Eur Neuropsychopharmacol 26:368–377

Li LM, Uehara K, Hanakawa T (2015) The contribution of interindividual factors to variability of response in transcranial direct current stimulation studies. Front Cell Neurosci 9:181

Liebetanz D, Koch R, Mayenfels S, Konig F, Paulus W, Nitsche MA (2009) Safety limits of cathodal transcranial direct current stimulation in rats. Clin Neurophysiol: Official J Int Fed Clin Neurophysiol 120:1161–1167

Lustenberger C, Boyle MR, Sankaraleengam A, Mellin J, Vaughn B, Flavio F (2016) Feedback-controlled transcranial alternating current stimulation reveals a functional role of sleep spindles in motor consolidation. Current Biology 26(16): 2127–2136

Marshall L, Born J (2011) Brain Stimulation During Sleep. Sleep Med Clin 6:10

Marshall L, Binder S (2013) Contribution of transcranial oscillatory stimulation to research on neural networks: an emphasis on hippocampo-neocortical rhythms. Front Human Neurosci 7:614

Marshall L, Mölle M, Hallschmid M, Born J (2004) Transcranial direct current stimulation during sleep improves declarative memory. J Neurosci: Official J Soc Neurosci 24:9985–9992

Marshall L, Helgadottir H, Mölle M, Born J (2006) Boosting slow oscillations during sleep potentiates memory. Nature 444:610–613

Marshall L, Kirov R, Brade J, Mölle M, Born J (2011) Transcranial electrical currents to probe EEG brain rhythms and memory consolidation during sleep in humans. PLoS ONE 6:e16905

Miranda PC, Lomarev M, Hallett M (2006) Modeling the current distribution during transcranial direct current stimulation. Clin Neurophysiol: Official J Int Fed Clin Neurophysiol 117: 1623–1629

Monai H, Ohkura M, Tanaka M, Oe Y, Konno A, Hirai H, Mikoshiba K, Itohara S, Nakai J, Iwai Y, Hirase H (2016) Calcium imaging reveals glial involvement in transcranial direct current stimulation-induced plasticity in mouse brain. Nat Commun 7:11100

Monte-Silva K, Kuo MF, Hessenthaler S, Fresnoza S, Liebetanz D, Paulus W, Nitsche MA (2013) Induction of late LTP-like plasticity in the human motor cortex by repeated non-invasive brain stimulation. Brain stimulation 6:424–432

Munz MT, Prehn-Kristensen A, Thielking F, Mölle M, Goder R, Baving L (2015) Slow oscillating transcranial direct current stimulation during non-rapid eye movement sleep improves behavioral inhibition in attention-deficit/hyperactivity disorder. Front Cell Neurosci 9:307

Nettersheim A, Hallschmid M, Born J, Diekelmann S (2015) The role of sleep in motor sequence consolidation: stabilization rather than enhancement. J Neurosci: Official J Soc Neurosci 35:6696–6702

Neuling T, Wagner S, Wolters CH, Zaehle T, Herrmann CS (2012) Finite-element model predicts current density distribution for clinical applications of tDCS and tACS. Front psychiatry 3:83

Niknazar M, Krishnan GP, Bazhenov M, Mednick SC (2015) Coupling of thalamocortical sleep oscillations are important for memory consolidation in humans. PLoS ONE 10:e0144720

Nitsche MA, Cohen LG, Wassermann EM, Priori A, Lang N, Antal A, Paulus W, Hummel F, Boggio PS, Fregni F, Pascual-Leone A (2008) Transcranial direct current stimulation: state of the art 2008. Brain stimulation 1:206–223

Nitsche MA, Jakoubkova M, Thirugnanasambandam N, Schmalfuss L, Hullemann S, Sonka K, Paulus W, Trenkwalder C, Happe S (2010) Contribution of the premotor cortex to consolidation of motor sequence learning in humans during sleep. J Neurophysiol 104:2603–2614

Ozen S, Sirota A, Belluscio MA, Anastassiou CA, Stark E, Koch C, Buzsáki G (2010) Transcranial electric stimulation entrains cortical neuronal populations in rats. The Journal of neuroscience: the official journal of the Society for Neuroscience 30:11476–11485

Pelletier SJ, Cicchetti F (2015) Cellular and molecular mechanisms of action of transcranial direct current stimulation: evidence from in vitro and in vivo models. Int J Neuropsychopharmacol 18

Pennartz CM, Lee E, Verheul J, Lipa P, Barnes CA, McNaughton BL (2004) The ventral striatum in off-line processing: ensemble reactivation during sleep and modulation by hippocampal ripples. The Journal of neuroscience: the official journal of the Society for Neuroscience 24:6446–6456

Pereira de Vasconcelos A, Cassel JC (2015) The nonspecific thalamus: A place in a wedding bed for making memories last? Neurosci Biobehav Rev 54:175–196

Prehn-Kristensen A, Munz M, Goder R, Wilhelm I, Korr K, Vahl W, Wiesner CD, Baving L (2014) Transcranial oscillatory direct current stimulation during sleep improves declarative memory consolidation in children with attention-deficit/hyperactivity disorder to a level comparable to healthy controls. Brain stimulation 7:793–799

Priori A (2003) Brain polarization in humans: a reappraisal of an old tool for prolonged non-invasive modulation of brain excitability. Clin Neurophysiol: Official J Int Fed Clin Neurophysiol 114:589–595

Radman T, Ramos RL, Brumberg JC, Bikson M (2009) Role of cortical cell type and morphology in subthreshold and suprathreshold uniform electric field stimulation in vitro. Brain stimulation 2:215–228

Rasch B, Born J (2013) About sleep's role in memory. Physiol Rev 93:681–766

Rolls A, Colas D, Adamantidis A, Carter M, Lanre-Amos T, Heller HC, de Lecea L (2011) Optogenetic disruption of sleep continuity impairs memory consolidation. Proc Natl Acad Sci USA 108:13305–13310

Savic B, Meier B (2016) How transcranial direct current stimulation can modulate implicit motor sequence learning and consolidation: a brief review. Front Human Neurosci 10:26

Stagg CJ, Best JG, Stephenson MC, O'Shea J, Wylezinska M, Kincses ZT, Morris PG, Matthews PM, Johansen-Berg H (2009) Polarity-sensitive modulation of cortical neurotransmitters by transcranial stimulation. J Neurosci: Official J Soc Neurosci 29:5202–5206

Stagg CJ, Lin RL, Mezue M, Segerdahl A, Kong Y, Xie J, Tracey I (2013) Widespread modulation of cerebral perfusion induced during and after transcranial direct current stimulation applied to the left dorsolateral prefrontal cortex. J Neurosci: Official J Soc Neurosci 33:11425–11431

Struber D, Rach S, Neuling T, Herrmann CS (2015) On the possible role of stimulation duration for after-effects of transcranial alternating current stimulation. Front Cell Neurosci 9:311

Tecchio F, Zappasodi F, Assenza G, Tombini M, Vollaro S, Barbati G, Rossini PM (2010) Anodal transcranial direct current stimulation enhances procedural consolidation. J Neurophysiol 104:1134–1140

Vossen A, Gross J, Thut G (2015) Alpha power increase after transcranial alternating current stimulation at alpha frequency (alpha-tACS) reflects plastic changes rather than entrainment. Brain stimulation 8:499–508

Wagner T, Fregni F, Fecteau S, Grodzinsky A, Zahn M, Pascual-Leone A (2007) Transcranial direct current stimulation: a computer-based human model study. NeuroImage 35:1113–1124

Wagner S, Rampersad SM, Aydin U, Vorwerk J, Oostendorp TF, Neuling T, Herrmann CS, Stegeman DF, Wolters CH (2014) Investigation of tDCS volume conduction effects in a highly realistic head model. J Neural Eng 11:016002

Weiss SA, Faber DS (2010) Field effects in the CNS play functional roles. Front Neural Circ 4:15

Westerberg CE, Florczak SM, Weintraub S, Mesulam MM, Marshall L, Zee PC, Paller KA (2015) Memory improvement via slow-oscillatory stimulation during sleep in older adults. Neurobiol Aging 36:2577–2586

Woods AJ, Antal A, Bikson M, Boggio PS, Brunoni AR, Celnik P, Cohen LG, Fregni F, Herrmann CS, Kappenman ES, Knotkova H, Liebetanz D, Miniussi C, Miranda PC, Paulus W, Priori A, Reato D, Stagg C, Wenderoth N, Nitsche MA (2016) A technical guide to tDCS, and related non-invasive brain stimulation tools. Clin Neurophysiol: Official J Int Fed Clin Neurophysiol 127:1031–1048

Memory Manipulation During Sleep: Fundamental Advances and Possibilities for Application

Lucia M. Talamini

Abstract Sleep is critically involved in cognitive functioning through content-specific information processing. Importantly, recent findings consistently show that these processes can be actively manipulated. For instance, by interfering with brain activity directly, or by presenting memory cues during sleep. This chapter will discuss recent advances in this field, considering basic research in both animals and human participants. Initial steps toward possible applications of sleep-related memory manipulations will also be discussed.

Keywords Targetted memory reactivation · Memory manipulation · Memory implantation

An increasing body of evidence suggests that sleep is not only essential for physical health, but also for a variety of mental functions. For instance, learning, memory consolidation and emotional coping all appear to benefit from sleep. The importance of sleep for mental function is not merely due to a general restorative effect of sleep. Rather, it seems to be related to brain-specific processes, whereby information acquired during the day is reprocessed and reorganised (Stickgold 2013). Pivotal contributions to our understanding of these processes, especially regarding the memory function of sleep, have come from studies using manipulations to directly influence these sleep-related processes. Some such manipulations have been aimed at neuronal activity, imposing artificial memories during sleep, some act on oscillatory population dynamics, boosting or interrupting particular EEG patterns, while still others have targeted the system at the sensory level, through presentation of memory cues during sleep. A recent development in this field

L.M. Talamini (✉)
Brain and Cognition Group, Department of Psychology,
University of Amsterdam, Amsterdam, The Netherlands
e-mail: L.M.Talamini@uva.nl

L.M. Talamini
Amsterdam Brain and Cognition (ABC),
University of Amsterdam, Amsterdam, The Netherlands

N. Axmacher and B. Rasch (eds.), *Cognitive Neuroscience of Memory Consolidation*, Studies in Neuroscience, Psychology and Behavioral Economics, DOI 10.1007/978-3-319-45066-7_19

concerns the use of closed-loop stimulation techniques, to precisely target stimuli at particular neural activity patterns. Studies of this type have generated strong support for many aspects of the 'sleep consolidation hypothesis', providing causal evidence for the role of different neural activity patterns in memory consolidation. On a more practical level, the observation that memory content can be manipulated during sleep has invited the exciting idea that such manipulations might be applied to practical purposes. This chapter will start with a brief introduction to sleep-related memory processing and its neural underpinnings. The sections thereafter will present an overview of the literature on memory manipulation during sleep.

Sleep-Related Memory Processing

Sleep-related neural processes supporting memory have been most directly investigated in rodents, using tasks recruiting the hippocampus (typically spatial, episodic or working memory tasks). In such studies, beneficial effects of sleep on memory performance were shown to involve the reactivation of previously encoded neuronal representations in hippocampo-(neo)cortical circuits (Peyrache et al. 2009; Ji and Wilson 2007; Atherton et al. 2015). While this so-called 'replay' activity has been observed in both rapid eye movement (REM) sleep (Louie and Wilson 2001) and nonREM sleep, the importance of sleep for declarative memory has, thus far, mostly been attributed to nonREM sleep (Born et al. 2006) and its signature oscillatory events, slow oscillations (Marshall et al. 2006; Marshall et al. 2011), sleep spindles (Gais et al. 2002; Clemens et al. 2005; Clemens et al. 2006; Schmidt et al. 2006; van der Helm et al. 2011a; Cox et al. 2012; Kaestner et al. 2013; Mednick et al. 2013; Cox et al. 2014a) and hippocampal sharp-wave ripples (O'Neill et al. 2010).

The large-scale coordination of neural activity, necessary for the putative reactivation of distributed memory representations, is thought to be reflected in the spatiotemporal coupling of these brain rhythms (Buzsaki 1996; Clemens et al. 2007; Holleman and Battaglia 2015). Specifically, reactivation of memory representations appears to occur during sharp wave ripple events in the hippocampus (O'Neill et al. 2010). These sharp wave ripples are time locked to the depolarised phase of thalamocortical spindles, which themselves are grouped into the up-state of slow oscillations (Mölle et al. 2002; Cox et al. 2014b; Staresina et al. 2015) (see also chapter by Bergmann and Staresina). The temporal coupling of sharp wave-ripples and spindles is thought to reflect the combined reactivation of hippocampal and (neo)cortical components of memory representations during sleep. The slow oscillations, in turn, have an important role in the functional coupling of cortical networks (Cox et al. 2014b) (Fig. 1). The combined dynamics are thought to underlie the spatiotemporal orchestration of sleep-related memory reactivation and sleep-related information processing in general.

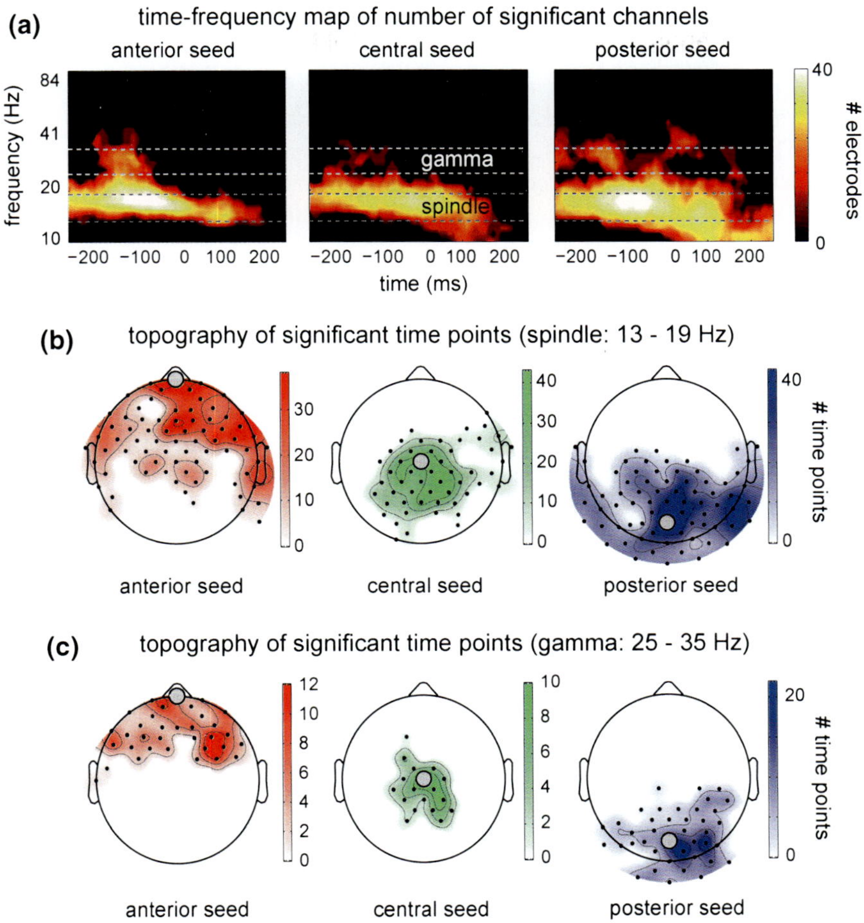

Fig. 1 High-density sleep-EEG (128 channels) was used to assess the spatiotemporal characteristics of oscillatory population dynamics during sleep. The figure shows the spatiotemporal extent of SO-phase effect on power. Three seed electrodes were used for SO detection. **a** Heat maps of clusters with significantly higher power in the up versus down state for an anterior (*left column*), central (*middle column*), and posterior seed electrode (*right column*). Indicated is the number of channels involved at every time–frequency point. Note the apparent presence of two distinct frequency ranges in these clusters, indicated between *dashed lines* and labelled "spindle" and "gamma." **b**, **c** Topographies of these frequency-specific effects (**b** spindles; **c** gamma) reveal that power modulations are highly localized. Channels closer to the seed electrode used for SO detection are involved in the significant cluster on more time points. Similarly, slow oscillations were found to modulate inter-site phase synchrony in the spindle range, as well as beta and gamma activity coupling to spindle phase (Cox et al. 2014b)

Manipulations of Oscillatory Sleep Patterns

Crucial insights regarding the neural underpinnings of sleep-related memory processing have come from studies involving precise manipulations of sleep physiology. The pertaining manipulations specifically enhance or suppress a particular neural activity pattern implicated in sleep-related memory processing. Importantly, studies using such methods have demonstrated the causal role of various sleep-related oscillations in memory consolidation.

A first such experiment applied transcranial, oscillating potentials (0.75 Hz) during early nocturnal non-REM sleep to boost slow oscillations in humans. The method was based on the more general finding that some endogenous brain rhythms can phase lock and resonate to external rhythms (Jefferys and Haas 1982; Hutcheon and Yarom 2000). Transcranial direct current stimulation (tDCS), indeed, led to increased slow oscillation power and slow spindle activity in short (1 min) intervals in between stimulations, and enhanced retention across sleep on a declarative memory task (Marshall et al. 2006; Marshall et al. 2004) (see chapter by Campos Beltran and Marshall). Later studies found that brief sound pulses applied upon detection of a slow oscillation negative half wave peak during nonREM sleep also led to boosting of slow oscillations and spindles, in a short period following the pulse (Ngo et al. 2013a, b) (Fig. 2). As in the study with tDCS, there was a concomitant declarative memory enhancement for material acquired prior to sleep. These studies support a causal involvement of slow oscillations in sleep-related memory processes.

Studies in rodents brought forth similar support for a causal role of spindles and sharp wave-ripples in memory processes (see also chapter by Maier and Kempter). For instance, pharmacological enhancement of spindle occurrence improved memory retention across sleep (Kaestner et al. 2013; Mednick et al. 2013) (see chapter by McDevitt, Krishnan, Bazhenov and Mednick), while interruption of sharp wave-ripples by closed-loop electrical stimulation interfered with rats' ability to recall spatial memory information (Girardeau et al. 2009; Ego-Stengel and Wilson 2010). Thus, a large body of physiological findings supporting the idea that memory consolidation is reflected in spatiotemporal sharp wave-ripple, spindle, slow oscillation coupling, was strengthened by evidence linking each of these three oscillatory phenomena causally to memory consolidation.

Targeted Memory Manipulation During Sleep

In the previous section, we saw that memory retention over sleep can be enhanced through manipulations of sleep physiology. A particularly exciting line of research concerns the possibility to target individual memories in the sleeping brain and manipulate these specifically. Indeed, according to recent studies, specific memories can be enhanced by presentation of memory cues during sleep. It is even possible to

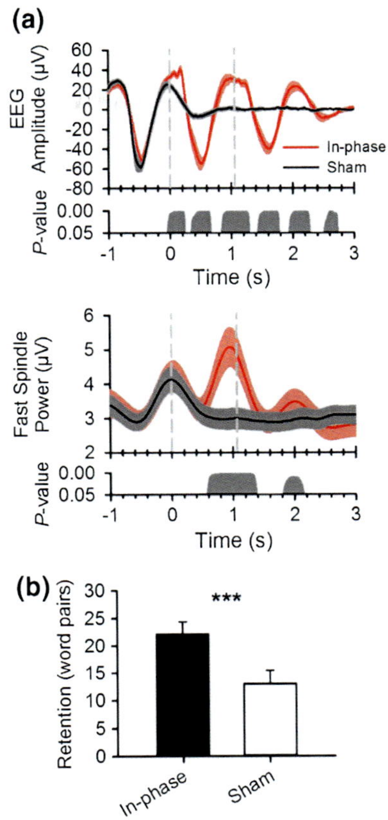

Fig. 2 Auditory stimulation in phase with slow oscillation up-states boosts sleep rhythms and promotes declarative memory. **a** During a so-called In-phase auditory stimulation condition (shown in *red*) participants were subjected to two brief clicks (50-ms *pink* noise) whenever online detection of a spontaneous slow oscillation occurred. These clicks were delivered such that they concurred with the next two up-coming slow oscillation up-states (*vertical grey lines*). This led to prolonging of the on-going slow oscillatory activity in comparison to a Sham control condition without stimulation (*black lines*, *upper panel*) and increased fast spindle power (12–15 Hz) phase-locked to the induced slow oscillation up-states (*lower panel*). *Bottom traces* indicate significant differences between conditions. **b** To assess effects on hippocampal memory consolidation subjects learned a declarative memory task of 120 paired associate words and performed cued recall before and after sleep. Retention (post-sleep minus pre-sleep recall performance) was distinctly higher during In-phase stimulation (*black bar*) than in the Sham condition (*empty bar*); *** $P < 0.001$

instil entirely new memories during sleep in the form of conditioned responses (Arzi et al. 2012). We might refer to such procedures, which target specific memory content during sleep, as 'targeted memory manipulations'.

The fact that sleep-related processing can be influenced by external cues may seem surprising, given the strongly reduced access of sensory input to thalamo-cortical circuits. However, despite this dampening, the sleeping brain maintains a lingering receptiveness to external stimuli. Obviously, external stimuli can still lead to arousal and awakening, provided they are sufficiently strong. More interestingly, the extent to which external stimuli will be perceived during sleep depends on stimulus properties like familiarity and personal meaning. For instance, irrelevant, familiar stimuli, like household or street noises, may not disturb a sound sleeper, while even soft crying of a baby may quickly awaken the parents. This suggests that at least some level of sensory stimulus processing persists during sleep.

Evidence for this notion comes from studies evaluating sleeping subjects' brain responses (typically evoked potentials) in response to simple auditory or tactile stimuli (Bastuji and García-Larrea 1999). The general tenet from this body of work is that, while the sleeping brain responds differently from wakefulness, it retains some residual capacity for performing simple processing relating to stimulus salience, novelty and significance. Recent findings suggest that the sleeping brain can respond in a stimulus-specific manner even to more complex stimuli, and that, using this propensity of the sleeping brain, information processing can be manipulated at the content level. For instance, in an early experiment, object locations were remembered better if an odour, present during learning, was also present during subsequent slow wave sleep (SWS) (Rasch et al. 2007) (see also chapter by Shanahan and Gottfried). This effect was later shown to be odour specific (Rihm et al. 2014). That is, the memory enhancing effect of odour presentation was dependent on the prior association of the odour with the items.

The choice for odour cues in these early experiments was guided by the fact that odour stimuli have a low chance of leading to arousals during sleep compared to, for instance, auditory stimuli. Furthermore, the olfactory system bypasses the thalamus, connecting directly to olfactory cortex, which, in turn, projects heavily to the hippocampus (see chapter by Wilson, Kondrakiewicz and Barnes). Thus odour cues, compared to other sensory cues, may have a better chance of reaching the hippocampus and leading to cued reactivation of hippocampus-dependent memory traces. Nonetheless, several studies have show that auditory stimuli can also be used as memory cues during sleep (Rudoy et al. 2009; Antony et al. 2012; Oudiette et al. 2013; Schreiner and Rasch 2014; Schreiner et al. 2015) (see chapter by Schreiner, Lehmann and Rasch). From an experimental perspective, the advantage of auditory cues is that they can more easily be used to create multiple unique cue-to-memory pairings, allowing experiments in which individual memories are specifically reactivated.

This property was exploited in an experiment in which subjects learned object-location associations while hearing characteristic object sounds (Rudoy et al. 2009). Part of these sounds were presented again during nonREM sleep. Upon waking, subjects recalled the locations associated to the sleep cues more accurately than other locations for which no cues were provided (Fig. 3). In another study, auditory cues presented during SWS were used to enhance skill learning (Antony et al. 2012). Specifically, the production, using a keyboard, of one of two practiced melodies, was enhanced by presenting that melody during a nap. Finally, two recent studies showed that verbal cueing during sleep can boost vocabulary learning (Schreiner and Rasch 2014; Schreiner et al. 2015). In the latter experiments sleeping

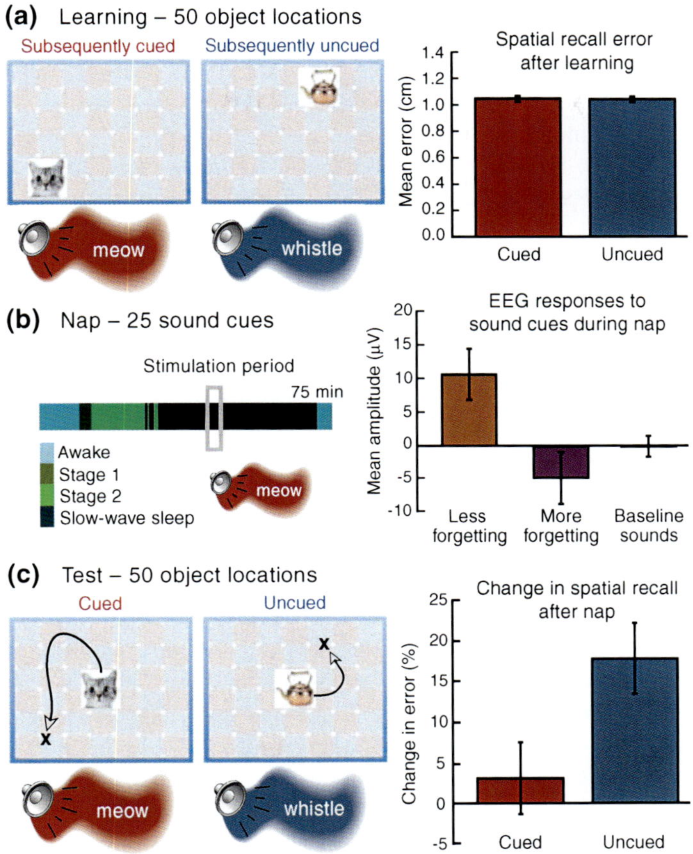

Fig. 3 Rudoy and colleagues (2009) manipulated memory processing during sleep using sounds. **a** In the initial stage of the experiment, individuals learned object-location associations on a grid while hearing the corresponding object sounds, using repeated presentations in a drop-out learning procedure. Accuracy was measured for each object as the distance between the remembered location and the original location, and accuracy at the conclusion of learning was matched for two sets of 25 objects, one set that was subsequently cued by the sounds during sleep and one that was not cued. **b** Sleep-staging data for a representative participant, showing the timing of the 3.5-min sequence of 25 sound cues that were presented at a very low level so as to not disturb ongoing sleep. Neural activity elicited by the sounds during sleep predicted later memory performance. EEG analyses showed that potentials at 400–800 ms after sound onset differed according to level of forgetting for corresponding object locations. **c** After the nap, individuals attempted to place each object in its correct location (*arrows* simulate motion of objects as individuals complete the task). Better spatial-location retention for cued compared to uncued objects was reflected by a smaller change in error, indicating that targeted memory reactivation through sounds presented during slow-wave sleep selectively improved spatial memory for corresponding objects

subjects were re-exposed, during nonREM sleep, to auditory presentations of foreign words learned prior to sleep. Upon waking, the correct translation of the foreign words to the mother tongue was more often recalled for sleep-cued respective to non-cued words (see Fig. 1 in the chapter of Schreiner, Lehman and Rasch).

The combined studies show that successful memory cueing during sleep can be achieved using different sensory cues, including olfactory and auditory ones. Such cueing can benefit memory performance over a variety of memory tasks, spanning the declarative and procedural memory domains. It might be noted that some of the above declarative memory studies administered memory cues in nonREM sleep (Rudoy et al. 2009; Schreiner and Rasch 2014; Schreiner et al. 2015), rather than SWS specifically (Rasch et al. 2007), which has more consistently been implicated in episodic/declarative memory consolidation (Stickgold 2013; Cox et al. 2012; Rasch and Born 2013; Sweegers and Talamini 2014). At the same time, the single study regarding skill learning (Antony et al. 2012) presented cues during SWS, which is not the sleep stage most strongly implicated in skill learning. It is as yet not clear whether cued memory reactivation during sleep benefits differently from cueing in different sleep stages and how this might depend on the type of cue and the type of memory task.

More in general, it might be noted that the size of the memory effects in these studies is typically modest and that memory benefits have not been observed in all studies. Interestingly, the larger part of studies showing benefits of sleep-cueing on post-sleep memory performance used declarative memory tasks with a clear semantic component to retrieval performance (Schreiner and Rasch 2014; Schreiner et al. 2015). This is also the case for studies showing enhanced memory retention following slow oscillation boosting (Marshall et al. 2006; Marshall et al. 2004; Ngo et al. 2013a, b). Examples of memory tasks used in these studies are paired associate tasks with semantically related associates and vocabulary learning tasks using related languages. On the other hand, studies using tasks relying more stringently on episodic memory did not always lead to behavioural memory improvement (Cox et al. 2014a; van Dongen et al. 2012). A study in our lab that used a strictly hippocampus-dependent word to location association task (similar to Cox et al. 2014a) and was optimized to detect possible effects on memory performance, also did not uncover such effects (Talamini and Cox, unpublished observations). While many experimental factors may underlie these differential outcomes, one possibility is that semantic memory effects may be induced through relatively local reactivations (e.g. within language networks for a verbal task), while episodic memory enhancements may require reactivation of larger scale networks, encompassing the hippocampus and widespread neocortical areas. Perhaps, effects based on relatively local processing are easier to achieve. Thus, while initial findings on targeted memory reactivation are highly exciting, many parameters determining the effectiveness of such procedures remain to be investigated.

Implanting New Memories

While the above studies regard the manipulation of existing memories during sleep, an equally intriguing question is whether novel information, processed during sleep, can leave lasting memory traces. A few older studies found either no evidence for sleep-learning of verbal material (Emmons and Simon 1956; Wood et al. 1992) or had methodological difficulties (Fox and Robbin 1952). A recent study, involving slow oscillation-upstate-locked presentation of real world sounds during sleep, also didn't find traces of lasting memory formation (Cox et al. 2014c). However, it appears that a simpler form of learning, classical conditioning, can take place during sleep. In a study with human participants, pleasant and unpleasant odours were paired with different tones during sleep. During ensuing wake, subjects displayed sniff responses to tones alone, suggesting an implicitly learned association with the odours (Arzi et al. 2012). Another study showed that conditioned responses can be extinguished during sleep (Hauner et al. 2013) (see also chapter by Shanahan and Gottfried). Human subjects underwent olfactory contextual fear conditioning during wake. Re-exposure to the odorant context during slow-wave sleep promoted stimulus-specific fear extinction, with parallel reductions of hippocampal activity and reorganization of amygdala ensemble patterns. This somewhat unintuitive finding may be understood considering sleep's role in dampening the emotions associated to a memory (Talamini et al. 2013; Gujar et al. 2011; van der Helm et al. 2011b; Yoo et al. 2007; Pace-Schott et al. 2011; Deliens et al. 2012; Wassing et al. 2016) (see also chapter by Cunningham and Payne). Sleep may thus benefit declarative consolidation, while reducing memories' emotional tone (Hofman et al. 2010).

A possible explanation for the differential results of studies investigating new learning during sleep may again be related to the neural networks underlying memory formation for the presented materials. While the successful conditioning and fear extinction attempts probably relied on subcortical and allocortical brain circuitry, the failure to establish memories for more complex stimuli, including verbal material and real world sounds, may be related to their dependence on neocortex.

Neural Mechanisms Underlying Targeted Memory Manipulation During Sleep

The results of targeted memory manipulations during sleep, suggest that at least part of sleep's benefits for memory are due to processes that act at the local level, on specific memories or parts thereof. Given replay findings in rodents (Peyrache et al. 2009; Ji and Wilson 2007; Atherton et al. 2015) and some limited evidence regarding synaptic plasticity during sleep (Sadowski et al. 2016; Frank 2015), a plausible hypothesis holds that the benefits arise through cue-induced reactivation of specific memory representations, leading to synaptic modifications that stabilise the pertaining ensembles. Direct evidence for this putative mechanism has been notoriously difficult to obtain. However, recent experiments, some using advanced experimentation techniques, have now provided convincing support.

In one of these studies, rats were trained on a sound to location association task, while neurons in the hippocampus were being recorded. Presentation of the sounds during subsequent sleep, biased reactivation events toward replaying the spatial representation associated with the pertaining cues. These results showed that the content of neuronal sleep replay can be manipulated, in a stimulus-specific manner, by external stimulation (Bendor and Wilson 2012).

A study in our lab on human subjects addressed this same issue at the macroscopic level (Cox et al. 2014a). During learning of word–location associations, words presented in the *left* and *right* visual hemifields were paired with different odours. Presentation of a single odour during a subsequent nap, aimed to selectively reactivate the subset of the studied material presented in the associated hemifield. We found topographically restricted fast spindle responses to memory cues, over posterior parietal areas contralateral to the cued hemifield, i.e. the areas where the visuospatial location of the reactivated material is know to be represented (Fig. 4). These results showed that memory cues can specifically reactivate associated memories, reflected in amplified fast spindling in the cortical area where the memory is stored. The combined experiments provide strong evidence that external cues can indeed reactivate specific, cue-associated memory representations at the level of both the hippocampus and neocortex.

While the two aforementioned studies link sleep-cueing to memory reactivation, two other studies, in rats, addressed the next step in the hypothesised sleep consolidation mechanism, asking whether sleep-reactivation of specific memories can lead to their modification (Barnes and Wilson 2014; de Lavilleon et al. 2015). In one of these studies, hippocampal CA1 was recorded while rats were exploring a maze. During subsequent sleep, the spontaneous activity of a well-identified place cell was used to trigger stimulations of the medial forebrain bundle (MFB) and, therewith, a neurochemical reward state (de Lavilleon et al. 2015). This closed-loop procedure created a strong post-sleep place preference for the associated place field in the maze. The study thus showed that place cells' reactivation activity during sleep still conveys relevant spatial information, which is functionally significant for navigation. Moreover, the findings suggest that existing place memories can undergo lasting modification during sleep. This latter point should be considered with some caution, as the artificial medial forebrain bundle stimulation, which likely increased acetylcholine and dopamine transmission, may induce a plasticity state that does not naturally occur during sleep.

The other study used olfactomimetic electrical stimulation of olfactory neurons in piriform cortex to create artificial odour memories (Barnes and Wilson 2014; see Fig. 5 of the chapter by Wilson, Kondrakiewicz and Barnes). Two different stimulation patterns were used as CS+ and CS−, respectively, in a fear conditioning procedure. Imposed replay of artificial olfactory memories during post-training SWS enhanced the strength of fear conditioning upon subsequent awake testing, suggesting that the cue reactivated the concomitant memory of the unconditioned fear stimulus and the association was strengthened. On the other hand identical replay during waking, as expected, induced fear extinction. Interestingly, imposed SWS replay of either the CS− or a random olfactomimetic stimulus, induced

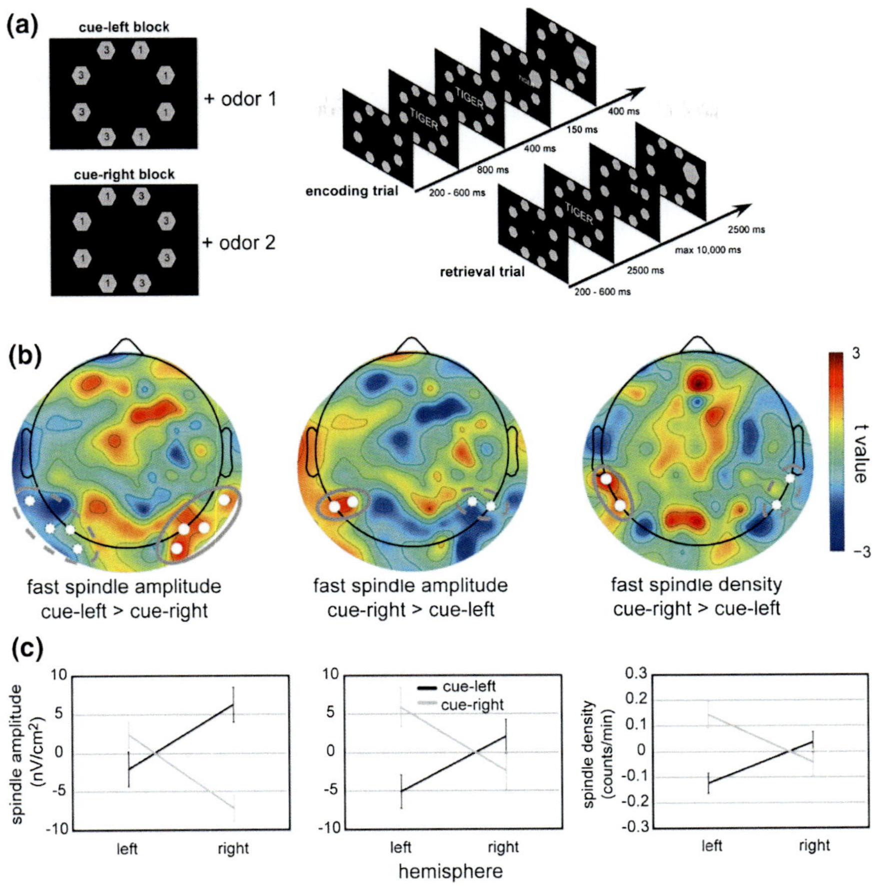

Fig. 4 High-density sleep-EEG with 128 channels was applied to show local sleep spindle modulations in relation to reactivation of specific memories (Cox et al. 2014a). **a** During learning of word-location associations, words presented in the left and right visual hemifields were paired with different odours. During a subsequent nap, a single odour was presented to selectively reactivate a subset of the studied material in sleeping subjects. *Left panel*: Schematic representation of word-location assignments for cue-left (*top*) and cue-right (*bottom*) word blocks. The numbers in each of the 8 locations indicate how many words were associated with that position. A block consisted of 16 words in total and was consistently paired with one odour. *Right panel*: Encoding (*top*) and retrieval (*bottom*) trial timing. **b–c** The reactivation of *left-sided* and *right-sided* stimuli resulted in differential fast spindle modulation topographies, with increased fast spindle amplitude (*left* and *middle panel*) and fast spindle density (*right panel*) in parieto-occipital areas contralateral to cueing side. **b** T maps showing significant cueing side effects. *Grey solid ovals mark* significant clusters; significant electrodes in each cluster are shown as *white dots*. *Grey dashed ovals* depict contralateral 'mirror' clusters used for assessing hemisphere-dependence of cueing side effects in (**c**). **c** Significant crossover interactions between cueing side and hemisphere for all clusters indicate bilaterally symmetrical spindle modulations in response to memory cues

generalisation of the fear memory to those artificial patterns. This is in line with the notion that information can be recombined during sleep replay (Gupta et al. 2010), possibly promoting new insights, problem solving and other creative processes. Of note, the main finding of this experiment appears opposite to that of the olfactory contextual fear conditioning experiment in humans, described in the previous section (Hauner et al. 2013). While the reason for this discordance is not apparent, it might be noted that the main effect in the rat study is considerably larger than that in the human study.

Targeting Memory Manipulations to Receptive Sleep Windows

As indicated earlier, some of the strongest evidence regarding episodic memory consolidation points to SWS and, at a more fine grained level, to temporally coupled slow oscillations (Marshall et al. 2006; Marshall et al. 2011), sleep spindles (Gais et al. 2002; Clemens et al. 2005; Clemens et al. 2006; Schmidt et al. 2006; van der Helm et al. 2011a; Cox et al. 2012; Kaestner et al. 2013; Mednick et al. 2013), and sharp-wave ripples. At first glance, signs of overall synchronised neural activity and reduced functional connectivity in cortical networks during SWS, seem in discordance with the idea that coordinated, large scale neural processes could be taking place. However, studies with higher temporal resolution suggest that neural networks during deep sleep fluctuate between two states, regulated by the phase of slow oscillations. Indeed, membrane potentials of neocortical circuits alternate between depolarized up-states and hyperpolarized down-states. The slow oscillation up-states hold most of the faster activity (Mölle et al. 2002; Cox et al. 2014b; Valderrama et al. 2012), are related to higher excitability (Bergmann et al. 2012) and plasticity (Hoffman et al. 2007), and feature complex interregional functional coupling of cortical networks (Cox et al. 2014b) (Fig. 1). On the other hand, all these characteristics are strongly reduced during the hyperpolarised down-states. In view of the above, slow wave up-states may represent windows of opportunity for memory consolidation and perhaps other forms of higher order information processing, while down-states reflect a deeply inhibited network state. As such, memory manipulations during sleep may work best if they are targeted to slow wave up-states.

To investigate this intriguing possibility, we recently developed a closed-loop procedure to target stimuli to any selected phase of on-going slow oscillations (Fig. 5) (Cox et al. 2014c). This procedure can be used for precise manipulation of the sleeping brain, while simultaneously brain activity is recorded at high spatial and temporal resolution through high-density EEG polysomnography. Phase targeting allows stimuli to be presented repeatedly, while being consistently time-locked to a specific oscillatory phase. This enables new and sophisticated experimentation. For instance, if indeed slow oscillations regulate windows of opportunity for neocortical information processing, a memory cue for neocortically-based information may need

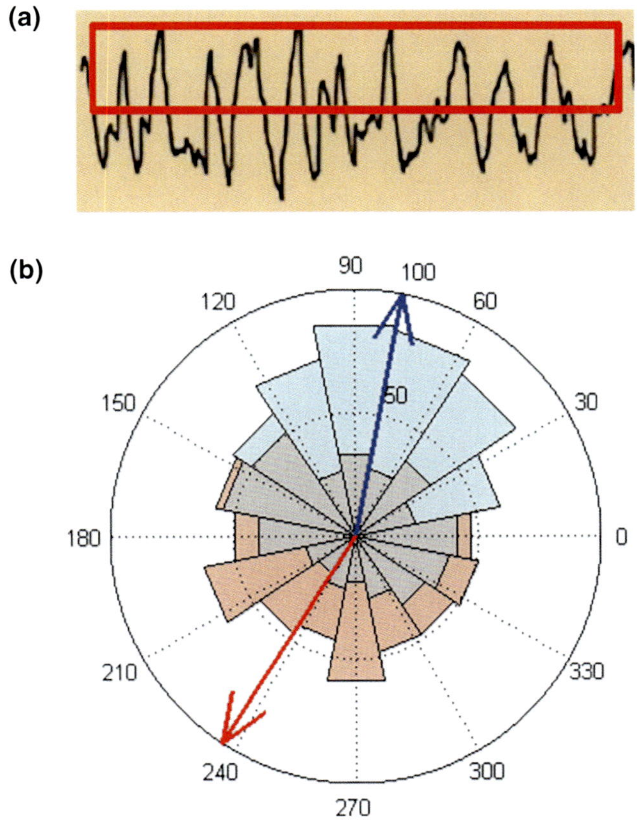

Fig. 5 Cox and colleagues developed a technique for presenting stimuli phase-locked to ongoing EEG oscillations (Cox et al. 2014c). **a** The procedure was used to present real world auditory stimuli phase-locked to either slow oscillation up-states (*red box*) or down-states, to test differential information processing in these two states (**b**) Phase targeting performance for up-state (*blue*) and down-state (*red*) of slow oscillations is shown in a rose plot (averaged over 12 subjects; N = 171). *Arrows* indicate average phase angle; 90° corresponds to the peak of the up state, 270° to the trough of the down state. Significant clustering of auditory stimulus presentations to the up-state (*blue*) or down-state (*red*) of slow oscillations was demonstrated (Rayleigh Test for 'non-uniformity' $*P < 1e{-}13$)

to arrive sufficiently often in the right slow oscillation phase in order to boost consolidation of the cued memory. After all, the build-up of lasting memory representations is a function of the number of times a representation has been activated. Similarly, repeated phase-optimized stimulus presentation could benefit memory modification or the induction of new learning. An ulterior benefit of phase-locked stimulation is that phase effects on neural processing can be analysed much more effectively than with random stimulation and post hoc stimulus sorting, as trials will be more tightly grouped in the phase(s) of interest and arbitrary stimulus sorting criteria are avoided. Furthermore, targeting of stimuli to optimized receptive

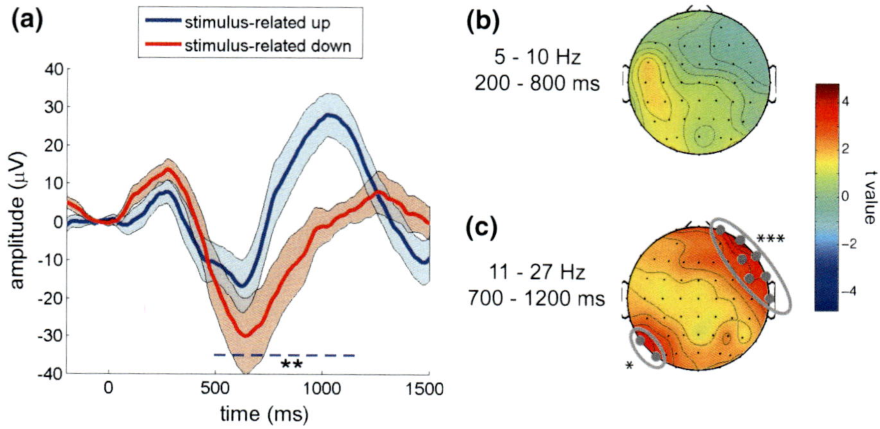

Fig. 6 Slow oscillation phase-dependent stimulus processing. **a** Differential stimulus-evoked waveforms for up (*blue*) and down (*red*) state-presented sound stimuli for frontal channel Fz. **b** Early stimulus-evoked theta power (coinciding with induced down-states) did not differ reliably between up- and down-targeted stimuli. **c** Late spindle/beta power (coinciding with induced up-states) was higher for up-targeted sounds than for down-targeted stimuli across the entire scalp, reaching significance in a right fronto-temporal area (electrodes Fp2, F8, FC6, T8, AF8, F6 and FT8), and a left parietal region (P7 and PO7). Reliable differences are indicated with: * $P < 0.025$; ** $P < 0.01$; *** $P < 0.001$. (From Cox et al. 2014c)

windows avoids unnecessary sleep disturbance with stimuli that have a poor chance of leading to memory change. Thus, stimulus locking to slow oscillation phase maximizes the opportunity to demonstrate slow oscillation phase-dependent learning and memory processing.

In a first study using this technique, we examined how cortical networks respond to real-world sound stimuli as a function of slow oscillation phase (Cox et al. 2014c). The sounds were repeatedly presented to sleeping subjects, targeted to either slow oscillation up or down-states, with a consistent sound to phase relation. Brain-wide responses to up and down-state stimuli, corrected for spontaneous slow oscillation dynamics, were evaluated. Up-state sounds induced a second up-state that occurred sooner in time, had higher amplitude, and featured enhanced spindle and beta activity (Fig. 6). Responses were relatively widespread, but largest over frontal cortex, the area that also provided the signal for phase targeting (Fz). These findings suggest enhanced stimulus processing with up-state-targeted stimulation.

Of note, both up- and down-state stimuli evoked K-complex-like responses, i.e. high-amplitude EEG patterns that are in many ways similar to slow oscillations. Compared to a no-stimulation condition, stimulation enhanced theta power around 300 ms after stimulation onset, corresponding to the sharp K-complex 'down-state', and induced a spindle/beta boost during the subsequent up-state, from 500 ms onwards (Fig. 7). In line with these findings, two other studies also found that the presentation of stimuli during sleep, be they simple tones or meaningful stimuli, was accompanied by a temporary rise of power in the theta (Schreiner and Rasch 2014)

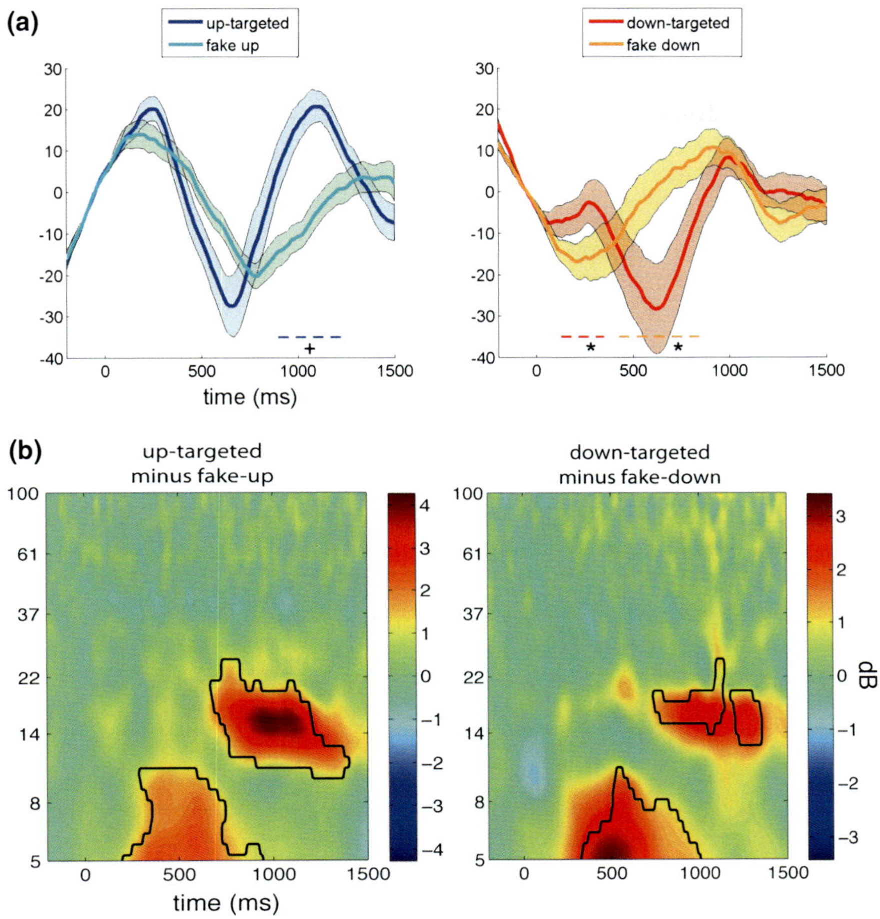

Fig. 7 Grand-average event-related potentials (**a**) and time-frequency power plots (**b**) for slow oscillation up-state and down-state-targeted auditory stimuli during sleep for frontal channel Fz. **a** Up-targeted (*left panel*) and down-targeted (*right panel*) events are shown against 'fake up' and 'fake-down' events, respectfully. The fake events are non-stimulated slow oscillations, where the target phases are matched to those of the stimulation events. *Error shading* indicates standard error of the mean. *Dashed coloured lines* near *bottom* signify time period of significant difference at cluster-level, *with colour* indicating the more positive waveform (+$P < 0.05$; * $P < 0.025$; ** $P < 0.01$; *** $P < 0.001$). **b** Time-frequency power difference plots corresponding to the *upper ERP panels*. Both up-targeted stimulus delivery compared with fake-up events (*left*) and down-targeted sounds compared to fake down events (*right*) elicit theta and spindle/beta activity relative to fake-down events. The spindle response is larger for the up-state stimulation condition. (From Cox et al. 2014c)

and spindle/gamma range (Ngo et al. 2013b; Schreiner and Rasch 2014). The induced theta power (in either phase condition) may directly reflect the induction of the K-complex like down-state, as the frequency of this relatively sharp deflection encompasses the low theta range.

Fig. 8 **a** Experiment time line. **b** (i) The main experimental protocol. Olfactory aversive ▶ partial-reinforcement trace conditioning between cigarette odour (Cig) and unpleasant odours. Stimuli were generated in blocks of 30 trials: 10 reinforced trials with unpleasant odour of ammonium sulphide (AS) (*yellow*), 10 reinforced trials with unpleasant odour of rotten fish (RF) (*brown*) and 10 non-reinforced trials (cigarette odour alone) (*grey*). (ii) The non-conditioned control protocol. Cigarette and unpleasant odour administration in randomized order such that the cigarette odour and unpleasant odours were non-conditioned. **c** (i) Percent change in smoked cigarettes in the first (days 1–3) and second half (days 5–7) of the experiment following conditioning during stage 2 sleep (*black*), REM sleep (*grey*), and wake (*outline*). (ii) Percent change in smoked cigarettes in the first (days 1–3) and second half (days 5–7) of the experiment following olfactory aversive conditioning (*black*), and non-conditioned odours (*striped*) administration during Stage 2 sleep. * $p < 0.05$, ** $p < 0.01$, *** $p < 0.005$. (From Arzi et al. 2014)

Although up-state stimuli in our study induced a network state that should be conducive to information processing and plasticity, enduring memory for the presented stimuli was not observed. Thus, it may not be possible to induce entirely new declarative memories during sleep. Alternatively, the limited number of presentations per stimulus (±3) in this nap study may not have been sufficient to effectuate notable memory build up. Current closed-loop experiments in our lab employ more frequent stimulation and use the procedure to address various aspects of sleep-related memory processes, including the possibility to enhance or depress sleep-related memory consolidation through phase-locked presentation of reactivation cues.

Closed-loop procedures, such as the one described above, present new and powerful experimental tools that are generating considerable interest in the field of sleep research. In general, closed-loop procedures allow the presentation of stimuli in alignment with specific patterns in on-going biophysical signals. The methods typically involve fast algorithms performing near real-time analysis of the pertaining signal, to detect the pattern of interest, and fast hardware/software loops to deliver stimuli in temporal concordance with a detected pattern.

In sleep and cognition studies, the first closed-loop methods were used to interrupt sharp wave-ripples in rodents (Girardeau et al. 2009; Ego-Stengel and Wilson 2010). Next, procedures in humans were developed to present auditory stimuli to particular slow oscillation phases (Ngo et al. 2013a, b; Cox et al. 2014c; Bergmann et al. 2012; Santostasi et al. 2016). The first of these phase-targeting methods was based on a simple, inflexible, procedure that does not adapt to ongoing fluctuations in slow oscillation frequency (Ngo et al. 2013a, b; Bergmann et al. 2012) and has not been validated in terms of phase targeting accuracy. Our own (Cox et al. 2014c) and another, very recently reported, method (Santostasi et al. 2016) are based on modelling and predicting on-going slow oscillation activity. These methods do adapt to the natural fluctuations in slow oscillation frequency and have, at least to some extent, been validated. Ongoing developments in our lab have recently resulted in a new, thoroughly validated, closed-loop stimulation procedure for oscillatory phase targeting, which is faster and more accurate than any previously reported method (Talamini et al. 2016). The new procedure enables interference with oscillatory dynamics much faster than the slow oscillation, opening up new and exciting research possibilities both within and beyond the field of sleep.

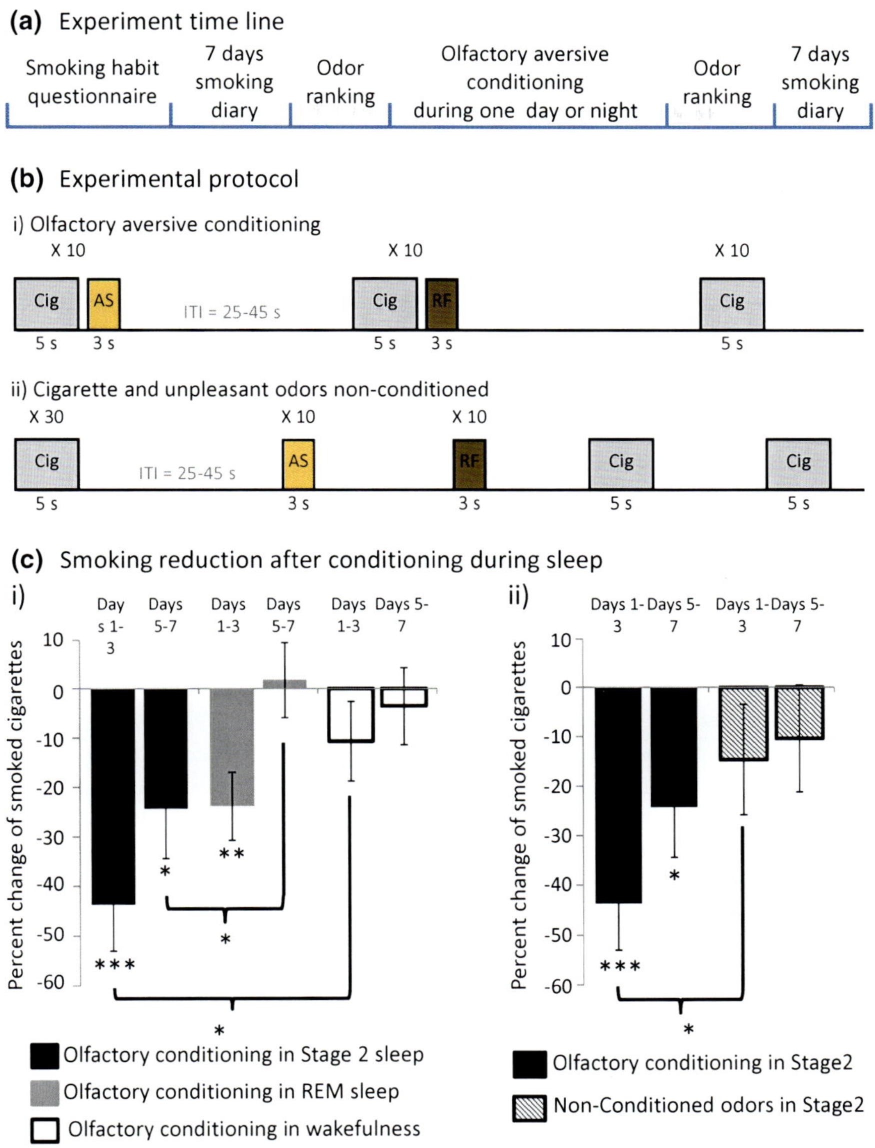

Applications

As discussed in previous sections, sleep-related memory manipulations have brought forth crucial evidence regarding the role of sleep in cognition. However, they also instigated the exiting idea that sleep's hidden processing potential might be harnessed for clinical and educational purposes. It might, for instance, be possible to modify traumatic memories, extinguish unadaptive behaviours, or even instil entirely new thoughts and ideas.

Concrete steps toward clinical application regard the use of aversive conditioning during sleep to reduce addictive behaviour in cigarette smokers (Fig. 8) (Arzi et al. 2014). Specifically, cigarette smoke odour was paired with a highly aversive odour in a partial conditioning protocol during either wake, stage 2 sleep or REM sleep. Conditioning during sleep, in particular during stage two, reduced cigarette smoking in the week after the conditioning night by approximately 30%, on average. Conditioning during REM was less effective (11% reduction) and conditioning during wake was not effective at all. The effects of conditioning tapered off in the course of the post-conditioning week. Nevertheless, these initial observations following just one conditioning night mark sleep-related conditioning as a promising strategy in the treatment of addiction.

Other possible clinical applications of sleep-related memory manipulation regard the modification of trauma memories. Perhaps the presentation, during sleep, of cues associated to beneficial therapy sessions, or 'safe circumstances', could strengthen these memories and benefit recovery. Alternatively, it might be possible to selectively 'erase' maladaptive memories, or to restore, sleep's normal role in the reduction of memories' emotional tone. Reports on sleep-related memory manipulation thus far mainly concern neutral memory. However, studies evaluating emotional memory manipulation during sleep are underway and will be informative as to possible applications in the emotional realm.

A different type of clinical application regards the possible treatment of sleep problems with external stimulation. Sleep problems have very high prevalence in modern societies, and many sleep disorders feature reduced SWS. While pharmacological treatments are available, these do not work in all people or in all conditions. To give just one example, no treatments with proven efficacy currently exist for PTSD-related sleep disorder. Moreover, sleep-promoting drugs, even when they do improve sleep, often do not restore normal sleep physiology. The benzodiazepines, for instance, have REM supressing effects. Finally, as with all pharmaceutical drugs, sleep medication has side effects, including paradoxical effects on sleep and risk of addiction. A new, non-pharmacological form of treatment would therefore be of considerable societal interest.

The applicability of external stimulation to this kind of goal remains to be seen. A recent study indicates that the responsiveness of healthy sleep to up-state targeted stimulation is self-limiting (Ngo et al. 2015). That is, boosting, in terms of slow oscillation amplitude and high frequency power content, only works for one or two oscillations in a row and then quickly fades out. In line with this finding, previously reported effects of external stimulations on sleep were all observed closely following stimulation; no studies have reported alterations in the amount of SWS or other sleep macrostructural parameters. It should, however, be considered that these findings were achieved with random stimulation or using an unvalidated slow oscillation phase targeting method with uncertain accuracy (Ngo et al. 2015). It will be interesting to see whether methods with demonstrated phase-targeting precision will lead to more favourable outcomes. Also, stimuli besides auditory ones, for instance electromagnetic fields, could be considered for this type of development.

Moving from clinical to educational possibilities, an application that seems close at hand concerns the use of sleep manipulations to aid learning. In particular, the existing literature on sleep-related memory manipulations suggests it should be possible to enhance the retention of information studied prior to sleep. A consideration here is that the memory effects in the studies thus far were typically small and memory benefits were not observed in all studies. It thus remains to be seen whether procedures can be developed in which the size and robustness of effects will be sufficient to be of practical interest. Also in this case, the targeting of stimuli to potentially receptive sleep windows, such as the slow oscillation up-state, may increase the effectiveness of sleep-related memory manipulations. As such, possible applications towards the enhancement of learning certainly merit investigation.

Concluding Remarks

Approaches in which the candidate process components of memory consolidation have been actively manipulated, have led to great advances in the field of sleep and memory. On this journey, sophisticated techniques have been developed to overcome the challenges posed by investigating brain activity in a state that is devoid of behavioural output. Clearly, the field is as yet at an early stage and many unknowns regarding sleep-related memory processing remain. This is especially the case considering the broader role of sleep in cognition, which, besides memory consolidation, involves more flexible information processing (Gupta et al. 2010) leading to insights (Wagner et al. 2004; Cai et al. 2009; Beijamini et al. 2014), as well as a central role in emotional memory regulation (Talamini et al. 2013; Gujar et al. 2011; van der Helm et al. 2011b; Yoo et al. 2007; Pace-Schott et al. 2011; Deliens et al. 2012; Wassing et al. 2016). Similarly, the multiple factors determining the success of memory manipulations during sleep have yet to be clarified. However, the tools, basic knowledge and broad interest from the research community are in place to support fast progress in this field, both at the level of fundamental advance and in terms of applications.

References

Antony JW, Gobel EW, O'Hare JK, Reber PJ, Paller KA (2012) Cued memory reactivation during sleep influences skill learning. Nat Neurosci 15(8):1114–1116

Arzi A et al (2012) Humans can learn new information during sleep. Nat Neurosci 15(10):1460–1465

Arzi A et al (2014) Olfactory aversive conditioning during sleep reduces cigarette-smoking behavior. J Neurosci: Official J Soc Neurosci 34(46):15382–15393

Atherton LA, Dupret D, Mellor JR (2015) Memory trace replay: the shaping of memory consolidation by neuromodulation. Trends Neurosci 38(9):560–570

Barnes DC, Wilson DA (2014) Slow-wave sleep-imposed replay modulates both strength and precision of memory. J Neurosci: Official J Soc Neurosci 34(15):5134–5142
Bastuji H, García-Larrea L (1999) Evoked potentials as a tool for the investigation of human sleep. Sleep Med Rev 3:23–45
Beijamini F, Pereira SIR, Cini FA, Louzada FM (2014) After being challenged by a video game problem, sleep increases the chance to solve it. PLoS ONE 9(1):e84342
Bendor D, Wilson MA (2012) Biasing the content of hippocampal replay during sleep. Nat Neurosci 15(10):1439–1444
Bergmann TO et al (2012) EEG-guided transcranial magnetic stimulation reveals rapid shifts in motor cortical excitability during the human sleep slow oscillation. J Neurosci: Official J Soc Neurosci 32(1):243–253
Born J, Rasch B, Gais S (2006) Sleep to remember. Neuroscientist 12(5):410–424
Buzsaki G (1996) The hippocampo-neocortical dialogue. Cereb Cortex 6(2):81–92
Cai DJ, Mednick SA, Harrison EM, Kanady JC, Mednick SC (2009) REM, not incubation, improves creativity by priming associative networks. Proc Natl Acad Sci 106(25): 10130–10134
Clemens Z, Fabo D, Halasz P (2005) Overnight verbal memory retention correlates with the number of sleep spindles. Neuroscience 132(2):529–535
Clemens Z, Fabo D, Halasz P (2006) Twenty-four hours retention of visuospatial memory correlates with the number of parietal sleep spindles. Neurosci Lett 403(1–2):52–56
Clemens Z et al (2007) Temporal coupling of parahippocampal ripples, sleep spindles and slow oscillations in humans. Brain 130(Pt 11):2868–2878
Cox R, Hofman WF, Talamini LM (2012) Involvement of spindles in memory consolidation is slow wave sleep-specific. Learn Mem 19(7):264–267
Cox R, Hofman WF, de Boer M, Talamini LM (2014a) Local sleep spindle modulations in relation to specific memory cues. Neuroimage 99:103–110
Cox R, van Driel J, de Boer M, Talamini LM (2014b) Slow oscillations during sleep coordinate interregional communication in cortical networks. J Neurosci: Official J Soc Neurosci 34 (50):16890–16901
Cox C, Korjoukov I, de Boer M, Talamini LM (2014c) Sound asleep: Processing and retention of slow oscillation phase-targeted stimuli. PLoS ONE 9(7):e101567
de Lavilleon G, Lacroix MM, Rondi-Reig L, Benchenane K (2015) Explicit memory creation during sleep demonstrates a causal role of place cells in navigation. Nat Neurosci 18(4): 493–495
Deliens G, Gilson M, Schmitz R, Peigneux P (2012) Sleep unbinds memories from their emotional context. Cortex J Devoted Study Nerv Syst Behav
Ego-Stengel V, Wilson MA (2010) Disruption of ripple-associated hippocampal activity during rest impairs spatial learning in the rat. Hippocampus 20(1):1–10
Emmons WH, Simon CW (1956) The non-recall of material presented during sleep. Am J Psychol 69(1):76–81
Fox BH, Robbin JS (1952) The retention of material presented during sleep. J Exp Psychol 43 (1):75–79
Frank MG (2015) Sleep and synaptic plasticity in the developing and adult brain. Curr Topics Behav Neurosci 25:123–149
Gais S, Molle M, Helms K, Born J (2002) Learning-dependent increases in sleep spindle density. J Neurosci: Official J Soc Neurosci 22(15):6830–6834
Girardeau G, Benchenane K, Wiener SI, Buzsaki G, Zugaro MB (2009) Selective suppression of hippocampal ripples impairs spatial memory. Nat Neurosci 12(10):1222–1223
Gujar N, Yoo SS, Hu P, Walker MP (2011) Sleep deprivation amplifies reactivity of brain reward networks, biasing the appraisal of positive emotional experiences. J Neurosci: Official J Soc Neurosci 31(12):4466–4474
Gupta AS, van der Meer MA, Touretzky DS, Redish AD (2010) Hippocampal replay is not a simple function of experience. Neuron 65(5):695–705

Hauner KK, Howard JD, Zelano C, Gottfried JA (2013) Stimulus-specific enhancement of fear extinction during slow-wave sleep. Nat Neurosci 16(11):1553–1555

Hoffman KL et al (2007) The upshot of up states in the neocortex: from slow oscillations to memory formation. J Neurosci: Official J Soc Neurosci 27(44):11838–11841

Hofman WF, Cox R, Talamini LM (2010) Effects of an emotional film on sleep EEG: relation with emotional attenuation over sleep. J Sleep Res 19(Suppl. 2):134

Holleman E, Battaglia FP (2015) Memory consolidation, replay, and cortico-hippocampal interactions. Analysis and modeling of coordinated multi-neuronal activity, Springer, Berlin, pp 207–221

Hutcheon BY, Yarom Y (2000) Resonance, oscillation and the intrinsic frequency preferences of neurons. Trends Neurosci 23:216–222

Jefferys JGH, Haas HL (1982) Synchronized bursting of CA1 hippocampal pyramidal cells in the absence of synaptic transmission. Nature 300:448–450

Ji D, Wilson MA (2007) Coordinated memory replay in the visual cortex and hippocampus during sleep. Nat Neurosci 10(1):100–107

Kaestner EJ, Wixted JT, Mednick SC (2013) Pharmacologically increasing sleep spindles enhances recognition for negative and high-arousal memories. J Cogn Neurosci

Louie K, Wilson MA (2001) Temporally structured replay of awake hippocampal ensemble activity during rapid eye movement sleep. Neuron 29:145–156

Marshall L, Molle M, Hallschmid M, Born J (2004) Transcranial direct current stimulation during sleep improves declarative memory. J Neurosci: Official J Soc Neurosci 24(44):9985–9992

Marshall L, Helgadottir H, Mölle M, Born J (2006) Boosting slow oscillations during sleep potentiates memory. Nature 444(7119):610–613

Marshall L, Kirov R, Brade J, Mölle M, Born J (2011) Transcranial electrical currents to probe EEG brain rhythms and memory consolidation during sleep in humans. PLoS ONE 6(2): e16905

Mednick SC et al (2013) The critical role of sleep spindles in hippocampal-dependent memory: a pharmacology study. J Neurosci: Official J Soc Neurosci 33(10):4494–4504

Mölle M, Marshall L, Gais S, Born J (2002) Grouping of spindle activity during slow oscillations in human non-rapid eye movement sleep. J Neurosci: Official J Soc Neurosci 22(24):10941–10947

Ngo HV, Claussen JC, Born J, Molle M (2013a) Induction of slow oscillations by rhythmic acoustic stimulation. J Sleep Res 22(1):22–31

Ngo HV, Martinetz T, Born J, Molle M (2013b) Auditory closed-loop stimulation of the sleep slow oscillation enhances memory. Neuron 78(3):545–553

Ngo HV et al (2015) Driving sleep slow oscillations by auditory closed-loop stimulation-a self-limiting process. J Neurosci: Official J Soc Neurosci 35(17):6630–6638

O'Neill J, Pleydell-Bouverie B, Dupret D, Csicsvari J (2010) Play it again: reactivation of waking experience and memory. Trends Neurosci 33(5):220–229

Oudiette D, Antony JW, Creery JD, Paller KA (2013) The role of memory reactivation during wakefulness and sleep in determining which memories endure. J Neurosci: Official J Soc Neurosci 33(15):6672–6678

Pace-Schott EF et al (2011) Napping promotes inter-session habituation to emotional stimuli. Neurobiol Learn Mem 95(1):24–36

Peyrache A, Khamassi M, Benchenane K, Wiener SI, Battaglia FP (2009) Replay of rule-learning related neural patterns in the prefrontal cortex during sleep. Nat Neurosci 12(7):919–926

Rasch B, Born J (2013) About sleep's role in memory. Physiol Rev 93(2):681–766

Rasch B, Buchel C, Gais S, Born J (2007) Odor cues during slow-wave sleep prompt declarative memory consolidation. Science 315(5817):1426–1429

Rihm JS, Diekelmann S, Born J, Rasch B (2014) Reactivating memories during sleep by odors: odor specificity and associated changes in sleep oscillations. J Cogn Neurosci 26(8):1806–1818

Rudoy JD, Voss JL, Westerberg CE, Paller KA (2009) Strengthening individual memories by reactivating them during sleep. Science 326(5956):1079

Sadowski JH, Jones MW, Mellor JR (2016) Sharp-wave ripples orchestrate the induction of synaptic plasticity during reactivation of place cell firing patterns in the hippocampus. Cell reports 14(8):1916–1929

Santostasi G et al (2016) Phase-locked loop for precisely timed acoustic stimulation during sleep. J Neurosci Methods 259:101–114

Schmidt C et al (2006) Encoding difficulty promotes postlearning changes in sleep spindle activity during napping. J Neurosci: Official J Soc Neurosci 26(35):8976–8982

Schreiner T, Rasch B (2014) Boosting vocabulary learning by verbal cueing during sleep. Cereb Cortex

Schreiner T, Lehmann M, Rasch B (2015) Auditory feedback blocks memory benefits of cueing during sleep. Nat Commun 6:8729

Staresina BP et al (2015) Hierarchical nesting of slow oscillations, spindles and ripples in the human hippocampus during sleep. Nat Neurosci 18(11):1679–1686

Stickgold R (2013) Parsing the role of sleep in memory processing. Curr Opin Neurobiol 23 (5):847–853

Sweegers CC, Talamini LM (2014) Generalization from episodic memories across time: a route for semantic knowledge acquisition. Cortex J Devoted Study Nerv Syst Behav 59:49–61

Talamini LM, Bringmann LF, De Boer M, Hofman WF (2013) Sleeping worries away or worrying away sleep? Physiological evidence on sleep-emotion interactions. PLoS ONE 8(5):1–10

Talamini LM, van Poppel EAM, Korjoukov I (2016) Hitting the right spot: a new closed-loop stimulation procedure for oscillatory phase targeting. ESRS congress, Bologna, 13–16 Sept 2016

Valderrama M et al (2012) Human gamma oscillations during slow wave sleep. PLoS ONE 7(4): e33477

van der Helm E, Gujar N, Nishida M, Walker MP (2011a) Sleep-dependent facilitation of episodic memory details. PLoS ONE 6(11):e27421

van der Helm E et al (2011b) REM sleep depotentiates amygdala activity to previous emotional experiences. Curr Biol 21(23):2029–2032

van Dongen EV et al (2012) Memory stabilization with targeted reactivation during human slow-wave sleep. Proc Natl Acad Sci USA 109(26):10575–10580

Wagner U, Gais S, Haider H, Verleger R, Born J (2004) Sleep inspires insight. Nature 427 (6972):352–355

Wassing R et al (2016) Slow dissolving of emotional distress contributes to hyperarousal. Proc Natl Acad Sci USA 113(9):2538–2543

Wood JM, Bootzin RR, Kihlstrom JF, Schacter DL (1992) Implicit and explicit memory for verbal information presented during sleep. Am J Psychol 76–81

Yoo SS, Gujar N, Hu P, Jolesz FA, Walker MP (2007) The human emotional brain without sleep–a prefrontal amygdala disconnect. Curr Biol 17(20):R877–R878

Scents and Reminiscence: Olfactory Influences on Memory Consolidation in the Sleeping Human Brain

Laura K. Shanahan and Jay A. Gottfried

Abstract Most animals sleep, even though the loss of vigilance associated with the sleep state can have serious consequences for survival. Why animals need to sleep is unclear, and the precise functions of sleep are not well understood. Recent studies have converged on the idea that sleep plays an active role in supporting memory consolidation. In the past decade, researchers have developed a unique method to manipulate specific memories during sleep. Known as targeted memory reactivation, this procedure involves presentation of a sensory cue during an initial encoding task, and then re-presentation of the same cue during slow-wave sleep, resulting in memory enhancement for the cue-associated items in the wake state. Odor stimuli have proven to be an effective entry point for reactivating memories in the sleeping human brain: the olfactory pathway projects directly to limbic networks supporting memory and emotion, and odors can be delivered with minimal risk of waking the sleeping subject. Here we review the human literature on olfactory targeted memory reactivation, detailing its initial applications and more recent progress in the field. Although the underlying neural mechanisms remain elusive, the capacity of sleep-borne odors to selectively target memories has important basic and translational research implications for shaping the consolidation of both declarative and emotional memories.

Keywords Sleep · Targeted memory reactivation · Olfaction · Memory consolidation

L.K. Shanahan (✉) · J.A. Gottfried
Department of Neurology, Northwestern University Feinberg School of Medicine, Chicago, IL, USA
e-mail: lshanahan4@gmail.com

J.A. Gottfried
e-mail: j-gottfried@northwestern.edu

N. Axmacher and B. Rasch (eds.), *Cognitive Neuroscience of Memory Consolidation*, Studies in Neuroscience, Psychology and Behavioral Economics, DOI 10.1007/978-3-319-45066-7_20

Odors and Memory Consolidation

The unique relationship between odors and memory has been celebrated creatively by druids, artists, philosophers, poets, and writers for centuries. Perhaps most famously, Marcel Proust described an olfactory memory in which the smell and taste of a French madeleine triggers a distant childhood memory, spawning a magnum opus across 7 novels (Proust 1913–1931). In a rather more contemporary twist, the writer Tom Robbins spun a deliciously olfactory tale in the book Jitterbug Perfume, where one of the chief characters, Marcel LeFever, possesses “a gift to detect odors too faint to register in others’ snouts.” That LeFever shares a first name with Proust is likely no accident. In an impassioned monologue, Marcel (of Jitterbug not madeleines) states: “Of our five senses, the one most connected to memory is the sense of smell. Although man has become increasingly visual in his orientations, although his olfactory receptor has shrunk until it is no larger than an American dime, sight simply cannot compete with smell when it comes to the ability to awaken memory. Memories associated with scent are invariably more immediate and more vivid than those associated solely with visual imagery or sound” (Robbins 1984).

Given the literary connections between olfaction and memory, it is perhaps unsurprising that odors have been scientifically shown to exert a powerful influence on the consolidation of associated memory traces. This chapter will focus specifically on the interaction between olfactory sensation and memory consolidation in the sleeping human brain, and thus complements other chapters in this volume on odor processing and memory consolidation in sleeping non-human animals (see also chapter by Wilson, Kondrakiewicz and Barnes) (Fig. 1).

Research Origins of Olfactory Targeted Memory Reactivation

Rasch and colleagues were the first to formally investigate the influence of odors on memory consolidation during human sleep (Rasch et al. 2007) (Fig. 2) (see also chapter by Talamini and by Schreiner, Lehmann and Rasch). In their landmark study, a rose odor (phenylethyl alcohol) was paired with a visuospatial learning task, in which subjects were required to learn the locations of several card pairs (similar to the card game “Concentration”). After learning, subjects went to sleep, and either the same rose odor or an odor vehicle was presented (in an alternating 30 s-on/30 s-off pattern to prevent habituation). Upon waking, subjects were asked to recall the card locations from pre-sleep learning. Intriguingly, subjects performed better on the memory post-test when they had been exposed to rose odor (vs. odor vehicle) during slow-wave sleep. The experimenters determined that this memory-enhancing effect was specific to slow-wave sleep, since delivering rose odor during wake or rapid eye movement (REM) sleep did not affect memory

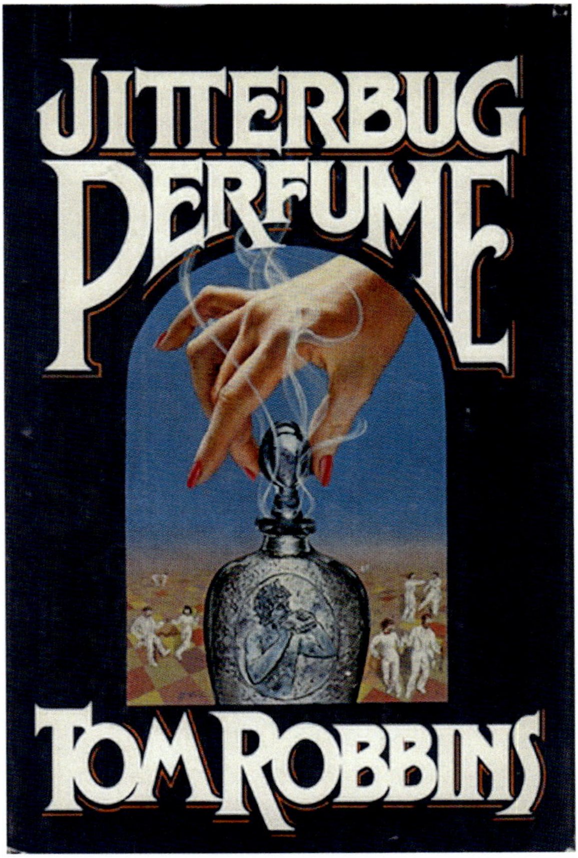

Fig. 1 Front cover art for the novel *Jitterbug Perfume* by Robbins (1984). Reproduced with permission from Random House

performance. Moreover, rose odor only improved recall if it had been previously paired with learning. Together, these behavioral findings suggested that odors can enhance the consolidation of associated declarative memories during sleep, without awareness of the sleeping subject.

In a follow-up experiment from the same paper, functional magnetic resonance imaging (fMRI) was used to investigate hippocampal activation during the odor reactivation task. Again, subjects learned card locations in the presence of rose odor, which was then presented during slow-wave sleep or wakefulness. Presentation of rose odor was shown to activate the hippocampus to a greater extent during slow-wave sleep than it did during wakefulness. The investigators concluded that the hippocampus is particularly receptive to olfactory sensory modulation during slow-wave sleep. It is worth noting that the original imaging paradigm did not include a non-reactivation odor (i.e., something other than rose odor) in either sleep or wake conditions, leaving open the possibility that odors activate the hippocampus to a greater extent during slow-wave sleep in general, and not as a consequence of prior odor-paired learning per se. However, in a follow-up control study from the same laboratory, presentation of a non-reactivation rose odor, which

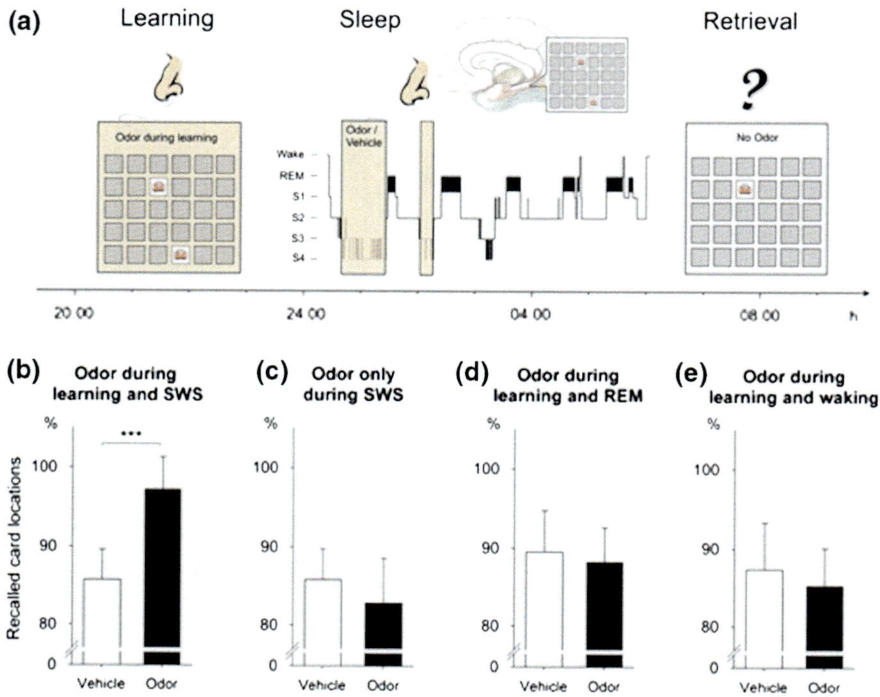

Fig. 2 The first olfactory targeted memory reactivation study. **a** Subjects learned card locations in the presence of rose odor. The same rose odor or an odorless vehicle (control) was then presented during sleep. Memory for card locations was tested upon waking, in absence of odor. **b** Odor presentation during learning and subsequent slow-wave sleep improved recall for card locations upon waking. **c** Odor presentation during slow-wave sleep did not affect recall if odor was not also delivered during prior learning. There was also no effect on recall when odor was presented during learning and again during subsequent REM sleep (**d**) or subsequent wakefulness (**e**). Reproduced from Rasch et al., Science 2007, with permission from AAAS

had not accompanied prior learning, did not activate the hippocampus (Rasch and Born 2007), suggesting that the association between odor and learning is the critical factor in driving these changes in the sleeping brain. This important study established that odor stimuli have the ability to actively promote activity in brain areas implicated in memory processing in the sleeping human brain, contradicting notions that sleep represents a passive state of sensory deafferentation (e.g., Ellenbogen et al. 2006, p. 717).

The practice of using sensory cues to influence memory consolidation has since been termed "targeted memory reactivation", or TMR (Oudiette and Paller 2013) (see also chapter by Talamini). Generally, memory consolidation is thought to occur through reactivation, where memory-related neural activity is replayed to facilitate its integration into existing cortical networks (Rasch and Born 2013). Researchers speculate that TMR appropriates the brain's natural reactivation process, biasing replay events toward associated memory traces, and resulting in privileged reactivation of those memories that have been "targeted". In theory, memories that are reactivated preferentially are more likely to be integrated into cortical networks for long-term storage, and are therefore stronger and more easily recalled. Although some mechanistic work has been done to support this notion (see next section), further research is needed to establish definitively that TMR evokes memory replay during sleep.

Recent studies have built on findings from the Rasch study, lending to a more complete understanding of TMR and its effects, both behavioral and neurobiological. The next sections of this chapter will review the human olfactory TMR literature, highlighting recent advances in the field and identifying outstanding questions.

Reactivation and Declarative Memory Consolidation

In 2011, the Rasch group conducted a second olfactory TMR study that refined the original experiment, specifically to investigate the role of reactivation in a memory paradigm incorporating memory interference (Diekelmann et al. 2011). As in their first TMR study, subjects learned card pair locations in the presence of an odor, and the same odor was presented during subsequent wake or slow-wave sleep. Following reactivation, subjects learned to associate cards from pre-sleep learning with novel locations (interference learning). Finally, subjects were tested on their knowledge of the original pre-sleep card pair locations. The main hypothesis was that odor presentation in either wake or sleep would yield memories more susceptible to interference. Such a prediction was consistent with reconsolidation theory, which posits that memories can re-enter an active state when reactivated (Nader and Hardt 2009) (see also chapter by Kessler, Blackwell and Kehyayan). Interestingly, and in contrast to the central hypothesis, olfactory TMR yielded disparate effects when the reactivation odor was presented during wake compared to slow-wave sleep. Subjects that underwent olfactory TMR (vs. sham reactivation with odor vehicle) during slow-wave sleep demonstrated attenuated memory decline for pre-sleep card locations following interference learning. These finding suggest that presenting the reactivation odor during slow-wave sleep stabilizes targeted memories, fortifying them against interference. However, when subjects underwent reactivation during wake (vs. sham), memory for the original object locations declined markedly following interference learning, suggesting that reactivation during wake renders targeted memories labile and vulnerable to interference.

In a complementary imaging experiment, a reactivation odor was delivered while subjects slept deeply (group 1, slow-wave sleep) or remained awake (group 2) during fMRI scanning. These data revealed that presenting the reactivation odor (vs. a non-reactivation odor) during slow-wave sleep activated left hippocampus (Fig. 3) as well as neocortical regions presumably involved in memory processing (e.g., temporal cortex), whereas presenting the reactivation odor during wake evoked activity in a more anterior brain region. Combined, these behavioral and imaging results suggest that olfactory TMR selectively activates different brain

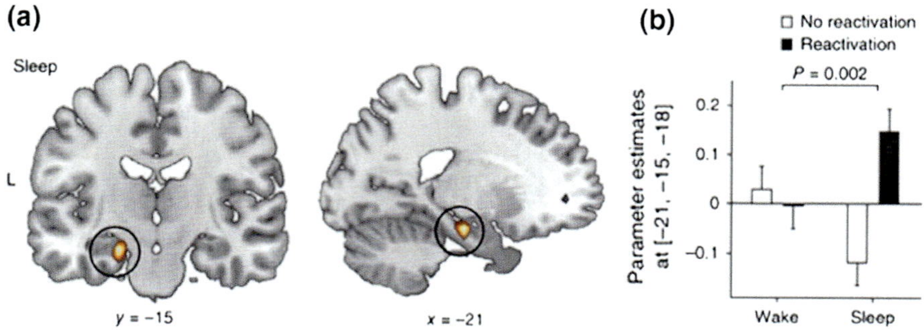

Fig. 3 **a** Presentation of a reactivation odor during slow-wave sleep activated *left* hippocampus. **b** Parameter estimates for the peak voxel in *left* hippocampus in response to the reactivation odor or a non-reactivation odor for wake and sleep conditions. Modified from Diekelmann et al. (2011)

regions depending on whether the subject is awake or asleep, rendering targeted memories more or less vulnerable to interference respectively. However, it should be noted that researchers did not test memory retention following reactivation in the MRI scanner. Thus, future research will be necessary to determine whether changes in brain activity during olfactory TMR are predictive of subsequent targeted changes in memory performance.

Diekelmann and colleagues conducted an additional olfactory TMR study in an effort to determine the extent to which sleep with reactivation had greater impact on subsequent memory performance than sleep per se (Diekelmann et al. 2012). By employing the same card location paradigm, researchers found that olfactory TMR during a 40-min nap improved memory consolidation to the same degree as did a 90-min nap without reactivation. However, a 40-min nap without reactivation was not adequate to enhance memory performance. These results suggest that olfactory TMR can enhance the efficiency with which memories are consolidated during sleep, even when sleep periods are restricted to shorter naps, though it is evident from these findings that sleep timing and duration play an important role in modulating these effects. An additional study assessing the stimulus specificity of olfactory TMR involved presentation of a target odor during learning, followed by presentation of either the target odor or a different (non-target) odor during slow-wave sleep (Rihm et al. 2014). As predicted, olfactory TMR only benefited memory outcomes in the target odor condition. Of note, presenting the target odor during slow-wave sleep altered electroencephalogram (EEG) activity, in which spectral power was increased in the 1.5–4 Hz delta frequency band over frontal electrodes, effectively increasing the negative-to-positive slope for slow oscillations. In addition, target odor presentation enhanced 13–15 Hz spindle power across parietal electrodes. These findings led the investigators to speculate that olfactory TMR might improve memory outcomes by influencing EEG events previously implicated in memory consolidation, such as sleep spindles and delta activity (Diekelmann and Born 2010).

Taken together, the above experiments provide solid support for the efficacy of olfactory TMR, given that each of them was able to independently replicate the original memory-enhancing effect found in the 2007 olfactory TMR study. However, it should be noted that each of these studies employed the same card location memory paradigm, raising questions in terms of the generalizability of the results. In 2014, Cox and colleagues conducted an olfactory TMR study using a unique memory paradigm involving a word-location association task (Cox et al. 2014). In their experiment, subjects learned to associate words with particular locations on a screen in the presence of two distinct odors. One half of the words was presented with odor A, while the other half was presented with odor B. Words paired with odor A had a leftward bias, meaning that they were presented mainly in leftward screen locations, while words paired with odor B had a rightward bias. During reactivation, one of the two odors was presented during non-REM sleep, to selectively reactivate words with either a leftward or rightward bias. They found that reactivation modulated spindle activity in a lateralized fashion during sleep. For instance, presenting rightward or leftward biased odors resulted in higher-amplitude fast spindles in contralateral brain regions (e.g., presenting the leftward biased odor increased fast spindle amplitude over right parieto-occipital regions). Although reactivation did not result in enhanced memory performance for targeted word locations, this may have been because subjects were over-trained on the task before sleep, in an effort to maximize chances of revealing spindle effects. As a result, it is not possible to draw conclusions about the relationship between the amount of local spindle activity and memory outcomes. Still, these results mesh well with those from the Rihm study, as they reinforce the idea that olfactory TMR is odor-specific, and can modulate spindle events during sleep.

Though the majority of olfactory TMR research has focused on reactivating memories during non-REM sleep stages, reactivation during REM sleep has been attempted as well. In 2014, the Rasch group used their card location memory paradigm with memory interference (as in Diekelmann et al. 2011) to demonstrate that reactivation during REM sleep has no effect on memory stability. These findings are in sharp contrast with those obtained by reactivating memories during slow-wave sleep (Diekelmann et al. 2011), and suggest that the efficacy of olfactory TMR (at least in the context of declarative memory paradigms) may be highly dependent on sleep stage. This is not surprising, as the physiology underlying slow-wave sleep and REM sleep is dramatically different: slow-wave sleep is characterized by high-amplitude, low-frequency EEG activity, while REM sleep is characterized by low-amplitude, high frequency EEG activity similar to that observed during the wake state (Silber et al. 2007). Moreover, while slow-wave sleep has been consistently shown to support declarative memory function, REM sleep has been implicated more in procedural and emotional memory consolidation (Rasch and Born 2013; Walker and van der Helm 2009). Whether olfactory reactivation during REM sleep might influence memory consolidation in an alternative paradigm is not known.

Reactivation Beyond Declarative Memory Consolidation

Almost all of the olfactory TMR literature is centered on the influence of reactivation on declarative memory consolidation, likely because the first olfactory TMR study was declarative in nature. However, a handful of olfactory TMR studies demonstrate that odors can have a broader influence on learning and memory, with implications for creative problem solving and fear learning. In 2012, Ritter and colleagues measured the impact of olfactory TMR on creativity (Ritter et al. 2012). In this study, researchers presented a problem requiring a creative solution in the presence of an odor, and then they presented the same odor (vs. no odor, or a different odor) while subjects slept. They found that subjects in the target odor condition generated more creative solutions to the proposed problem upon waking. However, one limitation of this study is that EEG activity was not monitored, precluding the possibility of presenting odors during particular sleep stages. Moreover, odor delivery was achieved via an odor diffuser (vs. a standard laboratory olfactometer). Robust evidence for this effect thus remains to be shown.

In 2013, Hauner and colleagues utilized olfactory TMR to modulate emotional memory consolidation during slow-wave sleep (Hauner et al. 2013) (Fig. 4). In a fear conditioning paradigm, subjects learned to associate pictures of two faces (conditioned stimuli, CS+) with electrical shocks (unconditioned stimulus, US) in the presence of two distinct contextual odors during fMRI scanning. The main hypothesis was that delivery of one of the contextual odors during slow-wave sleep (outside the MRI scanner) would selectively reactivate a subset of the acquired fear memories. Upon waking, subjects reentered the MRI scanner, where they were re-exposed to both CS+ face pictures. From pre- to post-conditioning, the physiological expression of fear (as measured by skin conductance response) declined in response to the targeted face CS+ to a greater extent than to the non-targeted face CS+. These findings suggest that odors can promote fear extinction during sleep in a stimulus-specific way, and could have important therapeutic implications for anxiety disorders such as PTSD. In parallel to the physiological effects, there was a reduction in mean fMRI activity in response to the targeted (vs. non-targeted) face CS+ from pre- to post-sleep in regions implicated in memory processing and fear memory expression, such as hippocampus, anterior cingulate cortex, and insula. Finally, reactivation disrupted multivariate ensemble patterns of fMRI activity in the amygdala from pre- to post-sleep (Fig. 4d), supporting the idea that in promoting fear extinction, reactivation induces a qualitative reorganization of the fear memory trace, rather than a quantitative diminution or elimination of the trace.

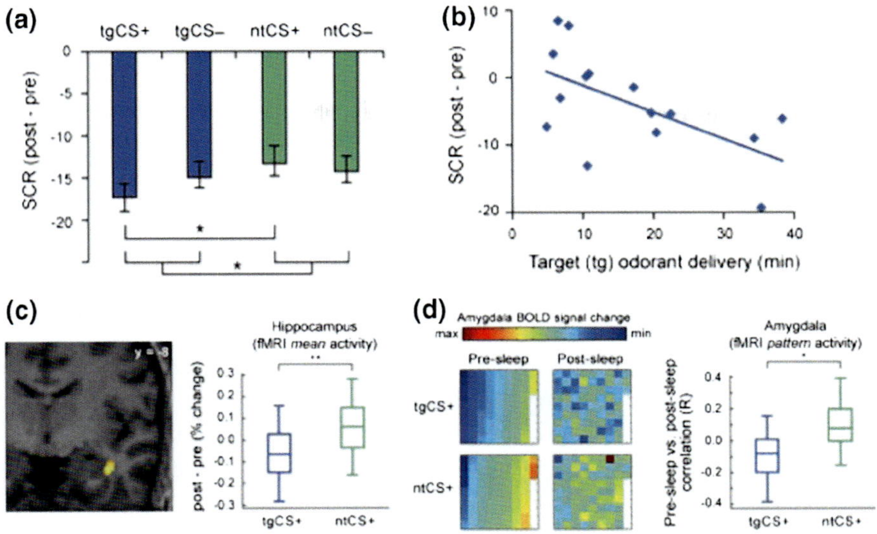

Fig. 4 Effects of olfactory TMR on fear memory consolidation. **a** Skin conductance response (SCR) decreased more for the targeted CS+ (tgCS+) from pre-sleep to post-sleep compared to the non-targeted CS+ (ntCS+). $*P \leq 0.05$, one-tailed. Error bars, s.e.m. **b** Extent of SCR reduction from pre-sleep to post sleep was correlated with duration of odor delivery during slow-wave sleep ($r = -0.61$, $P = 0.02$). **c** Mean fMRI activity in the anterior hippocampus exhibited greater decline in response from pre-sleep to post-sleep for the tgCS+ than for the ntCS+. $**P < 0.001$. **d** Ensemble maps of activation in *left* amygdala in a single subject (*left*) demonstrate that response patterns evoked by tgCS+ were more decorrelated from pre-sleep to post-sleep compared to those evoked by ntCS+. Each *square* represents signal intensity from a different voxel ($n = 75$), organized in columns from *top left* to *bottom right* in ascending order of tgCS+-evoked signal intensity in the pre-sleep condition. Across all 15 subjects (*right*), pre- and post-sleep pattern ensembles in amygdala became more distinct (less correlated) for tgCS+ than for ntCS +. $*P < 0.05$. Modified from Hauner et al. (2013)

TMR: A Broader View

In the first TMR experiments, olfactory stimuli were chosen as reactivation cues for a number of reasons. First, olfactory and limbic systems share significant anatomical overlap (Gottfried 2010), and odors have been shown to be particularly effective as memory cues (Chu et al. 2000). In addition, the olfactory system is unique compared to other sensory modalities in that it lacks a thalamic relay. Because the thalamus performs a sensory gating function (McCormick and Bal 1994), it follows that odors may enjoy privileged access to the medial temporal lobe, even during periods of sleep (Shanahan and Gottfried 2014). On a more practical level, multiple olfactory studies have demonstrated that odors can be presented unobtrusively during sleep without waking the sleeper (Carskadon and Herz 2004; Stuck et al. 2007).

Targeted memory reactivation is not an exclusive property of odor cues. Such manipulations can also be implemented successfully using auditory cues, such as tones or short sound clips that carry semantic meaning (e.g., a cat meowing, or a phone ringing). While olfactory cues are typically used more broadly to reactivate memory tasks during sleep (e.g., rose odor reactivates visuospatial learning), auditory cues have been employed to target particular task components. For instance, in the first auditory TMR experiment, pictures were paired with semantically related sound cues (e.g., picture of bell + clanging sound) during visuospatial learning, and then half of the sound cues were delivered during non-REM sleep to selectively reactivate half of the picture locations (Rudoy et al. 2009). As predicted, researchers found that memory performance was enhanced for targeted pictures (vs. non-targeted pictures). In the context of TMR, auditory cues offer certain advantages over olfactory ones. Namely, they can be delivered in a controlled way without the use of sophisticated machinery (i.e., an olfactometer), and human subjects can easily identify an extensive catalog of sounds, allowing researchers to target memories in a highly specific way. However, the significant advantages of olfactory TMR noted in the previous paragraph should not be discounted.

Finally, though the majority of TMR research thus far has focused on human subjects, several olfactory and auditory TMR studies have been recently conducted in rodents (Bendor and Wilson 2012; Rolls et al. 2013; Barnes and Wilson 2014). A recent study even demonstrated that olfactory TMR can enhance memory performance in honey bees trained on a classical conditioning task (Zwaka et al. 2015). Further discussion of auditory TMR, or of TMR in non-human species, is beyond the scope of this chapter, but more detailed consideration of these topics can be found in the chapters by Schreiner, Lehmann and Rasch and by Talamini of this volume.

Conclusion

Over the past decade, olfactory TMR researchers have demonstrated repeatedly that odors can be used as powerful tools to shape memory consolidation, representing a rare opportunity to gain insight into the enigma that is the sleeping human brain. Thus far, odors have been shown to enhance declarative memories, quell fear memories, and perhaps even incite creativity, but olfactory TMR research is still in its infancy. Future research should focus on the development of novel memory paradigms to test the generalizability of olfactory TMR across broader memory outcomes. In addition, more work must be done to determine whether odors can influence particular task components (as in auditory TMR), and whether olfactory TMR can be employed successfully during REM sleep in altered memory paradigms. Perhaps most importantly, the neurobiological mechanisms underlying the efficacy of olfactory TMR remain unclear. Prior studies have shown that in the context of olfactory TMR, odors impact sleep physiology (e.g., spindle power) and brain activity in the hippocampus and other regions, but future studies should go further, perhaps using more sophisticated fMRI analysis, or imaging techniques with optimized spatial and temporal resolution such as electrocorticography (ECoG), to directly address the theory that odors spur memory replay in the sleeping human brain (see also chapter by Zhang, Deuker and Axmacher).

That odors have the capacity to evoke the most vivid of episodic memories has been explored anecdotally for many years, as shown through the work of writers like Proust and Robbins. The recent scientific discovery that odors can be used to influence memories during sleep positively affirms these literary conceits and introduces an exciting twist on the same theme. Continued investigation of olfactory TMR and its effects on memory outcomes promises to lead to a more complete understanding of the fascinating link that exists between scents and reminiscence, and of how the central nervous system orchestrates this mnemonic jitterbug between the external world and our internal brain states.

References

Barnes DC, Wilson DA (2014) Slow-wave sleep-imposed replay modulates both strength and precision of memory. J Neurosci 34:5134–5142

Bendor D, Wilson MA (2012) Biasing the content of hippocampal replay during sleep. Nat Neurosci 15:1439–1444

Carskadon MA, Herz RS (2004) Minimal olfactory perception during sleep: why odor alarms will not work for humans. Sleep 27:402–405

Chu S, Downes JJ, Building ER (2000) Odour-evoked autobiographical memories: psychological investigations of proustian phenomena. Chem Senses 25:111–116

Cox R, Hofman WF, de Boer M, Talamini LM (2014) Local sleep spindle modulations in relation to specific memory cues. NeuroImage 99:103–110

Diekelmann S, Born J (2010) The memory function of sleep. Nat Rev Neurosci 11:114–126

Diekelmann S, Büchel C, Born J, Rasch B (2011) Labile or stable: opposing consequences for memory when reactivated during waking and sleep. Nat Neurosci 14:381–386

Diekelmann S, Biggel S, Rasch B, Born J (2012) Offline consolidation of memory varies with time in slow wave sleep and can be accelerated by cuing memory reactivations. Neurobiol Learn Mem 98:103–111

Ellenbogen JM, Payne JD, Stickgold R (2006) The role of sleep in declarative memory consolidation: passive, permissive, active or none? Curr Opin Neurobiol 16:716–722

Gottfried JA (2010) Central mechanisms of odour object perception. Nat Rev Neurosci 11: 628–641

Hauner KK, Howard JD, Zelano C, Gottfried JA (2013) Stimulus-specific enhancement of fear extinction during slow-wave sleep. Nat Neurosci 16:1553–1555

McCormick DA, Bal T (1994) Sensory gating mechanisms of the thalamus. Curr Opin Neurobiol 4:550–556

Nader K, Hardt O (2009) A single standard for memory: the case for reconsolidation. Nat Rev Neurosci 10:224–234

Oudiette D, Paller KA (2013) Upgrading the sleeping brain with targeted memory reactivation. Trends cogn sci 17:142–149

Proust M (1913–1931). In search of lost time. Grasset and Gallimard

Rasch B, Born J (2007) Maintaining memories by reactivation. Curr Opin Neurobiol 17:698–703

Rasch B, Born J (2013) About sleep's role in memory. Physiol Rev 93:681–766

Rasch B, Büchel C, Gais S, Born J (2007) Odor cues during slow-wave sleep prompt declarative memory consolidation. Science 315:1426–1429

Rihm J, Diekelmann S, Born J, Bjorn R (2014) Reactivating memories during sleep by odors: odor specificity and associated changes in sleep oscillations. J Cogn Neurosci 8:1806–1818

Ritter SM, Strick M, Bos MW, van Baaren RB, Dijksterhuis A (2012) Good morning creativity: task reactivation during sleep enhances beneficial effect of sleep on creative performance. J Sleep Res 21:643–647

Robbins T (1984) Jitterbug Perfume. Bantam USA, New York

Rolls A, Makam M, Kroeger D, Colas D, de Lecea L, Heller HC (2013) Sleep to forget: interference of fear memories during sleep. Molecular Psychiatry 18:1166–1170

Rudoy JD, Voss JL, Westerberg CE, Paller KA (2009) Strengthening individual memories by reactivating them during sleep. Science 326:1079

Shanahan LK, Gottfried JA (2014) Olfactory insights into sleep-dependent learning and memory. In: Barkai E, Wilson DA (eds) Odor memory and perception. Elsevier, Oxford, UK, pp 309–344

Silber MH, Ancoli-Israel S, Bonnet MH, Chokroverty S, Grigg-Damberger MM, Hirshkowitz M, Kapen S, Keenan SA, Kryger MH, Penzel T, Pressman MR, Iber C (2007) The visual scoring of sleep in adults. J Clin Sleep Med 3:121–131

Stuck BA, Stieber K, Frey S, Freiburg C, Hörmann K, Maurer JT, Hummel T (2007) Arousal responses to olfactory or trigeminal stimulation during sleep. Sleep 30:506–510

Walker MP, van der Helm E (2009) Overnight therapy? The role of sleep in emotional brain processing. Psychol Bull 135:731–748

Zwaka H, Bartels R, Franck V, Culo A, Menzel R (2015) Context odor presentation during sleep enhances memory in honeybees. Curr Biol 25:2869–2874

Reinforcing Language Learning During Sleep

Thomas Schreiner, Mick Lehmann and Björn Rasch

Keywords Sleep · Language learning · Memory · Reactivation

Language is a core ability of humans that enables us to communicate with each other. Using a restricted set of symbols to express an unlimited number of ideas requires the learning and storage of new words as well as the abstraction and generalization of linguistic rules. Sleep is essential for consolidating and integrating new memories into long-term memory, probably due to spontaneous reactivation of newly acquired memories during offline periods. Furthermore, empirical evidence consistently demonstrates that reactivations can be induced experimentally by presenting memory cues during sleep, resulting in enhanced memory performance the next day. In this chapter, we will summarize evidence in support for a beneficial role of memory consolidation processes occurring during sleep for learning a new language. Furthermore, we will review recent findings showing that central aspects

Thomas Schreiner and Mick Lehmann contributed equally to this work.

T. Schreiner · B. Rasch (✉)
Department of Psychology, Division of Cognitive Biopsychology and Methods, University of Fribourg, Rue P.-A. de Faucigny 2, 1701 Fribourg, Switzerland
e-mail: Bjoern.Rasch@unifr.ch

T. Schreiner · M. Lehmann · B. Rasch
Zurich Center for Interdisciplinary Sleep Research (ZiS), Zurich, Switzerland

M. Lehmann
Institute of Pharmacology and Toxicology, University of Zurich, Zurich, Switzerland

M. Lehmann
Department of Psychiatry, Psychotherapy and Psychosomatics, Psychiatric University Hospital Zurich, Zurich, Switzerland

T. Schreiner
Donders Institute for Brain, Cognition and Behaviour, Radboud University, Nijmegen, The Netherlands

N. Axmacher and B. Rasch (eds.), *Cognitive Neuroscience of Memory Consolidation*, Studies in Neuroscience, Psychology and Behavioral Economics, DOI 10.1007/978-3-319-45066-7_21

of language learning can be improved by targeted memory reactivation during sleep. Finally, we will discuss the potential underlying oscillatory mechanisms and propose a working model of successful memory reactivation and consolidation during sleep.

Memory and Sleep

Learning new words is crucial to develop language skills. During our infancy, the rate of word acquisition increases steadily and reaches a peak between the ages of 7 and 16, when we learn thousands of new words every year. Across the life span, we expand our vocabulary and learn foreign or second languages in parallel. The single-word knowledge is central for the subsequent emergence for syntactic and semantic structures (see Barrett 1989, for a review). "Language acquisition" is closely linked to basic memory processing, which is commonly divided into three consecutive steps: encoding, consolidation and retrieval (Gabrieli 1998). The encoding of a new lexical item leads to an initially labile memory trace that becomes less susceptible to interference during subsequent memory consolidation (McGaugh 2000). This process goes beyond classical consolidation and seeks on the one hand to preserve memories in their original form and, on the other hand, to store them into a more generalized representation. They become gradually transferred from a temporary store, mainly represented by the hippocampus, into pre-existing cortical knowledge networks for a stabilized long-term storage (Born and Wilhelm 2012) (see also the chapters by Cheng and Fernandez). While encoding and retrieval are optimally performed in the state of wakefulness, it has been consistently shown that sleep after learning is highly beneficial for consolidation mechanisms, resulting in an improved retrieval performance for various types of memory, inter alia declarative episodic memories including vocabulary learning (see Rasch and Born 2013, for a comprehensive review) (see also chapter by Schönauer and Gais). Moreover, sleep enhances logical reasoning and abstractive abilities, i.e. detecting general rules from individual experiences (see Walker and Stickgold 2010, for a review) (see also chapter by Rauss and Born). Both are cognitive capabilities that are indispensable in the manifold process leading to successful language acquisition.

The putative processes underlying a beneficial effect of sleep on memory have been described and conceptualized within the active system consolidation hypothesis (Diekelmann and Born 2010). This model for memory consolidation during sleep attempts to integrate a variety of findings and refers to the standard two-stage model of memory (Marr 1971; McClelland et al. 1995), a general concept that distinguishes between a fast-learning and a slow-learning store, which are represented by the hippocampus and the neocortex, respectively. Initial encoding critically relies on a fast-learning system, thus on the hippocampus. However, during system consolidation, memory representations are redistributed into a slow-learning memory system in cortical brain areas where they are integrated into larger neocortical networks and lose their dependency on hippocampal structures during retrieval (e.g. Frankland and Bontempi 2005) (see the chapter by Cheng for an alternative view). A central assumption of the active system consolidation is that this gradual redistribution process is driven by spontaneous and repeated reactivations of previously encoded memories during sleep such that synaptic connections within the neocortex are strengthened and form a more persistent memory representations (Rasch and Born 2013) (see also chapter by Zhang, Deuker and Axmacher).

There is now compelling evidence from rodents as well as humans that patterns of neuronal firing particular in the hippocampus that were present during encoding are replayed during subsequent sleep in the same sequential order (O'Neill et al. 2010; Pavlides and Winson 1989; Peigneux et al. 2004; Peyrache et al. 2009). However, reactivations are not limited to the hippocampal formation but occur in other neural circuitries relevant for memory consolidation as well, e.g. thalamus, striatum and prefrontal cortex (Ji and Wilson 2007; Pennartz et al. 2004; Peyrache et al. 2009) indicating that memory representations become gradually redistributed and integrated into larger neural networks. The active system consolidation theory postulates that a precise temporal coupling of slow oscillations, spindle and sharp-wave ripple activity is underlying the spontaneous reactivation processes during slow wave sleep (SWS) (Clemens et al. 2007) (see also chapter by Bergmann and Staresina). Slow oscillations originate in neocortical regions and activate neuronal assemblies in distant brain regions to become synchronized (Timoveef and Chauvette 2011). They are characterized by periods of synchronous neuronal firing ("up-states"), alternating with neuronal quiescence ("down-state") (Steriade 2006). Slow oscillations function as a pace maker synchronizing hippocampal memory reactivations with thalamo-cortical spindle activity and leading to the formation of spindle-ripple events (Eschenko et al. 2008), which have been proposed to promote long-lasting plastic changes in neocortical areas that is fundamental for memory formation (Rosanova and Ulrich 2005; Sadowski et al. 2016) (see also chapter by Maier and Kempter).

Sensory Processing and Semantic Discrimination During Sleep

The active system consolidation hypothesis provides a process-oriented model for the consolidation of recently acquired memories and their integration into pre-existing neural networks during sleep (Rasch and Born 2013). According to its central assumption that spontaneous reactivations during sleep are crucial for the memory's fate, inducing such reactivations experimentally should improve the consolidation process and thereby affect subsequent recall performance (Diekelmann 2014). This can be achieved by associating the learning environment or material with cues (e.g. an odour or a sound), which are presented during subsequent periods of sleep (Oudiette and Paller 2013) (see also chapter by Talamini, Shanahan and Gottfried). The rationale behind this experimental approach is that merely presenting the cue triggers the entire memory reactivation (see chapter for a detailed description of targeted memory reactivation (TMR)). A prerequisite for the successful application of this technique is that (a) sensory information reaches sensory, associative, and mnemonic areas during sleep and (b) semantic categories or specific characteristics of stimuli are successfully discriminated during sleep. We will discuss evidence for both aspects in the following section.

Our responsiveness to external stimulation is reduced during sleep, however, the extent of isolation from the environmental inputs is a current matter of debate. An increasing number of studies, ranging from animal (e.g. Edeline et al. 2000; Edeline 2003) to human EEG (e.g. Bastuji and García-Larrea 1999) and neuroimaging studies (e.g. Portas et al. 2000), has clearly shown that the sleeping brain responds to sensory stimulation. Already a half a century ago, Reivich et al. (1968) demonstrated that the local blood flow in the auditory cortex during sleep is significantly increased. In a more recent study, neurons in the primary and secondary auditory cortex even displayed similar firing patterns during sleep and wakefulness (Issa and Wang 2008). Furthermore, several studies reported that the ability to discriminate deviant auditory stimuli in an oddball paradigm persists in the sleeping state (for a review see Bastuji and García-Larrea 1999): Deviant tones elicited K-complexes (KC) of higher amplitude as compared to standard tones during sleep (Karakaş et al. 2007; Strauss et al. 2015). In a recent study using simultaneous EEG/fMRI, it has been shown that the presentation of tones during sleep was associated with significant activation of brain areas involved in auditory processing, in particular when auditory stimulations were followed by a KC (Dang-Vu et al. 2011). Interestingly, the neural response was not observable when the tone was played during a sleep spindle, implying that ongoing spontaneous brain activity profoundly influence the responsiveness of the brain during sleep.

With respect to semantic discrimination during sleep, early studies have already demonstrated that the presentation of predefined target words (Shanon 1979) or the participant's own name during NREM sleep (Oswald et al. 1960) elicits higher amplitude KCs as compared to the presentation of other names or tones (see also e.g. Perrin et al. 1999; Portas et al. 2000; Pratt et al. 1999). Most recently, a study reported a stronger neuronal response for angry versus neutral voices (Blume et al. 2016), implying that stimulus properties such as self-relevance and emotional prosody render stimuli salient even during NREM. Furthermore, the presentation of semantically unrelated word pairs (Brualla et al. 1998; Perrin et al. 2002) and incongruent sentences (Ibáñez et al. 2006) led to a significantly more negative ERP component, which is elicited by the same mechanisms as the N400 during wakefulness and thought to be a marker for verbal discordance.

Altogether these findings strongly suggest that the brain's ability to detect auditory stimuli and their intrinsic semantic content persists during sleep. This provides an essential prerequisite for triggering memory reactivations during sleep by presenting associated cues.

Inducing Reactivations by Cueing During Sleep

Numerous studies have now successfully demonstrated that targeted memory reactivation can indeed improve memory consolidation (Diekelmann et al. 2011; Rasch et al. 2007; Rihm et al. 2014; Schönauer et al. 2013; Schreiner et al. 2015a, b; Schreiner and Rasch 2015a, b; Lehmann et al. 2016; for a review see Oudiette and Paller 2013). Moreover, studies using trial-unique auditory cues improved memory for a selection of learned associations (e.g. Rudoy et al. 2009). The memory benefit was specifically observed for cued associations, suggesting that it is possible to externally trigger reactivations for selected experiences or learning material. It is noteworthy that creating an association between learning and cues is a requirement for effective TMR. Applying the cues alone, without a previous association with learning, does not affect memory performance (Cox et al. 2014; Rasch et al. 2007). Moreover, it is assumed that the beneficial memory effect depends on the prior performance level (Cairney et al. 2016; Creery et al. 2015). Furthermore, temporal lobe epileptic patients with uni- or bilateral hippocampal sclerosis showed a gradual decrease in cueing benefits with higher degrees of damaged hippocampal tissue, indicating that structural integrity of the hippocampus is critical for successful memory reactivation during sleep (Fuentemilla et al. 2013) (see also chapter by Baker and Zeman on memory consolidation deficits in patients). In the following sections, we illustrate how sleep and TMR can potentially further improve our existing capacities of language learning.

Sleep and Language Learning

Language is largely defined as a coordinated system of arbitrary signals and rule-directed structures, allowing us to communicate (Brandone et al. 2006). Accordingly, during the process of language learning, we basically give meaning to symbols and sounds, thereby enabling us to express our thoughts and describe the environment (Pinker 2000).

As pointed out above, sleep plays a crucial role in reinforcing memories. An increasing body of research has begun to unravel the role of the sleeping brain in supporting different aspect of language learning and has provided first evidence for the beneficial impact of sleep on expanding the "mental lexicon". The strengthening effect of sleep on newly acquired memories has been repeatedly demonstrated for new word learning.

Sleep has been proven to be beneficial not only during the process of language acquisition in 16 month old infants (Horvath et al. 2015) but also during second language learning in children and adolescents (Gais et al. 2006; Henderson et al. 2012). Furthermore sleep enhances recall performance for novel pseudo-words in 7 and 12 year olds (Brown et al. 2012) and promotes the learning of unknown words of a native language in adults (Kurdziel and Spencer 2015). Thus, the well-established link between sleep and the strengthening and stabilization of new declarative memories has been also proven to be valid for the learning of new words.

A series of recent studies tested if—apart from its strengthening effect—sleep would actively support the integration of newly learned words into pre-existing knowledge networks. Those studies investigated whether the acquisition of newly spoken word forms (e.g. "cathedruke") would interfere with the recognition performance for well-known words (e.g. "cathedral"), particularly when learning was followed by sleep (Dumay and Gaskell 2007; Tamminen et al. 2010; Tham et al. 2015). The rationale of this approach is that only when words are successfully integrated into existing knowledge networks, the novel word is capable of causing lexical competition during spoken word recognition (Gaskell and Dumay 2003). Thus, the emergence of such an inhibitory effect during later recognition testing would indicate that memory integration already occurred.

Consistent with the general notion that sleep might underlie the process of integration of new memories (Rasch and Born 2013), recognition of familiar words was inhibited only when learning was followed by an interval filled with sleep but not after wakefulness (Dumay and Gaskell 2007; Tham et al. 2015), indicating an active role of sleep regarding the integration of newly acquired words into pre-existing vocabulary. Interestingly, sleep spindles during the retention interval (Tamminen et al. 2010) and theta activity during later recognition testing (Bakker et al. 2015) were predictive for successful integration.

However, sleep not only integrates new words but also supports the mental lexicon by promoting the generalization of word meanings in both, toddlers (Friedrich et al. 2015) and adults (Lau et al. 2011). In addition, sleep benefits perceptual generalization in speech recognition (Fenn et al. 2003). Accordingly, sleep actively supports the reorganization of recent memories, by creating superordinate, semantic knowledge derived from individual episodic experiences (see chapter by Cheng on the relationship between consolidation and semantic learning).

Within this context, Davis and Gaskell (2009) proposed a complementary learning systems (CLS) consolidation model specifically for word learning. In this model, being closely tied to the active system consolidation theory described above, the learning and lexicalization of new words starts with an initial familiarization phase followed by slow lexical consolidation processes. The hippocampus plays a central role during the formation of an initially weak memory representation, which is subsequently replayed offline during NREM sleep. Those reactivation processes, originating from the hippocampus, lead to an improved integration of lexical representation in the pre-existing vocabulary. Thereby, the discriminatory precision between overlapping representations during later speech perception is improved and facilitates automatic word recognition (Davis and Gaskell 2009).

Reactivation processes during sleep that underlie the strengthening and integration of newly learned words might not be exclusively important for the human being but play a similarly role in the animal kingdom as well. Even though the underlying consolidation mechanisms might vary, replay activity seems to be equally important when it comes to song learning in birds (Dave and Margoliash 2000). Thus, the reactivation of previously learned words or sounds during sleep may be a critical factor for language learning in general.

The majority of studies investigating the role of sleep in language learning have focused mainly on the impact of sleep on the acquisition of new words and their integration into pre-existing knowledge networks. However, when acquiring a new language, one must uncover the complex grammatical system, comprising semantic, syntactic morphological and phonological rules. It requires the ability to extract and generalize them. As sleep has been proven to support processes of abstraction (Fischer et al. 2006; Stickgold and Walker 2013; Wagner et al. 2004) it seems plausible that grammar learning might as well benefit from consolidation processes during sleep.

In accordance with this assumption, infants were capable of abstracting the general grammatical pattern of a shortly presented artificial language when they were allowed to sleep after learning (Gómez et al. 2006), while remembering the overall grammatical pattern of the language even 24 h later (Hupbach et al. 2009). Moreover, sleep benefits grammar learning not only in infants but also adults show enhanced rule abstraction to an artificial grammar (Nieuwenhuis et al. 2013) and learning of hidden linguistic rules after sleep (Batterink et al. 2014).

Hence, these studies could convincingly demonstrate that apart from its influence on the mental lexicon, sleep also supports grammar learning, possibly by unfolding its impact on the generalization and abstraction of prior learned memories (e.g. Stickgold and Walker 2013).

The attempt has been made to explain specifically how reactivation processes during sleep could lead to improved abilities for abstraction and generalization, which might also be applicable for grammar learning (Lewis and Durrant 2011; Stickgold and Walker 2013). In particular, it is assumed that abstraction and generalization originate from the repeated reactivation of overlapping memory representations, going along with the selective strengthening of common elements. However, for both, grammar and word learning, more experimental evidence for the pivotal role of replay activity during sleep seems to be essential in order to draw stronger conclusions, which we will describe in the next section.

Improving Language Learning by Cueing During Sleep

As outlined above, sleep improves diverse aspects of language learning and it has been proposed that reactivation processes during sleep might underlie these beneficial effects. If replay activity during sleep is actually beneficial for language learning, inducing it experimentally during sleep should result in a specific memory benefit. Furthermore, this should be true for diverse aspects of language learning, e.g. vocabulary and grammar learning or rule generalization.

A recent study investigated whether the replay of verbal cues during sleep benefits vocabulary learning (Schreiner and Rasch 2015a). In the evening, German-speaking participants learned Dutch words and the corresponding German translations. During subsequent NREM sleep, the reactivation of previously learned associations was experimentally induced by solely replaying half of the Dutch words (cued words), while the other half was not replayed and was used as control stimuli (uncued words, see Fig. 1a, for a summary of the procedure). In accordance with the proposed models, cueing words during sleep enhanced memory

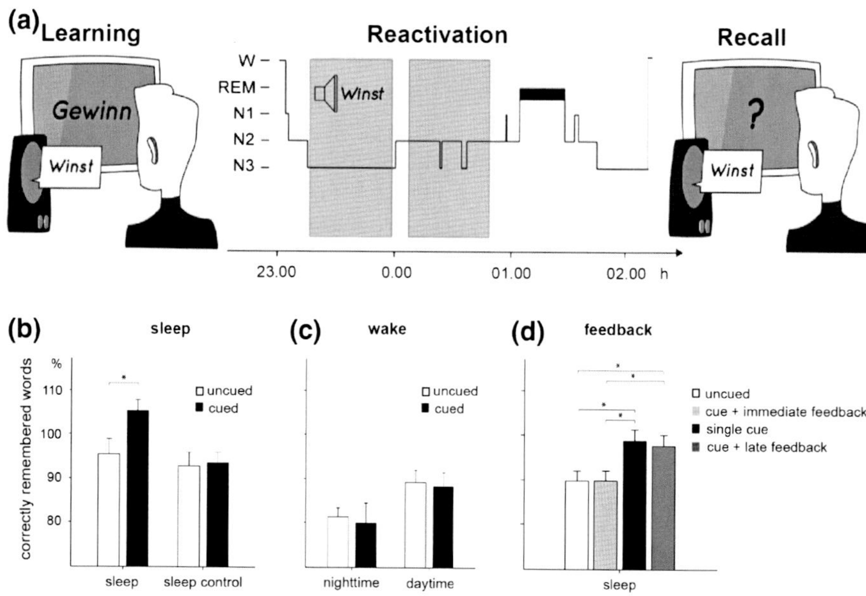

Fig. 1 Replaying foreign vocabulary during sleep enhances consolidation processes. **a** Procedures: Participants studied in the evening 120 Dutch-German word pairs. Learning was followed by a 3 h retention interval either filled with wakefulness or sleep. During the time of the retention interval some of the Dutch words were replayed for 90 min. A cued recall procedure was performed after the retention interval, testing the participant's memory for the German translations. **b** In the cueing sleep group, memory for 'cued' words (*black bar*) was enhanced as compared to 'uncued' words (*white bar*) Memory performance for uncued words was almost identical to recall performance in the control sleep group, where no cues were delivered during sleep **c** As compared to NREM sleep, cueing had no beneficial effect on later memory performance when performed during day- and night-time wakefulness. **d** In case that Dutch words were immediately followed by correct and false verbal feedback (i.e., the correct or wrong German translation, respectively, *light grey bar*), the cueing related effects on memory vanished, as there was no difference in memory performance to uncued words (*white bar*). Replaying single Dutch words (*black bar*) again improved memory performance. The same effect emerged when Dutch words were followed by delayed correct feedback (1500 ms instead of 200 ms). Retrieval performance is indicated as percentage of recalled German translations with performance before sleep set to 100%. Values are means ± s.e.m. $*P \leq 0.05$

performance for the correct German translation, whereas the recall performance for uncued words was not different from that of a group receiving no verbal cues during sleep at all (Fig. 1b). When participants did not sleep but stayed awake after learning in the evening, cueing did not improve the recall performance (Fig. 1c). This pattern of results implies that the benefits of auditory cueing exceeds the normal consolidation effects of sleep on memory performance and provides evidence for the sleep specificity of the TMR approach.

While controlling for circadian influences on cognitive states, fatigue during learning and retrieval might have influenced the effects of cueing on memory performance. To exclude this potential confounding factor, the authors investigated additionally the effect of vocabulary cueing during daytime wakefulness, using the same learning material (Schreiner and Rasch 2015b). However, cueing did not increase the number of recalled word pairs although the subjects were well rested during retrieval testing.

In both studies only the Dutch words were replayed without the respective German translation. The rationale behind this approach is that presenting solely the memory cue (the Dutch word) should trigger the reactivation of the memory in its entirety (Dutch-German association). It is, however, not known if replaying the whole word-pair (Dutch word followed by its German translation) during sleep results in an additional memory benefit.

When the Dutch word was followed after 200 ms by the correct German translation, the beneficial cueing effects vanished completely and memory performance did not differ from uncued word pairs (Schreiner et al. 2015a, b, see Fig. 1d). The presentation of the Dutch word followed by a false German translation or a pure tone, thus a stimulus without any prior learned content, also abolished the cueing related memory benefits. Interestingly, the time lag between the presentation of the two words seems to be critical. When the correct German translation followed by an extended inter-stimulus interval of 1500 ms, the beneficial cueing effect was restored again. Thus, the abolition of the cueing effect seems to be independent from the second words content. Rather the timing of the second stimulus, thus the tight time lag between the Dutch word and the subsequent German stimulus might have led to the extinction of cueing related benefits.

In sum, these studies provide first evidence that presenting prior learned foreign vocabularies during subsequent sleep effectively induces reactivation processes and thereby enhances memory performance for the respective translation. The question remains, however, whether other language-related aspects similarly profit from the TMR approach. Batterink and Paller (2015) successfully filled this gap, by demonstrating that grammatical generalization can be boosted by cueing during sleep as well. As in prior studies on sleep and grammar learning, subjects learned grammatical rules of an artificial language. Subsequently, phrases from the previously learned language were repeatedly presented during an afternoon nap. When re-exposed to the phrases during sleep, participants showed enhanced grammatical generalization after the nap, indicating that grammar learning may profit from cueing during sleep.

Oscillatory Correlates of Vocabulary Cueing During Sleep

Our understanding of processes supporting enduring plastic changes after reactivation has been improved by the description of the underlying oscillatory mechanisms. The exact contribution of specific frequency bands and their interplay, however, remain to be identified.

Oscillatory activity in the theta (4–7 Hz) and gamma (25–100 Hz) range during encoding and recall testing have been persistently associated with successfull memory consolidation (for reviews see Düzel et al. 2010; Nyhus and Curran 2010). Synchronized neural activity in the theta and gamma frequency range have been linked to synaptic processes and induced long term potentiation (LTP) in the hippocampus, thereby enabling memory formation (Axmacher et al. 2006; Huerta and Lisman 1995; Hyman et al. 2003). In addition, the interplay of theta and gamma is thought to bind individual memory components and their temporal ordering (Lisman and Jensen 2013). However, during retrieval of declarative memories, theta and gamma oscillations could cause the reinstatement of the entire memory representations in the cortex via feedback projections from the hippocampus to the cortex (Nyhus and Curran 2010). Moreover, theta activity seems to reflect successful integration of newly learned words in the pre-existing vocabulary (Bakker et al. 2015) and the strength of episodic memory traces (Klimesch et al. 2006).

Accordingly, power in theta and gamma activity is usually increased during successful encoding and the correct recognition of words (Mormann et al. 2005; Osipova et al. 2006; Trimper et al. 2014). In line with these findings, enhanced theta and gamma activity has been observed when Dutch words were correctly identified during a recognition test and the elicited theta activity was even stronger for words that were cued during prior sleep (Schreiner et al. 2015a, b), which possibly reflects the preferential integration of replayed words into the mental lexicon (Bakker et al. 2015) and memory trace strength (Klimesch et al. 2006).

Apart from wakefulness, successful cueing during NREM sleep might be similarly associated with elevated theta activity (Schreiner and Rasch 2015a; Schreiner et al. 2015a, b; Lehmann et al. 2016).

Cueing of Dutch words that was most beneficial for later recall (i.e., gained words: not remembered before sleep, but successfully recalled after sleep) was associated with the strongest increases in induced theta power and sleep spindle activity (Fig. 2c).

When a tone, the correct or an incorrect German translation was presented immediately after the Dutch word, this characteristic increase in theta and spindle activity almost completely vanished (Schreiner et al. 2015a, b; Fig. 2d). Notably, blocking oscillatory activity in those frequency bands was paralleled by a corresponding abolition of memory benefits during later recall. Accordingly, when the verbal feedback was presented with a delay of 1500 ms, both, cueing related increases in theta and spindle oscillation and behavioural memory benefits

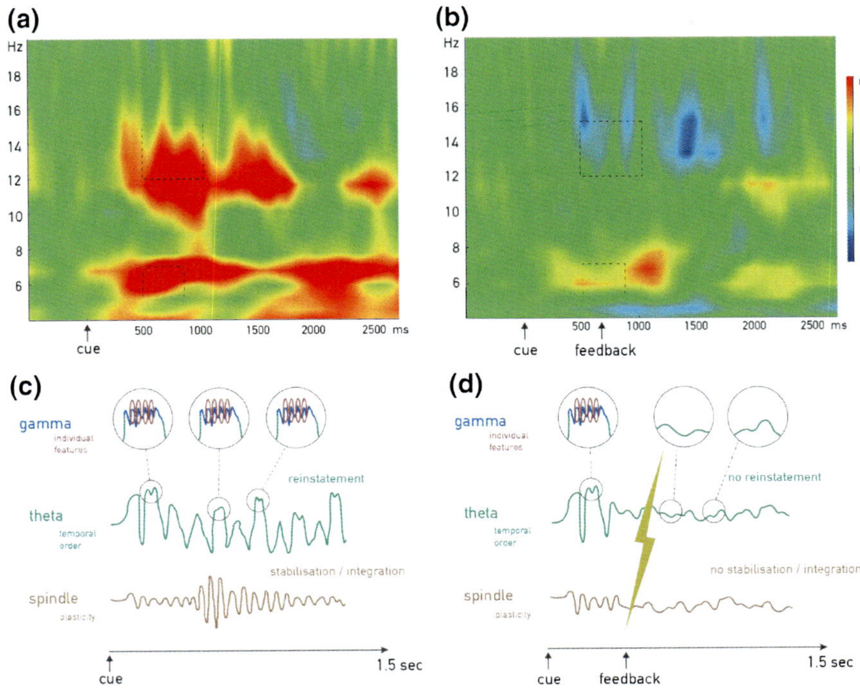

Fig. 2 Working model and empirical findings for neural oscillatory mechanisms underlying the memory benefits of cueing during sleep. **a** In a recent study (Schreiner et al. 2015b), successful cueing of Dutch words during NREM sleep was associated with enhanced power in the theta and spindle band (representative electrode F3). Verbal cues were presented at time 0 ms. **b** The differences in the theta and spindle band vanished when Dutch cues were immediately followed by verbal feedback. Importantly this effect was paralleled by a blockade of memory benefits of cueing on the behavioural level, indicating the critical role of theta and spindle activity for successful stabilization of memories by cueing during sleep. Values are increase in oscillatory power in %. **c** The presentation of a memory-related cue during NREM causes the hippocampus dependent reinstatement of the entire memory representation in the cortex, which is coded in theta and gamma activity. The theta-gamma interaction (in *green* and *blue*, respectively) provides the means for the temporal ordering and binding of single memory features. Spindle activity is critical for facilitating local plastic processes underlying the stabilization and integration of reactivated memories into long-term knowledge networks. According to our model, the successful reinstatement (reflected by increase theta power on the surface and probably associated with hippocampal sharp-wave ripple activity) together with the simultaneous or slightly delayed emergence of sleep spindles (*brown*) is a crucial prerequisite to successfully stabilize memories during sleep upon their reactivation. **d** Disturbing these combined oscillatory signals by presenting further auditory stimuli in close proximity disrupts the reinstatement (theta, gamma) and stabilization/integration (spindle) process and blocks the beneficial effect of cueing during sleep. Delaying further auditory stimulation by a least 1.5 s after the cue is sufficient for a successful completion of the stabilization and integration process (not shown)

re-emerged, indicating that a critical time window exists for the occurrence of oscillatory mechanisms that are relevant for successful memory cueing during sleep.

Given the crucial role of slow oscillations in models of memory consolidation during sleep (Rasch and Born 2013), it is assumed that slow oscillations precisely orchestrate reactivation processes, driving the dialogue between subcortical structures and the neocortex (Diekelmann and Born 2010). Considering that a slow-oscillation (0.5–1.5 Hz) might be required to complete one cycle after the presentation of a memory related cue in order to improve memory strength, the critical time window would be between 0.75 and 2 s, which corresponds exactly to the observation.

In agreement with this interpretation, there is ample evidence that slow oscillations facilitate the temporally grouping of spindle and ripple activity (Clemens et al. 2007; Staresina et al. 2015) (for details, see chapter by Bergmann and Staresina). Slow oscillations might have a similar grouping effect on theta activity, with theta activity being primarily bundled in the depolarized up-states. After presentation of a memory cue during sleep, the co-occurrence of a slow oscillations up-states and related oscillatory processes (endogenous reactivation signals, SW-R, spindles/theta oscillations etc.) might be needed in order to let memory benefits unfold. Accordingly, immediate presentation of a second stimulus might interfere with the formation of a synchronized up-state and thereby block the memory enhancing effects of cueing.

Overall, successful cueing of vocabulary during sleep is closely tied to increased activity in the theta and sleep spindle range, suggesting that their interplay with slow oscillations is critical for stabilizing memories traces during NREM sleep.

A Working Model for Stabilizing Reactivated Memories During Sleep

In contrast to slow oscillations and sleep spindles, oscillations in the theta band are currently not embedded in models of memory processes acting during NREM sleep. Nonetheless, a growing number of findings indicate that theta oscillations during NREM sleep play a substantial role in the processing of memories.

In patients suffering from Alzheimer's disease and amnestic mild cognitive impairment, and similarly in healthy subjects enhanced theta power and additionally faster theta frequency during NREM sleep predicts better memory performance (Hot et al. 2011; Schabus et al. 2005; Westerberg et al. 2012). Moreover, processing of rare stimuli in the context of a mismatch paradigm is associated with enhanced theta activity (Karakaş et al. 2007) and salient stimulus presentation during sleep is similarly linked to stronger theta synchronization (Blume et al. 2016), implicating an involvement of theta activity in auditory related sensory-memory processes during sleep. In addition, it has been shown that spike timing during theta oscillations, nested in the delta rhythm, might supervise reciprocal interactions between superficial and deep cortical layers, thereby mimicking the cortical dynamics with regards to memory and sensory processes during wakefulness (Carracedo et al. 2013).

Based on the above presented findings, one might conclude that theta activity during NREM sleep is similarly involved in sleep-related reactivation and stabilization processes. However, the precise involvement of the theta rhythm in consolidation mechanisms during sleep and the possible interaction with slow oscillations and sleep spindles have to be further elucidated.

Here, we suggest in a working model, that theta activity might represent the cueing related reinstatement of a memory representation (see Fig. 2). Accordingly, given that the connection between the cue and the associated memory is sufficiently strong and the cue is replayed in a specific oscillatory state (e.g. up-state of the SO), power in the theta band increases after the presentation of the cue, reinstating the associated memory trace. We suggest that a reinstatement-related enhancement of theta synchrony occurs regardless of the global brain state and can therefore occur during waking, NREM and REM sleep.

In parallel to a model derived from waking studies (Nyhus and Curran 2010), we suggest that theta and gamma oscillations provoke the hippocampus-related reinstatement of memory representations in cortical areas, not only during wakefulness but as well during NREM. Beyond that, theta activity might initiate synaptic potentiation (Larson et al. 1986) and coordinate activity between the prefrontal cortex and the hippocampus (Benchenane et al. 2010).

The fate of reactivated memories is most likely not ultimately determined by the occurrence or absence of changes in theta activity but possibly also by the brains macro state: reactivated memories during waking might be bolstered (in case that they are actively and repeatedly encoded or that feedback is given), unaffected or actually destabilized (when interfering input engages after reactivation) (Nader and Hardt 2009). In the context of NREM sleep, however, reactivation processes always lead to a strengthening of associated memories in case that stabilization processes may unfold without interruption (this question is also discussed in the chapter by Zhang, Deuker and Axmacher). We propose that an increase in theta power together with the simultaneous or slightly delayed emergence of sleep spindles is a crucial prerequisite to facilitate this stabilizing effect. According to the active system consolidation hypothesis, sleep spindles play a pivotal role in the transfer of reactivated memories from the hippocampus to neocortical sites (Born and Wilhelm 2012): specifically, hippocampal signals of reactivation are assumed to be nested in individual troughs of spindles (Mölle et al. 2009; Staresina et al. 2015; Siapas and Wilson 1998; Sirota and Buzsáki 2005) (see chapter by Bergmann and Staresina). Furthermore, sleep spindles are thought to prime and maintain LTP in cortical circuits by provoking Ca^{2+} influx for successive plastic processes (Contreras et al. 1997).

Interestingly, it has been demonstrated in rodents that reactivation processes seem to occur briefly before spindles emerge (Peyrache et al. 2009) and that sleep spindles are not inevitably associated with hippocampal activity (Andrade et al. 2011). Moreover, thalamic and hippocampal inputs to the cortex might even be inhibited during sleep spindles (Peyrache et al. 2011), suggesting that sleep spindles might be involved in supporting local plasticity of previously reactivated memories rather than reactivation and redistribution of memories itself (Genzel et al. 2014). In line with this assumption, not sleep spindles per se but theta activity might reflect more directly the successful reactivation and reinstatement of a memory trace, possibly with a strong association with hippocampal sharp-wave ripple activity. Thus, a successful reinstatement (reflected by theta oscillations) accompanied or immediately followed by spindle oscillations might be required to successfully cause plastic changes that underlie stabilization and integration of memories during sleep (see Fig. 2a). Interrupting the interplay of those oscillatory mechanisms by e.g. replaying auditory cues following effective reactivations interrupts associated reinstatement and stabilization (theta/spindle) processes and thereby disables the beneficial effects of TMR during NREM sleep (Fig. 2b). In support for the functionally different role of theta and spindle activity, it has been shown that memory reactivations in the context of REM sleep similarly result in increased theta synchronization (successful reinstatement of the related memory), but might not stabilize memory traces to same extent due to absence of oscillatory activity in the spindle range and thus reactivations result not in enhanced memory consolidation (Lehmann et al. 2016).

The active system consolidation theory ascribes slow oscillations the role of a time-giving pace maker, which orchestrates neural processes associated with memory reactivation. Whereas successfully cued memories are more often succeeded by slow waves, not remembered word pairs were as well followed by slow oscillations (Schreiner and Rasch 2015a, b; Schreiner et al. 2015a, b). Likewise, enhancing solely slow wave activity by pharmaceuticals without enhancing spindle activity fails to increase consolidation processes during sleep (Feld et al. 2013), while the concurrent enhancement of slow wave and sleep spindle activity by pharmaceutical, electric or auditory stimulation has proven to be effective (Marshall et al. 2006; Mednick et al. 2013; Ngo et al. 2013, 2015) (see chapter by Campos Beltran and Marshall). Accordingly, slow oscillations are not exclusively effective, but might constitute a relevant precondition for successful cueing by regulating the temporal succession of reactivation related processes and moreover the exchange between subcortical and neocortical structures. Within a certain time window, theta and spindle activities might be critical for the reinstatement of memories and their subsequent stabilization following cueing during sleep. However, the question remains, whether presenting auditory cues during the up-states of slow oscillations, which are reflecting higher synchronicity in neural firing (Destexhe et al. 2007), would further increase cueing-related improvement of language learning.

Conclusion

Summing up, the research reviewed in this chapter provides evidence for the key role of sleep in the context of language learning. It has been demonstrated that the sleeping brain fosters several facets of language learning, be it the learning of new words or generalization and abstraction processes with regards to grammar acquisition (Batterink et al. 2014; Henderson et al. 2012). Thus, sleep might represent a favourable neuronal state to expedite and alleviate those learning processes.

First evidence, that TMR during sleep can be used to improve foreign vocabulary and grammar learning crucially expands the scope of research on sleep and language learning. Findings, reviewed in this chapter set the stage for upcoming advances in the field, ranging from a better understanding concerning the memory function of sleep to the development of technical applications and their implementation in educational settings, in order to facilitate language learning.

Acknowledgements This work was supported by a grant from the Swiss National Science Foundation (SNSF) (100014_162388) and the research pool of the University of Fribourg. Thomas Schreiner is supported by a grant of the SNSF (P2ZHP1_164994) and Mick Lehmann is supported by the Forschungskredit of the University of Zurich, grant no. FK-16-044.

References

Andrade KC, Spoormaker VI, Dresler M, Wehrle R, Holsboer F, Sämann PG, Czisch M (2011) Sleep spindles and hippocampal functional connectivity in human NREM sleep. J Neurosci 31(28):10331–10339. doi:10.1523/JNEUROSCI.5660-10.2011

Axmacher N, Mormann F, Fernández G, Elger CE, Fell J (2006) Memory formation by neuronal synchronization. Brain Res Rev 52(1):170–182. doi:10.1016/j.brainresrev.2006.01.007

Bakker I, Takashima A, van Hell JG, Janzen G, McQueen JM (2015) Changes in theta and beta oscillations as signatures of novel word consolidation. J Cogn Neurosci 27(7):1286–1297

Barrett M (1989) Early language development. In: Slater A, Bremner G (eds) Infant development, Lawrence Erlbaum Associates Ltd, Hove, UK

Bastuji H, García-Larrea L (1999) Evoked potentials as a tool for the investigation of human sleep. Sleep Med Rev 3(1):23–45

Batterink LJ, Oudiette D, Reber PJ, Paller KA (2014) Sleep facilitates learning a new linguistic rule. Neuropsychologia 65C:169–179. doi:10.1016/j.neuropsychologia.2014.10.024

Batterink LJ, Paller KA (2015) Sleep-based memory processing facilitates grammatical generalization: evidence from targeted memory reactivation. Brain Lang. doi:10.1016/j.bandl.2015.09.003

Benchenane K, Peyrache A, Khamassi M, Tierney PL, Gioanni Y, Battaglia FP, Wiener SI (2010) Coherent theta oscillations and reorganization of spike timing in the hippocampal-prefrontal network upon learning. Neuron 66(6):921–936. doi:10.1016/j.neuron.2010.05.013

Blume C, del Giudice R, Lechinger J, Wislowska M, Heib DP, Hoedlmoser K, et al (2016) Preferential processing of emotionally and self-relevant stimuli persists in unconscious N2 sleep. Brain Lang doi:10.1016/j.bandl.2016.02.004.

Born J, Wilhelm I (2012) System consolidation of memory during sleep. Psychol Res 76(2): 192–203. doi:10.1007/s00426-011-0335-6

Brandone AC, Salkind SJ, Golinkoff RM, Hirsh-Pasek K (2006) Language development. Children's needs III: development, prevention, and intervention. Washington, DC, US, National Association of School Psychologists, pp 499–514

Brown H, Weighall A, Henderson LM, Gaskell MG (2012) Enhanced recognition and recall of new words in 7- and 12-year-olds following a period of offline consolidation. J Exp Child Psychol 112(1):56–72. doi:10.1016/j.jecp.2011.11.010

Brualla J, Romero MF, Serrano M, Valdizán JR (1998) Auditory event-related potentials to semantic priming during sleep. Electroencephalogr Clin Neurophysiol 108:283–290. doi:10.1016/S0168-5597(97)00102-0

Cairney SA, Lindsay S, Sobczak JM, Paller KA, Gaskell MG (2016) The benefits of targeted memory reactivation for consolidation in sleep are contingent on memory accuracy and direct cue-memory associations. Sleep 39(5):1139–1150

Carracedo LM, Kjeldsen H, Cunnington L, Jenkins A, Schofield I, Cunningham MO, Whittington MA et al (2013) A neocortical delta rhythm facilitates reciprocal interlaminar interactions via nested theta rhythms. J Neurosci 33(26):10750–10761. doi:10.1523/JNEUROSCI.0735-13.2013

Clemens Z, Molle M, Eross L, Barsi P, Halasz P, Born J (2007) Temporal coupling of parahippocampal ripples, sleep spindles and slow oscillations in humans. Brain 130(11): 2868–2878. doi:10.1093/brain/awm146

Contreras D, Destexhe A, Steriade M (1997) Intracellular and computational characterization of the intracortical inhibitory control of synchronized thalamic inputs in vivo. J Neurophysiol 78(1):335–350

Cox R, Korjoukov I, de Boer M, Talamini LM (2014) Sound asleep: processing and retention of slow oscillation phase-targeted stimuli. PLoS ONE 9(7):e101567. doi:10.1371/journal.pone.0101567

Creery JD, Oudiette D, Antony JW, Paller KA (2015) Targeted memory reactivation during sleep depends on prior learning. Sleep 38(5):755–763. doi:10.5665/sleep.4670

Dang-Vu TT, Bonjean M, Schabus M, Boly M, Darsaud A, et al (2011) Interplay between spontaneous and induced brain activity during human non-rapid eye movement sleep. Proc Natl Acad Sci USA 108:15438–15443

Dave A, Marholiash D (2000) Song replay during sleep and computational rules for sensorimotor vocal learning. Science 290(5492):812–816. doi:10.1126/science.290.5492.812

Davis MH, Gaskell MG (2009) A complementary systems account of word learning: neural and behavioural evidence. Philos Trans R Soc Lond B Biol Sci 364(1536):3773–3800. doi:10.1098/rstb.2009.0111

Destexhe A, Hughes SW, Rudolph M, Crunelli V (2007) Are corticothalamic "up"states fragments of wakefulness? Trends Neurosci. doi:10.1016/j.tins.2007.04.006

Diekelmann S (2014) Sleep for cognitive enhancement. Front Syst Neurosci 8:46. doi:10.3389/fnsys.2014.00046

Diekelmann S, Born J (2010) The memory function of sleep. Nat Rev Neurosci 11(2):114–126. doi:10.1038/nrn2762

Diekelmann S, Büchel C, Born J, Rasch B (2011) Labile or stable: opposing consequences for memory when reactivated during waking and sleep. Nat Neurosci 14(3):381–386

Dumay N, Gaskell MG (2007) Sleep-associated changes in the mantal representation of spoken words. Psychol Sci 18(1):35–39

Düzel E, Penny WD, Burgess N (2010) Brain oscillations and memory. Curr Opin Neurobiol. doi:10.1016/j.conb.2010.01.004

Edeline JM (2003) The thalamo-cortical auditory receptive fields: Regulation by the states of vigilance, learning and the neuromodulatory systems. In: Experimental brain research, vol. 153, pp. 554–572. doi:10.1007/s00221-003-1608-0

Edeline JM, Manunta Y, Hennevin E (2000) Auditory thalamus neurons during sleep: changes in frequency selectivity, threshold, and receptive field size. J Neurophysiol 84(2):934–952

Eschenko O, Ramadan W, Mölle M, Born J, Sara SJ (2008) Sustained increase in hippocampal sharp-wave ripple activity during slow-wave sleep after learning. Learn Mem, 15:222–228, Cold Spring Harbor, NY. doi:10.1101/lm.726008

Feld GB, Wilhelm I, Ma Y, Groch S, Binkofski F, Mölle M, Born J (2013) Slow wave sleep induced by GABA agonist tiagabine fails to benefit memory consolidation. Sleep 36(9): 1317–1326. doi:10.5665/sleep.2954

Fenn KM, Nusbaum HC, Margoliash D (2003) Consolidation during sleep of perceptual learning of spoken language. Nature, vol. 425

Fischer S, Drosopoulos S, Tsen J, Born J (2006) Implicit learning—explicit knowing: a role for sleep in memory system interaction. J Cogn Neurosci 18(3):311–319. doi:10.1162/jocn.2006.18.3.311

Frankland PW, Bontempi B (2005) The organization of recent and remote memories. Nat Rev Neurosci 6(2):119–130. doi:10.1038/nrn1607

Friedrich M, Wilhelm I, Born J, Friederici AD (2015) Generalization of word meanings during infant sleep. Nat Commun 6:6004. doi:10.1038/ncomms7004

Fuentemilla L, Miró J, Ripollés P, Vilà-Balló A, Juncadella M, Castañer S, Rodríguez-Fornells A (2013) Hippocampus-dependent strengthening of targeted memories via reactivation during sleep in humans. Curr Biol 23(18):1769–1775. doi:10.1016/j.cub.2013.07.006

Gabrieli JD (1998) Cognitive neuroscience of human memory. Annu Rev Psychol 49:87–115. doi:10.1146/annurev.psych.49.1.87

Gais S, Lucas B, Born J (2006) Sleep after learning aids memory recall. Learn Mem, 13(3):259–262, Cold Spring Harbor, NY. doi:10.1101/lm.132106

Gaskell MG, Dumay N (2003) Lexical competition and the acquisition of novel words. Cognition 89(2):105–132. doi:10.1016/S0010-0277(03)00070-2

Genzel L, Kroes MCW, Dresler M, Battaglia FP (2014) Light sleep versus slow wave sleep in memory consolidation: a question of global versus local processes? Trends Neurosci 37(1):9–10. doi:10.1016/j.tins.2013.10.002

Gómez RL, Bootzin RR, Nadel L (2006) Naps promote abstraction in language-learning infants. Psychol Sci 17(8):670–674. doi:10.1111/j.1467-9280.2006.01764.x

Henderson LM, Weighall AR, Brown H, Gareth Gaskell M (2012) Consolidation of vocabulary is associated with sleep in children. Dev Sci 15(5):674–687. doi:10.1111/j.1467-7687.2012.01172.x

Horvath K, Myers K, Foster R, Plunkett K (2015) Napping facilitates word learning in early lexical development. J Sleep Res 24(5):503–509. doi:10.1111/jsr.12306

Hot P, Rauchs G, Bertran F, Denise P, Desgranges B, Clochon P, Eustache F (2011) Changes in sleep theta rhythm are related to episodic memory impairment in early Alzheimer's disease. Biol Psychol 87(3):334–339

Huerta PT, Lisman JE (1995) Bidirectional synaptic plasticity induced by a single burst during cholinergic theta oscillation in CA1 in vitro. Neuron 15:1053–1063. doi:10.1016/0896-6273(95)90094-2

Hupbach A, Gomez RL, Bootzin RR, Nadel L (2009) Nap-dependent learning in infants. Dev Sci 12(6):1007–1012. doi:10.1111/j.1467-7687.2009.00837.x

Hyman JM, Wyble BP, Goyal V, Rossi CA, Hasselmo ME (2003) Stimulation in hippocampal region CA1 in behaving rats yields long-term potentiation when delivered to the peak of theta and long-term depression when delivered to the trough. J Neurosci 23, 11725–11731 (23/37/11725 [pii])

Ibáñez A, López V, Cornejo C (2006) ERPs and contextual semantic discrimination: degrees of congruence in wakefulness and sleep. Brain Lang 98(3):264–275

Issa EB, Wang X (2008) Sensory responses during sleep in primate primary and secondary auditory cortex. J Neurosci 28:14467–14480. doi:10.1523/JNEUROSCI.3086-08.2008

Ji D, Wilson MA (2007) Coordinated memory replay in the visual cortex and hippocampus during sleep. Nat Neurosci 10(1):100–107. doi:10.1038/nn1825

Karakaş S, Cakmak ED, Bekçi B, Aydin H (2007) Oscillatory responses representing differential auditory processing in sleep. Int J Psychophysiol: Official J Int Organ Psychophysiol 65(1):40–50. doi:10.1016/j.ijpsycho.2007.02.010

Klimesch W, Hanslmayr S, Sauseng P, Gruber W, Brozinsky CJ, Kroll NEA, Doppelmayr M et al (2006) Oscillatory EEG correlates of episodic trace decay. Cereb Cortex 16:280–290. doi:10.1093/cercor/bhi107

Kurdziel LBF, Spencer RMC (2015) Consolidation of novel word learning in native English-speaking adults. Memory, 8211(April): 1–11, Hove, England. doi:10.1080/09658211.2015.1019889

Larson J, Wong D, Lynch G (1986) Patterned stimulation at the theta frequency is optimal for the induction of hippocampal long-term potentiation. Brain Res 368(2):347–350. doi:10.1016/0006-8993(86)90579-2

Lau H, Alger SE, Fishbein W (2011) Relational memory: a daytime nap facilitates the abstraction of general concepts. PLoS ONE, 6(11). doi:10.1371/journal.pone.0027139

Lehmann M, Schreiner T, Seifritz E, Rasch B (2016) Emotional arousal modulates oscillatory correlates of targeted memory reactivation during NREM, but not REM sleep. Sci Rep 6:39229. 10.1038/srep39229

Lewis PA, Durrant SJ (2011) Overlapping memory replay during sleep builds cognitive schemata. Trends Cogn Sci 15(8):343–351. doi:10.1016/j.tics.2011.06.004

Lisman JE, Jensen O (2013) The theta-gamma neural code. Neuron 77(6):1002–1016. doi:10.1016/j.neuron.2013.03.007

Marr D (1971) Simple memory: a theory for archicortex. Philos Trans R Soc Lond Series B Biol Sci 262:23–81. doi:10.1098/rstb.1971.0078

Marshall L, Helgadóttir H, Mölle M, Born J (2006) Boosting slow oscillations during sleep potentiates memory. Nature 444, Nature Publishing Group

McClelland JL, McNaughton BL, O'Reilly RC (1995) Why there are complimentary learning systems in the hippocampus and neocortex: Insights from the successes and failures of connectionist models of learning and memory. Psychol Rev 102:419–457

McGaugh JL (2000) Memory–a century of consolidation. Science 287(5451):248–251, New York, NY

Mednick SC, McDevitt EA, Walsh JK, Wamsley E, Paulus M, Kanady JC, Drummond SPA (2013) The critical role of sleep spindles in hippocampal-dependent memory: a pharmacology study. J Neurosci 33(10):4494–4504. doi:10.1523/JNEUROSCI.3127-12.2013

Mölle M, Eschenko O, Gais S, Sara SJ, Born J (2009) The influence of learning on sleep slow oscillations and associated spindles and ripples in humans and rats. Eur J Neurosci 29:1071–1081. doi:10.1111/j.1460-9568.2009.06654.x

Mormann F, Fell J, Axmacher N, Weber B, Lehnertz K, Elger CE, Fernández G (2005) Phase/amplitude reset and theta-gamma interaction in the human medial temporal lobe during a continuous word recognition memory task. Hippocampus 15(7):890–900. doi:10.1002/hipo.20117

Nader K, Hardt O (2009) A single standard for memory: the case for reconsolidation. Nat Rev Neurosci 10(3):224–234. doi:10.1038/nrn2590

Ngo HVV, Martinetz T, Born J, Mölle M (2013) Auditory closed-loop stimulation of the sleep slow oscillation enhances memory. Neuron 78(3):545–553. doi:10.1016/j.neuron.2013.03.006

Ngo H-VV, Miedema A, Faude I, Martinetz T, Molle M, Born J (2015) Driving sleep slow oscillations by auditory closed-loop stimulation–a self-limiting process. J Neurosci 35 (17):6630–6638. doi:10.1523/JNEUROSCI.3133-14.2015

Nieuwenhuis ILC, Folia V, Forkstam C, Jensen O, Petersson KM (2013) Sleep promotes the extraction of grammatical rules. PLoS ONE 8. doi:10.1371/journal.pone.0065046

Nyhus E, Curran T (2010) Functional role of gamma and theta oscillations in episodic memory. Neurosci Biobehav Rev 34(7):1023–1035. doi:10.1016/j.neubiorev.2009.12.014

O'Neill J, Pleydell-Bouverie B, Dupret D, Csicsvari J (2010) Play it again: reactivation of waking experience and memory. Trends Neurosci 33(5):220–229. Retrieved from http://www.ncbi.nlm.nih.gov/pubmed/20207025

Osipova D, Takashima A, Oostenveld R, Fernández G, Maris E, Jensen O (2006) Theta and gamma oscillations predict encoding and retrieval of declarative memory. J Neurosci 26 (28):7523–7531. doi:10.1523/JNEUROSCI.1948-06.2006

Oswald I, Taylor AM, Treisman M (1960) Discriminative responses to stimulation during human sleep. Brain 83(3):440–453. doi:10.1093/brain/83.3.440

Oudiette D, Paller KA (2013) Upgrading the sleeping brain with targeted memory reactivation. Trends Cogn Sci (3):142–149

Pavlides C, Winson J (1989) Influences of hippocampal place cell firing in the awake state on the activity of these cells during subsequent sleep episodes. J Neurosci 9(8):2907–2918

Peigneux P, Laureys S, Fuchs S, Collette F, Perrin F, Reggers J, Maquet P et al (2004) Are spatial memories strengthened in the human hippocampus during slow wave sleep? Neuron 44(3):535–545. doi:10.1016/j.neuron.2004.10.007

Pennartz CMA, Lee E, Verheul J, Lipa P, Barnes CA, McNaughton BL (2004) The ventral striatum in off-line processing: ensemble reactivation during sleep and modulation by hippocampal ripples. J Neurosci 24:6446–6456. doi:10.1523/JNEUROSCI.0575-04.2004

Perrin F, Bastuji H, Garcia-Larrea L (2002) Detection of verbal discordances during sleep. Neuroreport 13(10):1345–1349

Perrin F, Garcia-Larrea L, Mauguiere F, Bastuji H (1999) A differential brain response to the subject's own name persists during sleep. Clin Neurophysiol 110(12):2153–2164. doi:10.1016/S1388-2457(99)00177-7

Peyrache A, Battaglia FP, Destexhe A (2011) Inhibition recruitment in prefrontal cortex during sleep spindles and gating of hippocampal inputs. Proc Natl Acad Sci USA 108(41). 10.1073/pnas.1103612108

Peyrache A, Khamassi M, Benchenane K, Wiener SI, Battaglia FP (2009) Replay of rule-learning related neural patterns in the prefrontal cortex during sleep. Nat Neurosci 12:919–926. doi:10.1038/nn.2337

Pinker S (2000) Survival of the clearest. Nature 404:441–442

Portas CM, Krakow K, Allen P, Josephs O, Armony JL, Frith CD (2000) Auditory processing across the sleep-wake cycle: simultaneous EEG and fMRI monitoring in humans. Neuron 28:991–999. doi:10.1016/S0896-6273(00)00169-0

Pratt H, Berlad I, Lavie P (1999) "Oddball" event-related potentials and information processing during REM and non-REM sleep. Clin Neurophysiol 110(1):53–61

Rasch B, Born J (2013) About sleep's role in memory. Physiol Rev 93(2):681–766. doi:10.1152/physrev.00032.2012

Rasch B, Büchel C, Gais S, Born J (2007) Odor cues during slow-wave sleep prompt declarative memory consolidation. Science 315(5817):1426–1429. Retrieved from http://www.ncbi.nlm.nih.gov/pubmed/17347444

Reivich M, Isaacs G, Evarts E, Kety S (1968) The effect of slow wave sleep and REM sleep on regional cerebral blood flow in cats. J Neurochem 15(4):301–306

Rihm JS, Diekelmann S, Born J, Rasch B (2014) Reactivating memories during sleep by odors: odor specificity and associated changes in sleep oscillations. J Cogn Neurosci 26:1806–1818. doi:10.1162/jocn_a_00579

Rosanova M, Ulrich D (2005) Pattern-specific associative long-term potentiation induced by a sleep spindle-related spike train. J Neurosci 25(41):9398–9405. doi:10.1523/JNEUROSCI.2149-05.2005

Rudoy JD, Voss JL, Westerberg CE, Paller KA (2009) Strengthening individual memories by reactivating them during sleep. Science 326(5956):1079. Retrieved from http://www.ncbi.nlm.nih.gov/pubmed/19965421

Sadowski JHLP, Jones MW, Mellor JR (2016) Sharp-wave ripples orchestrate the induction of synaptic plasticity during reactivation of place cell firing patterns in the hippocampus. Cell Reports. doi:10.1016/j.celrep.2016.01.061

Schabus M, Hoedlmoser K, Pecherstorfer T, Kloesch G (2005) Influence of midday naps on declarative memory performance and motivation. Somnologie 9:148–153

Schönauer M, Geisler T, Gais S (2013) Strengthening procedural memories by reactivation in sleep. J Cogn Neurosci 1–11. doi:10.1162/jocn_a_00471

Schreiner T, Rasch B (2015a) Boosting vocabulary learning by verbal cueing during sleep. Cerebral Cortex 25(11):4169–4179, New York, NY: 1991. doi:10.1093/cercor/bhu139

Schreiner T, Rasch B (2015b) Cueing vocabulary in awake subjects during the day has no effect on memory. Somnologie - Schlafforschung Und Schlafmedizin 19(2):133–140. doi:10.1007/s11818-015-0005-9

Schreiner T, Göldi M, Rasch B (2015a) Cueing vocabulary during sleep increases theta activity during later recognition testing. Psychophysiology 52(11):1538–1543. doi:10.1111/psyp.12505

Schreiner T, Lehmann M, Rasch B (2015b) Auditory feedback blocks memory benefits of cueing during sleep. Nat Commun 6:8729. doi:10.1038/ncomms9729

Shanon B (1979) Semantic processing during sleep. Bull Psychon Soc 14:382. doi:10.3758/BF0332948

Siapas AG, Wilson MA (1998) Coordinated interactions between hippocampal ripples and cortical spindles during slow-wave sleep. Neuron 21:1123–1128. doi:10.1016/S0896-6273(00)80629-7

Sirota A, Buzsáki G (2005) Interaction between neocortical and hippocampal networks via slow oscillations. Thalamus Relat Syst 3(04):245. doi:10.1017/S1472928807000258

Staresina BP, Bergmann TO, Bonefond M, van der Meij R, Jensen O, Deuker L, Elger CE, Axmacher N, Fell, J (2015) Hierarchical nesting of slow oscillations, spindles and ripples in the human hippocampus during sleep. Nat Neurosci 18:1679–1686. doi:10.1038/nn.4119

Steriade M (2006) Grouping of brain rhythms in corticothalamic systems. Neuroscience 137:1087–1106. doi:10.1016/j.neuroscience.2005.10.029

Stickgold R, Walker MP (2013) Sleep-dependent memory triage: evolving generalization through selective processing. Nat Neurosci 16(2):139–145. doi:10.1038/nn.3303

Strauss M, Sitt JD, King J-R, Elbaz M, Azizi L, Buiatti M, Dehaene S et al (2015). Disruption of hierarchical predictive coding during sleep. Proc Natl Acad Sci 201501026. doi:10.1073/pnas.1501026112

Tamminen J, Payne JD, Stickgold R, Wamsley EJ, Gaskell MG (2010) Sleep spindle activity is associated with the integration of new memories and existing knowledge. J Neurosci 30:14356–14360. doi:10.1523/JNEUROSCI.3028-10.2010

Tham EKH, Lindsay S, Gaskell MG (2015) Markers of automaticity in sleep-associated consolidation of novel words. Neuropsychologia 71:146–157. doi:10.1016/j.neuropsychologia.2015.03.025

Timoveef I, Chauvette S (2011) Thalamocortical oscillations: local control of EEG slow waves. Curr Top Med Chem 11(19):2457–2471

Trimper JB, Stefanescu RA, Manns JR (2014) Recognition memory and theta-gamma interactions in the hippocampus. Hippocampus 24(3):341–353. doi:10.1002/hipo.22228

Wagner U, Gais S, Haider H, Verleger R, Born J (2004) Sleep inspires insight. Nature 427:21–24

Walker MP, Stickgold R (2010) Overnight alchemy: sleep-dependent memory evolution. Nat Rev Neurosci 11(3):218. http://doi.org/10.1038/nrn2762-c1

Westerberg CE, Mander BA, Florczak SM, Weintraub S, Mesulam M-M, Zee PC, Paller KA (2012) Concurrent impairments in sleep and memory in amnestic mild cognitive impairment. J Int Neuropsychol Soc JINS 18(3):1–11. doi:10.1017/S135561771200001X

Part V
Clinical Translation

Reconsolidation and Trauma Memory

Henrik Kessler, Simon E. Blackwell and Aram Kehyayan

Abstract This chapter highlights how advances in reconsolidation research could lead to the development of novel treatments to influence traumatic memories. Patients with Posttraumatic Stress Disorder (PTSD) suffer from recurrent intrusive memories of the traumatic event(s). The concept of reconsolidation implies that already consolidated memories can be changed after reactivating them and applying an update to the original memory trace. This chapter briefly reviews the relevant literature on (mainly human) reconsolidation research. Afterwards, PTSD is explained as a disorder of memory. Finally, first approaches are described that make use of postulated reconsolidation-update mechanisms in order to change traumatic memories.

Keywords PTSD · Memory reconsolidation · Innovative treatments · Translational medicine

Complementary to the main focus of this book, memory consolidation, this chapter briefly discusses the concept of *reconsolidation* and its possible application in the treatment of patients suffering from traumatic memories. The basic idea has two premises: (1) A previously consolidated memory can be rendered malleable via reactivation with a retrieval cue, followed by a process of restabilisation and potential update of the memory trace (Sara 2000; Nader et al. 2000a).

H. Kessler (✉)
Department of Psychosomatic Medicine and Psychotherapy,
LWL-University Hospital, Ruhr-Universität Bochum, Bochum, Germany
e-mail: Henrik.Kessler@ruhr-uni-bochum.de

S.E. Blackwell
Mental Health Research and Treatment Center, Department of Psychology,
Ruhr-Universität Bochum, Massenbergstraße 9-13, 44787 Bochum, Germany
e-mail: Simon.Blackwell@ruhr-uni-bochum.de

A. Kehyayan
Department of Psychosomatic Medicine and Psychotherapy,
LWL-University Hospital, Ruhr-Universität Bochum, Bochum, Germany

N. Axmacher and B. Rasch (eds.), *Cognitive Neuroscience of Memory Consolidation*, Studies in Neuroscience, Psychology and Behavioral Economics, DOI 10.1007/978-3-319-45066-7_22

An intervention post-reactivation could influence this "reconsolidation" process and thus modify the memory. (2) Posttraumatic Stress Disorder (PTSD) can be understood as a disorder of memory (Brewin 2014). A hallmark symptom of PTSD is "recurrent, involuntary and intrusive distressing memories of the traumatic event (s)" (APA 2013), and such memories are the main target of effective treatments. Following these two premises, reconsolidation-update mechanisms may provide a framework for the modification of symptomatic trauma memories. This is an example of how advances in the science of memory could lead to novel treatments for mental disorders (Holmes et al. 2014).

It has in fact been suggested that memory reconsolidation could be the key factor in understanding the changes brought about by psychotherapy across many mental disorders (Lane et al. 2015). This chapter, however, focuses on the idea that reconsolidation-update mechanisms may be effective in the treatment of trauma memories that typically haunt patients with PTSD. First, a brief overview of the history and concept of reconsolidation will be provided, emphasising the literature on human reconsolidation. Second, the idea of PTSD as a disorder of memory will be explained. Third, studies applying the concept of reconsolidation to the modification of traumatic memories in healthy participants and patients with PTSD will be reviewed.

The History and Concept of Reconsolidation

The idea that old memories can be modified following reactivation was first demonstrated empirically in animal studies in the 1960s, where e.g. electroconvulsive shock led to amnesia for passive-avoidance memories after their retrieval (Misanin et al. 1968). At the time, this process was termed "cue-dependent amnesia", and it has only later been referred to as reconsolidation. Despite those interesting early studies, over the following decades memory research focused on the concept of consolidation, leading to the traditional view of an unidirectional process with increasing stability leading to a "fixed" memory (challenged e.g. in Nader and Hardt 2009; Nadel et al. 2012). The concept of reconsolidation was rediscovered around the turn of the century via animal studies in which, for example, fear memories were weakened with pharmacological agents (Nader et al. 2000b). This rediscovery and re-conceptualisation (Sara 2000; Nader et al. 2000a) led to a plethora of studies demonstrating reconsolidation in animals (reviewed e.g. in Nader and Hardt 2009; Nadel et al. 2012). However, most agents applied in animal studies are toxic in humans, and thus investigation of reconsolidation in humans required different experimental procedures and lagged behind the animal work. Consequently, the evidence for reconsolidation in humans is therefore more limited, albeit increasing (Finnie and Nader 2012; Besnard et al. 2012; Schwabe et al. 2014). The experimental protocol to investigate reconsolidation, derived from the first animal studies, can be outlined as follows (Lewis 1969): (1) Consolidated memories have to be *reactivated* via a retrieval cue; (2) The memory-changing intervention has to be administered *afterwards*; (3) Effects on memory should be tested after short-term effects of the intervention have vanished as reconsolidation should update *long-term* memory.

Important human reconsolidation studies have picked up on one approach from the animal studies, conditioned fear responses, and investigated *fear memories*. Kindt and colleagues demonstrated that the behavioral expression of conditioned fear responses to stimuli (startle response) in healthy participants could be erased by administering the beta-adrenergic receptor antagonist propranolol prior to reactivation of the fear memory, 24 h after initial conditioning (Kindt et al. 2009). Importantly, both memory reactivation and propranolol were required for this effect; either component alone (propranolol only, or reactivation plus placebo) resulted in no disruption of fear memory (Kindt et al. 2009). In a follow-up study (Soeter and Kindt 2010), the authors showed that this effect persisted one month after the intervention. Interestingly, propranolol decreased only the fear startle response, leaving declarative memory intact. In yet another study by this lab, previously conditioned startle fear responses were selectively "neutralized" by the disruption of reconsolidation with propranolol (Soeter and Kindt 2011). Because the intervention (propranolol) in the above mentioned studies was administered *before* memory reactivation and hence does not strictly follow Lewis' protocol, doubts have been raised about whether the mechanism responsible for this effect was in fact reconsolidation (Schiller and Phelps 2011). However, a response to this has been proposed (Brunet et al. 2011a, b): Due to its pharmacokinetic properties, propranolol must be administered in advance in order to be effective within the time window in which most of the reconsolidation process occurs (2 h in animals; Przybyslawski et al. 1999). Propranolol used in the context of episodic memories and patient studies will be further addressed below.

Nevertheless, as a reaction to the uncertainties surrounding the use of propranolol within a strict reconsolidation protocol, Schiller and colleagues' own study applied an alternative experimental approach to investigate reconsolidation in humans, using a non-invasive protocol (Schiller et al. 2010). Reactivated fear memories were updated with non-fearful information during the time-window of reconsolidation, leading to a reduction in fear responses. This effect lasted a year and was apparent only for reactivated memories. Another study confirmed that fear memories can be erased by disrupting reconsolidation with extinction training and also showed changes in amygdala-based fear networks (Agren et al. 2012). Additionally, this study showed that a distinct reconsolidation time-window might also exist in humans: the extinction training of fear-conditioned stimuli was only effective if performed within 6 h of memory reactivation. The extinction protocol applied in Schiller's study was successfully replicated with auditory aversive stimuli (Oyarzun et al. 2012). However, there have also been studies failing to replicate the effect initially reported by Schiller and colleagues (e.g. Soeter and Kindt 2011; Golkar et al. 2012). This discrepancy challenges the robustness of the effect of the extinction protocol and underlines the need to further investigate the exact parameters of the experimental setting, e.g. differences in time-windows, and hence between extinction and reconsolidation (Schwabe et al. 2014).

Another stream of research in human reconsolidation is concerned with *episodic memories*. Interestingly, the roots of this approach date back to William James' notion that memories are changed according to the cognitive context in which they are retrieved (James 1892) and Bartlett's experiments showing changes in memory according to cultural expectations (Bartlett 1932). The claim that memories change over time according to retrieval contexts and are hence fundamentally dynamic in nature is in fact well-established in cognitive psychology, without using the term "reconsolidation" (Loftus 2005). In a test of reconsolidation for episodic memories, healthy participants learned a list of objects and were asked to learn a second list of objects 24 h later (Hupbach et al. 2007). Only participants who were reminded of the first list (reactivation of consolidated memory) prior to learning the second list later intertwined items of list 2 when asked about list 1 on day three. This study supports the constructive nature of memory with new information being incorporated into old memories (reconsolidation *update*). In a series of experiments using a retrieval-relearning technique, episodic memories of a fictional movie were impaired when false information was provided during the relearning phase, after reactivation of the initially learned memories (Chan and LaPaglia 2013). Episodic memories have also been impaired in humans via reactivation followed by electroconvulsive therapy (Kroes et al. 2014). Interestingly, consolidated memories have been shown to be strengthened if there is *no* intervention after their reactivation (Forcato et al. 2011). Changing episodic memories with supposed reconsolidation mechanisms has also been investigated via application of pharmacological agents (reviewed in Lonergan et al. 2013). In a study mainly focusing on effects of cortisol and propranolol on the recall of emotionally-valenced words, de Quervain (2007) found no effect of propranolol on long-term recall of previously learned words. Similar results were obtained in another study (Tollenaar et al. 2009a), which found no immediate or prolonged effects of propranolol on recognition of previously learned words. Cortisol administration, on the other hand, led to impaired recognition. However, propranolol was found to significantly reduce sympathetic arousal. In contrast, there are also studies that have observed an effect of propranolol on declarative memory: Kroes et al. (2010) report that propranolol administered before a cued recall task significantly impaired recall of negatively valenced (but not neutral) words 24 h later, compared to placebo. Another study using neutral and emotionally negative pictures as stimulus material found that propranolol administered before memory reactivation specifically impaired subsequent recognition of negative, but not neutral, pictures (Schwabe et al. 2012) (see also chapter by Meir Drexler and Wolf). Both memory reactivation without propranolol and propranolol administered without memory reactivation had no effect on subsequent declarative memory. Tollenaar and colleagues also conducted a study investigating the effects of cortisol and propranolol on psychophysiological responding to negative autobiographical memories in healthy subjects, using a script-driven imagery procedure (Tollenaar et al. 2009b). Participants prepared a script of a negative life event, and were later asked to imagine this event as a memory reactivation while under the influence of cortisol or propranolol. The same procedure was repeated after a 1 week wash-out period.

This study focused on psychophysiological reactions that accompany negative episodic memories (often strongly pronounced in PTSD patients) and closely resembled the studies by Brunet and colleagues in clinical populations (see below). Interestingly, the authors found no effect of either cortisol or propranolol on skin conductance reactivity.

In sum, episodic memories can be influenced by reconsolidation mechanisms in humans, but the direction of change—impairment, strengthening or update—varies. All three possibilities may have a function in human life: important information needs to be strengthened, new information should lead to an update, and some memories are better weakened. The disentanglement of the three processes in experimental settings via consideration of context variables and procedural constraints could be a fruitful scientific venture. It should also be noted that fear conditioning and episodic memory tasks involve two fundamentally different types of emotional memory with different underlying neural mechanisms (Lonergan et al. 2013). Disentangling the two mechanisms warrants further enquiry.

PTSD as a Disorder of Memory

That dysfunctions in memory comprise a core component of PTSD psychopathology is clear from a consideration of its basic symptoms: It is *memories* of traumatic events from the past that cause a "reliving" of those events even when the actual threat is no longer present. The historical roots of this view date at least back to Pierre Janet's and early Sigmund Freud's observations of patients suffering from "hysteria": Those patients experienced traumatic situations in their childhood and the repressed memories of these were thought to cause their symptoms (Janet 1894; Freud and Breuer 1895). Later on, Pitman conceptualized pathologically enhanced memory for traumatic events in PTSD as an "overconsolidation" of trauma memories due to hyperactivity of stress hormones during traumatic events (Pitman 1989). Recent descriptions of PTSD emphasize three core symptoms: intrusions, hyperarousal and avoidance (APA 2013). Intrusions (or flashbacks in the most extreme form) are essentially intrusive *memories* of past events that patients experience as if happening in the present. Hyperarousal is typically accompanied by enhanced activity in the sympathetic nervous system (increased heart frequency, blood pressure, muscle tone, transpiration, etc.) in reaction to triggers that remind patients of traumatic events. Most of those bodily reactions are components of what comprises conditioned fear responses in experimental research. Finally, avoidance is a logical consequence of the former two symptoms, since patients tend to avoid triggers, contexts and situations that remind them of traumatic events, essentially (sometimes desperately) trying to reduce intrusions and hyperarousal. Therefore, two of three core symptoms in PTSD are the result of pathological memory processes, whether it is concerning episodic memories of traumatic events (intrusions) or conditioned fear responses (hyperarousal).

Those compelling facts have led many to view PTSD as characterized by—amongst other factors—disturbances in memory (Ehlers and Clark 2000; Brewin 2003, 2014; McNally 2003; van der Kolk 2007; Axmacher et al. 2010). This view is also reflected in the relative effectiveness of different treatment modalities for PTSD. The most successful treatment components include some sort of exposure to traumatic memories with the aim to change and/or reprocess them (NICE 2005). In a well-established therapy for intrusive images, Eye Movement Desensitization and Reprocessing (EMDR), patients are asked to reactivate their traumatic memories, while rapidly moving their eyes horizontally (Shapiro 2001). This eye movement has been thought to cause a reprocessing of traumatic memories allowing them to be better integrated in the patient's biography; recent accounts suggest that taxing working memory (via eye movements) while the memory is reactivated results in a weakening of the memory trace (van den Hout and Engelhard 2012). Another successful treatment, cognitive behaviour therapy, involves the patients recounting and reliving the traumatic event (exposure), and reframing it via cognitive techniques (Ehlers and Clark 2000). Within cognitive behaviour therapy, "imagery rescripting" may also be used to reprocess memories of trauma (e.g. Arntz 2012). Following Lane's et al. (2015) influential article, all these treatments seeking to modify traumatic memories may tap into processes of reconsolidation and trigger update-mechanisms. The key question guiding the studies reviewed in the remainder of this chapter is: How can we make use of reconsolidation-update mechanisms in order to change trauma memories effectively using interventions that are more easily available than individual psychotherapy sessions?

Reconsolidation-Update and the Modification of Traumatic Memories

There has been only limited application of the concept of reconsolidation to the modification of traumatic memories in healthy participants and patients with PTSD. In essence, basic science studies investigating the modification of "trauma" memories via reconsolidation-update mechanisms have used the trauma film paradigm as an experimental trauma analogue to create intrusions in otherwise healthy participants (Marks et al. 2014; James et al. 2015). Patient studies have largely used propranolol to interfere with reconsolidation of previously reactivated actual trauma memories (Giustino et al. 2016).

With regards to basic science studies with healthy participants, researchers have aimed to "induce" PTSD-like symptoms (i.e. intrusive visual memories) in healthy volunteers using the trauma film paradigm (James et al. 2016). After watching a distressing film depicting scenes with highly negative and arousing content, participants reliably report intrusions of these scenes over the following days. In a study by Marks et al. (2014), healthy participants watched such a distressing film, and on the next day completed an extinction training (repeated presentations of film segments). Frequency and intensity of intrusions was assessed 24 h after extinction training. Memory was either reactivated before extinction training (extinction within reconsolidation time window), after extinction training (extinction outside reconsolidation time window), or extinction was performed without memory reactivation. Surprisingly, participants who received the extinction training within the time window of reconsolidation reported *more* intrusions than subjects in the other two groups. While the authors interpret their result as being consistent with reconsolidation theory (as the intervention did have an impact on later intrusions), the reason for the unexpected increase in intrusions remains elusive.

In another basic science study, 24 h after watching a trauma film, healthy participants had their (already consolidated) memories of the film reactivated, and then played the computer game *Tetris* (James et al. 2015). This game was chosen as an intervention to disrupt reconsolidation as it recruits visuospatial resources and should compete with the processing of (visual) intrusive memories. Only the combination of prior memory reactivation and *Tetris* gameplay, but not reactivation or *Tetris* alone, caused a reduction in intrusions from the trauma film compared to a no-task control group.

Only a limited number of studies have tried to apply interventions based on reconsolidation in *clinical populations*, and these have mostly used propranolol. Brunet and colleagues conducted 3 studies with PTSD patients (Brunet et al. 2011a, b). Patients received propranolol weekly over the course of 6 weeks prior to treatment sessions in which they were asked to read aloud self-written accounts of their traumatic event or, in one study, re-narrate this event. In all 3 studies, PTSD symptom severity decreased significantly over the course of time, and 51–71% of patients no longer met the full criteria for PTSD at follow up 6 months post-treatment. However, only one of these studies involved a control group, who refused treatment sessions and only took part in PTSD-symptoms assessment at time points corresponding to pre-treament, post-treatment, and follow-up in the treated group. There was no control group receiving e.g. reactivation of trauma memory without propranolol, and thus, it could not be shown that treatment effects were specifically due to reconsolidation blockade. Alternatively, treatment effects could have been caused by extinction processes or habituation induced by repeated exposure to traumatic material, which constitutes the basis of many exposure-based PTSD interventions. In another study with PTSD patients (Brunet et al. 2008), the authors showed that propranolol administered after retrieval of trauma-related memories resulted in lower physiological responses during a script-driven mental imagery procedure involving the same trauma-related material one week later, compared to placebo. Further, they showed perpetuation of this effect 4 months post-treatment (Brunet et al. 2014).

In their critical review of studies applying propranolol in PTSD patients, Giustino et al. (2016) discuss the question of what memory processes are modulated by propranolol. They review rodent as well as human studies and discuss whether propranolol acts primarily through the disruption of memory reconsolidation (potentially by indirectly inhibiting protein synthesis), or serves as an enhancer of extinction learning (potentially by reducing psychological stress and helping to restore an optimal level of noradrenergic signalling; see also chapter of Meir-Drexler and Wolf). Reconsolidation and extinction have been shown to be mutually exclusive and to rely on different molecular mechanisms in rodents (Merlo et al. 2014). Giustino and colleagues state that many studies appear to suggest that propranolol blocks the reconsolidation of fear memories in healthy volunteers (see above) as well as in PTSD patients. However, certain limitations may apply that could restrict the effectiveness of reconsolidation blockade via propranolol in PTSD patients: first, there is evidence that only weak and recent memories are susceptible to erasure (Milekic and Alberini 2002; Suzuki et al. 2004). Second, memories might not necessarily undergo reconsolidation unless there is new information to be encoded (Sevenster et al. 2012). Finally, over-consolidation through frequent reactivation of fear memories in PTSD could make them more resistant to reconsolidation blockade (Pitman and Delahanty 2005). Additionally, the authors criticize the study procedures used in many reconsolidation paradigms as not suited to differentiate reliably between effects of reconsolidation blockade, or extinction enhancement. They conclude that an ambiguity remains as to which memory mechanisms are modulated by propranolol, and that current evidence rather seems to support maximum effectiveness of propranolol when administered shortly after trauma and coupled with extinction training.

The only study—to our knowledge—trying a different approach to disrupt reconsolidation in PTSD patients is the case report by Gahr et al. (2014). They describe the effect of 8 sessions of electroconvulsive therapy (ECT) in a patient with severe, chronic, therapy-refractory PTSD (caused by multiple traumatic events) and a major depressive episode. ECT was administered shortly after the patient was asked to remember and describe verbally one of his traumatic events, and resulted in decrease in PTSD and depressive symptoms. PTSD symptoms related to the targeted event subsided completely. Further, the patient even reported that he could hardly remember the targeted event, suggesting a change not only in intrusive memories (e.g. flashbacks) but also in voluntarily accessible episodic memory. The authors speculate that the observed effect could be due to reconsolidation modifications caused by ECT.

In summary, applying the theory of memory reconsolidation to a conceptualisation of PTSD as a disorder of memory leads to the suggestion that reconsolidation-update mechanisms could provide a route for effective treatment. Studies in both animals and human participants provide evidence for the possibility of memory reconsolidation, and the view of PTSD as a disorder of memory is borne out by consideration of its symptomatology and the central components of successful treatment approaches. While research applying the ideas of reconsolidation to PTSD phenomena is in its infancy, the potential for development of efficient and accessible treatment approaches derived from basic memory science makes this a compelling area for clinical translational research.

References

Agren T, Engman J, Frick A, Björkstrand J, Larsson E-M, Furmark T, Fredrikson M (2012) Disruption of reconsolidation erases a fear memory trace in the human amygdala. Science 337 (6101):1550–1552. doi:10.1126/science.1223006

American Psychiatric Association (2013) Diagnostic and statistical manual of mental disorders. 5th edn. American Psychiatric Publishing

Arntz A (2012) Imagery rescripting as a therapeutic technique: review of clinical trials, basic studies, and research agenda. J Exp Psychopathol 3(2):189–208

Axmacher N, Do Lam AT, Kessler H, Fell J (2010) Natural memory beyond the storage model: repression, trauma, and the construction of a personal past. Front Hum Neurosci 4:211

Bartlett FC (1932) Remembering: a study in experimental and social psychology. Cambridge University Press, Cambridge

Besnard A, Caboche J, Laroche S (2012) Reconsolidation of memory: a decade of debate. Prog Neurobiol 99:61–80

Brewin CR (2003) Posttraumatic stress disorder: malady or myth?. Yale University Press, New Haven

Brewin CR (2014) Episodic memory, perceptual memory and their interaction: foundations for a theory of posttraumatic stress disorder. Psychol Bull 140:69–97. doi:10.1037/a0033722

Brunet A, Orr SP, Tremblay J, Robertson K, N K, Pitman RK (2008) Effect of post-retrieval propranolol on psychophysiologic responding during subsequent script-driven traumatic imagery in post-traumatic stress disorder. J Psychiatr Res 42(6):503–506. doi:10.1016/j.jpsychires.2007.05.006

Brunet A, Ashbaugh AR, Saumier D, Nelson M, Pitman RK, Tremblay J et al (2011a) Does reconsolidation occur in humans: a reply. Front Behav Neurosci 5:74

Brunet A, Poundja J, Tremblay J, Bui E, Thomas E, Orr SP et al (2011b) Trauma reactivation under the influence of propranolol decreases posttraumatic stress symptoms and disorder: 3 open-label trials. J Clin Psychopharmacol 31(4):547–550. doi:10.1097/JCP.0b013e318222f360

Brunet A, Thomas É, Saumier D, Ashbaugh AR, Azzoug A, Pitman RK et al (2014) Trauma reactivation plus propranolol is associated with durably low physiological responding during subsequent script-driven traumatic imagery. Can J Psychiatry (Revue Canadienne de Psychiatrie) 59(4):S228–232

Chan JC, LaPaglia JA (2013) Impairing existing declarative memory in humans. Proc Natl Acad Sci U S A 110:9309–9313

de Quervain D (2007) Preventive effect of β-adrenoceptor blockade on glucocorticoid-induced memory retrieval deficits. Am J Psychiatry 164(6):967. doi:10.1176/appi.ajp.164.6.967

Ehlers A, Clark DM (2000) A cognitive model of posttraumatic stress disorder. Behav Res Ther 38:319–345

Finnie PSB, Nader K (2012) The role of metaplasticity mechanisms in regulating memory destabilization and reconsolidation. Neurosci Biobehav Rev 36:1667–1707

Forcato C, Rodriguez ML, Pedreira ME (2011) Repeated labilization-reconsolidation processes strengthen declarative memory in humans. PLoS ONE 6:e23305

Freud S, Breuer J (1895) Studien über hysterie. Fischer, Frankfurt a. M

Gahr M, Schonfeldt-Lecuona C, Spitzer M, Graf H (2014) Electroconvulsive therapy and posttraumatic stress disorder: first experience with conversation-based reactivation of traumatic memory contents and subsequent ECT-mediated impairment of reconsolidation. J Neuropsychiatry Clin Neurosci 26(3):E38–9. doi:10.1176/appi.neuropsych.13070159

Giustino TF, Fitzgerald PJ, Maren S (2016) Revisiting propranolol and PTSD: memory erasure or extinction enhancement? Neurobiol Learn Mem 130:26–33. doi:10.1016/j.nlm.2016.01.009

Golkar A, Bellander M, Olsson A, Ohman A (2012) Are fear memories erasable?-reconsolidation of learned fear with fear-relevant and fear-irrelevant stimuli. Front Behav Neurosci 6:80

Holmes EA, Craske MG, Graybiel AM (2014) A call for mental-health science. Clinicians and neuroscientists must work together to understand and improve psychological treatments [Comment]. Nature 511:287–289. doi:10.1038/511287a

Hupbach A, Gomez R, Hardt O, Nadel L (2007) Reconsolidation of episodic memories: a subtle reminder triggers integration of new information. Learn Mem 14:47–53

James W (1892) Psychology: the briefer course. Harper, Torchbooks, New York

James EL et al (2015) Computer game play reduces intrusive memories of experimental trauma via reconsolidation update mechanisms. Psychol Sci 1–15. doi:10.1177/0956797615583071 (OnlineFirst)

James EL, Lau-Zhu A, Clark IA, Visser RM, Hagenaars MA, & Holmes EA (2016) The trauma film paradigm as an experimental psychopathology model of psychological trauma: intrusive memories and beyond. Clin Psychol Rev 47:106–142. doi:10.1016/j.cpr.2016.04.010

Janet P (1894) Der Geisteszustand der Hysterischen. Deuticke, Leipzig/Wien

Kindt M, Soeter M, Vervliet B (2009) Beyond extinction: erasing human fear responses and preventing the return of fear. Nat Neurosci 12:256–258. doi:http://www.nature.com/neuro/journal/v12/n3/full/nn.2271.html

Kroes MCW, Strange BA, Dolan RJ (2010) Beta-adrenergic blockade during memory retrieval in humans evokes a sustained reduction of declarative emotional memory enhancement. J Neurosci Official J Soc Neurosci 30(11):3959–3963. doi:10.1523/JNEUROSCI.5469-09.2010

Kroes MCW et al (2014) An electroconvulsive therapy procedure impairs reconsolidation of episodic memories in humans. Nat Neurosci 17:204–2016. doi:10.1038/nn.3609

Lane RD, Ryan L, Nadel L, & Greenberg L (2015) Memory reconsolidation, emotional arousal, and the process of change in psychotherapy: new insights from brain science. Behav Brain Sci, 38, e1

Lewis DJ (1969) Sources of experimental amnesia. Psychol Rev 76:461–472

Loftus EF (2005) Planting misinformation in the human mind: a 30-year investigation of the malleability of memory. Learn Mem 12:361–366

Lonergan MH, Olivera-Figueroa LA, Pitman RK, Brunet A (2013) Propranolol's effects on the consolidation and reconsolidation of long-term emotional memory in healthy participants: a meta-analysis. J Psychiatry Neurosci JPN 38(4):222–231. doi:10.1503/jpn.120111

Marks EH, Zoellner LA (2014) Attenuating fearful memories: effect of cued extinction on intrusions. Emotion (Washington, DC) 14(6):S1143–1154. doi:10.1037/a0037862

McNally RJ (2003) Remembering trauma. Harvard University Press. Traumatic experiences are unforgettable, Cambridge, MA

Merlo E, Milton AL, Goozée ZY, Theobald DE, Everitt BJ (2014) Reconsolidation and extinction are dissociable and mutually exclusive processes: behavioral and molecular evidence. J Neurosci Official J Soc Neurosci 34(7):2422–2431. doi:10.1523/JNEUROSCI.4001-13.2014

Milekic MH, Alberini CM (2002) Temporally graded requirement for protein synthesis following memory reactivation. Neuron 36(3):521–525. doi:10.1016/S0896-6273(02)00976-5

Misanin JR, Miller RR, Lewis DJ (1968) Retrograde amnesia produced by electroconvulsive shock after reactivation of a consolidated memory trace. Science 160:554–555

Nadel L, Hupbach A, Gomez R, Newman-Smith K (2012) Memory formation, consolidation and transformation. Neurosci Biobehav Rev 36:1640–1645

Nader K, Hardt O (2009) A single standard for memory: the case for reconsolidation. Nat Rev Neurosci 10:224–234

Nader K, Schafe G, LeDoux J (2000a) The labile nature of consolidation theory. Nat Rev Neurosci 1:216–219

Nader K, Schafe GE, LeDoux JE (2000b) Fear memories require protein synthesis in the amygdala for reconsolidation after retrieval. Nature 406:722–726

National Institute for Health and Clinical Excellence (2005) Post-traumatic stress disorder (PTSD): the management of PTSD in adults and children in primary and secondary care. Report No. CG026. National Institute for Health and Clinical Excellence, London

Oyarzun JP, Lopez-Barroso D, Fuentemilla L, Cucurell D, Pedraza C, Rodriguez-Fornells A, de Diego-Balaguer R (2012) Updating fearful memories with extinction training during reconsolidation: a human study using auditory aversive stimuli. PLoS ONE 7(6):e38849

Pitman RK (1989) Post-traumatic stress disorder, hormones, and memory. Biol Psychiatry 26:221–223

Pitman RK, Delahanty DL (2005) Conceptually driven pharmacologic approaches to acute trauma. CNS Spectr 10(2):99–106

Przybyslawski J, Roullet P, Sara SJ (1999) Attenuation of emotional and nonemotional memories after their reactivation, role of beta adrenergic receptors. J Neurosci 19:6623–6628

Sara SJ (2000) Retrieval and reconsolidation: toward a neurobiology of remembering. Learn Mem 7:73–84

Schiller D, Phelps EA (2011) Does reconsolidation occur in humans? Front Behav Neurosci 5:24

Schiller D et al (2010) Preventing the return of fear in humans using reconsolidation update mechanisms. Nature 463:49–53

Schwabe L, Nader K, Wolf OT, Beaudry T, Pruessner JC (2012) Neural signature of reconsolidation impairments by propranolol in humans. Biol Psychiatry 71(4):380–386. doi:10.1016/j.biopsych.2011.10.028

Schwabe L, Nader K, Pruessner JC (2014) Reconsolidation of human memory: Brain mechanisms and clinical relevance. Biol Psychiatry 76(4):274–280

Sevenster D, Beckers T, Kindt M (2012) Retrieval per se is not sufficient to trigger reconsolidation of human fear memory. Neurobiol Learn Mem 97(3):338–345. doi:10.1016/j.nlm.2012.01.009

Shapiro F (2001) Eye movement desensitization and reprocessing: Basic principles, protocols and procedures, 2nd edn. Guilford, New York

Soeter M, Kindt M (2010) Dissociating response systems: erasing fear from memory. Neurobiol Learn Mem 94(1):30–41. doi:10.1016/j.nlm.2010.03.004

Soeter M, Kindt M (2011) Disrupting reconsolidation: pharmacological and behavioral manipulations. Learn Mem 18(6):357–366

Suzuki A, Josselyn SA, Frankland PW, Masushige S, Silva AJ, Kida S (2004) Memory reconsolidation and extinction have distinct temporal and biochemical signatures. J Neurosci Official J Soc Neurosci 24(20):4787–4795. doi:10.1523/JNEUROSCI.5491-03.2004

Tollenaar MS, Elzinga BM, Spinhoven P, Everaerd W (2009a) Immediate and prolonged effects of cortisol, but not propranolol, on memory retrieval in healthy young men. Neurobiol Learn Mem 91(1):23–31. doi:10.1016/j.nlm.2008.08.002

Tollenaar MS, Elzinga BM, Spinhoven P, Everaerd W (2009b) Psychophysiological responding to emotional memories in healthy young men after cortisol and propranolol administration. Psychopharmacology 203(4):793–803. doi:10.1007/s00213-008-1427-x

van den Hout M, Engelhard IM (2012) How does EMDR work? J Exp Psychopathol 3(5):724–738

Van der Kolk BA (2007) The history of trauma in psychiatry. In: Friedman MJ, Keane TM, Resick PA (eds) Handbook of PTSD: science and practice. Guilford Press, New York, pp 19–36

Sleep-Related Interventions to Improve Psychotherapy

Christoph Nissen, Marion Kuhn, Elisabeth Hertenstein and Nina Landmann

Abstract Mental disorders are among the most prevalent and impairing disorders worldwide. According to the WHO (World Health Organization), about a third of the population will suffer from a mental disorder across the lifespan. Current guidelines include psychotherapy, or at least psychoeducation, as a first-line treatment component for all mental disorders. Psychotherapy represents an interpersonal process that involves the modification of cognition, emotion and behavior. As such, it can, at least in part, be conceptualized as the quantitative strengthening and qualitative reorganization of novel and more adaptive memory representations. Given that sleep substantially modulates learning, memory and underlying neural refinements, this chapter centers on the idea that sleep-related interventions can be used to augment the treatment effects of psychotherapy. The first part identifies basic memory processes with particular relevance for psychotherapy. The second part evaluates the potential of sleep-related interventions prior to and after psychotherapy, as well as the modulation of distinct aspects of sleep to augment these memory processes in psychotherapy and discusses further directions.

Keywords Psychotherapy · Sleep · Memory · Strengthening · Reorganization

Introduction

Psychotherapy can be Conceptualized as Memory Formation

Within the past decades, the neurosciences have identified distinct types of learning and underlying neural refinements with potential relevance for the strengthening and reorganization of novel memories during psychotherapy (Ramirez et al. 2015; Dunsmoor et al. 2015). Yet for a long time, ideological and methodical disputes

C. Nissen (✉) · M. Kuhn · E. Hertenstein · N. Landmann
Department of Psychiatry and Psychotherapy, University Medical Center Freiburg, Hauptstraße 5, 79104 Freiburg, Germany
e-mail: christoph.nissen@uniklinik-freiburg.de

N. Axmacher and B. Rasch (eds.), *Cognitive Neuroscience of Memory Consolidation*, Studies in Neuroscience, Psychology and Behavioral Economics, DOI 10.1007/978-3-319-45066-7_23

have characterized the interplay between psychotherapy and the neurosciences. This has resulted in a variety of psychotherapy constructs, such as the transference-hypothesis (Freud 1939), with poorly defined neural correlates, as well as in a variety of basic neuroscience findings, such as the description of synaptic long-term potentiation (LTP) (Duffy et al. 1981), that are not systematically addressed in psychotherapy.

In the following section, we identify basic learning processes with, at least partly, unraveled neural correlates, which are of particular relevance for psychotherapy. This aims to pave the way for a more systematic neurobiological augmentation of psychotherapy. Of note, we assume that the described basic learning processes are important for all psychotherapy schools, such as cognitive behavior therapy or psychoanalysis, since treatment effects most likely emerge from common molecular pathways in the brain. Still at this point, the current chapter has the closest link to cognitive behavior therapy that is explicitly based on concepts of learning and memory.

Memory Processes in Psychotherapy

Memory formation comprises the quantitative strengthening and qualitative reorganization of memories. Strengthening refers to the veridical stabilization or enhancement of previously encoded memories, whereas reorganization denotes the emergence of qualitatively new representations, which have not been directly encoded. Figure 1 translates this concept to psychotherapy. As depicted, we selected a memory perspective as the main structure of this chapter and tried to match associated psychotherapy processes. This perspective does not directly address other important aspects of psychotherapy, such as the therapist-patient relationship and many others. However, it might facilitate the translation of basic science knowledge to the augmentation of psychotherapy, including sleep-related interventions.

A major mechanism for memory formation is synaptic LTP. LTP is characterized by several properties, namely cooperativity, input specificity, and associativity (Citri and Malenka 2008). Cooperativity means that the induction of LTP requires coincident activation of multiple excitatory afferent fibers converging on a given neuron. Input specificity indicates that LTP is restricted to stimulated afferent fibers, and associativity means that a weak input that is not strong enough to elicit LTP can be potentiated when it is activated in association with a stronger input. These properties make LTP an attractive cellular mechanism for learning and memory. LTP has been described in different neural networks, including the hippocampus, amygdala and the neocortex, that are of relevance for distinct types of learning (Malenka and Bear 2004).

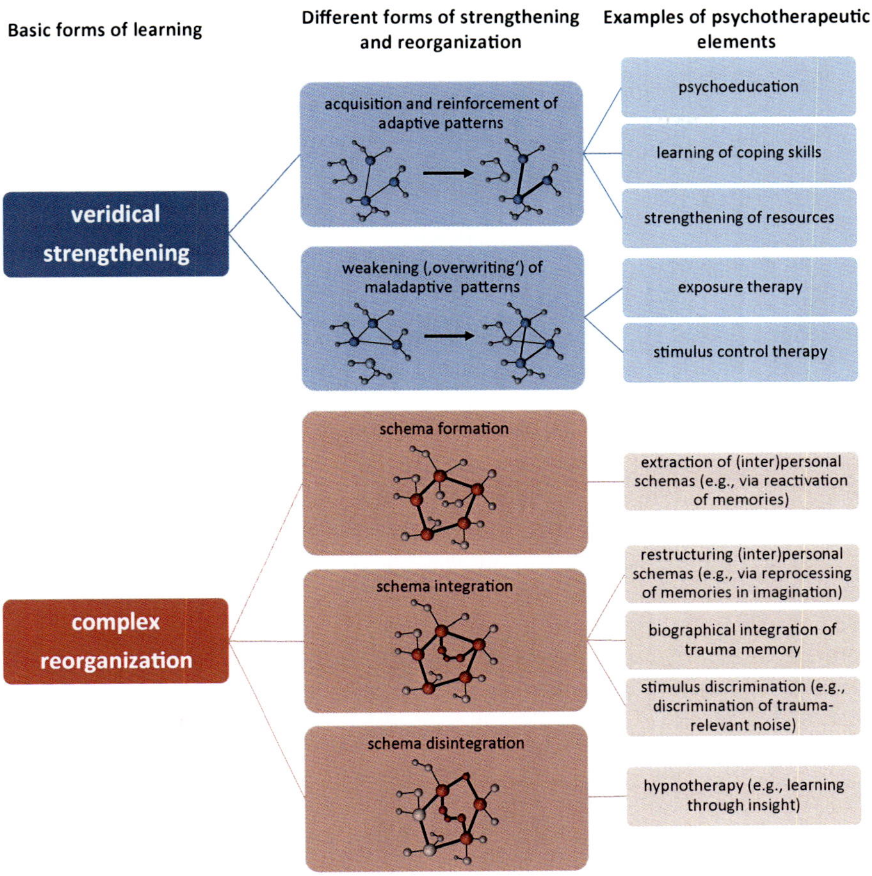

Fig. 1 Memory processes in psychotherapy. Psychotherapy aims at promoting adaptive and weakening non-adaptive patterns of cognition, emotion and behavior. It is important to note that, on a neural level, these changes appear to primarily emerge from the acquisition and strengthening of novel adaptive memories that might complement and, on a behavioral level, even erase non-adaptive patterns. Yet on a neural level, the entire deletion of consolidated memory traces appears to not or rarely occur. The figure denotes basic memory processes and gives examples of related psychotherapy processes

Veridical Strengthening

Acquisition and Reinforcement of Adaptive Patterns

Psychotherapy aims at promoting adaptive patterns of cognition, emotion and behavior. The acquisition and reinforcement of adaptive patterns comprises, for instance, psychoeducation about mental disorders. Particularly, the acquisition of factual knowledge about the pathophysiology and treatment options can provide hope, foster self-efficacy and reduce stigmatization. Other examples include the acquisition of new coping skills (e.g., tension regulation skills and relaxation techniques) and the reinforcement of individual resources (e.g., reactivating exercise habits or social contacts).

Weakening of Maladaptive Patterns

Psychotherapy aims at weakening maladaptive patterns of cognition, emotion and behavior. Importantly, this weakening on a behavioral level is not equivalent to the weakening or deletion of a memory trace on the neural level, but rather requires the acquisition of novel and more adaptive memory representations, which overbalance the initial trace. For instance, exposure therapy with response prevention, the state-of-the-art treatment in patients with anxiety disorders, can result in extinction of fear on the behavioral level. Yet on a neural level, fear extinction requires the acquisition of a novel safety memory trace (Hartley and Phelps 2010). More specifically, in exposure therapy, the patient is gradually confronted with fear-provoking stimuli along with response prevention. The aim is the experience of habituation and the reduction of excessive fear. Another example for weakening of a maladaptive behavior is stimulus control therapy, an evidence-based treatment of addictive behavior. Here for instance, patients with alcohol dependence are instructed to identify and avoid triggers for alcohol abuse (e.g., specific locations or persons). This also includes the acquisition of new and more adaptive memories of alternative behavior.

Reorganization

Memory reorganization refers to the formation of qualitatively new memory content, which has not directly been encoded. Critically, the human brain does not memorize exact recollections of events, but actively constructs and modifies memories (Nadel et al. 2007) (see also chapters by Cheng and by Rauss and Born). Reorganization can be subdivided into schema formation, schema integration and schema disintegration (Landmann et al. 2014). Here, schemas refer to overarching constructs containing a set of related information (see also chapter by Fernandez). For example, the schema 'cat' contains all information necessary to recognize a cat and distinguish it from other animals such as dogs. Schemas facilitate the recollection and usage of memories. Specifically, schema formation refers to the abstraction of new rules and schemas from a set of information. Schema integration means that new information is included into an already existing schema, e.g. the integration of a new word into a mental lexicon. Interestingly, schema integration can also lead to memory distortion and false memories, i.e. the recollection of events which actually never happened (Bergman and Roediger 1999). If too much schema-inconsistent information is received, existing schemas need to be abandoned or changed to enable new perspectives and creativity (schema disintegration and reintegration).

Schema Formation

Schema formation, e.g. the abstraction of communalities and the extraction of rules, is often required in psychotherapy. For example, patients with major depression are motivated to engage in rewarding activities. Whereas such activities can have a short-term antidepressant effect, long-term improvements depend on the formation of adaptive cognitive schemas such as increased expectation of self-efficacy. Such adaptive schemas are often under-represented in patients with depression and might be established in the course of therapy.

Schema Integration

In schematherapy, 'schema' refers to a maladaptive pattern of thoughts, experiences and behavior, which has been developed in childhood or youth and is modified in the course of therapy (Kellogg and Young 2006). An example for a schema is 'emotional inhibition', referring to the exaggerated inhibition of spontaneous emotional expressions, such as anger. Restructuring this schema may rely on schema integration, assuming that the initial schema cannot and maybe should not be erased, but complemented by new and more adaptive information. Another example of schema integration is the restructuring of memory in patients with post-traumatic stress disorder (PTSD) (see also chapter by Kessler, Blackwell and Kehyayan). Here, psychotherapy aims to create a coherent trauma narrative (instead of fragmented memory) and to attenuate excessive emotional responses. During trauma exposure, new adaptive memory traces are included into the memory representation (e.g., 'Then I was in deadly terror, but now I know that I have survived'). Another example for schema integration is stimulus discrimination. Patients with PTSD often experience excessive fear in response to trauma-related, but actually neutral stimuli, such as sirens in road accident survivors. The goal of stimulus discrimination is to add new and safe associations to the stimulus, such as 'The fire brigade is coming to help a cat descending from a tree' or 'Perfect fourth', which reminds me of the beginning of the song 'Amazing Grace'.

Schema Disintegration

Some problems, including intrapsychic and interpersonal problems discussed in psychotherapy, cannot be solved by schema formation or schema integration, because existing ways of thinking reach their limits (patients become 'stuck' on a problem during repeated unsuccessful search for a solution). These problems require the disintegration and recombination of schemas in the form of associative thinking and creativity. An example for schema disintegration in psychotherapy is creative problem solving, which is often used in hypnotherapy. Hypnotic states can be induced by certain instructions (hypnotic suggestions, see Oakley and Halligan 2009). The hypnotic state is characterized by physiological changes (relaxation, changes in body scheme and body perception), and changes in information processing (narrowing of the attentional focus, decreased influence of verbal/logic control mechanisms, increased influence of emotion). The rationale of hypnotherapy is that the induced state can be used to facilitate access to suppressed content (difficult experiences as well as resources) and enable new, unconventional solutions to problematic situations. Some studies indicate that a brain activation pattern which is similar to REM sleep might promote creativity (for overview, Landmann et al. 2015). For example, it has been shown that creativity is facilitated directly after awakening from REM sleep, assuming that REM-like brain activity persists into the wake state for a certain period (Walker et al. 2002). Brain activity during REM sleep is characterized by selective reactivation of cholinergic neurotransmission compared to slow wave sleep (SWS), leading to increased activation of limbic and paralimbic structures and a deactivation of the frontal cortex (Pace-Schott and Hobson 2002). This brain activation pattern may suppress a 'top down' problem solving style while simultaneously promoting unconventional and creative solutions.

Sleep-Related Interventions to Improve Psychotherapy

Sleep has been shown to modify both the quantitative strengthening and qualitative reorganization of memory (Rasch and Born 2013; Landmann et al. 2014) (see also chapters by Schönauer and Gais and by Rauss and Born). Therefore, targeted manipulations of sleep could be used to increase the efficiency of psychotherapy.

The aim of the following section is to evaluate the potential of sleep-related interventions to augment memory processes in psychotherapy. Of note at this point, very few studies have directly investigated the effects of sleep-related interventions on psychotherapy. Therefore, the current chapter rather provides a theoretical framework and research agenda, than a systematic review of completed studies. To provide a structure with direct potential for clinical implementation, we decided to organize the following sections into sleep prior to and after psychotherapy sessions. In a final section, potential manipulations of sleep will be discussed.

Sleep Prior to Psychotherapy

Healthy sleep is important for preparing the brain for subsequent acquisition of novel memory (Maquet 2001). To date to our knowledge, no study has directly manipulated distinct aspects of sleep prior to psychotherapy in patients with mental disorders. Therefore, the following paragraphs refer to results from healthy humans and animals with the aim to derive hypotheses for future psychotherapy research.

Strengthening

Sleep loss and especially sleep deprivation have been demonstrated to impair the acquisition and reinforcement of both, neutral declarative and emotional memories. After 36 h of sleep deprivation, the formation of temporal memory (recency discrimination) is significantly impaired in healthy humans, even if unspecific effects of sleepiness are reduced by the intake of caffeine (Harrison and Horne 2000). Functional magnetic resonance imaging (fMRI) showed that the deficit in the formation of new episodic memories after sleep deprivation is associated with reduced hippocampal and prefrontal activity (Yoo et al. 2007b).

In the emotional memory system, some, although not conclusive evidence supports the notion that rapid eye movement (REM) sleep prior to learning might be particularly important for the acquisition of emotional memory content. After 5 h of REM sleep deprivation, rats showed considerable deficits in the encoding of an active avoidance conditioning task with a reduction in the number of avoidances compared to a group without REM sleep deprivation, and these impairments could not be compensated by continued practice during the learning session (Gruart-Masso et al. 1995). Also in humans, REM sleep prior to learning might be relevant for the acquisition of emotional memory. Here, after REM sleep deprivation, the encoding of positive and neutral words was selectively impaired, leading to a relative shift in the encoding of negative words (Walker and van der Helm 2009). At the functional neuroanatomical level, an enhanced activity of the amygdala to negative emotional stimuli has been observed after sleep deprivation in healthy humans (Yoo et al. 2007a).

On the neural level, encoding deficits after sleep deprivation have been proposed to emerge from synaptic saturation processes after prolonged wake-related synaptic up-scaling leading to reduced inducibility of synaptic LTP in rodent hippocampus (Vyazovskiy et al. 2008). We recently observed in healthy humans that one night of sleep deprivation leads to increased cortical excitability, presumably related to increased net synaptic strength, along with decreased inducibility of LTP-like plasticity, based on electroencephalographic (EEG) and transcranial magnetic stimulation (TMS) studies (Kuhn et al. in press *Nat Comm*).

These findings from animals and healthy humans suggest that a sufficient sleep duration and quality prior to psychotherapy sessions are important to allow for optimal encoding of novel memory content and treatment efficacy. Of particular note, disruptions of sleep continuity represent a transdiagnostic syndrome across a broad variety of mental disorders, such as major depression, anxiety disorders, obsessive-compulsive disorders or schizophrenia (Baglioni et al. 2016). Here, the treatment of sleep disturbances might improve the encoding of novel declarative information during psychotherapy sessions (e.g., education about the disorder and treatment options) or attenuate negatively tinted emotional memory acquisition related to poor sleep. However, these hypotheses remain to be further tested. For these studies, it is important to note that some interventions that can, at least on the short run, improve sleep continuity, such as benzodiazepines or benzodiazepine receptor agonists, might disrupt critical functions of sleep, such as synaptic refinements as a prerequisite for optimal memory encoding and consolidation (Hall-Porter et al. 2014; Roth et al. 1984).

Besides the possibility to treat sleep disturbances in order to avoid deficits in memory formation due to sleep loss, the implementation of additional sleep periods immediately before learning promotes subsequent memory acquisition (Van Der Werf et al. 2009; Mander et al. 2011). For instance, performance on a declarative learning task was improved in healthy humans after a nap with 100 min sleep opportunity, compared to those without the nap intervention (Mander et al. 2011). On the neural level, this finding can be explained by the synaptic homeostasis hypothesis proposing a desaturation of synapses during sleep (synaptic downscaling), which is thought to result in a renewed capacity for the encoding of new information through associative synaptic plasticity (Tononi and Cirelli 2014).

Future studies are needed to systematically test the hypothesis that additional periods of sleep, most likely in the form of naps, before a psychotherapy session promote the encoding of novel adaptive memories during psychotherapy. A broad variety of homeostatic (interval between sleep and psychotherapy) and circadian (time-of-day) aspects might be tested. However, it should be noted that under some circumstances additional periods of sleep might also exert adverse effects. Particularly, in patients with major depressive disorder, daytime naps can have substantially detrimental effects, perhaps related to differences in homeostatic and associative processes of synaptic plasticity (Kuhn et al. 2016; Wolf et al. 2015).

Reorganization

The effects of sleep prior to a psychotherapy session on reorganization of memory during or after this session remain to be further elucidated.

Sleep After Psychotherapy

Strengthening

Sleep after learning has been shown to foster the quantitative strengthening of newly encoded memories (Rasch and Born 2013). Underlying mechanisms include a replay of initially weak memory traces, which is best described for a hippocampal-neocortical network during SWS, contributing to the transformation into long-term memory (Rasch and Born 2013). Since many psychotherapy interventions require a strengthening of memories (Fig. 1), sleep after psychotherapy may augment treatment effects. The best researched example is exposure with response prevention (ERP), a treatment for anxiety disorders. ERP is based on fear conditioning, a basal emotional learning paradigm. It has been shown that sleep is important for fear conditioning. Preliminary evidence suggests that this concept can be translated to clinical application, i.e. ERP can be augmented with naps after psychotherapy sessions.

During fear acquisition, a previously neutral stimulus (then conditioned stimulus, CS) is paired with an unconditional, fear eliciting stimulus (UCS). After successful fear acquisition, the CS alone elicits the fear response (e.g., startle response). During fear extinction, the acquired association between CS and UCS is attenuated through repeated presentations of the CS without the UCS. As outlined above, on the neural level however, the fear response is not erased, but is actively complemented by a new memory trace (Hartley and Phelps 2010), implicating that previously extinguished fear responses can recover (Brooks and Bouton 1993; Rescorla 2004). This is why the reduction or extinction of fear on a behavioral level can be subsumed under memory strengthening on a neural level (Fig. 1). Synaptic LTP in the lateral amygdala has been identified as a key mechanism of fear acquisition and extinction (Rogan et al. 1997). For instance, in auditory fear conditioning (i.e. when a tone is paired with a fear-eliciting stimulus such as an electrical shock), an input signal is transferred from the medial geniculate nucleus of the auditory thalamus to the lateral amygdala. This transmission is largely mediated through N-methyl-D-aspartate (NMDA et al. 1997). The signal is then transferred to the central nucleus of the amygdala and from there to other brain centers responsible for the execution of the behavioral fear response. In rodents, selective antagonism of NMDA-receptors in the lateral amygdala has been demonstrated to block both the local induction of LTP and fear conditioning, showing that local LTP in the lateral amygdala underlies fear conditioning (Maren 1999).

Animal as well as human research suggest that sleep modulates fear conditioning (Pace-Schott et al. 2015). In humans, total sleep deprivation after initial fear acquisition significantly impairs the later recall of learned fear (Menz et al. 2013) and the generalization of fear to a different context (Kuriyama et al. 2010). Sleep deprivation after fear extinction impairs generalization (Pace-Schott et al. 2009). These studies investigated the effects of total sleep deprivation and thus do not allow for conclusions on whether SWS, REM sleep, or both are important for the observed effects. Other studies, suggest that selective REM sleep deprivation blocks fear extinction in rats and humans (Spoormaker et al. 2012; Fu et al. 2007), that REM sleep disruption is associated with impaired consolidation of extinction (Spoormaker et al. 2010), and that the impairment of fear conditioning is correlated with REM sleep in humans (Menz et al. 2013).

Furthermore, REM sleep seems to play a role in safety learning, i.e. learning that a specific stimulus will never be paired with an electric shock. In a study comparing napping versus quiet rest after exposition and habituation to aversive pictures, SWS was associated with better between-session habituation (Pace-Schott et al. 2011). In summary, on the behavioral level, there is strong evidence that sleep deprivation impairs fear acquisition, the generalization of fear extinction, and other types of emotional learning, which in turn indicates that sleep is necessary for these processes (see also chapter by Cunningham and Payne on emotional memory consolidation).

A few recent studies have transferred this basic knowledge to the context of psychotherapy. Two studies have investigated the effect of sleep after ERP sessions on treatment outcome in patients with specific phobia. In the first study, young women with spider phobia watched short videos of spiders (exposure treatment) (Pace-Schott et al. 2012). All participants had to review the same videos as well as videos of another spider after a certain delay. Participants were randomized to four delay-groups: (i) 12 h delay with normal nighttime sleep, (ii) 12 h delay with sleep deprivation, (iii) 2 h delay with waking in the morning, or (iv) 2 h delay with waking in the evening. Subjective fear ratings, skin conductance response and heart rate acceleration decreased across the first session in all groups (successful extinction). In the wake conditions, fear ratings and objective fear measures increased across the delay (loss of extinction). The group that slept showed decreased skin conductance response across the delay (extinction augmentation) and decreased reactivity to the novel spider (extinction generalization). The results indicate that sleep after exposure promotes retention and generalization of extinction memory and argue against circadian explanations. In the second study (Kleim et al. 2014), 40 participants with spider phobia according to DSM IV criteria participated in a virtual reality exposure session. Participants were randomized to a 90 min daytime nap or wakefulness following exposure. One week after the session, reductions in self-reported fear and phobic cognitions when approaching a living spider were significantly larger in the sleep group. Both measures of symptom reduction were associated with greater percentages of stage 2 sleep during the nap, supporting the notion that specific NREM-related brain activity is responsible for sleep-dependent memory enhancement.

These studies suggest that the idea of using sleep for the augmentation of fear extinction and extinction generalization can be transferred to practical application in the context of psychotherapy. However, spider phobia is a simple phobia with relatively low severity and complexity, which often does not considerably interfere with daily life.

First and preliminary evidence for the hypothesis that sleep might influence therapy outcome also in more complex disorders was observed in patients with generalized social phobia. One hundred and sixty nine patients received twelve weeks of group CBT (Zalta et al. 2013). Poor sleep quality at baseline was associated with slower improvement in the course of therapy and higher symptom severity after treatment. Moreover, subjective restfulness of sleep after a therapy session was associated with lower symptom levels the next session (controlled for symptom severity in the previous session). These results are consistent with the notion that poor sleep quality decreases the effects of CBT. However, due to the observational rather than experimental design, causal conclusions cannot be drawn. Future research is needed to investigate the efficacy of sleep for the augmentation of exposure treatment in more complex disorders, and to investigate the effects of sleep on other forms of memory strengthening in the context of psychotherapy, e.g. the acquisition of new skills or cognitive restructuring (Fig. 1).

Critical Time Windows for Sleep After Psychotherapy

Animal research suggests that sleep after the encoding of novel information needs to be placed within a critical time window in order to boost memory consolidation. In rats, sleep deprivation up to five hours after acquisition impaired memory consolidation in a spatial learning task (Morris Water Maze Task) (Graves et al. 2003). In contrast, sleep deprivation 5–10 h after acquisition did not have a significant effect on later performance. However, due to methodological limitations, it cannot be ruled out that the lack of a significant effect in the 5–10 h group was due to a statistical artifact (substantially smaller sample size in the late compared to the early sleep deprivation group with numerically similar performance in both groups). In addition, this finding has not yet been replicated in humans. If a critical time window for sleep's effect on memory consolidation exists, this knowledge could be transferred into clinical practice of psychotherapy, e.g. by implementing brief periods of sleep directly after a psychotherapy session.

Reorganization

Some, but not unequivocal evidence suggests that sleep also promotes the qualitative reorganization of memories (Landmann et al. 2014) (see chapters by Schönauer and Gais, by Rauss and Born, and by Cheng). Yet up to now, no studies on memory reorganization in the context of psychotherapy have been conducted. One area of particular interest is the reorganization of memory in trauma therapy (see also chapter by Kessler, Blackwell and Kehyayan). Here, trauma-related memory traces are reactivated during exposure therapy, and specific interventions are used with the aim of reorganizing trauma memory, e.g. attenuating excessive emotional reactions while preserving factual memory, or integrating dissociated memories.

In addition, therapeutic sleep deprivation could be used for reduced memory consolidation, e.g. after a traumatic event. A recent study in healthy humans found that sleep deprivation, compared to sleep, following an experimental analogue traumatic event (trauma film) reduced the emotional effect and the amount of intrusive memories (Porcheret et al. 2015). If this translates to real life traumatic events, sleep deprivation directly after the event might reduce the consolidation of trauma-related memories and the risk for the development of PTSD. However, at this point it is unclear how targeted periods of sleep deprivation after traumatic events might affect distinct aspects of memory formation, including explicit and implicit emotional aspects, and how this might reduce or even increase the long-term risk for trauma-associated health problems.

Another example for memory reorganization in the context of psychotherapy is Imagery Rehearsal Therapy (IRT) for the treatment of nightmare disorder, in which patients are instructed to rewrite the frightening content in a less anxiety provoking way (e.g., with a more favorable ending) (Krakow and Zadra 2006). The rationale of this treatment is that new memory traces are created with repeated rehearsal of the modified version and reactivated during sleep, leading to an alleviation of the nightmare.

Beyond the context of psychotherapy, sleep-related interventions could also be used to augment other treatments related to neuroplasticity. Specifically, brain lesions after stroke or traumatic injury require a reorganization of cortical networks, e.g. one intact brain region taking over the functions of a lesioned one (Elbert and Rockstroh 2004). It is plausible that sleep can promote such neuroplastic processes. However, sleep disturbances are highly prevalent in stroke survivors (Sterr et al. 2008). Moreover, the setting in intensive care units typically include a 24 h exposure to bright light and high levels of noise, often leading to disturbed sleep. Improving sleep might have the potential to improve the rehabilitation process.

Manipulation of Sleep in Combination with Psychotherapy

In addition to the targeted implementation of sleep periods in combination with psychotherapy, the manipulation of sleep-specific processes might improve cognitive performance and learning (Diekelmann 2014). Current approaches include (1) the reactivation of memory during sleep using cues, (2) the modulation of sleep-specific brain activity using non-invasive brain stimulation techniques and hypnosis, and (3) the pharmacological manipulation of specific neurotransmitter systems. Some of these approaches has been translated to psychotherapy research.

The Reactivation of Memory During Sleep Using Olfactory and Auditory Cues

An interesting non-invasive approach is the reactivation of memory content during sleep using olfactory or auditory cues (Rasch et al. 2007; Rudoy et al. 2009) (see chapters by Schreiner, Lehmann and Rasch and by Talamini). In a seminal study, participants learned a visuospatial object-location task in the evening with time-locked administration of an odor to the learning stimuli (Rasch et al. 2007). Recall performance in the morning after one night of sleep was significantly enhanced after re-exposure of the odor during SWS. In contrast, odor re-exposure was ineffective during REM sleep or wakefulness. Further studies demonstrated that a selective reactivation of memory content during sleep using odor exposure promotes resistance to interference (Diekelmann et al. 2011) and enhances memory consolidation (Diekelmann et al. 2012). These effects likely emerge from cue-induced reactivation of task-related memory networks during sleep as indicated by a reactivation of hippocampal and neocortical regions in fMRI studies (Rasch et al. 2007; Diekelmann et al. 2011) (see also chapter by Zhang, Deuker and Axmacher). In a study investigating 60 medication-free volunteers with spider-phobia, olfactory cues were used with the aim of increasing the beneficial effect of sleep (Rihm et al. in press). Memories of subjective therapy success were verbalized after exposure in vivo and paired with a background odor. Afterwards, patients either stayed awake or had a 90 min daytime nap. The nap group was divided into two subgroups that received either a re-presentation of the odor or a presentation of an odorless vehicle during NREM sleep. Re-presentation of the odor did not improve treatment outcome, but increased left-lateralized frontal slow spindle (11.0–13.0 Hz) and right-lateralized parietal fast spindle (13.0–15.0 Hz) activity, which might be indicative of a successful reactivation of therapy-related memories during sleep.

The reactivation of memory during sleep cannot only strengthen desired memories, but could also help to weaken undesired memory traces. In a contextual fear conditioning procedure in human participants, pictures of faces were paired with mild electric shocks while odors were presented as contextual background (Hauner et al. 2013) (see chapter by Shanahan and Gottfried). The re-presentation of one of the contextual odor cues during subsequent SWS resulted in enhanced fear extinction, i.e. reduced fear responses to the faces in the next morning. In a similar study in mice, an odor was used as conditioned stimulus that was paired with foot shocks (Rolls et al. 2013). During post-training sleep, mice were re-exposed to the odor. Interestingly, when tested during subsequent periods of wakefulness, fear responses to the odor were increased. However, the injection of a protein synthesis inhibitor into the basolateral amygdala before the re-exposure of the odor during sleep resulted in fear extinction, i.e. reduced fear responses to the conditioned odor. These findings demonstrate that subtle methodological variations determine whether the reactivation of fear memories during sleep leads to a strengthening or weakening on the behavioral level.

The described approaches might be further transferred to psychotherapy. However, it is to note that, at this point in time, it is largely unclear, how exactly complex psychotherapy contents are processed and how selective enhancement of distinct memory aspects might promote or decrease therapeutic effects.

The Modulation of Sleep-Specific Brain Activity Using Non-invasive Brain Stimulation Techniques and Hypnosis

Another non-invasive approach is the manipulation of specific aspects of sleep using brain stimulation techniques. The enhancement of slow wave activity by transcranial application of oscillating potentials (0.75 Hz) during early nocturnal non-REM sleep increases the consolidation of hippocampus-dependent declarative memory in young adults (Marshall et al. 2006) (see chapter by Campos Beltran and Marshall). These results could be replicated to some extent in patients with schizophrenia (Göder et al. 2013), but not in elderly adults (Eggert et al. 2013). The same effect was observed in healthy participants using closed-loop auditory stimulation during sleep that enhanced slow wave activity and the consolidation of declarative memory (Ngo et al. 2013). We recently provided evidence that total sleep time can be manipulated using transcranial direct current stimulation (tDCS) prior to sleep (Frase et al. 2016), presumably mediated by prolonged changes of cortical arousal that via 'top-down' pathways appears to extend to subcortical sleep-wake regulation networks (Krone et al. 2016). Yet at this point, we were only able to increase cortical arousal and reduce total sleep time, but unable to extend total sleep time in healthy participants. Increasing arousal and reducing total sleep time might be relevant for numerous clinical conditions characterized by reduced arousal and hypersomnia, such as observed after various types of brain lesion (e.g. Frase et al. 2015). Another recent study showed that transcranial alternating current stimulation in the lower gamma band frequency during REM sleep can induce lucid dreams—dreams with self-reflective awareness (Voss et al. 2014). The induction of lucid dreams might represent an opportunity to practice new behavior or modify the content of repetitive dreams, thereby augmenting the effectiveness of psychotherapy (see also chapter by Schredl).

The induction of slow oscillations during sleep might not only enhance the consolidation of previously encoded memory content, but also increase the capacity for encoding new information during subsequent wakefulness. Transcranial slow oscillation stimulation (tSOS) during an afternoon nap in healthy humans improved the encoding of hippocampus-dependent declarative material (pictures, word-pairs, word list) compared to a nap without stimulation (Antonenko et al. 2013). Intensifying slow oscillations might promote synaptic downscaling in saturated neuronal networks and reinstate the capacity to encode new information in healthy humans. However, in patients with mental disorders, the enhancement of synaptic downscaling during naps might lead to adverse effects. For instance, patients with major depressive disorder might exhibit a reduced level of overall synaptic strength compared to healthy humans (Wolf et al. 2015). During the normal waking period, they possibly do not reach the optimal window of associative plasticity. Additional enhancement of synaptic downscaling during naps using tSOS might even pronounce this impairment and detrimentally affect psychotherapy.

To date, to our knowledge, no study tried to augment the effects of psychotherapy using current stimulation techniques during sleep. An ongoing study tries to augment CBT with prefrontal tDCS during wakefulness in patients with major depression. In this study, tDCS is applied during group psychotherapy sessions and the effects of psychotherapy are compared to a group with sham stimulation and a group without stimulation (German Brain Stimulation Center).

A potential alternative to non-invasive brain stimulation for increasing SWS and improving cognition is hypnotic suggestion. Two studies indicate that an auditory hypnotic suggestion to 'sleep deeper' increases the percentage of SWS in females suggestible to hypnosis. The increase in SWS during a daytime nap with hypnotic suggestion, compared to a control tape (within-subject comparison with an interval of 7 days between sessions), was 57% in older females (Cordi et al. 2015) and 81% in young females (Cordi et al. 2014). In low suggestible women, no such effect of the hypnotic suggestion was observed. In the study investigating older women, cognitive functioning was assessed with a semantic verbal fluency task before and after the nap. In high suggestible women, verbal fluency increased significantly more after the nap with hypnotic suggestion (and more SWS) compared to the control condition. Such an improvement was not observed in low suggestible women (verbal fluency was numerically better after the control tape), suggesting that these benefits were effects of the hypnotic suggestion and the associated increase in SWS. A limitation of the studies is that low suggestible women had more SWS in the control condition compared to high suggestible women. This may have limited their potential for an increase in SWS and performance. In summary, hypnotic suggestion appears to have the potential to increase SWS during a nap and increase verbal fluency in selected individuals.

Pharmacological Interventions

The strengthening and reorganization of memories is modulated by a number of different neurotransmitters and hormones that are differentially regulated during wakefulness and sleep. Therefore, the manipulation of sleep by pharmacological substances represents another method to influence processes of memory formation and, potentially, the effects of psychotherapy.

SWS is characterized by low levels of noradrenaline, acetylcholine and cortisol. Increasing noradrenergic availability using the noradrenaline reuptake inhibitor reboxetine enhances memory retention in an odor recognition task (Gais et al. 2011). In contrast, the reduction of noradrenergic activity during sleep by the administration of the $\alpha 2$-autoreceptor agonist clonidine impairs subsequent memory performance. Furthermore, SWS-related consolidation of declarative memories is blocked when the central cholinergic tone is increased by administration of physostigmine during a period of SWS-rich sleep (Gais and Born 2004). Similar to acetylcholine, the increase of cortisol levels during SWS blocks the enhancement of sleep-related memory consolidation (Plihal and Born 1999; Wilhelm et al. 2011).

Another interesting finding emphasizing the role of glutamatergic neurotransmission for sleep-dependent offline consolidation of memories is the observation that the NMDA receptor coagonist D-cycloserine (DCS) administered during retention sleep facilitates consolidation of declarative memory (word pairs), whereas the administration of DCS during a wake interval remained without effect on retention of word pairs (Feld et al. 2013).

A study examining the augmentation of psychotherapy demonstrated that long-term effects of CBT in patients with insomnia disorder (12 months follow-up) are more robust without co-administration of zolpidem, a benzodiazepine receptor agonist (Morin et al. 2009). Benzodiazepine receptor agonism transiently improves sleep duration, but has been shown to impair sleep-associated processes of learning and neuronal plasticity (Silva et al. 2003), which might contribute to the observed worse long-term outcome. However, possible mechanisms are complex and include further aspects like conditioning of drug intake and self-efficacy.

Other studies have investigated the effects of M1 muscarinic acetylcholine receptor (mAChR) activation on sleep (Nissen et al. 2006a) and sleep related memory formation (Nissen et al. 2006b) demonstrating that the direct M1 mAChR agonist RS-86 reduces REM sleep latency and SWS duration in healthy humans but has no clinically relevant effect on pre–post sleep consolidation of declarative or procedural memories.

Conclusion and Directions

As outlined in this chapter, further translation of basic science knowledge on memory formation and its modulation by sleep appears to have the potential to promote the effects of psychotherapy. Given the high prevalence of mental disorders and still limited treatment effects, this would represent a substantial benefit to public health. It appears to be a timely opportunity to systematically investigate the effects of sleep prior to and after distinct types of memory formation during psychotherapy sessions, as well as those of targeted manipulations of sleep, on short- and long-term treatment effects.

Of note, the chapter centered on the potential of augmenting treatment effects in individuals with clinical disorders. However, the discussed approaches might not only promote therapy effects, but also decrease them, or worse, even harm patients, e.g. by facilitating the implementation of false memories or by affecting complex biography concepts. These aspects and the potential use for neuroenhancement of healthy functioning require a careful scientific and ethical evaluation.

References

Antonenko D, Diekelmann S, Olsen C, Born J, Molle M (2013) Napping to renew learning capacity: enhanced encoding after stimulation of sleep slow oscillations. Eur J Neurosci 37 (7):1142–1151. doi:10.1111/ejn.12118

Baglioni C, Nanovska S, Regen W, Spiegelhalder K, Feige B, Nissen C et al (2016) Sleep and mental disorders: a meta-analysis of polysomnographic research. Psychol Bull in press

Bergman ET, Roediger HL (1999) Can Bartlett's repeated reproduction experiments be replicated? Mem Cogn 27(6):937–947. http://www.ncbi.nlm.nih.gov/pubmed/?term=Bergman+and+Roediger+1999. Checked 19 Nov 2015

Brooks DC, Bouton ME (1993) A retrieval cue for extinction attenuates spontaneous recovery. J Exp Psychol Anim Behav Process 19(1):77–89

Citri A, Malenka RC (2008) Synaptic plasticity: multiple forms, functions, and mechanisms. Neuropsychopharmacol Official Publ Am Coll Neuropsychopharmacol 33(1):18–41. doi:10.1038/sj.npp.1301559

Cordi MJ, Schlarb AA, Rasch B (2014) Deepening sleep by hypnotic suggestion. Sleep 37:1143–1152

Cordi MJ, Hirsinger S, Mérillat S, Rasch B (2015) Improving sleep and cognition by hypnotic suggestion in the elderly. Neuropsychologica 69:176–182

Diekelmann S (2014) Sleep for cognitive enhancement. Front Syst Neurosci 8:46. doi:10.3389/fnsys.2014.00046

Diekelmann S, Büchel C, Born J, Rasch B (2011) Labile or stable: opposing consequences for memory when reactivated during waking and sleep. Nat Neurosci 14(3):381–386. doi:10.1038/nn.2744

Diekelmann S, Biggel S, Rasch B, Born J (2012) Offline consolidation of memory varies with time in slow wave sleep and can be accelerated by cuing memory reactivations. Neurobiol Learn Mem 98(2):103–111. doi:10.1016/j.nlm.2012.07.002

Duffy C, Teyler TJ, Shashoua VE (1981) Long-term potentiation in the hippocampal slice: evidence for stimulated secretion of newly synthesized proteins. Science (New York, NY) 212 (4499):1148–1151

Dunsmoor JE, Murty VP, Davachi L, Phelps EA (2015) Emotional learning selectively and retroactively strengthens memories for related events. Nature 520(7547):345–348. doi:10.1038/nature14106

Eggert T, Dorn H, Sauter C, Nitsche MA, Bajbouj M, Danker-Hopfe H (2013) No effects of slow oscillatory transcranial direct current stimulation (tDCS) on sleep-dependent memory consolidation in healthy elderly subjects. Brain Stimulation 6(6):938–945. doi:10.1016/j.brs.2013.05.006

Elbert T, Rockstroh B (2004) Reorganization of human cerebral cortex: the range of changes following use and injury. Neuroscientist 10(2):129–141

Feld GB, Lange T, Gais S, Born J (2013) Sleep-dependent declarative memory consolidation–unaffected after blocking NMDA or AMPA receptors but enhanced by NMDA coagonist D-cycloserine. Neuropsychopharmacol Official Publ Am Coll Neuropsychopharmacol 38 (13):2688–2697. doi:10.1038/npp.2013.179

Frase L, Maier JG, Zittel S, Freyer T, Riemann D, Normann C et al (2015) Bifrontal anodal transcranial direct current stimulation (tDCS) improves daytime vigilance and sleepiness in a patient with organic hypersomnia following reanimation. Brain Stimulation 8(4):844–846. doi:10.1016/j.brs.2015.05.009

Frase L, Piosczyk H, Zittel S, Jahn F, Selhausen P, Krone L et al (2016) Modulation of total sleep time by transcranial direct current stimulation (tDCS). Neuropsychopharmacol Official Publ Am Coll Neuropsychopharmacol. doi:10.1038/npp.2016.65

Freud S (1939) Die Traumdeutung. Franz Deuticke, Leipzig und Wien

Fu J, Li P, Ouyang X, Gu C, Song Z, Gao J et al (2007) Rapid eye movement sleep deprivation selectively impairs recall of fear extinction in hippocampus-independent tasks in rats. Neuroscience 144(4):1186–1192. doi:10.1016/j.neuroscience.2006.10.050

Gais S, Born J (2004) Low acetylcholine during slow-wave sleep is critical for declarative memory consolidation. Proc Natl Acad Sci U S A 101(7):2140–2144. doi:10.1073/pnas.0305404101

Gais S, Rasch B, Dahmen JC, Sara S, Born J (2011) The memory function of noradrenergic activity in non-REM sleep. J Cogn Neurosci 23(9):2582–2592. doi:10.1162/jocn.2011.21622

Göder R, Baier PC, Beith B, Baecker C, Seeck-Hirschner M, Junghanns K, Marshall L (2013) Effects of transcranial direct current stimulation during sleep on memory performance in patients with schizophrenia. Schizophrenia Res 144(1–3):153–154. doi:10.1016/j.schres.2012.12.014

Graves LA, Heller EA, Pack AL, Abel T (2003) Sleep deprivation selectively impairs memory consolidation for contextual fear conditioning. Learn Mem 10(3):168–176

Gruart-Masso A, Nadal-Alemany R, Coll-Andreu M, Portell-Cortes I, Marti-Nicolovius M (1995) Effects of pretraining paradoxical sleep deprivation upon two-way active avoidance. Behav Brain Res 14(72):181–183

Hall-Porter JM, Schweitzer PK, Eisenstein RD, Ahmed HA, Walsh JK (2014) The effect of two benzodiazepine receptor agonist hypnotics on sleep-dependent memory consolidation. J Clin Sleep Med JCSM Official Publ Am Acad Sleep Med 10(1):27–34. doi:10.5664/jcsm.3352

Harrison Y, Horne JA (2000) Sleep loss and temporal memory. Q J Exp Psychol 53(1):271–279

Hartley CA, Phelps EA (2010) Changing fear: the neurocircuitry of emotion regulation. Neuropsychopharmacol 35(1):136–146

Hauner KK, Howard JD, Zelano C, Gottfried JA (2013) Stimulus-specific enhancement of fear extinction during slow-wave sleep. Nat Neurosci 16(11):1553–1555. doi:10.1038/nn.3527

Kellogg SH, Young JE (2006) Schema therapy for borderline personality disorder. J Clin Psychol 62(4):445–458. doi:10.1002/jclp.20240

Kleim B, Wilhelm FH, Temp L, Margraf J, Wiederhold BK, Rasch B (2014) Sleep enhances exposure therapy. Psychol Med 44(7):1511–1519

Krakow B, Zadra A (2006) Clinical management of chronic nightmares: imagery rehearsal therapy. Behav Sleep Med 4(1):45–70

Krone L, Frase L, Piosczyk H, Selhausen P, Zittel S, Jahn F et al (2016) Top-down control of arousal and sleep: fundamentals and clinical implications. Sleep Med Rev. doi:10.1016/j.smrv.2015.12.005

Kuhn M, Wolf E, Maier J, Mainberger F, Feige B, Schmid H et al (in press) Sleep recalibrates homeostatic and associative synaptic plasticity in the human cortex. Nat Commun

Kuhn M, Mainberger F, Feige B, Maier JG, Mall V, Jung NH et al (2016) State-dependent partial occlusion of cortical LTP-like plasticity in major depression. Neuropsychopharmacol Official Publ Am Coll Neuropsychopharmacol 41(6):1521–1529. doi:10.1038/npp.2015.310

Kuriyama K, Soshi T, Kim Y (2010) Sleep deprivation facilitates extinction of implicit fear generalization and physiological response to fear. Biol Psychiatry 68:991–998

Landmann N, Kuhn M, Piosczyk H, Feige B, Baglioni C, Spiegelhalder K et al (2014) The reorganisation of memory during sleep. Sleep Med Rev 18(6):531–541. doi:10.1016/j.smrv.2014.03.005

Landmann N, Kuhn M, Maier J-G, Spiegelhalder K, Baglioni C, Frase L et al (2015) REM sleep and memory reorganization: potential relevance for psychiatry and psychotherapy. Neurobiol Learn Mem 122:28–40. doi:10.1016/j.nlm.2015.01.004

Malenka RC, Bear MF (2004) LTP and LTD: an embarrassment of riches. Neuron 44(1):5–21. doi:10.1016/j.neuron.2004.09.012

Mander BA, Santhanam S, Saletin JM, Walker MP (2011) Wake deterioration and sleep restoration of human learning. Curr Biol CB 21(5):R183–4. doi:10.1016/j.cub.2011.01.019

Maquet P (2001) The role of sleep in learning and memory. Science 294(5544):1048–1052

Maren S (1999) Long-term potentiation in the amygdala: a mechanism for emotional learning and memory. Trends Neurosci 22(12):561–567

Marshall L, Helgadottir H, Mölle M, Born J (2006) Boosting slow oscillations during sleep potentiates memory. Nature 444(7119):610–613

Menz MM, Rihm JS, Salari N, Born J, Kalisch R, Pape HC et al (2013) The role of sleep and sleep deprivation in consolidating fear memories. NeuroImage 75:87–96. doi:10.1016/j.neuroimage.2013.03.001

Morin CM, Vallières A, Guay B, Ivers H, Savard J, Mérette C et al (2009) Cognitive behavioral therapy, singly and combined with medication, for persistent insomnia: a randomized controlled trial. JAMA 301(19):2005–2015. doi:10.1001/jama.2009.682

Nadel L, Campbell J, Ryan L (2007) Autobiographical memory retrieval and hippocampal activation as a function of repetition and the passage of time. Neural Plast 90472. http://www.ncbi.nlm.nih.gov/pubmed/18274617. Checked 19 Nov 2015

Ngo H-VV, Martinetz T, Born J, Mölle M (2013) Auditory closed-loop stimulation of the sleep slow oscillation enhances memory. Neuron 78(3):545–553. doi:10.1016/j.neuron.2013.03.006

Nissen C, Nofzinger EA, Feige B, Waldheim B, Radosa MP, Riemann D, Berger M (2006a) Differential effects of the muscarinic M1 receptor agonist RS-86 and the acetylcholine-esterase inhibitor donepezil on REM sleep regulation in healthy volunteers. Neuropsychopharmacol Official Publ Am Coll Neuropsychopharmacol 31(6):1294–1300. doi:10.1038/sj.npp.1300906

Nissen C, Power Ann E, Power AE, Nofzinger Eric A, Nofzinger EA, Feige Bernd et al (2006b) M1 muscarinic acetylcholine receptor agonism alters sleep without affecting memory consolidation. J Cogn Neurosci 18(11):1799–1807. doi:10.1162/jocn.2006.18.11.1799

Oakley DA, Halligan PW (2009) Hypnotic suggestion and cognitive neuroscience. Trends Cogn Sci 13:264–270

Pace-Schott EF, Germain A, Milad MR (2015) Effects of sleep on memory for conditioned fear and fear extinction. Psychol Bull 141(4):835–57

Pace-Schott EF, Hobson JA (2002) The neurobiology of sleep: genetics, cellular physiology and subcortical networks. Nat Rev Neurosci 3(8):591–605. doi:10.1038/nrn895

Pace-Schott EF, Milad MR, Orr SP, Rauch SL, Stickgold R, Pitman RK (2009) Sleep promotes generalization of extinction of conditioned fear. Sleep 32:19–26

Pace-Schott EF, Shepherd E, Spencer RM, Marcello M, Tucker M, Propper RE, Stickgold R (2011) Napping promotes inter-session habituation to emotional stimuli. Neurobiol Learn Mem 95(1):24–36

Pace-Schott EF, Verga PW, Bennett TS, Spencer RMC (2012) Sleep promotes consolidation and generalization of extinction learning in simulated exposure therapy for spider fear. J Psychiatr Res 46(8):1036–1044. doi:10.1016/j.jpsychires.2012.04.015

Plihal W, Born J (1999) Memory consolidation in human sleep depends on inhibition of glucocorticoid release. NeuroReport 10(13):2741–2747

Porcheret K, Holmes EA, Goodwin GM, Foster RG, Wulff K (2015) Psychological effect of an analogue traumatic event reduced by sleep deprivation. Sleep 38:1017–1025

Ramirez S, Liu X, MacDonald CJ, Moffa A, Zhou J, Redondo RL, Tonegawa S (2015) Activating positive memory engrams suppresses depression-like behaviour. Nature 522(7556):335–339. doi:10.1038/nature14514

Rasch B, Born J (2013) About sleep's role in memory. Physiol Rev 93(2):681–766. doi:10.1152/physrev.00032.2012

Rasch B, Büchel C, Gais S, Born J (2007) Odor cues during slow-wave sleep prompt declarative memory consolidation. Science 315(5817):1426–1429

Rescorla RA (2004) Spontaneous recovery. Learn Mem 11(5):501–509

Rihm JS, Sollberger SB, Soravia LM (2016) Re-presentation of olfactory exposure therapy success cues during non-rapid eye movement sleep did not increase therapy outcome but inreased sleep. Front Human Neurosci (in press)

Rogan MT, Stäubli UV, LeDoux JE (1997) Fear conditioning induces associative long-term potentiation in the amygdala. Nature 390(6660):604–607

Rolls A, Makam M, Kroeger D, Colas D, de Lecea L, Craig HH (2013) Sleep to forget: interference of fear memories during sleep. Mol Psychiatry 18(11):1166–1170. doi:10.1038/mp.2013.121

Roth T, Roehrs T, Wittig R, Zorick F (1984) Benzodiazepines and memory. Br J Clin Pharmacol 18(Suppl 1):45S–49S

Rudoy JD, Voss JL, Westerberg CE, Paller KA (2009) Strengthening individual memories by reactivating them during sleep. Science (New York, NY) 326(5956):1079. doi:10.1126/science.1179013

Silva A, Collao A, Orellana M, Melendez J, Caviedes P, Cardenas Am (2003) Zopiclone, but not brotizolam, impairs memory storage during sleep. Neurosci Res 47(2):241–243

Spoormaker VI, Sturm A, Andrade KC, Schröter MS, Goya-Maldonado R, Holsboer F et al (2010) The neural correlates and temporal sequence of the relationship between shock exposure, disturbed sleep and impaired consolidation of fear extinction. J Psychiatr Res 44(16):1121–1128. doi:10.1016/j.jpsychires.2010.04.017

Spoormaker VI, Schröter MS, Andrade KC, Dresler M, Kiem SA, Goya-Maldonado R et al (2012) Effects of rapid eye movement sleep deprivation on fear extinction recall and prediction error signaling. Hum Brain Mapp 33(10):2362–2376. doi:10.1002/hbm.21369

Sterr A, Herron K, Dijk D-J, Ellis J (2008) Time to wake-up: sleep problems and daytime sleepiness in long-term stroke survivors. Brain Inj 22(7–8):575–579. doi:10.1080/02699050802189727

Tononi G, Cirelli C (2014) Sleep and the price of plasticity: from synaptic and cellular homeostasis to memory consolidation and integration. Neuron 81(1):12–34. doi:10.1016/j.neuron.2013.12.025

Van Der Werf YD, Altena E, Schoonheim MM, Sanz-Arigita EJ, Vis JC, de Rijke W, Van Someren EJW (2009) Sleep benefits subsequent hippocampal functioning. Nat Neurosci 12(2):122–123. doi:10.1038/nn.2253

Voss U, Holzmann R, Hobson A, Paulus W, Koppehele-Gossel J, Klimke A, Nitsche MA (2014) Induction of self awareness in dreams through frontal low current stimulation of gamma activity. Nat Neurosci 17(6):810–812. doi:10.1038/nn.3719

Vyazovskiy VV, Cirelli C, Pfister-Genskow M, Faraguna U, Tononi G (2008) Molecular and electrophysiological evidence for net synaptic potentiation in wake and depression in sleep. Nat Neurosci 11(2):200–208. doi:10.1038/nn2035

Walker MP, van der Helm E (2009) Overnight therapy? The role of sleep in emotional brain processing. Psychol Bull 135(5):731–748. doi:10.1037/a0016570

Walker MP, Liston C, Hobson JA, Stickgold R (2002) Cognitive flexibility across the sleep-wake cycle: REM-sleep enhancement of anagram problem solving. Brain Res Cogn Brain Res 14(3):317–324

Wilhelm I, Wagner U, Born J (2011) Opposite effects of cortisol on consolidation of temporal sequence memory during waking and sleep. J Cogn Neurosci 23(12):3703–3712. doi:10.1162/jocn_a_00093

Wolf E, Kuhn M, Norman C, Mainberger F, Maier JG, Maywald S et al (2015) Synaptic plasticity model of therapeutic sleep deprivation in major depression. Sleep Med Rev 30:53–62. doi:10.1016/j.smrv.2015.11.003

Yoo SS, Gujar N, Hu P, Jolesz FA, Walker MP (2007a) The human emotional brain without sleep-a prefrontal amygdala disconnect. Curr Biol 23(20):R877–8

Yoo SS, Hu PT, Gujar N, Jolesz FA, Walker M (2007b) A deficit in the ability to form new human memories without sleep. Nat Neurosci 10(3):385–392

Zalta AK, Dowd S, Rosenfield D, Smits JAJ, Otto MW, Simon NM et al (2013) Sleep quality predicts treatment outcome in CBT for social anxiety disorder. Depress Anxiety 30(11):1114–1120. doi:10.1002/da.22170

Accelerated Long-Term Forgetting in Epilepsy—and Beyond

John Baker and Adam Zeman

Abstract Accelerated long-term forgetting (ALF) is the excessively rapid loss of information over intervals longer than those typically used in neuropsychological assessment, most often 30 min. It has been described primarily in people with epilepsy, but it may occur in other contexts, for example preclinical Alzheimer's disease and mild cognitive impairment. We review the methodological aspects of the assessment of long-term forgetting; evidence relevant to the interval over which forgetting occurs in ALF; its relationship to sleep, occurence in children and approaches to treatment. Although ALF undoubtedly occurs in clinical practice, there is continuing uncertainty about two theoretical issues: (i) whether it reflects an impairment of memory acquisition, memory consolidation or a combination of the two; (ii) whether it results from structural or physiological changes in the brain or from a combination of both.

Keywords Accelerated long term forgetting · Transient epileptic amnesia · Memory · Epilepsy

Epilepsy can be associated with a variety of cognitive symptoms and impairments. Over the last decade a particular pattern of epilepsy-associated memory impairment has attracted increasing attention. Accelerated Long-term Forgetting (ALF) is the abnormally rapid forgetting of memories over intervals longer than the 30 min or so traditionally used to measure 'delayed recall' during a neuropsychological assessment. People with ALF often report that memories 'leak away' more quickly than would be expected, and some studies have reported a correlation between subjective complaints of memory impairment and objective measures of ALF. In this chapter we will review the evolution of this concept, clarifying its clinical presentation and associations, underlying pathophysiology, and treatment options. We will review recent evidence that ALF may not be confined to patients with epilepsy.

J. Baker · A. Zeman (✉)
Cognitive and Behavioural Neurology, University of Exeter Medical School, St Luke's Campus, Magdalen Road, Exeter EX1 2LU, UK
e-mail: A.Zeman@exeter.ac.uk

N. Axmacher and B. Rasch (eds.), *Cognitive Neuroscience of Memory Consolidation*, Studies in Neuroscience, Psychology and Behavioral Economics, DOI 10.1007/978-3-319-45066-7_24

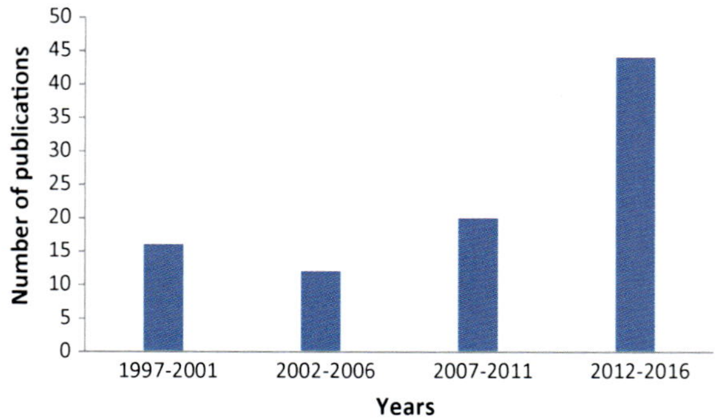

Fig. 1 Number of publications on ALF 1996–2016

We performed a literature search using PUBMED to identify all relevant articles. A search using the terms 'accelerated AND forgetting' yielded 114 results. Of these, 44 were published in the last 5 years and 64 in the last 10 years. This provides a sense of the relatively recent development of this term and its proliferation over the last decade (Fig. 1). The results are summarised in Table 1.

Defining ALF

It has long been recognised that TLE is often associated with memory impairment, but it has generally been assumed that standard testing intervals of half an hour or so are adequate to assess long term retention and forgetting rates. The possibility that accelerated forgetting might become apparent in some patients over intervals longer than those usually studied was first raised by a series of case reports (Kapur et al. 1997; O'Connor et al. 1997; Lucchelli and Spinnler 1998; Mayes et al. 2003; Cronel-Ohayon et al. 2006).

Subsequent publications have focused on group studies, predominantly in patients with temporal lobe epilepsy (TLE) (see Table 1). Whereas the early case reports involved patients with epilepsy secondary to a previous brain injury, the majority of the group studies have reported ALF in patients with no visible abnormality on MRI scan, in common with the majority of patients with TLE (Manes et al. 2005; Butler et al. 2007). A number of these single case reports and case series have been summarised previously (Bell and Giovagnoli 2007; Butler and Zeman 2008; Zeman et al. 2013; Elliott et al. 2014).

Patients with TLE show a range of memory impairments, and not all of them will exhibit ALF. Some will have deficits on standard memory test intervals (Mameniskiene et al. 2006); while others will show no impairment even over longer delays (Giovagnoli et al. 1995; Bell et al. 2005; Bell 2006). However, a number of studies have found evidence of normal or relatively normal memory retention after a 30 min delay (Blake et al. 2000; Butler et al. 2007) in association with accelerated forgetting after that interval (Martin et al. 1991).

Table 1 Summary of ALF studies

Author	Year	Number of patients	Age (mean)	Diagnosis	Test intervals
Giovagnoli	1995	24	38	TLE	1, 24 h, 3, 6, 13 d
Kapur	1997	1	62	Complex partial seizures	30 m, 6 w
O'Connor	1997	1	42	Complex partial seizures	2, 24, 48, 72 h, 1 w
Lucchelli	1998	1	65	Complex partial seizures	10 m, 60 m, 24 h, 1 w, 41 d
Blake	2000	21	34	TLE	30 m, 8 w
Mayes	2003	1	46	Complex partial seizures	30 m, 3 w
Bell	2005	42	37	TLE	30 m, 24 h
Jokeit	2005	162	38	epilepsy	30 m
Manes	2005	7	57	TEA	30 m, 6 w
Bell	2006	25	39	TLE	30 m, 2 w
Cronel-Ohayon	2006	1	18	Complex partial seizures	60 m, 1 w, 29 d
Mameniskiene	2006	70	33	TLE	30 m, 4 w
Butler	2007	24	68	TEA	30 m, 1, 3 w
Davidson	2007	21	12	IGE	30 m, 1 w
Manes	2008	10	64	subjective memory complaints	30 m, 6 w
Jansari	2010	1	63	TLE	30 m, 24 h, 1, 2, 4 w
Muhlert	2010	11	69	TEA	40 s, 30 m, 24 h, 1, 3 w
Deak	2011	7	44	TLE	30 m, 12 h
Galassi	2011	1	58	TLE	30 m, 1 w
Tramoni	2011	5	43	TLE	1 h, 6 w
Barkas	2012	12		TLE	3–6 w
Butler	2012	22	66	TEA	30 m, 1 w
Narayanan	2012	14	34	TLE	30 m, 4 w
Wilkinson	2012	27	37	TLE	1 h, 6 w

(continued)

Table 1 (continued)

Author	Year	Number of patients	Age (mean)	Diagnosis	Test intervals
Fitzgerald	2013	39		epilepsy	30 m, 24 h, 4 d
Hoefeijzers	2013	17	66	TEA	30 m, 1, 3 w
Mary	2013	32		healthy adults - sleep	30 m, 1 w
McGibbon	2013	1	68	TLE	5, 30, 55 m, 4, 24 h
Atherton	2014	11	68	TEA	30 m, 12 h
Evans	2014	7	40	TLE	30 m, 1 w
Gascoigne	2014	23	13	TLE	30 m, 1 w
Hoefeijzers	2015	11	59	TEA	30 m, 3, 8, 24 h, 1 w
Tu	2014	7	51	Thalamic stroke	1, 24 h, 1, 2, 4 w
Walsh	2014	15	75	MCI	30 m, 1 w
Landowsky-Brooks	2015	42	43	25 with head injury	30 m, 4 h
Miller	2015	7	45	Epilepsy	30 m, 1 w
Witt	2015	1	35	anti-GAD encephalitis	30 m, 1 w
Cassel	2016	18	39	TLE	10 m, 1 d, 1 w
Zeman	2016	1	52	Spinal injury/baclofen infusion	30 m, 1 w
Weston	2016	21	38	AD mutation carriers	30 m, 1 w
Bell	2007	Review article			
Butler	2008	Review article			
Zeman	2013	Review article			
Elliott	2014	Review article			
Guerts	2015	Review article			

TLE Temporal lobe epilepsy, *IGE* Idiopathic generalised epilepsy, *TEA* Transient epileptic amnesia, *MCI* Mild cognitive impairment, *AD* Alzheimer's disease, GAD Glutamic acid decarboxylase, *h* hours, *m* minutes, *d* days, *w* weeks

Table 2 TEA diagnostic criteria (from Butler and Zeman 2008)

Definition of TEA (from Butler and Zeman 2008)
(1) a history of recurrent witnessed episodes of transient amnesia
(2) cognitive functions other than memory judged to be intact during typical episodes by a reliable witness
(3) evidence for a diagnosis of epilepsy based on one or more of the following: (a) epileptiform abnormalities on electroencephalography (b) the concurrent onset of other clinical features of epilepsy (e.g. lip-smacking, olfactory hallucinations) (c) a clear-cut response to anticonvulsant therapy

ALF is particularly common in association with transient epileptic amnesia (TEA), a subtype of TLE (Table 2). Patients with TEA typically report three distinct forms of memory impairment: amnestic seizures, retrograde memory impairment (the inability to evoke autobiographical memories from the past, often combined with impairment of remote topographical memory), and ALF (Zeman et al. 2013). 44% of patients with TEA report symptoms suggestive of accelerated forgetting (Butler et al. 2007). Moreover, Butler et al. reported a correlation between subjective memory complaints and measures of accelerated forgetting, but not with measures of memory obtained at standard intervals (Butler et al. 2009).

If it is accepted that some patients with epilepsy perform normally on memory tests at standard intervals, but show impairment at extended ones, what kinds of learning and memory are affected? The phenomenon has been described in the context of declarative rather than non-declarative or procedural memory (Muhlert et al. 2010), and may especially affect the recall of context-rich episodic memories (Jansari et al. 2010; Tramoni et al. 2011). In the following section we consider the optimal methodology for the assessment of ALF.

Measuring ALF

A careful recent review of the literature by Elliott et al. (2014) concluded that ALF is a distinctive and robust clinical phenomenon but also identified a range of methodological variations in previous studies which sometimes hampers the interpretation of results. The sources of variation included:

i. **Selection of control participants**: studies have varied in the care with which control participants have been matched to patients on measures of general cognitive functioning, such as IQ, educational background and age. These may all be relevant to long term memory retention and forgetting.
ii. **Test material and procedures**:
    a. verbal vs visual material: most but not all studies have compared the learning and forgetting of verbal and visual material: this is desirable, in case there are material specific effects.

b. assessment procedure: most but not all studies have used measure of recall and recognition: this, again, is desirable given evidence that these measures may tap into partially separable processes (Aggleton and Brown 1999). Elliott et al. (2014) review evidence that subtle differences in procedures can affect test results.
c. ceiling and floor effects: these have been present in some studies, limiting their ability to detect and/or compare forgetting in patients and/or controls.
d. matching initial learning: some but not all studies have succeeded in matching initial learning. Where this is not achieved, the interpretation of forgetting curves, from different points of departure, is controversial. Elliott et al. (2014) discuss a range of techniques used to match initial learning.
e. rehearsal effects: there is a risk, in studies of long-term retention, that participants may rehearse the material they have learned consciously. Differential rehearsal across groups could confound the intended comparison. It is not clear that this is a serious practical problem, but some researchers have used material that would be difficult to rehearse to overcome this obstacle.
f. Short-term memory (STM) influence: some but not all studies have included a distraction procedure before the measure of immediate recall to obtain a result that is not contaminated by working memory.

iii. **Analysing of forgetting rates**: As mentioned above, where levels of initial learning differ the comparison of forgetting curves is problematic and there is a variety of potential approaches to data analysis.

To minimise these discrepancies, this paper concludes with a series of considerations which it suggests should be applied to future studies in this area (Table 3).

There is a need to develop a reliable approach to the clinical identification of ALF. Miller et al. (2015) recently assessed 60 healthy control subjects using three standardised measures of memory (Rey Auditory Verbal Learning (RAVLT), Logical Memory (LM), and Aggie Figures) with recall delays of 30 min and 7 days in order to establish normative values. 15 patients with focal epilepsy were studied using the same tasks. Seven of the patients showed ALF. Although this is a small

Table 3 Methodological guidance for ALF studies (from Elliott et al. 2014)

Methodological guidance for further studies in ALF (from Elliott et al. 2014)
1. Patient and control groups should be matched, at least for age and intellectual ability
2. Ideally, both verbal and non-verbal test material should be used
3. Ideally, forgetting should be measured using both recall and recognition tests
4. Ceiling and floor effects should be avoided as far as possible
5. The potential for rehearsal and repeated recall should be avoided as far as possible
6. The immediate delay period should be long enough to ensure information is stored in LTM and retrieval is not reliant on STM processes
7. Effort should be made to equate initial learning (whilst avoiding overlearning)

study the suggestion is that this triad of tests, with a recall period of 7 days, may be an effective means of identifying ALF in patients with epilepsy who complain of memory impairment.

The Forgetting Interval in ALF

The formation of memories is a highly complex, time-demanding process involving a series of biological steps and anatomical regions (Fig. 2). Disruption of any of these steps can impair memory (Kopelman 2002). Initial memory acquisition requires that information gains access to the relevant memory system ('encoding') with rapid associated changes in synaptic strength ('early' consolidation). Over time, the fragile early memory trace is strengthened ('late consolidation'), at least in part through processes of 'replay' (see chapter by Zhang, Deuker and Axmacher). Memories must then be stored, and retrieved when required. There is evidence that they remain labile during storage, especially at times of retrieval, when they may be strengthened or weakened by a process of 'reconsolidation' (see also chapter by Kessler, Blackwell and Kehyayan). The complexity of these processes challenges the simple, traditional distinction in cognitive psychology between 'short' and 'long term' memory, and predicts the existence of forms of amnesia corresponding to disruption of these processes. The locus of the memory process disturbed in ALF remains unclear but studies of its time course are relevant to determining this: if ALF is, in fact, predictable from the point of memory acquisition then it may well be due to a pathology of encoding or early acquisition. If, however, at least in some instances, ALF only becomes apparent sometime after acquisition, then later processes of consolidation may be involved, though, as we mention below, behavioural evidence alone may never be sufficient to establish this firmly.

A variety of different delays has been investigated, from several hours to several weeks (see Table 1). The onset of ALF over shorter delays has been examined by Hoefeijzers et al. (2015). This study investigates the recall of word lists in TEA patients and controls at intervals of 30 min, 3, 8 and 24 h. Although the TEA patients were not significantly different from controls at the immediate and 30 min intervals, there were significant differences between 30 min and 8 h. There was a reduction in recall between 30 min and 3 h but this did not reach significance. No further forgetting was observed over the first night in either group (i.e., between 8 h and 24 h).

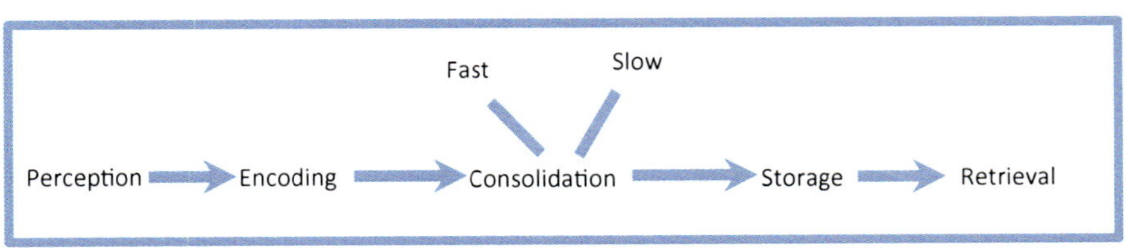

Fig. 2 Representation of multiple stages in memory process

Other studies have identified evidence of ALF just one hour after acquisition (Wilkinson et al. 2012; McGibbon and Jansari 2013). Wilkinson et al. (2012) reports accelerated forgetting of both verbal and non-verbal material in a cohort of 27 TLE patients. However, the pattern of forgetting was dependent on the different materials learnt and on the lateralisation of hippocampal sclerosis. Those with left sided abnormalities demonstrated accelerated forgetting of verbal material at the 1 h delay, whereas the right-sided group showed normal retention at this interval with accelerated forgetting over a 6-week delay. In contrast, patients with right-sided hippocampal sclerosis showed faster forgetting over 1 h of the non-verbal task when compared to those with left sided abnormalities.

The time course of forgetting has recently been examined in detail by Cassel et al. (2016). In this paper, to clarify the interval after learning at which ALF can be said to start, 14 TLE patients were examined using a story and a route learning task. Recall was tested at intervals of 30 s (following a distractor task), 10 min, one day and one week. It was found that patients' and controls' performance did not differ at 30 s on either task. However, TLE patients required more learning trials to reach criterion. In the story task, accelerated forgetting was progressive from 30 s onwards, although differences in forgetting rate only became statistically significant after one week. In the route learning task, patients showed faster forgetting over the first 10 min with comparable forgetting rates thereafter. These data suggest that in this group of patients accelerated forgetting occurs as a consequence of impaired acquisition/early consolidation.

In contrast to these findings, in a reanalysis of data from the study of Butler et al. (2007), Hoefeijzers examined the fate of words which had received the same number of learning trials and retrievals in patients with TEA and control participants in a world list learning study. There was no significant difference between the two groups on an initial recall test at 30 min. However, recall performance was significantly lower for TEA patients when compared to controls after 1 week, suggesting a disruption of 'late' consolidation.

As an aside, we note three relevant complexities: first, it may well be that earlier and later processes interact in such a way that an early defect is amplified by its effects at later stages: for example, recurrent rounds of 'replay' may give rise to a widening difference between memories differing slightly in strength at or soon after acquisition (Zeman et al. 2016). Second, it is not clear that purely behavioural data will ever be sufficient to resolve the question of the locus of the underlying impairment in ALF, as we currently have no absolutely reliable behavioural measure of memory strength: it is likely, therefore, that settling this question will require direct measures of brain activity associated with the various stages of memory processing. Third, patients describe the loss of memories over varied time-scales, from days to months: it may be that different mechanisms underlie ALF in different clinical situations.

ALF and Sleep

Theories of memory consolidation have placed particular emphasis on the role of sleep in the long-term retention of declarative memories (see chapter by Schönauer and Gais). Given the possibility that ALF may be a disorder of memory consolidation, its connection with sleep has been investigated. The inter-relationship of sleep and memory consolidation is reviewed elsewhere in this book. In a study of sleep quality and in particular its association to ALF, Mary et al. (2013) found that increased sleep fragmentation ('intra-sleep awakenings') was associated with ALF. However, this study also suggests that this sleep fragmentation was not enough by itself to explain the phenomenon and that therefore memory consolidation processes cannot depend exclusively on sleep quality after learning. In support of this a number of studies have shown that ALF can occur without an intervening period of sleep (see previous section).

The frequent occurrence of amnesic attacks on awakening in patients with TEA (Butler et al. 2007), raised a suspicion that ALF in this condition might be related to disruption of sleep-related memory processing. However, Atherton et al. (2014) found ALF patients derived the same benefit from sleep as controls. In this study, TEA patients, tested after an 8 h interval, showed ALF after a day awake but not following a comparable period of overnight sleep. This suggests that ALF in TEA may be related more to the effects of retroactive interference from novel information than to a disruption of sleep-related consolidation. This view is supported by other studies examining TEA and TLE which have also shown ALF during waking hours which is not worsened by an intervening period of sleep (Fitzgerald et al. 2013; Deak et al. 2011; Hoefeijzers et al. 2015).

ALF in Children with Epilepsy

A number of studies have attempted to identify ALF in children with epilepsy (Cronel-Ohayon et al. 2006; Davidson et al. 2007; Gascoigne et al. 2014). Assessing ALF in children is not straightforward. However, given the long term implications of an accelerated rate of forgetting for young people in full-time education, it is certainly clinically important. Although temporal lobe epilepsy is the most common form of adult-onset epilepsy, seizures involving the temporal lobes are also common in children. Gascoigne et al. (2014) identify twenty-three children between the ages of 6 and 16 years of age with temporal lobe epilepsy. 7 of these patients had had a temporal lobe resection. ALF was identified in this group for verbal, but not visual information. No correlation was identified between epilepsy severity and ALF. ALF has also been observed in children with generalised epilepsy (Davidson et al. 2007).

Treatment of ALF in Epilepsy

Several studies have investigated the response to treatments ranging from repeated prompting (McGibbon and Jansari 2013) and medication (Midorikawa and Kawamura 2007; Razavi et al. 2010; Barkas et al. 2012) to surgery (Galassi et al. 2011; Evans et al. 2014).

At the more conservative end of this spectrum, McGibbon and Jansari (2013) examined the role of repeated recall in reducing the effects of ALF. They found that repeated recall has a protective effect against the delayed forgetting seen in ALF. Tramoni et al. (2011) reported normal performance over 6 weeks in a task in which facts were taught initially to a criterion of 90% correct in patients who showed ALF for episodic information. Further work examining the value and limits of such strategies in patients with ALF would be valuable

If ALF is a manifestation of ongoing epileptic activity it is reasonable to suppose that anticonvulsant medications may have a role in reducing these memory symptoms. Midorikawa and Kawamura (2007) showed improvement of ALF but not retrograde amnesia in a patient with TEA following the initiation of anticonvulsant medication. Razavi et al. (2010) also found resolution of memory symptoms in their TEA patient following the introduction of carbamazepine. However Jansari et al. (2010) did not find a significant improvement in the rate of long-term forgetting in their patient after lamotrigine had been started.

In cases of medically intractable temporal lobe epilepsy, surgery remains the best option for seizure control. A number of studies have looked at the role of temporal lobectomy in relation to ALF. Results have been mixed. In their case report Galassi et al. (2011) show that a left temporal polectomy reduced seizure frequency and ALF in a 58 yr old male patient with a 20 year history of medically intractable epilepsy. Evans et al. (2014) report a cohort of seven patients with temporal lobe epilepsy. A longitudinal design was used to assess ALF pre- and post-epilepsy surgery. Visual and verbal materials were used with recall and recognition tests. ALF was confirmed prior to surgery. The study identified a degree of impairment of initial learning in the group post-surgery, in keeping with hippocampal resection, which complicated the interpretation of further forgetting. However, retention was unimpaired between the 30 min and one week delays on all 8 subtests following surgery. It is also worth noting that at the time of testing in these patients, none had had a change in their epilepsy medication, precluding this from being a contributing factor in their improvement. Given that the indication for surgery in these patients was medically intractable epilepsy, the study concludes that persistent and recurrent seizure activity may have had a causative role in the pattern of forgetting before surgery.

Some novel therapeutic approaches have also been investigated. This includes the use of the selective serotonin reuptake inhibitor fluoxetine (Barkas et al. 2012). In this paper it was shown that patients with hippocampal sclerosis show impairments of acquisition for a spatial task with accelerated forgetting of this task once learned. Administration of fluoxetine reversed the learning deficit, but left the pattern of accelerated forgetting unchanged.

ALF in Other Conditions

While ALF has been described predominantly in epilepsy, a growing literature is exploring the possibility that it can occur in other contexts. We highlight recent papers describing ALF in cases with limbic encephalitis, stroke, subjective memory complaints, mild cognitive impairment (MCI) and Alzheimer's disease (Manes et al. 2005; Walsh et al. 2014; Weston et al. 2016) and during intrathecal therapy with the GABA (B) receptor agonist, Baclofen.

In recent years a growing variety of types of autoimmune limbic encephalitis have been described, typically presenting with memory disturbance, seizures, emotional symptoms and personality change. Witt et al. (2015) describes a case of ALF associated with glutamic acid decarboxylase (GAD) antibody related limbic encephalitis. This patient, a 35 year-old male, complained of severe anterograde and retrograde memory deficits characterized by accelerated long-term forgetting. Video EEG monitoring confirmed a left temporal epileptic focus and subclinical seizure activity but no overt seizures at the time of initial presentation. Cognitive testing identified normal learning and initial recall at 30 min but significantly impaired recall for information at 1 week, in keeping with the pattern seen in other ALF patients. He was treated with monthly steroid pulses and anticonvulsant treatment and reported an improvement in his anterograde memory.

ALF has also been demonstrated in patients with thalamic stroke (Tu et al. 2014). In this study, using a visual task, 7 patients with previous thalamic strokes were tested at intervals of: 1, 24 h, 1, 2, and 4 weeks. Accelerated forgetting of newly acquired contextual information was identified in patients within 24 h when compared to healthy controls.

It is not surprising, given the clinical features and demographics of reported cases of ALF, that there has been interest in its potential relationship to MCI, particularly as ALF has been shown to correspond well to subjective memory concerns. Manes et al. (2005) in a relatively early paper on the topic of ALF showed that accelerated forgetting was identifiable in patients attending a memory clinic with subjective memory concerns who performed normally in standard neuropsychological tests. In this study of 10 individuals with complaints of memory loss but normal cognitive evaluation, 7 patients with MCI and 9 healthy controls, recall of both verbal and non-verbal information was tested immediately and then subsequently at intervals of 3 min and 6 weeks. There was no significant difference between the control group and the group with memory complaints on logical memory or the Rey complex figure at the immediate and 30 min intervals, but a significant difference had developed by 6 weeks. At this time point the scores of the cognitively normal with memory complaints group approached those of the MCI group, with no significant difference between them. The relationship between MCI and ALF has since been explored further by Walsh et al. (2014). In this study, although MCI subjects had an increased rate of forgetting within the first 30 min, a greater rate of forgetting was also identified between the 30 min and 1 week recall sessions. This result is important as it shows that the standard tests performed in the memory clinic, which typically involve a delay of 30 min, may under-estimate the deficits experienced by this MCI cohort.

A very recent study Weston et al. (2016) investigated whether ALF may be an early feature of Alzheimer's disease. In a study of 21 carriers of pathological, AD-causing mutations and 11 control patients, matched for age and performance on standard memory tests, accelerated long-term forgetting for both verbal and non-verbal material was found in the mutation carriers at 7 days. These pre-symptomatic AD patients were tested on average 7.2 years before their predicted symptom onset. Patients and controls showed similar performance at an initial recall interval of 30 min. It is therefore suggested that ALF may be an early, pre-symptomatic feature of familial Alzheimer's disease, indeed perhaps the earliest feature of AD-related cognitive decline.

A recent review by Guerts et al. (2015), while highlighting the possible occurrence of ALF in early AD, as just discussed, noted a lack of evidence for ALF in 8/11 studies ranging over Korsakoff's syndrome, depression (with or without ECT), traumatic brain injury and multiple sclerosis.

Finally, ALF has also been described as side effect of medication. Baclofen is a GABA (B) receptor agonist. It is use widely for conditions causing increased muscle tone and spasticity, particularly when associated with pain. In a recent paper ALF was identified in a patient receiving treatment with an intrathecal baclofen pump. During a period of dose escalation, the patient reported a constellation of memory disorders very similar to those seen in TEA: amnestic episodes, ALF and autobiographical amnesia. These were confirmed on objective testing. As the baclofen dose was reduced the amnestic episodes and ALF resolved while the autobiographical amnesia persisted. While it is possible that baclofen therapy, given at unusually high doses in this case, induced TEA, the authors also raise the possibility that signalling at the GABA (B) receptor may play a specific role in ALF.

Pathophysiology of ALF

Given that ALF appears to be distinctive clinical phenomenon, two key questions about its nature come to the fore: first, is it a disorder of memory acquisition and early consolidation or of later phases of consolidation? This has already been discussed above, in Sect. The Forgetting Interval in ALF. Second, is it a consequence of disturbed physiology, disordered anatomy or some additional factors such as mood or drug treatment (Table 4)?

Table 4 possible mechanisms for ALF in epileptic patients

Possible mechanisms for ALF in epileptic patients
1. Clinical or subclinical seizure activity
2. Structural brain pathology
3. Side effect of anti-convulsant therapy
4. Psychological mechanisms

In the context of epilepsy, might ALF be a direct consequence of seizures? They do not appear to be *required* for ALF as ALF has been demonstrated in their absence, for example in patients with TEA (see e.g. Butler et al. 2007) and some authors, for example Blake et al. (2000), found no relationship between overt seizure frequency and memory performance. However, Mameniskiene et al. (2006) reported a positive correlation between long-term forgetting and seizures during their experimental period. They also identified a relationship between *subclinical* epileptiform EEG activity and long-term forgetting. This relationship was confirmed by Fitzgerald et al. (2013) who report evidence that inter-ictal discharges disrupt memory consolidation. The reduction of ALF by treatment in at least some patients also points to a role for disordered physiology in its causation.

Material-specific differences in learning rates could potentially arise from variations in the locations of an epileptogenic focus, and the origin of the epileptic activity. However, previous data have not consistently shown that hemispheric differences cause material-specificity in ALF. Blake et al. (2000) using a test of verbal memory, identified ALF in patients with TLE originating from the left but not the right hemisphere. However, in an earlier investigation, Martin et al. (1991) were not able to find a hemispheric effect when testing verbal memory. Other studies have also shown ALF for both verbal and non-verbal information regardless of the site of activity (Mameniskiene et al. 2006; Butler et al. 2007).

There is, therefore, some tentative evidence that disordered physiology contributes to ALF in at least some patients. Does disturbed anatomy also play a role? Several of the early case studies documented ALF in patients with focal pathology. Subsequent group studies have identified abnormalities within and beyond the hippocampus in patients with Transient Epileptic Amnesia (Butler et al. 2007; Butler and Zeman 2008; Mosbah et al. 2014). Tramoni et al. (2011) identified metabolic changes in the temporal lobes using PET and magnetic resonance spectroscopy in patients with temporal lobe epilepsy and ALF. The occurrence of ALF in patients with preclinical AD and MCI which ultimately give rise to structural pathology in the temporal lobes may also point to a role for structural pathology. However, this evidence is equivocal as anatomical and physiological disturbance go hand in hand in AD (Vossel et al. 2013), and, in general, attempts to correlate the degree of structural change with the extent of ALF have been unsuccessful. Thus the role of structural pathology in ALF remains uncertain.

It has also been suggested that the pattern of memory impairment seen in ALF among patients with epilepsy could be a consequence of anticonvulsant medication. This seems unlikely, given that patients often report ALF prior to the initiation of anticonvulsant therapies and, as we have seen, often report an improvement in memory following the introduction of medications to reduce seizure frequency (Galassi et al. 2011; Butler et al. 2007). However, there is certainly some evidence that antiepileptic drugs can negatively affect cognition (Ortinski and Meador 2004; Jokeit et al. 2005). These negative effects of anticonvulsant medications are most commonly seen with higher doses and the use of polypharmacy. Jokeit et al. (2005) investigated this question. It was shown that high serum levels of carbamazepine, phenobarbital or phenytoin in patients with refractory temporal lobe epilepsy were associated with impaired performance in verbal and nonverbal memory retention tasks when compared to patients with lower levels. However, neither high anti-convulsant doses nor the use of multiple medications are common among patient with epilepsy and ALF. TEA patients, in particular, are typically responsive to low doses of a single epilepsy medication (Butler and Zeman 2008). Moreover, it is rare for patients with ALF to be treated with the older anti-convulsants, which are more prone to cognitive side-effects.

Historically, discrepancies between subjective reports of memory impairment and normal performance on neuropsychological testing have often been attributed to low mood or poor self-esteem (Giovagnoli et al. 1997; Elixhauser et al. 1999). The development of reliable paradigms for identifying and diagnosing ALF has shown that mood is unlikely to play a major causal role in this condition. Several studies have reported an absence of correlation between ratings of mood and ALF (Blake et al. 2000; Mameniskiene et al. 2006; Butler et al. 2007). If ALF were a manifestation of low mood per se we would expect to see an accelerated rate of forgetting in patients with depression: this has not been shown on previous studies (Lewis and Kopelman 1998).

Conclusions

Since early case reports documented a form of long-term forgetting which occurred at delays longer than those utilised as part of standard memory testing, the literature on accelerated long-term forgetting has grown substantially. There is now a significant body of evidence that ALF occurs as a clinical phenomenon, especially in the context of epilepsy. There is also growing evidence that this pattern of forgetting occurs in conditions other than temporal lobe epilepsy, although this remains the most common cause. Recent studies of ALF in amnestic MCI and pre-symptomatic Alzheimer's disease patients will no doubt stimulate further interest in the topic.

A range of important questions about ALF await a definitive answer. There is continuing disagreement and uncertainty about whether ALF results from a deficit of memory acquisition/early consolidation or of later phases of consolidation. Similarly, the roles of physiological and structural disturbance in causing ALF remain uncertain. These various possibilities are not mutually exclusive: It may well be that early and late consolidation, physiological and structural pathologies are all involved, to varying degrees across differing clinical contexts. Finally, further work is required to develop reliable methods for eliciting ALF in everyday practice.

References

Aggleton JP, Brown M (1999) Episodic memory, amnesia, and the hippocampal-anterior thalamic axis. Behav Brain Sci 22:425–489

Atherton K, Noble A et al (2014) Sleep-dependent memory consolidation and accelerated forgetting. Cortex 54:92–105

Barkas L, Redhead E et al (2012) Fluoxetine restores spatial learning but not accelerated forgetting in mesial temporal lobe epilepsy. Brain 135:2358–2374

Bell B (2006) WMS-III logical memory performance after a two-week delay in temporal lobe epilepsy and control groups. J Clin Exp Neuropsychol 28:1435–1443

Bell B, Fine J et al (2005) Temporal lobe epilepsy and the selective reminding test: the conventional 30-minute delay suffices. Psychol Assess 17:103–109

Bell B, Giovagnoli A (2007) Recent innovative studies of memory in temporal lobe epilepsy. Neuropsychol Rev 17:455–476

Blake RV, Wroe SJ et al (2000) Accelerated forgetting in patients with epilepsy: evidence for an impairment in memory consolidation. Brain 123:472–483

Butler C, Graham K et al (2007) The syndrome of transient epileptic amnesia. Ann Neurol 61: 587–598

Butler C, Zeman A (2008) Recent insights into the impairment of memory in epilepsy: transient epileptic amnesia, accelerated long-term forgetting and remote memory impairment. Brain 131:2243–2263

Butler CR, Bhaduri A et al (2009) Transient epileptic amnesia: regional brain atrophy and its relationship to memory deficits. Brain 132:357–368

Cassel A, Morris R et al (2016) Forgetting in temporal lobe epilepsy: when does it become accelerated? Cortex 1683

Cronel-Ohayon S, Zesiger P et al (2006) Deficit in memory consolidation (abnormal forgetting rate) in childhood temporal lobe epilepsy: pre and postoperative long-term observation. Neuropediatrics 37:317–324

Davidson M, Dorris L et al (2007) Memory consolidation and accelerated forgetting in children with idiopathic generalised epilepsy. Epilepsy Behav 11:394–400

Deak M, Stickgold R et al (2011) The role of sleep in forgetting in temporal lobe epilepsy: a pilot study. Epilepsy Behav 21(4):462–466

Elixhauser A, Leidy N et al (1999) The relationship between memory performance, perceived cognitive function, and mood in patients with epilepsy. Epilepsy Res 37:13–24

Elliott G, Isaac C, Muhlert N (2014) Measuring forgetting: a critical review of accelerated long-term forgetting studies. Cortex 54:16–32

Evans S, Elliott G et al (2014) Can temporal lobe epilepsy surgery ameliorate accelerated long-term forgetting? Neuropsychologia 53:64–74

Fitzgerald Z, Mohamed A et al (2013) Accelerated long-term forgetting: a newly identified memory impairment in epilepsy. J Clin Neurosci 20:1486–1491

Galassi R, Sambati L et al (2011) Accelerated long-term forgetting in temporal lobe epilepsy: evidence of improvement after left temporal pole lobectomy. Epilepsy Behav 22:793–795

Gascoigne M, Smith M et al (2014) Accelerated long-term forgetting in children with temporal lobe epilepsy. Neuropsychologia 59:93–102

Giovagnoli A, Casazza M et al (1995) Visual learning on a selective reminding procedure and delayed recall in patients with temporal lobe epilepsy. Epilepsia 36:704–711

Giovagnoli AR, Mascheroni S et al (1997) Self-reporting of everyday memory in patients with epilepsy: relation to neuropsychological, clinical, pathological and treatment factors. Epilepsy Res 28:119–128

Guerts S, van der Werf S, Kessels R (2015) Accelerated forgetting? An evaluation on the use of long-term forgetting rates in patients with memory problems. Front Psychol 6:article 752

Hoefeijzers S et al (2015) Accelerated long-term forgetting can become apparent within 3–8 hours of wakefulness in patients with transient epileptic amnesia. Neuropsychology 29(1):117–125

Hoefeijzers S, Dewar M et al (2013) Accelerated long-term forgetting in transient epileptic amnesia: an acquisition of consolidation deficit. Neuropsychologia 51:1549–1555

Jansari A, Davis K et al (2010) When 'long-term memory' no longer means 'forever': analysis of accelerated long-term forgetting in a patient with temporal lobe epilepsy. Neuropsychologia 48:1707–1715

Jokeit H, Kramer G, Ebner Al (2005) Do antiepileptic drugs accelerate forgetting? Epilepsy Behav 6:430–432

Kapur N, Millar J et al (1997) Very long-term amnesia in association with temporal lobe epilepsy: evidence for multiple-stage consolidation processes. Brain Cogn 35:58–70

Kopelman M (2002) Disorders of memory. Brain 125:2152–2190

Landowsky-Brooks R (2015) Four-hour delayed memory recall for stories: theoretical and clinical implication of measuring accelerated long-term forgetting. Appl Neuropsychol Adult 00:1–8

Lewis P, Kopelman M (1998) Forgetting rates in neuropsychiatric disorders. JNNP 65:890–898

Lucchelli F, Spinnler H (1998) Ephemeral new traces and evaporated remote engrams: a form of neocortical temporal lobe amnesia? A preliminary case report. Neurocase 4:447–459

Mameniskiene R, Jatuzis D et al (2006) The decay of memory between delayed and long-term recall in patients with temporal lobe epilepsy. Epilepsy Behav 8:278–288

Manes F, Graham K et al (2005) Autobiographical amnesia and accelerated forgetting in transient epileptic amnesia. J Neurol Neurosurg Psychiatry 76:1387–1391

Martin RC, Loring DW, et al. (1991) Impaired long-term retention despite normal verbal learning in patients with temporal lobe dysfunction. Neuropsychology 5:3–12

Manes F, Serrano C et al (2008) Accelerated forgetting in subjects with memory complaints: a new form of mild cognitive impairment? J Neurol 255(7):1067–1070

Mary A, Schreiner S et al (2013) Accelerated long-term forgetting in aging and intra-sleep awakenings. Front Psychol 4:750

Mayes A, Isaac C et al (2003) Long-term amnesia: a review and detailed illustrative case study. Cortex 39:567–603

McGibbon T, Jansari A (2013) Detecting the onset of accelerated long-term forgetting: evidence from temporal lobe epilepsy. Neuropsychologia 51(1):114–122

Midorikawa A, Kawamura M (2007) Recovery of long-term anterograde amnesia, but not retrograde amnesia, after initiation of an anti-epileptic drug in a case of transient epileptic amnesia. Neurocase 13(5):385–389

Miller L, Flanagan E et al (2015) Old does with new tricks: detecting accelerated long-term forgetting by extending traditional measures. Epilepsy Behav 45:205–211

Mosbah A, Tramoni E et al (2014) Clinical, neuropsychological, and metabolic characteristics of transient epileptic amnesia syndrome. Epilepsia 55(5):699–706

Muhlert N, Milton F et al (2010) Accelerated forgetting of real-life events in transient epileptic amnesia. Neuropsychologia 48(11):3235–3244

Narayanan J, Duncan R et al (2012) Accelerated long-term forgetting in temporal lobe epilepsy: verbal, nonverbal and autobiographical memory. Epilepsy Behav 25(4):622–630

O'Connor M, Sieggreen M et al (1997) Accelerated forgetting in association with temporal lobe epilepsy and paraneoplastic encephalitis. Brain Cogn 35:71–84

Ortinski P, Meador K (2004) Cognitive side effects of anti-epileptic drugs. Epilepsy Behav 5: 60–65

Razavi M, Barrah J, Paradiso S (2010) A longitudinal study of transient epileptic amnesia. Cogn Behav Neurol 23(2):142–145

Tramoni E, Felician O et al (2011) Long-term consolidation of declarative memory: insight from temporal lobe epilepsy. Brain 134(3):816–831

Tu S, Miller L et al (2014) Accelerated forgetting of contextual details due to focal medio-dorsal thalamic lesion. Front Behav Neurosci 8:320

Vossel K et al (2013) Seizures and epileptiform activity in the early stages of Alzheimer's disease. JAMA Neurol 70(9):1158–1166

Walsh C, Wilkins S et al (2014) Memory consolidation in aging and MCI after 1 week. Neuropsychology 28(2):273–280

Weston P, Liang Y et al (2016) Accelerated long-term forgetting in pre-symptomatic familial Alzheimer's disease (accepted for presentation at Association of British Neurologists Annual Meeting 2016 ahead of publication)

Wilkinson H, Holdstock J et al (2012) Long-term accelerated forgetting of verbal and non-verbal information in temporal lobe epilepsy. Cortex 48(93):317–332

Witt J-A, Vogt V et al (2015) Loss of autonoetic awareness of recent autobiographical episodes and accelerated long-term forgetting in a patient with previously unrecognized glutamic acid decarboxylase antibody related limbic encephalitis. Front Neurol 6:130

Zeman A, Butler C et al (2013) Novel forms of forgetting in temporal lobe epilepsy. Epilepsy Behav 26(3):335–342

Zeman A, Hoefeijzers S et al (2016) The GABAB receptor agonist, baclofen, contributes to three distinct varieties of amnesia in the human brain—a detailed case report. Cortex 74(9):9–19